**EMS Textbooks in Mathematics**

*EMS Textbooks in Mathematics is* a book series aimed at students or professional mathematicians seeking an introduction into a particular field. The individual volumes are intended to provide not only relevant techniques, results and their applications, but afford insight into the motivations and ideas behind the theory. Suitably designed exercises help to master the subject and prepare the reader for the study of more advanced and specialized literature.

Jørn Justesen and Tom Høholdt, *A Course In Error-Correcting Codes*
Markus Stroppel, *Locally Compact Groups*
Peter Kunkel and Volker Mehrmann, *Differential-Algebraic Equations*
Dorothee D. Haroske and Hans Triebel, *Distributions, Sobolev Spaces, Elliptic Equations*

Thomas Timmermann

# An Invitation to Quantum Groups and Duality

## From Hopf Algebras to Multiplicative Unitaries and Beyond

European Mathematical Society

Author:

Thomas Timmermann
Fachbereich Mathematik und Informatik
Westfälische Wilhelms-Universität Münster
Einsteinstraße 62
48149 Münster
Germany

E-Mail: timmermt@math.uni-muenster.de

2000 Mathematical Subject Classification (primary; secondary): 16W30, 46-01, 46L55; 22A22, 22D25, 22D35, 46L10

ISBN 978-3-03719-043-2

The Swiss National Library lists this publication in The Swiss Book, the Swiss national bibliography, and the detailed bibliographic data are available on the Internet at http://www.helveticat.ch.



Contact address:

European Mathematical Society Publishing House
Seminar for Applied Mathematics
ETH-Zentrum SEW A27
CH-8092 Zürich
Switzerland

Email: info@ems-ph.org
Homepage: www.ems-ph.org

Typeset using the authors' TEX files: I. Zimmermann, Freiburg
Printing: Printforce, Alphen aan den Rijn (NL)

9 8 7 6 5 4 3 2

*To my parents*
*Bettina and Werner Timmermann*

# Preface

The original aim of this book was to present some results of my PhD thesis on quantum groupoids in the setting of $C^*$-algebras. But soon I realized that it would be more useful to provide an introduction to the mathematical background of the thesis than to focus on my own special results.

This book is not written by an expert and does not aim at experts; rather, it is addressed at graduate students and non-experts from other fields. It shall provide an introduction to quantum groups in the setting of $C^*$-algebras and von Neumann algebras, and enable the reader to proceed to advanced topics and research articles. Roughly, I tried to write the book that I missed when I started to learn the theory.

Much of the material presented in this book is scattered over many research articles and was not yet covered in introductory texts. I tried to select the most important approaches, to present the main results of several foundational articles in a coherent manner and from one common perspective, and to explain the context and the interrelations of the individual approaches. Apart from the last chapter, which summarizes some of the main results of my PhD thesis, little in this book is original. The presentation and the choice of topics is, of course, strongly influenced by my personal view and limited by my personal knowledge. Several omissions had to be made in order to finish this book in finite time and space.

Naturally, it is difficult to avoid misprints and minor mistakes – I hope that only few serious errors remained. Certainly, there are places where the presentation could be improved, where references should be added, or where other corrections could and should be made. I am grateful for every hint, correction, or comment that is send to the author or the publisher. An up-to-date table of corrections can be found at the following web address:

`http://www.math.uni-muenster.de/~timmermt/quantum-groups.html`

The suggestion to write this book came from my PhD supervisor Joachim Cuntz, who also provided the contact to Manfred Karbe of the EMS Publishing House. I would like to take this opportunity to thank Joachim Cuntz for this initiative, and for his generous support during the last years.

I would like to thank the EMS Publishing House, in particular Manfred Karbe and Irene Zimmermann, for the friendly cooperation, and Stefaan Vaes for comments on the book and many helpful suggestions. Following his advice, I included the examples of quantum groups presented in Chapter 6 and Section 8.4 – much to the benefit of the reader and the book, I think.

This book was written at the SFB 478 "Mathematische Strukturen in der Mathematik" in Münster. I would like to thank the SFB, and the Deutsche Forschungsgemeinschaft who funds this SFB, for support, and for the ideal working environment.

Furthermore, I would like to thank the members of our research group "Funktionalanalysis, Operatoralgebren und nichtkommutative Geometrie", in particular Alcides Buss, Siegfried Echterhoff, and Walther Paravicini, for many interesting suggestions and discussions.

Finally, I would like to thank my wife Kristina Timmermann for her kind support during the last years, and my father Werner Timmermann, who carefully read the manuscript and improved it by innumerable helpful suggestions, corrections, and hints.

Münster, January 2008 Thomas Timmermann

# Contents

# Introduction

The aim of this book is to give an introduction to the duality of quantum groups and to quantum groups in the setting of $C^*$-algebras and von Neumann algebras.

Roughly, a Hopf algebra or quantum group is the natural generalization of a group within the setting of non-commutative geometry: following the general principle of non-commutative geometry, the underlying space of the group is replaced by an algebra, and the group operations are replaced by additional structure maps on this algebra.

In the setting of $C^*$-algebras and von Neumann algebras, the term "quantum group" refers to generalizations of locally compact groups. In other fields of mathematics, the term "quantum group" is usually applied to a wide range of mathematical objects, which are studied by quite different methods. Therefore it seems appropriate to give an overview before we outline the approach adopted in this book.

## Hopf algebras and quantum groups in the algebraic setting

Initially, Hopf algebras were studied in a purely algebraic setting. The first examples appeared in the following situations:

**Algebraic topology.** In the study of the cohomology ring $H^*(G)$ of a compact Lie group $G$, Hopf investigated the map $\Delta\colon H^*(G) \to H^*(G) \otimes H^*(G)$ induced by the multiplication map $G \times G \to G$, and used algebraic properties of this map to determine the structure of $H^*(G)$. More generally, if $X$ is "a group up to homotopy", more precisely, a Hopf space, then $H^*(X)$ is a Hopf algebra, and this algebraic structure can be used to show that $X$ has the same cohomology like a product of spheres [66].

**Affine algebraic groups.** A basic principle in algebraic geometry is to describe an affine space $X$ in terms of the algebra of regular functions $\mathcal{O}(X)$. Now, almost by definition, algebraic group structures on $X$ correspond bijectively with Hopf algebra structures on $\mathcal{O}(X)$. Applications of the Hopf algebra point of view are given, for example, in [1].

**Representation theory of groups.** Further natural examples of Hopf algebras are the group algebra $\Bbbk G$ of a finite group $G$ and the universal enveloping algebra $U(\mathfrak{g})$ of a Lie algebra $\mathfrak{g}$.

These examples fall into two classes: the first two examples of Hopf algebras are commutative, and the last examples satisfy a cocommutativity condition that is, in some sense, dual to commutativity. In particular, both classes are closely related to classical groups.

The theory of Hopf algebras received strong new impulses when new classes of examples were constructed that were neither commutative nor cocommutative:

**Deformations.** One of the most influential developments in the theory of Hopf algebras was the introduction of $q$-deformations of universal enveloping algebras associated to certain Lie algebras. First examples were constructed by Faddeev and the Leningrad school in connection with work on the quantum inverse scattering method; later, Drinfeld and Jimbo produced a $q$-deformed Hopf algebra for every semisimple complex Lie algebra [37], where the deformation is related to a certain Poisson structure on the initial Lie algebra. These Drinfeld–Jimbo Hopf algebras and their representation theory are very well understood, see, for example, [23], [24], [68], [79], [80], [84], [103], [140].

**Knot invariants and the Yang–Baxter equation.** There exists an intriguing connection between physics, low-dimensional topology, and the corepresentation theory of certain Hopf algebras. The starting point is that the category of corepresentations of a Hopf algebra carries a natural tensor product, very much like the category of representations of a group. This tensor product is symmetric only if the Hopf algebra is cocommutative. But for certain Hopf algebras called braided or triangular, there exists a braiding, which is an isomorphism $c_{V,W} : V \otimes W \to W \otimes V$, natural in the corepresentations $V$ and $W$. The coherence constraints on such a braiding can be related to planar braid diagrams and to the quantum Yang–Baxter equation known from physics. In particular, one can construct knot invariants and solutions of the Yang–Baxter equation out of braided Hopf algebras. Conversely, solutions of the quantum Yang–Baxter equation give rise to bialgebras, and, in special cases, to Hopf algebras. A very nice account of these topics can be found in [79].

Unlike the first commutative and cocommutative examples of Hopf algebras, the new examples listed above are no longer directly related to classical groups; therefore they are usually called quantum groups.

**Algebraic quantum groups and their duality.** An algebraic framework for the study of quantum groups and their duality was developed by Van Daele [174], [177]. In his theory, a quantum group is a non-unital Hopf algebra equipped with an integral, which is an analogue of the Haar measure of a locally compact group, and to every such quantum group, one can associate a dual quantum group.

## Quantum groups in the setting of $C^*$-algebras and von Neumann algebras

A major motivation for the introduction of quantum groups in the setting of $C^*$-algebras and von Neumann algebras was the generalization of Pontrjagin duality to non-abelian locally compact groups:

**Kac algebras and generalized Pontrjagin duality.** For every locally compact abelian group $G$, the set of characters $\widehat{G}$ is a locally compact abelian group again, and the Pontrjagin–van Kampen theorem says that $\widehat{\widehat{G}} \cong G$ (see Section 1.1). For a non-abelian locally compact group, a generalized dual can no longer be defined in the form of a group, and one has to look for a larger category (of "quantum groups") that includes both locally compact groups and their generalized duals. This problem was solved by Vainerman and Kac [167], [168], and by Enock and Schwartz [47]: they defined the notion of a Kac algebra, which is a von Neumann algebra equipped with similar structure maps like a Hopf algebra, and constructed for every Kac algebra $A$ a dual Kac algebra $\widehat{A}$, such that $\widehat{\widehat{A}} \cong A$. An important rôle in their theory is played by the analogue of the Haar measure of a locally compact group, which is part of the structure of a Kac algebra. A $C^*$-algebraic counterpart of the theory was developed by Vallin and Enock [49], [170].

The concept of a Kac algebra, however, turned out to be too restrictive to include all interesting examples of quantum groups in the setting of $C^*$-algebras:

**Compact quantum groups.** Woronowicz developed a general theory of compact quantum groups in the setting of $C^*$-algebras [193], [202], which contains examples that do not satisfy all axioms of a Kac algebra. This theory is very appealing: the definition of a compact quantum group is concise, the existence of a Haar measure on every compact quantum group can be deduced from the axioms, and the corepresentation theory of every such quantum group is very similar to the representation theory of a compact group.

A new perspective on quantum groups in the setting of $C^*$-algebras and von Neumann algebras was introduced by Baaj and Skandalis:

**Multiplicative unitaries.** Examples of multiplicative unitaries were used for a long time in the theory of quantum groups, till Baaj and Skandalis put them center-stage, formulated an abstract definition, and gave a comprehensive treatment [7]. Roughly, a multiplicative unitary simultaneously encodes a quantum group and the dual of that quantum group; conversely, to every "reasonable" quantum group, one can associate a multiplicative unitary.

Finally, comprehensive theories of locally compact quantum groups were developed, which cover all known examples:

**Locally compact quantum groups / weighted Hopf algebras.** The theories developed by Vaes and Kustermans [91], [93], and Masuda, Nakagami, and Woronowicz [110], seem to give a definite answer to the question "What is a locally compact quantum group in the setting of $C^*$-algebras/von Neumann algebras?".

## Organization of the book

The aim of this book, as stated above, is to give an introduction to the duality of quantum groups, and to quantum groups in the setting of $C^*$-algebras and von Neumann algebras.

One possible approach would be to start immediately with a study of locally compact quantum groups, which form the most general framework. For someone who is familiar with Hopf algebras and with the high-level analytic techniques used in the theory of locally compact quantum groups, this is probably the best choice. In this book, however, we shall adopt another approach, which may be better suited for graduate students and researchers from other fields.

**Contents of the book.** Part I of this book provides an introduction to quantum groups in a purely algebraic setting. After a review of Hopf algebras and their duality (Chapter 1), we discuss the duality of algebraic quantum groups developed by Van Daele (Chapter 2). This theory provides a very nice model for the generalizations of Pontrjagin duality that will be considered in Part II and yields many fundamental formulas. Finally, we investigate algebraic compact quantum groups (Chapter 3). This class of quantum groups can be studied not only in an algebraic, but also in a $C^*$-algebraic setting, and will serve us as a bridge for the passage to the setting of $C^*$-algebras.

In Part II, we turn to quantum groups in the setting of $C^*$-algebras and von Neumann algebras. First, we discuss the problems that arise in the definition of a Hopf $C^*$-algebra or Hopf–von Neumann algebra, consider examples related to locally compact groups, and list the existing approaches (Chapter 4). Then, we present Woronowicz's theory of $C^*$-algebraic compact quantum groups (Chapter 5), which is particularly accessible and close to the algebraic setting, and consider important examples (Chapters 6). Closely related to quantum groups in the setting of $C^*$-algebras and von Neumann algebras are multiplicative unitaries, which are studied subsequently (Chapter 7). Part II ends with an overview of the theory of locally compact quantum groups and some examples (Chapter 8). We focus on motivation, which can often be found in the theory of algebraic quantum groups, and explain the central analytic tools, but do not give any proof.

Part III of this book is devoted to selected topics. First, we discuss coactions of quantum groups on $C^*$-algebras, reduced crossed products, and a generalization of the Takesaki–Takai duality theorem for group actions (Chapter 9). The crossed product construction and the duality theorem make essential use of multiplicative unitaries, and are due to Baaj and Skandalis. Next, we give an introduction to measurable quantum groupoids, or, more precisely, to pseudo-multiplicative unitaries on Hilbert spaces (Chapter 10). In particular, we present the relative tensor product of Hilbert spaces, which is also known as Connes' fusion. Finally, we summarize some results of the author's thesis on pseudo-multiplicative unitaries on $C^*$-modules and quantum groupoids in the setting of $C^*$-algebras (Chapter 11).

Frequently used notation and important terms used in this book are listed in separate indices, and some background is compiled in a short appendix.

**Prerequisites.** This book should be accessible to graduate students and researchers from other fields of mathematics. For Part I, no special background is needed. Part II assumes some familiarity with Hilbert spaces, $C^*$-algebras, and von Neumann algebras, as summarized in the appendix; at some points, we use the language of $C^*$-modules, which is also summarized in the appendix. Part III contains advanced topics and is addressed at readers with some background in the field of $C^*$-algebras or von Neumann algebras.

**Logical dependence of the chapters.** The logical dependence of the individual chapters of this book is sketched in the following diagram:

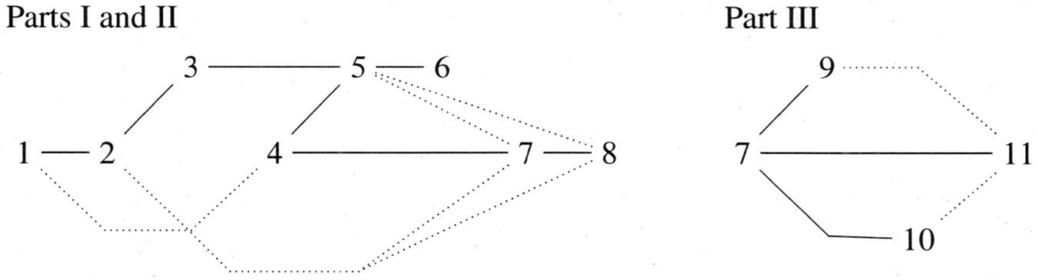


The dotted lines indicate that a chapter provides examples or motivation for the developments in a subsequent chapter, without that an understanding of the first chapter is needed for an understanding of the second one.

## Preliminaries and notation

Let us fix some notation and terminology.

As usual, the letters $\mathbb{N}, \mathbb{Z}, \mathbb{R}, \mathbb{C}$ denote the sets of natural, integer, real, and complex numbers, respectively. We put $\mathbb{T} := \{z \in \mathbb{C} : |z| = 1\}$. The letter $\mathbb{k}$ will stand for an arbitrary field.

We adopt the following convention. A *sesquilinear form* on a complex vector space $H$ is a map $\langle \cdot | \cdot \rangle \colon H \times H \to H$ that is conjugate-linear in the first variable and linear in the second variable. In particular, we apply this convention to inner products on Hilbert spaces.

The domain of definition of a map $\phi$ is denoted by $\mathrm{Dom}(\phi)$, and the image is denoted by $\mathrm{Im}(\phi)$. The identity map on a set $X$ is denoted by $\mathrm{id}_X$ or shortly by id.

We denote the set of all bounded linear operators from a Hilbert space $H_1$ to a Hilbert space $H_2$ by $\mathcal{L}(H_1, H_2)$, and the subset of all compact linear operators by

$\mathcal{K}(H_1, H_2)$. Furthermore, we use the ket-bra notation: for every element $\xi$ of a Hilbert space $H$, we define maps

$$|\xi\rangle\colon \mathbb{C} \to H,\ \lambda \mapsto \lambda\xi, \quad \text{and} \quad \langle\xi|\colon H \to \mathbb{C},\ \zeta \mapsto \langle\xi|\zeta\rangle.$$

Considering $\mathbb{C}$ as a Hilbert space, we have $|\xi\rangle \in \mathcal{K}(\mathbb{C}, H)$, $\langle\xi| \in \mathcal{K}(H, \mathbb{C})$, and $|\xi\rangle^* = \langle\xi|$.

Given a subset $X$ of a vector space $V$, we denote by span $X \subseteq V$ the linear span of $X$. If $V$ is a topological vector space, we denote by $\overline{\text{span}}\, X$ and $[X]$ the closed linear span of $X$. We say that $X \subseteq V$ is *linearly dense* in $Y \subseteq V$ if $\overline{\text{span}}\, X = Y$.

In Part II and III, we denote the algebraic tensor product by the symbol "$\odot$" to distinguish it from the minimal tensor product of $C^*$-algebras.

# Part I

# From groups to quantum groups

# Chapter 1
# Hopf algebras

This chapter gives a brief introduction to Hopf algebras. We focus on examples related to groups, on the axiomatics, and on the duality of Hopf algebras. The contents is standard and can be found in nearly every text on Hopf algebras or quantum groups, for example, in [23], [29], [79], [80], [111], [140]; two classical references are [1], [145].

## 1.1 Motivation: Pontrjagin duality

One of the motivations to study Hopf algebras is the question how the Pontrjagin duality of locally compact abelian groups can be extended to non-abelian groups. Let us briefly recall classical Pontrjagin duality. To each locally compact abelian group $G$, one can associate a *dual group* $\widehat{G}$, which is locally compact and abelian again, as follows. As a set, $\widehat{G}$ consists of all characters on $G$, that is, of all continuous group homomorphisms $G \to \mathbb{T}$; the group operation is given by pointwise multiplication of characters, and the topology is the topology of uniform convergence on compact subsets. The following standard examples are well-known from Fourier analysis:

$$\widehat{\mathbb{Z}} \cong \mathbb{T}, \quad \widehat{\mathbb{Z}/n\mathbb{Z}} \cong \mathbb{Z}/n\mathbb{Z}, \quad \widehat{\mathbb{R}} \cong \mathbb{R}, \quad \widehat{\mathbb{T}} \cong \mathbb{Z}.$$

Evidently, each element $x \in G$ determines a character on $\widehat{G}$ by evaluation at $x$. Denote this character by $\mathrm{ev}_x$.

**Theorem** (Pontrjagin duality). *For every locally compact abelian group $G$, the map $G \to \widehat{\widehat{G}}$ given by $x \mapsto \mathrm{ev}_x$ is an isomorphism of topological groups.*

This result can be proven via two different strategies:

- The classical approach [112], [122], [190] uses the structure theory of abelian groups. First, the duality is verified for special classes of groups like the examples listed above or all discrete and compact abelian groups. Then, the contravariant functor $G \mapsto \widehat{G}$ is shown to be compatible with direct sums, extensions, limits, and colimits. The structure theory says that these operations, applied to the special classes of groups considered before, yield *all* locally compact abelian groups, and so the theorem follows.

- The modern approach [20], [33], [150, Chapter VII, Section 3] uses the Fourier transformation and functional analysis in varying levels of abstraction.

Over the last decades, much work has been spent on generalizations of Pontrjagin duality to larger classes of groups, in particular to non-abelian locally compact groups. For such a group, characters retain too little information – they only capture abelian quotients of the group. A natural solution is to consider higher-dimensional representations as well. But then, the dual of a group can no longer be equipped with a group structure, and one has to look for a new category of "generalized groups" that contains locally compact groups and their duals. Roughly, one can distinguish two approaches:

**"The dual of a group is its representation theory".** This approach can be compared to Grothendieck's idea to replace a space by its category of sheaves; here, one replaces a group $G$ by the category of all representations of $G$ on vector spaces, the morphisms in this category being the intertwiners of representations. Depending on the context, additional requirements on the representations and intertwiners may be necessary, for example, continuity. Equipped with the natural tensor product of representations, this category becomes a symmetric monoidal or symmetric tensor category; the tensor subcategory of all one-dimensional representations corresponds precisely to the group of characters. One early achievement of this approach is the Tannaka–Krein duality theorem [22], [25], [62], [151], which says that a compact group can be reconstructed from the category of its representations. For a survey on this approach, see [73].

**"The dual of a space is its function algebra".** The idea is to encode the underlying space of a group $G$ by some "coordinate algebra" $A$ of functions on $G$, and the multiplication $G \times G \to G$ by a comultiplication $A \to A \otimes A$. This leads to the notion of a Hopf algebra, which, in various variants, is the central topic of this book. The dual of the group is encoded by the group algebra which can be thought of as the coordinate algebra of the dual group. For an abelian group, this interpretation can be made precise using the Fourier transform.

Of course, both approaches are intimately related. Roughly, the (representation theory of the) coordinate algebra of a group encodes the underlying space of the group, and the (representation theory of the) group algebra encodes the representation theory of the group.

For locally compact groups, the second approach has successfully been pursued in the setting of von Neumann algebras and $C^*$-algebras [47], [91], leading to a generalization of Pontrjagin duality which covers all locally compact groups. This generalization is outlined in Chapter 8.

## 1.2 The concept of a Hopf algebra

### 1.2.1 Definition

We fix the following notation. The letter $\Bbbk$ will always stand for a field. Given an algebra $A$, we denote by $m\colon A \otimes A \to A$ the multiplication map $a \otimes b \mapsto ab$, and, if $A$ is unital, by $\eta\colon \Bbbk \to A$ the *unit map* $\lambda \mapsto \lambda 1_A$.

**Definition 1.2.1.** A *Hopf algebra* is a unital algebra $A$ (over $\Bbbk$), equipped with

i) a unital homomorphism $\Delta\colon A \to A \otimes A$, called the *coproduct* or *comultiplication*, which is *coassociative* in the sense that the square

$$\begin{CD} A @>{\Delta}>> A \otimes A \\ @V{\Delta}VV @VV{\mathrm{id}\otimes\Delta}V \\ A \otimes A @>>{\Delta\otimes\mathrm{id}}> A \otimes A \otimes A \end{CD} \tag{1.1}$$

commutes, that is, $(\Delta \otimes \mathrm{id}_A) \circ \Delta = (\mathrm{id}_A \otimes \Delta) \circ \Delta$;

ii) a homomorphism $\epsilon\colon A \to \Bbbk$, called the *counit*, which makes the diagram

$$\begin{CD} A \otimes A @<{\Delta}<< A @>{\Delta}>> A \otimes A \\ @V{\epsilon\otimes\mathrm{id}}VV @| @VV{\mathrm{id}\otimes\epsilon}V \\ \Bbbk \otimes A @>>{\cong}> A @<<{\cong}< A \otimes \Bbbk \end{CD} \tag{1.2}$$

commute, that is, $(\epsilon \otimes \mathrm{id}_A) \circ \Delta = \mathrm{id}_A = (\mathrm{id}_A \otimes \epsilon) \circ \Delta$;

iii) a linear map $S\colon A \to A$, called the *antipode*, which makes the diagram

$$\begin{CD} A \otimes A @<{\Delta}<< A @>{\Delta}>> A \otimes A \\ @V{S\otimes\mathrm{id}}VV @VV{\eta\circ\epsilon}V @VV{\mathrm{id}\otimes S}V \\ A \otimes A @>>{m}> A @<<{m}< A \otimes A \end{CD} \tag{1.3}$$

commute, that is, $m \circ (S \otimes \mathrm{id}_A) \circ \Delta = \eta \circ \epsilon = m \circ (\mathrm{id}_A \otimes S) \circ \Delta$.

We shall always use the letters $\Delta, \epsilon$, and $S$ to denote the comultiplication, counit, and antipode of a Hopf algebra. When several Hopf algebras $A, B, \ldots$ are considered simultaneously, we index the structure maps by the respective Hopf algebras and write $\Delta_A, \epsilon_A, S_A, \Delta_B, \epsilon_B, S_B, \ldots$.

**Definition 1.2.2.** A *morphism of Hopf algebras* $A$ and $B$ is a unital algebra homomorphism $F\colon A \to B$ that is compatible with the structure maps in the sense that the squares

$$
\begin{array}{ccc}
A & \xrightarrow{\Delta_A} & A \otimes A \\
{\scriptstyle F}\downarrow & & \downarrow{\scriptstyle F\otimes F} \\
B & \xrightarrow[\Delta_B]{} & B \otimes B,
\end{array}
\qquad
\begin{array}{ccc}
A & \xrightarrow{\epsilon_A} & \Bbbk \\
{\scriptstyle F}\downarrow & & \| \\
B & \xrightarrow[\epsilon_B]{} & \Bbbk,
\end{array}
\qquad
\begin{array}{ccc}
A & \xrightarrow{S_A} & A \\
{\scriptstyle F}\downarrow & & \downarrow{\scriptstyle F} \\
B & \xrightarrow[S_B]{} & B
\end{array}
$$

commute, that is, $\Delta_B \circ F = (F \otimes F) \circ \Delta_A$, $\epsilon_B \circ F = \epsilon_A$, and $S_B \circ F = F \circ S_A$.

**Remark 1.2.3.** We shall see in Remark 1.3.23 and Proposition 1.3.17 that the definition of a Hopf algebra and the definition of a morphism of Hopf algebras can be weakened.

### 1.2.2 Examples related to groups

Let $G$ be a group. We denote by $\Bbbk(G)$ the algebra of all $\Bbbk$-valued functions on $G$, where the addition and multiplication are defined pointwise. The structure maps of $G$, that is,

| the multiplication | the inclusion of the unit | the inversion |
|---|---|---|
| $G \times G \to G,$ | $\{e\} \hookrightarrow G,$ | $G \to G,$ |
| $(x, y) \mapsto xy,$ | $e \mapsto e,$ | $x \mapsto x^{-1},$ |

induce the following algebra homomorphisms:

$$
\begin{array}{lll}
\Delta\colon \Bbbk(G) \to \Bbbk(G \times G), & \epsilon\colon \Bbbk(G) \to \Bbbk, & S\colon \Bbbk(G) \to \Bbbk(G), \\
(\Delta(f))(x, y) := f(xy), & \epsilon(f) := f(e), & (S(f))(x) := f(x^{-1}).
\end{array}
\tag{1.4}
$$

Equipped with these structure maps, $\Bbbk(G)$ almost forms a Hopf algebra – the only defect is that the target of the map $\Delta$ is $\Bbbk(G \times G)$ and not $\Bbbk(G) \otimes \Bbbk(G)$. We can identify $\Bbbk(G) \otimes \Bbbk(G)$ with a subspace of $\Bbbk(G \times G)$, but the image of $\Delta$ is not contained in this subspace unless $G$ is finite. However, each unital subalgebra $A \subseteq \Bbbk(G)$ that satisfies $\Delta(A) \subseteq A \otimes A$ and $S(A) \subseteq A$ *is* a Hopf algebra with respect to the restrictions of the maps $\Delta$, $\epsilon$, and $S$. This can be verified by direct calculations, or by comparing diagrams (1.1)–(1.3) with diagrams that express the group axioms.

**Example 1.2.4** (The function algebra of a finite group). Let $G$ be a finite group. Then the tensor product $\Bbbk(G) \otimes \Bbbk(G)$ can be identified with $\Bbbk(G \times G)$, and the algebra $\Bbbk(G)$, equipped with the maps $\Delta$, $\epsilon$, and $S$ defined in (1.4), forms a Hopf algebra.

Let us rewrite the structure maps of this Hopf algebra in terms of a canonical basis. For each $x \in G$, we define a function $\delta_x \in \Bbbk(G)$ by

$$\delta_x(y) := \delta_{x,y} = \begin{cases} 1, & x = y, \\ 0, & \text{otherwise.} \end{cases}$$

Then the family $(\delta_x)_{x \in G}$ is a basis of $\Bbbk(G)$, and for each $x \in G$,

$$\Delta(\delta_x) = \sum_{\substack{y,z \in G \\ yz = x}} \delta_y \otimes \delta_z, \quad \epsilon(\delta_x) = \delta_{x,e}, \quad S(\delta_x) = \delta_{x^{-1}}.$$

**Example 1.2.5** (The algebra of representative functions of a group). Let $G$ be a topological group. Recall that a continuous complex-valued function $f$ on $G$ is *representative* if and only if it satisfies the following equivalent conditions:

i) the linear span of all left- and right-translates of $f$, that is, of all functions of the form $f(z \cdot x)\colon y \mapsto f(zyx)$, where $x, z \in G$, has finite dimension;

ii) the linear span of all right-translates of $f$, that is, of all functions of the form $f(\cdot x)\colon y \mapsto f(yx)$, where $x \in G$, has finite dimension;

iii) the linear span of all left-translates of $f$, that is, of all functions of the form $f(z\,\cdot)\colon y \mapsto f(zy)$, where $z \in G$, has finite dimension;

iv) there exists a continuous representation $\pi$ of $G$ on some finite-dimensional complex vector space $V$ and elements $v \in V$, $\phi \in V' = \operatorname{Hom}_{\Bbbk}(V, \Bbbk)$ such that $f(x) = \phi(\pi(x)v)$ for all $x \in G$.

Let us prove the equivalence of these conditions. Clearly, i) implies ii) and iii). If $f$ satisfies condition ii), then right translation defines a continuous representation $\pi$ of $G$ on the space $V := \operatorname{span}\{f(\cdot x) \mid x \in G\}$; in detail, $\pi$ is given by $\pi(x)g := g(\cdot x)$ for all $g \in V$ and $x \in G$. Since $f(x) = (\pi(x)f)(e) = \epsilon(\pi(x)f)$ for all $x \in G$, the function $f$ satisfies condition iv). A similar argument shows that iii) implies iv). Finally, assume that $f$ has the form described in iv), and denote by $\mathcal{C}(\pi)$ the space of all functions of the form $x \mapsto \psi(\pi(x)w)$, where $w \in V$ and $\psi \in V'$. Then $\dim \mathcal{C}(\pi) \leq (\dim V)^2 < \infty$, and for all $x, z \in G$, the function $f(z \cdot x)$ belongs to $\mathcal{C}(\pi)$ because it can be written in the form $f(z \cdot x) = \phi(\pi(z \cdot x)v) = \psi(\pi(\cdot)w)$, where $\psi = \phi(\pi(z)\cdot)$ and $w = \pi(x)v$. Therefore, $f$ satisfies condition i).

Denote the space of all continuous representative functions on $G$ by $\operatorname{Rep}(G)$. A moment of consideration shows that this space is a unital subalgebra of $\mathbb{C}(G)$. We show that $\Delta(\operatorname{Rep}(G)) \subseteq \operatorname{Rep}(G) \otimes \operatorname{Rep}(G)$ and $S(\operatorname{Rep}(G)) \subseteq \operatorname{Rep}(G)$, which implies that Rep($G$), equipped with the restrictions of the maps $\Delta$, $\epsilon$, and $S$ defined in formula (1.4), forms a Hopf algebra.

So, consider a function $f \in \mathrm{Rep}(G)$ of the form $f(x) = \phi(\pi(x)v)$ as in condition iv) above. Choose a basis $(w_i)_i$ of $V$ and denote by $(\psi_i)_i$ the associated dual basis of $V'$. Then $w = \sum_i w_i \psi_i(w)$ for each $w \in V$, and hence

$$(\Delta(f))(x, y) = f(xy) = \phi(\pi(x)\pi(y)v) = \sum_i \phi(\pi(x)w_i) \cdot \psi_i(\pi(y)v)$$

for all $x, y \in G$. Therefore $\Delta(f) = \sum_i f_{1,i} \otimes f_{2,i}$, where $f_{1,i} = \phi(\pi(\cdot)w_i) \in \mathrm{Rep}(G)$ and $f_{2,i} = \psi_i(\pi(\cdot)v) \in \mathrm{Rep}(G)$ for all $i$, and $\Delta(f) \in \mathrm{Rep}(G) \otimes \mathrm{Rep}(G)$. Next, consider the contragredient representation $\pi'$ of $\pi$, which is the representation on $V'$ given by $\pi'(x)\psi = \psi(\pi(x)^{-1}\cdot)$ for all $\psi \in V'$ and $x \in G$. Denote by $\mathrm{ev}_v \in V''$ the functional given by $\psi \mapsto \psi(v)$. Then

$$(S(f))(x) = f(x^{-1}) = \phi(\pi(x^{-1})v) = \mathrm{ev}_v(\pi'(x)\phi) \quad \text{for all } x \in G,$$

and hence $S(f) \in \mathrm{Rep}(G)$.

Let us add two remarks:

i) For every group $G$ and every field $\Bbbk$, one can define a Hopf algebra $\mathrm{Rep}_{\Bbbk}(G) \subseteq \Bbbk(G)$ of representative functions which are defined by conditions i)–iv) as above, but without any continuity assumption on $f$ or on the representation $\pi$.

ii) If $G$ is compact, then Rep($G$) is dense in $C(G)$ by the Peter–Weyl-Theorem [22, Chapter III], [62, Chapter 7], so that the Hopf algebra Rep($G$) is sufficiently large to encode the group $G$.

**Example 1.2.6** (The coordinate algebra of a matrix group). Let $G$ be one of the groups $\mathrm{SL}_n(\Bbbk)$, $\mathrm{SO}_n(\Bbbk)$, or $\mathrm{Sp}_n(\Bbbk)$, where $n \in \mathbb{N}$. For $i, j = 1, \dots, n$, denote by $u_{ij} \colon M_n(\Bbbk) \to \Bbbk$ the function that maps each matrix to its $(i, j)$th entry. Consider the unital subalgebra $\mathcal{O}(G) \subseteq \Bbbk(G)$ generated by the restrictions of these functions. If the characteristic of $\Bbbk$ is 0, this algebra can easily be described in terms of generators and relations; for example, $\mathcal{O}(\mathrm{SL}_n(\Bbbk))$ is isomorphic to $\Bbbk[U_{11}, \dots, U_{nn}]/(\det -1)$, where $\det = \sum_\sigma \mathrm{sgn}(\sigma) \cdot U_{1\sigma(1)} \dots U_{n\sigma(n)}$, the sum being taken over all permutations of the set $\{1, \dots, n\}$.

The spaces $\Delta(\mathcal{O}(G))$ and $S(\mathcal{O}(G))$ are contained in $\mathcal{O}(G) \otimes \mathcal{O}(G)$ and $\mathcal{O}(G)$, respectively, which implies that the algebra $\mathcal{O}(G)$, equipped with the restrictions of the maps $\Delta$, $\epsilon$, and $S$ defined in (1.4), forms a Hopf algebra. Indeed, in terms of the generators $u_{ij}$, the comultiplication and the counit are given by

$$\Delta(u_{ij}) = \sum_k u_{ik} \otimes u_{kj} \quad \text{and} \quad \epsilon(u_{ij}) = \delta_{i,j} \quad \text{for all } i, j,$$

respectively, because $u_{ij}(cd) = \sum_k u_{ik}(c)u_{kj}(d)$ and $u_{ij}(1_n) = \delta_{i,j}$ for all $c, d \in M_n(\Bbbk)$. Using Cramer's rule, one can also express $S(u_{ij})$ in terms of the generators $u_{kl}$, where $k, l = 1, \ldots, n$.

The groups $\mathrm{SL}_n(\Bbbk)$, $\mathrm{SO}_n(\Bbbk)$, and $\mathrm{Sp}_n(\Bbbk)$ are particular examples of *linear algebraic groups* or *affine algebraic groups*. Every affine algebraic group over $\Bbbk$ can be encoded by a Hopf algebra:

**Example 1.2.7.** (The coordinate ring of an affine algebraic group) Let $G$ be an affine algebraic group over an algebraically closed field $\Bbbk$, that is, an affine algebraic variety over $\Bbbk$ whose set of points is equipped with the structure of a group, such that the group operations are morphisms of algebraic varieties (over $\Bbbk$). Then the coordinate ring $\mathcal{O}(G)$, which is the algebra of polynomial functions on the variety $G$, is a Hopf algebra with respect to the operations given in formula (1.4), see [19], [65], [143].

Almost by definition, the assignment $G \mapsto \mathcal{O}(G)$ defines a contravariant equivalence between the category of affine algebraic groups over $\Bbbk$ and the category of commutative Hopf algebras over $\Bbbk$ that are finitely generated and reduced, see [1, Chapter 4], [65], or [145, Chapter XV].

Let us consider two elementary examples. To simplify the discussion, we assume that the characteristic of $\Bbbk$ is 0. Let $n \in \mathbb{N}$.

- The coordinate Hopf algebra $\mathcal{O}(\Bbbk^n)$ of the additive group $\Bbbk^n$ is isomorphic the polynomial algebra $\Bbbk[X_1, \ldots, X_n]$, where $X_i$ corresponds to the $i$th coordinate function $x_i : (c_1, \ldots, c_n) \mapsto c_i$. In terms of the generators $X_i$, the comultiplication, counit, and antipode are given by

$$\Delta(X_i) = X_i \otimes 1 + 1 \otimes X_i, \quad \epsilon(X_i) = 0, \quad S(X_i) = -X_i \quad \text{for all } i,$$

  because $(\Delta(x_i))(c, d) = x_i(c + d) = c_i + d_i = x_i(c) + x_i(d)$, $\epsilon(x_i) = x_i(0) = 0$, and $(S(x_i))(c) = x_i(-c) = -c_i$ for all $c, d \in \Bbbk^n$.

- The multiplicative group $\Bbbk^n_\times := (\Bbbk \setminus \{0\})^n$ can be identified with the affine algebraic variety $\{(c, d) \in \Bbbk^n \times \Bbbk^n \mid c_i d_i = 1 \text{ for all } i\}$; its coordinate Hopf algebra $\mathcal{O}(\Bbbk^n_\times)$ is isomorphic to the algebra of Laurent polynomials $\Bbbk[X_1, X_1^{-1}, \ldots, X_n, X_n^{-1}]$, where the comultiplication, counit, and antipode are given by

$$\Delta(X_i) = X_i \otimes X_i, \quad \epsilon(X_i) = 1, \quad S(X_i) = X_i^{-1} \quad \text{for all } i.$$

The Hopf algebras presented next do not consist of functions on a group:

**Example 1.2.8** (The group algebra of a discrete group). Let $G$ be a discrete group. Recall that the *group algebra* $\Bbbk G$ of $G$ is the vector space of all finitely supported

$\Bbbk$-valued functions on $G$, equipped with the convolution product $(f * g)(x) := \sum_{yz=x} f(y)g(z)$. Equivalently, $\Bbbk G$ can be defined as the universal algebra generated by a family of elements $(U_x)_{x\in G}$, subject to the relation $U_y U_z = U_{yz}$ for all $y, z \in G$.

The group algebra $\Bbbk G$ is a Hopf algebra with respect to the maps

$$\begin{aligned} \Delta\colon \Bbbk G &\to \Bbbk G \otimes \Bbbk G, & \epsilon\colon \Bbbk G &\to \Bbbk, & S\colon \Bbbk G &\to \Bbbk G, \\ \Delta(U_x) &:= U_x \otimes U_x, & \epsilon(U_x) &:= 1, & S(U_x) &:= U_{x^{-1}}, \end{aligned}$$

where $\Delta$ and $\epsilon$ are multiplicative, and $S$ is antimultiplicative. Indeed, straightforward calculations show that for these maps, the diagrams (1.1)–(1.3) commute.

Let us assume that $G$ is abelian and $\Bbbk = \mathbb{C}$. Then the Hopf algebra $\Bbbk G = \mathbb{C}G$ can be described in terms of the dual group $\widehat{G}$: In that case, $\mathbb{C}G$ is isomorphic to the Hopf algebra $\mathrm{Rep}(\widehat{G})$ introduced in Example 1.2.5. Let us sketch the proof. First, we can identify $\widehat{G}$ with the group

$$\Big\{(z_x)_{x\in G} \in \textstyle\prod_{x\in G} \mathbb{T} \mid z_x z_y = z_{xy} \text{ for all } x, y \in G\Big\} \subseteq \displaystyle\prod_{x\in G} \mathbb{T}.$$

By Tychonoff's Theorem, $\prod_{x\in G} \mathbb{T}$ is compact, and hence $\widehat{G}$ is compact as well. Since $\widehat{G}$ is abelian, each of its continuous finite-dimensional representations is equivalent to a direct sum of continuous one-dimensional representations. Such representations correspond bijectively with characters on $\widehat{G}$, and hence, by Pontrjagin duality, with elements of $G$. Thus the algebra $\mathrm{Rep}(\widehat{G})$ is generated by the family of functions $(\mathrm{ev}_x)_{x\in G}$ given by $\mathrm{ev}_x(\chi) = \chi(x)$. To show that these functions are linearly independent, we use the Haar measure $\hat{\lambda}$ on $\widehat{G}$: the linear independence follows easily from the equation $\int_{\widehat{G}} \overline{\mathrm{ev}_x}\, \mathrm{ev}_y \, d\hat{\lambda} = \int_{\widehat{G}} \mathrm{ev}_{x^{-1}y} = 0$ for all $x \neq y$, which can be deduced from the relation $\mathrm{ev}_z(\chi) \int_{\widehat{G}} \mathrm{ev}_z \, d\hat{\lambda} = \int_{\widehat{G}} \mathrm{ev}_z(\chi\,\cdot) d\hat{\lambda} = \int_{\widehat{G}} \mathrm{ev}_z \, d\hat{\lambda}$, $z \in G$, $\chi \in \widehat{G}$. Thus $(\mathrm{ev}_x)_x$ is a basis of $\mathrm{Rep}(\widehat{G})$. Now the assignment $U_x \mapsto \mathrm{ev}_x$ defines an isomorphism of Hopf algebras $\mathbb{C}G \xrightarrow{\cong} \mathrm{Rep}(\widehat{G})$, because

$$\begin{gathered} (\mathrm{ev}_x \mathrm{ev}_y)(\chi) = \mathrm{ev}_x(\chi)\, \mathrm{ev}_y(\chi) = \chi(x)\chi(y) = \chi(xy) = \mathrm{ev}_{xy}(\chi), \\ (\Delta(\mathrm{ev}_x))(\chi, \zeta) = \mathrm{ev}_x(\chi\zeta) = (\chi\zeta)(x) = \chi(x)\zeta(x) = \mathrm{ev}_x(\chi)\, \mathrm{ev}_x(\zeta), \\ (S(\mathrm{ev}_x))(\chi) = \mathrm{ev}_x(\chi^{-1}) = \chi(x)^{-1} = \chi(x^{-1}) = \mathrm{ev}_{x^{-1}}(\chi), \\ \epsilon(\mathrm{ev}_x) = \mathrm{ev}_x(e) = 1 \quad \text{for all } \chi, \zeta \in \widehat{G} \text{ and } x, y \in G. \end{gathered}$$

In particular, the Hopf algebra $\mathbb{C}\mathbb{Z}^n$ is isomorphic to the Hopf algebra $\mathrm{Rep}(\mathbb{T}^n)$. Moreover, if the characteristic of $\Bbbk$ is 0, then the Hopf algebra $\Bbbk\mathbb{Z}^n$ is isomorphic to the Hopf algebra $\mathcal{O}(\Bbbk^n_\times)$ (see Example 1.2.7) via the map $U_k \mapsto X_1^{k_1} \dots X_n^{k_n}$, where $k = (k_1, \dots, k_n) \in \mathbb{Z}^n$.

**Example 1.2.9** (The universal enveloping algebra of a Lie algebra). Let $\mathfrak{g}$ be a Lie algebra over $\Bbbk$. Recall that the *universal enveloping algebra* $U(\mathfrak{g})$ of $\mathfrak{g}$ is the universal unital algebra generated by elements of $\mathfrak{g}$, subject to the relation $xy - yx = [x, y]$ for all $x, y \in \mathfrak{g}$. Equivalently, $U(\mathfrak{g})$ can be characterized by the following universal property: the algebra $U(\mathfrak{g})$ contains $\mathfrak{g}$ as a subspace, and for every unital algebra $A$ and every linear map $F: \mathfrak{g} \to A$ that satisfies $F([x, y]) = F(x)F(y) - F(y)F(x)$ for all $x, y \in \mathfrak{g}$, there exists a unique unital algebra homomorphism $U(\mathfrak{g}) \to A$ that extends $F$. This universal property implies that the linear maps

$$\begin{aligned} &\Delta: \mathfrak{g} \to U(\mathfrak{g}) \otimes U(\mathfrak{g}), && \epsilon: \mathfrak{g} \to \Bbbk, && S: \mathfrak{g} \to \mathfrak{g}, \\ &\Delta(x) := x \otimes 1 + 1 \otimes x, && \epsilon(x) := 0, && S(x) := -x, \end{aligned}$$

extend to unital algebra homomorphisms

$$\Delta: U(\mathfrak{g}) \to U(\mathfrak{g}) \otimes U(\mathfrak{g}), \quad \epsilon: U(\mathfrak{g}) \to \Bbbk, \quad S: U(\mathfrak{g}) \to U(\mathfrak{g})^{\mathrm{op}},$$

where $U(\mathfrak{g})^{\mathrm{op}}$ denotes the opposite algebra of $U(\mathfrak{g})$, that is, the algebra obtained by reversing the multiplication. Indeed, for all $x, y \in \mathfrak{g}$,

$$\begin{aligned} [\Delta(x), \Delta(y)] &= [x \otimes 1 + 1 \otimes x,\ y \otimes 1 + 1 \otimes y] \\ &= [x, y] \otimes 1 + 1 \otimes [x, y] = \Delta([x, y]), \\ [\epsilon(x), \epsilon(y)] &= 0 = \epsilon([x, y]), \\ [S(x), S(y)] &= (-y)(-x) - (-x)(-y) = -[x, y] = S([x, y]). \end{aligned}$$

Routine calculations show that these maps turn $U(\mathfrak{g})$ into a Hopf algebra. For further details, see [65, Chapter XVI], [56, Chapter 3], or [181, Sections 3.2, 3.4].

Let us give an almost trivial example. Consider $\Bbbk^n$ as a Lie algebra with trivial Lie bracket given by $[c, d] = 0$ for all $c, d \in \Bbbk^n$. A comparison with Example 1.2.7 shows that if the characteristic of $\Bbbk$ is 0, then $U(\Bbbk^n)$ is isomorphic to the Hopf algebra $\mathcal{O}(\Bbbk^n)$.

**Example 1.2.10.** Important examples of Hopf algebras which attracted much attention over the last decades are $q$-deformations of the coordinate Hopf algebra $\mathcal{O}(G)$ of a semisimple complex Lie group $G$, and of the universal enveloping algebra $U(\mathfrak{g})$ of a semisimple complex Lie algebra $\mathfrak{g}$. Comprehensive accounts of these $q$-deformed Hopf algebras can be found in many books, for example, in [23], [24], [68], [79], [80], [84], [103], [140]; see also the introduction to Chapter 6.

## 1.3 Axiomatics of Hopf algebras

To obtain a better understanding of the concept of a Hopf algebra, we shall take a closer look at the different algebraic structures that appear in the definition. We begin with the comultiplication, which we first consider separately and then in combination with the multiplication of the underlying algebra. Next, we turn to the antipode and deduce a list of standard relations which are permanently used in the theory of Hopf algebras. Finally, we characterize Hopf algebras in terms of two natural maps which are important for generalizations of Hopf algebras in the setting of non-unital algebras, of $C^*$-algebras, and of von Neumann algebras.

### 1.3.1 Coalgebras and bialgebras

At the beginning, the theory of Hopf algebras may be difficult to learn because it is based not only on the language of algebras and modules, but also on the less familiar language of coalgebras and comodules. For example, a Hopf algebra is a unital algebra and simultaneously a counital coalgebra. These structures are compatible in a sense that is made precise by the concept of a bialgebra, and the existence of an antipode can be characterized in terms of a convolution product which merges the two structure maps. In the following paragraphs, we explain these statements and the terms involved.

An algebra $A$ can be considered as a vector space equipped with a linear map $m\colon A \otimes A \to A,\ a \otimes b \mapsto ab$, which is associative in the sense that the square

$$\begin{array}{ccc} A \otimes A \otimes A & \xrightarrow{m\otimes \mathrm{id}} & A \otimes A \\ {\scriptstyle \mathrm{id}\otimes m}\big\downarrow & & \big\downarrow{\scriptstyle m} \\ A \otimes A & \xrightarrow[m]{} & A \end{array} \tag{1.5}$$

commutes. Each element $a \in A$ gives rise to a linear map $\eta\colon \Bbbk \to A,\ \lambda \mapsto \lambda a$, and the element $a$ is the unit for the multiplication if and only if the diagram

$$\begin{array}{ccccc} \Bbbk \otimes A & \xleftarrow{\cong} & A & \xrightarrow{\cong} & A \otimes \Bbbk \\ {\scriptstyle \eta\otimes \mathrm{id}}\big\downarrow & & \big\| & & \big\downarrow{\scriptstyle \mathrm{id}\otimes\eta} \\ A \otimes A & \xrightarrow[m]{} & A & \xleftarrow[m]{} & A \otimes A \end{array} \tag{1.6}$$

commutes. Reversing all arrows in this description of an algebra, we obtain the definition of a coalgebra:

**Definition 1.3.1.** A *coalgebra* (*over* $\Bbbk$) is a vector space $A$ equipped with a linear map $\Delta\colon A \to A \otimes A$ called the *coproduct* or *comultiplication* that is *coassociative* in the sense that $(\Delta \otimes \mathrm{id}_A) \circ \Delta = (\mathrm{id}_A \otimes \Delta) \circ \Delta$, or, equivalently, that diagram (1.1) on page 5 commutes.

Let $(A, \Delta)$ be a coalgebra. A linear map $\epsilon\colon A \to \Bbbk$ is a *counit* for $(A, \Delta)$ if $(\epsilon \otimes \mathrm{id}_A) \circ \Delta = \mathrm{id}_A = (\mathrm{id}_A \otimes \epsilon) \circ \Delta$, that is, if the diagram (1.2) on page 5 commutes. A coalgebra is called *counital* if it has a counit.

A *morphism of coalgebras* $(A, \Delta_A)$ and $(B, \Delta_B)$ is a linear map $F: A \to B$ that satisfies $\Delta_B \circ F = (F \otimes F) \circ \Delta_A$. The map $F$ is *counital* if $A$ and $B$ have counits $\epsilon_A$ and $\epsilon_B$, respectively, and if $\epsilon_B \circ F = \epsilon_A$.

If $(A, \Delta)$ is a coalgebra and the comultiplication $\Delta$ is understood, we freely speak of $A$ itself as a coalgebra.

**Remarks 1.3.2.** i) Every coalgebra has at most one counit. Indeed, if $\epsilon_1$ and $\epsilon_2$ are counits for a coalgebra $(A, \Delta)$, then

$$\epsilon_1 = \epsilon_1 \circ (\mathrm{id}_A \otimes \epsilon_2) \circ \Delta = (\epsilon_1 \otimes \epsilon_2) \circ \Delta = \epsilon_2 \circ (\epsilon_1 \otimes \mathrm{id}_A) \circ \Delta = \epsilon_2.$$

ii) Given coalgebras $(A, \Delta_A)$ and $(B, \Delta_B)$, we can construct the following new coalgebras:

*Coopposite coalgebra.* Denote by $\Sigma: A \otimes A \to A \otimes A$ the flip map $a \otimes b \mapsto b \otimes a$. Then $(A, \Delta_A)^{\mathrm{cop}} := (A, \Sigma \circ \Delta_A)$ is a coalgebra, called the *coopposite coalgebra of* $(A, \Delta_A)$. Evidently, a linear map $\epsilon: A \to \Bbbk$ is a counit for $(A, \Delta_A)$ if and only if it is a counit for $(A, \Delta_A)^{\mathrm{cop}}$. The coalgebra $(A, \Delta_A)$ is called *cocommutative* if $(A, \Delta_A)^{\mathrm{cop}} = (A, \Delta_A)$, that is, if $\Sigma \circ \Delta_A = \Delta_A$.

*Direct sum.* Denote by $\Delta_{A \oplus B}$ the composition of the map $\Delta_A \oplus \Delta_B: A \oplus B \to (A \otimes A) \oplus (B \otimes B)$ with the natural inclusion $(A \otimes A) \oplus (B \otimes B) \hookrightarrow (A \oplus B) \otimes (A \oplus B)$. Then $(A \oplus B, \Delta_{A \oplus B})$ is a coalgebra. If $A$ and $B$ possess counits $\epsilon_A$ and $\epsilon_B$, respectively, then the map $(a, b) \mapsto \epsilon_A(a) + \epsilon_B(b)$ is a counit for $(A \oplus B, \Delta_{A \oplus B})$.

*Tensor product.* Denote by $\Delta_{A \otimes B}$ the composition of the map $\Delta_A \otimes \Delta_B: A \otimes B \to A \otimes A \otimes B \otimes B$ with the isomorphism $A \otimes A \otimes B \otimes B \xrightarrow{\cong} A \otimes B \otimes A \otimes B$ given by $a_1 \otimes a_2 \otimes b_1 \otimes b_2 \mapsto a_1 \otimes b_1 \otimes a_2 \otimes b_2$. Then $(A \otimes B, \Delta_{A \otimes B})$ is a coalgebra. If $A$ and $B$ possess counits $\epsilon_A$ and $\epsilon_B$, respectively, then the map $a \otimes b \mapsto \epsilon_A(a)\epsilon_B(b)$ is a counit for $(A \otimes B, \Delta_{A \otimes B})$.

For calculations in coalgebras, the following *Sweedler notation* or *Sigma notation* is very useful.

**Notation 1.3.3.** Let $(A, \Delta)$ be a coalgebra and $a \in A$. Then $\Delta(a) \in A \otimes A$ can be written in the form $\Delta(a) = \sum_i a_{1,i} \otimes a_{2,i}$, where $a_{1,i}, a_{2,i} \in A$. To simplify the notation, we suppress the summation index $i$ and write

$$\Delta(a) = \sum_i a_{1,i} \otimes a_{2,i} =: \sum a_{(1)} \otimes a_{(2)}.$$

Here, the subscripts "(1)" and "(2)" indicate the order of the factors in the tensor product; thus, for example, $\Sigma(\Delta(a)) = \sum a_{(2)} \otimes a_{(1)}$. We extend this notation to iterated applications of $\Delta$ as follows. Since $\Delta$ is coassociative, the elements

$$(\mathrm{id}_A \otimes \Delta)(\Delta(a)) = \sum a_{(1)} \otimes \Delta(a_{(2)}) = \sum a_{(1)} \otimes (a_{(2)})_{(1)} \otimes (a_{(2)})_{(2)}$$

and

$$(\Delta \otimes \mathrm{id}_A)(\Delta(a)) = \sum \Delta(a_{(1)}) \otimes a_{(2)} = \sum (a_{(1)})_{(1)} \otimes (a_{(1)})_{(2)} \otimes a_{(2)}$$

are equal. We write this element as $\sum a_{(1)} \otimes a_{(2)} \otimes a_{(3)}$. More generally, consider the maps $\Delta^{(n)} \colon A \to A \otimes \cdots \otimes A$ ($n+1$ factors), inductively defined by

$$\Delta^{(0)} := \mathrm{id}_A \quad \text{and} \quad \Delta^{(n+1)} := (\Delta^{(n)} \otimes \mathrm{id}_A) \circ \Delta \quad \text{for } n \geq 0.$$

By coassociativity, every map $A \to A \otimes \cdots \otimes A$ that is obtained by $n$ successive applications of $\Delta$ to one factor of the intermediate tensor product $A \otimes \cdots \otimes A$ coincides with $\Delta^{(n)}$. We write

$$\Delta^{(n)}(a) =: \sum a_{(1)} \otimes \cdots \otimes a_{(n+1)}.$$

**Examples 1.3.4.** i) Let $(A, \Delta_A)$ and $(B, \Delta_B)$ be coalgebras. Then for all $a \in A$, $b \in B$,

$$\begin{aligned}
\Sigma(\Delta(a)) &= \sum a_{(2)} \otimes a_{(1)}, \\
\Delta_{A\oplus B}((a,b)) &= \sum (a_{(1)}, 0) \otimes (a_{(2)}, 0) + \sum (0, b_{(1)}) \otimes (0, b_{(2)}), \\
\Delta_{A\otimes B}(a \otimes b) &= \sum a_{(1)} \otimes b_{(1)} \otimes a_{(2)} \otimes b_{(2)}.
\end{aligned}$$

ii) In Sweedler notation, the axioms for the counit and the antipode of a Hopf algebra $A$ take the form

$$\sum \epsilon(a_{(1)})a_{(2)} = a = \sum a_{(1)}\epsilon(a_{(2)})$$

and

$$\sum S(a_{(1)})a_{(2)} = \eta(\epsilon(a)) = \sum a_{(1)}S(a_{(2)}) \quad \text{for all } a \in A.$$

A combination of these axioms yields the following useful formula:

$$\begin{aligned}
\sum S(a_{(1)})a_{(2)} \otimes a_{(3)} &= \sum S((a_{(1)})_{(1)})(a_{(1)})_{(2)} \otimes a_{(2)} \\
&= \sum \eta(\epsilon(a_{(1)})) \otimes a_{(2)} = \sum 1_A \otimes \epsilon(a_{(1)})a_{(2)} = 1_A \otimes a.
\end{aligned}$$

Next, we consider algebra structures and coalgebra structures that are compatible in a natural sense.

**Lemma 1.3.5.** *Let $A$ be a vector space equipped with the structure of an algebra and of a coalgebra. Then the following conditions are equivalent:*

i) *The comultiplication $\Delta\colon A \to A \otimes A$ is an algebra homomorphism.*

ii) *The multiplication $m\colon A \otimes A \to A$ is a morphism of coalgebras.*

iii) *The following diagram commutes:*

$$\begin{array}{ccccc} A \otimes A & & \xrightarrow{\quad m \quad} & & A \\ \downarrow{\scriptstyle \Delta\otimes\Delta} & & & & \downarrow{\scriptstyle \Delta} \\ A \otimes A \otimes A \otimes A & \xrightarrow{\mathrm{id}\otimes\Sigma\otimes\mathrm{id}} & A \otimes A \otimes A \otimes A & \xrightarrow[m\otimes m]{} & A \otimes A. \end{array} \tag{1.7}$$

*Proof.* Condition iii) is equivalent to i) and ii) because the multiplication and comultiplication of $A \otimes A$ are given by

$$(A \otimes A) \otimes (A \otimes A) \xrightarrow{\mathrm{id}\otimes\Sigma\otimes\mathrm{id}} A \otimes A \otimes A \otimes A \xrightarrow{m\otimes m} A \otimes A$$

and

$$A \otimes A \xrightarrow{\Delta\otimes\Delta} A \otimes A \otimes A \otimes A \xrightarrow{\mathrm{id}\otimes\Sigma\otimes\mathrm{id}} (A \otimes A) \otimes (A \otimes A)$$

respectively. □

**Definition 1.3.6.** A *bialgebra* (*over* $\Bbbk$) is a vector space $A$ equipped with the structure of an algebra and a coalgebra such that diagram (1.7) commutes. We shall usually not mention the multiplication map $m$ explicitly and refer to the pair $(A, \Delta)$ consisting of the algebra $A$ and the comultiplication $\Delta$ as a bialgebra.

A bialgebra is called *unital* if it is unital as an algebra and the comultiplication is a unital algebra homomorphism; it is called *counital* if it is counital as a coalgebra and the multiplication is a counital morphism of coalgebras.

A *morphism of bialgebras* $(A, \Delta_A)$ and $(B, \Delta_B)$ is a linear map $F\colon A \to B$ that is a morphism of algebras and of coalgebras. It is called *unital/counital* if it is unital/counital as a map of algebras/coalgebras.

**Remarks 1.3.7.** i) Often, bialgebras are assumed to be unital and counital. We explicitly state these assumptions whenever we impose them.

ii) For a unital/counital bialgebra $(A, \Delta)$, the compatibility conditions between the unit and the comultiplication / between the counit and the multiplication amount to the commutativity of the following squares:

$$\begin{CD} \Bbbk @>{\eta}>> A \\ @V{\cong}VV @VV{\Delta}V \\ \Bbbk\otimes\Bbbk @>>{\eta\otimes\eta}> A\otimes A \end{CD} \qquad \Big/ \qquad \begin{CD} A\otimes A @>{\epsilon\otimes\epsilon}>> \Bbbk\otimes\Bbbk \\ @V{m}VV @VV{\cong}V \\ A @>>{\epsilon}> \Bbbk. \end{CD} \tag{1.8}$$

In particular, $(A, \Delta)$ is counital if and only if the counit $\epsilon$ is multiplicative.

ii) Given bialgebras $(A, \Delta_A)$ and $(B, \Delta_B)$, we can construct the following new bialgebras:

*Opposite and coopposite bialgebra.* Reversing the multiplication, the comultiplication, or both, of $(A, \Delta_A)$, we obtain three new bialgebras. More precisely, denote by $\Sigma\colon A\otimes A \to A\otimes A$ the flip, and by $A^{\mathrm{op}}$ the opposite algebra of $A$. Then the pairs $(A, \Delta_A)^{\mathrm{op}} := (A^{\mathrm{op}}, \Delta_A)$, $(A, \Delta_A)^{\mathrm{cop}} := (A, \Sigma \circ \Delta_A)$, $(A, \Delta_A)^{\mathrm{op,cop}} := (A^{\mathrm{op}}, \Sigma \circ \Delta_A)$ are bialgebras again.

*Direct sum and tensor product.* The vector spaces $A \oplus B$ and $A \otimes B$ are bialgebras with respect to the usual algebra structure and the coalgebra structure defined in Remark 1.3.2 ii).

iv) Evidently, every Hopf algebra is a unital and counital bialgebra. Furthermore, every unital and counital bialgebra $(A, \Delta)$ admits at most one antipode that turns it into a Hopf algebra: if $S_1$ and $S_2$ are antipodes for $A$, then

$$\begin{aligned} S_1(a) = \sum \epsilon(a_{(1)})S_1(a_{(2)}) &= \sum S_2(a_{(1)})a_{(2)}S_1(a_{(3)}) \\ &= \sum S_2(a_{(1)})\epsilon(a_{(2)}) = S_2(a) \quad \text{for all } a \in A. \end{aligned}$$

Remarks 1.3.2 i) and 1.3.7 iv) imply that the counit and the antipode of a Hopf algebra are uniquely determined by the comultiplication. Therefore, we shall refer to a pair $(A, \Delta)$ as a Hopf algebra if $A$ is a Hopf algebra with comultiplication $\Delta$.

### 1.3.2 Convolution

Let $(A, \Delta)$ be a coalgebra and $B$ an algebra over $\Bbbk$. Then the space of linear maps $\mathrm{Hom}_{\Bbbk}(A, B)$ carries an important *convolution product*, defined by

$$f * g := m_B \circ (f \otimes g) \circ \Delta_A \quad \text{for all } f, g \in \mathrm{Hom}_{\Bbbk}(A, B),$$

that is,

$$(f * g)(a) := \sum f(a_{(1)})g(a_{(2)}) \quad \text{for all } f, g \in \mathrm{Hom}_{\Bbbk}(A, B), a \in A.$$

The convolution product is associative because

$$((f * g) * h)(a) = \sum f(a_{(1)})g(a_{(2)})h(a_{(3)}) = (f * (g * h))(a)$$

for all $f, g, h \in \mathrm{Hom}_{\Bbbk}(A, B)$ and $a \in A$. Thus $\mathrm{Hom}_{\Bbbk}(A, B)$ becomes an algebra.

Particularly interesting is the case $B = \Bbbk$: The convolution product turns $A' = \mathrm{Hom}_{\Bbbk}(A, \Bbbk)$ into an algebra. When we speak of $A'$ as an algebra, we always refer to this algebra structure.

**Remarks 1.3.8.** i) If $(A, \Delta)$ is a finite-dimensional bialgebra or Hopf algebra, then $A'$ can also be equipped with the structure of a bialgebra or Hopf algebra, see Theorem 1.4.1.

ii) In general, the convolution algebra $\mathrm{Hom}_{\Bbbk}(A, B)$ need not be unital. But if $(A, \Delta)$ has a counit $\epsilon_A$ and $B$ has a unit $\eta_B$, then the composition $\eta_B \circ \epsilon_A$ is a unit for the convolution algebra $\mathrm{Hom}_{\Bbbk}(A, B)$. Indeed,

$$((\eta_B \circ \epsilon_A) * f)(a) = \sum 1_B \epsilon_A(a_{(1)}) f(a_{(2)}) = \sum f(\epsilon_A(a_{(1)})a_{(2)}) = f(a)$$

and similarly $(f * (\eta_B \circ \epsilon_A))(a) = f(a)$ for all $f \in \mathrm{Hom}_{\Bbbk}(A, B)$ and $a \in A$.

iii) Every algebra homomorphism $F\colon B \to C$ induces an algebra homomorphism

$$F_*\colon \mathrm{Hom}_{\Bbbk}(A, B) \to \mathrm{Hom}_{\Bbbk}(A, C), \quad f \mapsto F \circ f.$$

Indeed, for all $f, g \in \mathrm{Hom}_{\Bbbk}(A, B)$ and $a \in A$,

$$\begin{aligned}((F_* f) * (F_* g))(a) &= \sum F(f(a_{(1)}))F(g(a_{(2)})) \\ &= \sum F(f(a_{(1)})g(a_{(2)})) = (F_*(f * g))(a).\end{aligned}$$

Likewise, if $(D, \Delta_D)$ is a coalgebra and $G\colon D \to A$ is a morphism of coalgebras, then

$$G^*\colon \mathrm{Hom}_{\Bbbk}(A, B) \to \mathrm{Hom}_{\Bbbk}(D, B), \quad g \mapsto g \circ G,$$

is an algebra homomorphism. The assignments $((A, \Delta), B) \mapsto \mathrm{Hom}_{\Bbbk}(A, B)$ and $(F, G) \mapsto F_* \circ G^* = G^* \circ F_*$ define a bifunctor from the categories of coalgebras and algebras to the category of algebras.

Using the convolution product, we can characterize Hopf algebras among bialgebras as follows:

**Remark 1.3.9.** Let $(A, \Delta)$ be a unital and counital bialgebra and $S\colon A \to A$ a linear map. Then diagram (1.3) on page 5 commutes if and only if $S$ is inverse to the identity map $\mathrm{id}_A$ in the convolution algebra $\mathrm{Hom}_{\Bbbk}(A, A)$, that is, if $S * \mathrm{id}_A = \eta \circ \epsilon = \mathrm{id}_A * S$. In particular, the bialgebra $(A, \Delta)$ can be equipped with the structure of a Hopf algebra if and only if the identity map $\mathrm{id}_A$ is invertible in the convolution algebra $\mathrm{Hom}_{\Bbbk}(A, A)$.

Let us turn to another important convolution product. Assume that $(A, \Delta)$ is a coalgebra. For each $a \in A$ and $f \in A'$, we define

$$f * a := (\mathrm{id}_A \otimes f)(\Delta(a)) = \sum a_{(1)} f(a_{(2)})$$

and

$$a * f := (f \otimes \mathrm{id}_A)(\Delta(a)) = \sum f(a_{(1)}) a_{(2)}.$$

**Lemma 1.3.10.** *If $(A, \Delta)$ is a bialgebra, then the maps*

$$A' \times A \to A, \ (f, a) \mapsto f * a, \quad A \times A' \to A, \ (a, f) \mapsto a * f,$$

*turn $A$ into a bimodule over the algebra $A'$.*

*Proof.* First, we show that the map $(f, a) \mapsto f * a$ turns $A$ into a left module over $A'$. For each $h \in A'$, put $\rho(h) := (\mathrm{id} \otimes h) \circ \Delta \colon A \to A$. We need to show that $\rho(f * g) = \rho(f)\rho(g)$ for all $f, g \in A'$. But

$$\begin{aligned}\rho(f * g) = (\mathrm{id} \otimes (f * g)) \circ \Delta &= (\mathrm{id} \otimes f \otimes g) \circ (\mathrm{id} \otimes \Delta) \circ \Delta \\ &= (\mathrm{id} \otimes f \otimes g) \circ (\Delta \otimes \mathrm{id}) \circ \Delta \\ &= (\mathrm{id} \otimes f) \circ \Delta \circ (\mathrm{id} \otimes g) \circ \Delta = \rho(f)\rho(g).\end{aligned}$$

A similar argument shows that the map $(a, f) \mapsto a * f$ turns $A$ into a right module over $A'$. Finally, these maps turn $A$ into a bimodule because for all $f, g \in A'$ and $a \in A$,

$$(f * a) * g = \sum f(a_{(1)}) a_{(2)} g(a_{(3)}) = f * (a * g). \qquad \square$$

**Remark 1.3.11.** If $(A, \Delta)$ is a coalgebra with counit $\epsilon$, then $\epsilon * a = a = a * \epsilon$ for all $a \in A$; if $(A, \Delta)$ is a Hopf algebra, then $S * a = \eta(\epsilon(a)) = a * S$ for all $a \in A$.

### 1.3.3 Properties of the antipode

The antipode of a Hopf algebra satisfies several fundamental relations that are not obvious from the definition. To some extent, the antipode of a Hopf algebra behaves like the inversion of a group: the inversion of a group is antimultiplicative, and the antipode of a Hopf algebra is both antimultiplicative and anticomultiplicative.

**Proposition 1.3.12.** *The antipode of a Hopf algebra $(A, \Delta)$ is a unital and counital morphism $(A, \Delta) \to (A, \Delta)^{\mathrm{op,cop}}$ of bialgebras, that is, the following conditions hold:*

$$\text{i) } S \circ m = m \circ \Sigma \circ (S \otimes S), \qquad \text{ii) } S \circ \eta = \eta,$$

$$\text{iii) } \Delta \circ S = (S \otimes S) \circ \Sigma \circ \Delta, \qquad \text{iv) } \epsilon \circ S = \epsilon.$$

*Equivalently, for all $a, b \in A$,*

i)′ $S(ab) = S(b)S(a)$, ii)′ $S(1_A) = 1_A$,

iii)′ $\sum (S(a))_{(1)} \otimes (S(a))_{(2)} = \sum S(a_{(2)}) \otimes S(a_{(1)})$, iv)′ $\epsilon(S(a)) = \epsilon(a)$.

*Proof.* i) Consider $A \otimes A$ as a coalgebra (see Remark 1.3.2 ii)) and $\mathrm{Hom}_{\Bbbk}(A \otimes A, A)$ as an algebra with respect to the convolution product (see Section 1.3.2). We claim:

$$(S \circ m) * m = \eta \circ \epsilon \circ m = m * (m \circ \Sigma \circ (S \otimes S)). \tag{1.9}$$

Since $\epsilon \circ m : A \otimes A \to \Bbbk$ is the counit of $A \otimes A$, the map $\eta \circ \epsilon \circ m : A \otimes A \to A$ is the unit in $\mathrm{Hom}_{\Bbbk}(A \otimes A, A)$ (see Remark 1.3.8 ii)). Thus, equation (1.9) implies that $m$ is invertible with respect to the convolution product, and that the inverse of $m$ is $S \circ m = m \circ \Sigma \circ (S \otimes S)$. Let us prove (1.9). For all $a, b \in A$,

$$\begin{aligned} ((S \circ m) * m)(a \otimes b) &= \sum S(m(a_{(1)} \otimes b_{(1)})) \cdot m(a_{(2)} \otimes b_{(2)}) \\ &= \sum S(a_{(1)} b_{(1)}) a_{(2)} b_{(2)} \\ &= \sum S((ab)_{(1)})(ab)_{(2)} = \eta(\epsilon(ab)), \end{aligned}$$

$$\begin{aligned} (m * (m \circ \Sigma \circ (S \otimes S)))(a \otimes b) &= \sum m(a_{(1)} \otimes b_{(1)}) \cdot m(S(b_{(2)}) \otimes S(a_{(2)})) \\ &= \sum a_{(1)} b_{(1)} S(b_{(2)}) S(a_{(2)}) \\ &= \sum a_{(1)} \eta(\epsilon(b)) S(a_{(2)}) = \eta(\epsilon(a)) \eta(\epsilon(b)). \end{aligned}$$

ii) We have $1_A = (\eta \circ \epsilon)(1_A) = (m \circ (S \otimes \mathrm{id}) \circ \Delta)(1_A) = S(1_A)1_A$.

iii) Consider $A \otimes A$ as an algebra and $\mathrm{Hom}_{\Bbbk}(A, A \otimes A)$ as an algebra with respect to the convolution product. We claim:

$$\Delta * (\Delta \circ S) = \Delta \circ \eta \circ \epsilon = ((S \otimes S) \circ \Sigma \circ \Delta) * \Delta. \tag{1.10}$$

Since $\Delta(1_A) = 1_A \otimes 1_A$, the map $\Delta \circ \eta : \Bbbk \to A \otimes A$ is the unit map of $A \otimes A$. Therefore $\Delta \circ \eta \circ \epsilon$ is the unit in $\mathrm{Hom}_{\Bbbk}(A, A \otimes A)$ (see Remark 1.3.8 ii)). Thus, equation (1.10) implies that $\Delta$ is invertible with respect to the convolution and that the inverse of $\Delta$ is $\Delta \circ S = (S \otimes S) \circ \Sigma \circ \Delta$. Let us prove (1.10). For all $a \in A$,

$$\begin{aligned}(\Delta * (\Delta \circ S))(a) &= \sum \Delta(a_{(1)})\Delta(S(a_{(2)})) \\ &= \sum \Delta(a_{(1)}S(a_{(2)})) = (\Delta \circ \eta \circ \epsilon)(a),\end{aligned}$$

$$\begin{aligned}(((S \otimes S) \circ \Sigma \circ \Delta) * \Delta)(a) &= \sum (S(a_{(2)}) \otimes S(a_{(1)})) \cdot (a_{(3)} \otimes a_{(4)}) \\ &= \sum S(a_{(2)})a_{(3)} \otimes S(a_{(1)})a_{(4)} \\ &= \sum 1_A \otimes \epsilon(a_{(2)})S(a_{(1)})a_{(3)} \\ &= \sum 1_A \otimes S(a_{(1)})a_{(2)} = (\Delta \circ \eta \circ \epsilon)(a).\end{aligned}$$

iv) For all $a \in A$,

$$\begin{aligned}\epsilon(S(a)) &= \sum \epsilon(S(a_{(1)}))\epsilon(a_{(2)}) \\ &= \sum \epsilon(S(a_{(1)})a_{(2)}) = (\epsilon \circ \eta \circ \epsilon)(a) = \epsilon(a).\end{aligned}$$ □

**Remarks 1.3.13.** i) In the proof above, the equations $(S \circ m) * m = \eta \circ \epsilon \circ m$ and $\Delta * (\Delta \circ S) = \Delta \circ \eta \circ \epsilon$ can also be deduced from Remark 1.3.8 iii): since $m$ is a morphism of coalgebras and $\Delta$ is an algebra homomorphism,

$$(S \circ m) * m = m^*(S) * m^*(\mathrm{id}_A) = m^*(S * \mathrm{id}_A) = m^*(\eta \circ \epsilon) = \eta \circ \epsilon \circ m,$$
$$\Delta * (\Delta \circ S) = \Delta_*(\mathrm{id}_A) * \Delta_*(S) = \Delta_*(\mathrm{id}_A * S) = \Delta_*(\eta \circ \epsilon) = \Delta \circ \eta \circ \epsilon.$$

ii) Assuming some familiarity with the Sweedler notation, we can verify conditions i)′ and ii)′ in Proposition 1.3.12 by direct calculations:

$$\begin{aligned}S(a)S(b) &= \sum S(a_{(1)})S(b_{(1)})\,\epsilon(b_{(2)})\epsilon(a_{(2)}) \\ &= \sum S(a_{(1)})S(b_{(1)})\epsilon(b_{(2)}a_{(2)}) \\ &= \sum S(a_{(1)})S(b_{(1)})(b_{(2)}a_{(2)})_{(1)}S((b_{(2)}a_{(2)})_{(2)}) \\ &= \sum S(a_{(1)})S(b_{(1)})b_{(2)}a_{(2)}S(b_{(3)}a_{(3)}) \\ &= \sum S(a_{(1)})a_{(2)}S(ba_{(3)}) = S(ba)\end{aligned}$$

and

$$\begin{aligned}\sum S(a_{(2)}) \otimes S(a_{(1)}) &= \sum (S(a_{(2)}) \otimes S(a_{(1)}))\Delta(\eta(\epsilon(a_{(3)}))) \\ &= \sum (S(a_{(2)}) \otimes S(a_{(1)}))\Delta(a_{(3)}S(a_{(4)})) \\ &= \sum (S(a_{(2)})a_{(3)} \otimes S(a_{(1)})a_{(4)})\Delta(S(a_{(5)})) \\ &= \sum (\eta(\epsilon(a_{(2)})) \otimes S(a_{(1)})a_{(3)})\Delta(S(a_{(4)})) \\ &= \sum (1_A \otimes S(a_{(1)})a_{(2)})\Delta(S(a_{(3)})) = \Delta(S(a)).\end{aligned}$$

The previous proposition implies that for every Hopf algebra, the square of the antipode is a morphism of the Hopf algebra. But unlike the inversion of a group, the antipode of a Hopf algebra need not be involutive, that is, $S^2$ need not be equal to the identity – the antipode need not even bijective.

**Proposition 1.3.14.** *For every Hopf algebra $(A, \Delta)$, the following conditions are equivalent:*

i) *The antipode $S$ of $(A, \Delta)$ is bijective.*

ii) *The bialgebra $(A, \Delta)^{\mathrm{op}}$ is a Hopf algebra.*

iii) *The bialgebra $(A, \Delta)^{\mathrm{cop}}$ is a Hopf algebra.*

*If these conditions hold, then $S^{-1}$ is the antipode of $(A, \Delta)^{\mathrm{op}}$ and of $(A, \Delta)^{\mathrm{cop}}$.*

The proof involves the following evident statement:

**Lemma 1.3.15.** *Let $(A, \Delta)$ be a Hopf algebra and $T : A \to A$ a linear map. Then the following conditions are equivalent:*

i) *The bialgebra $(A, \Delta)^{\mathrm{op}}$ is a Hopf algebra with antipode $T$.*

ii) $m \circ \Sigma \circ (T \otimes \mathrm{id}) \circ \Delta = \eta \circ \epsilon = m \circ \Sigma \circ (\mathrm{id} \otimes T) \circ \Delta$.

iii) $\sum a_{(2)}T(a_{(1)}) = \eta(\epsilon(a)) = \sum T(a_{(2)})a_{(1)}$ *for all $a \in A$.*

iv) $m \circ (\mathrm{id} \otimes T) \circ \Sigma \circ \Delta = \eta \circ \epsilon = m \circ (T \otimes \mathrm{id}) \circ \Sigma \circ \Delta$.

v) *The bialgebra $(A, \Delta)^{\mathrm{cop}}$ is a Hopf algebra with antipode $T$.* □

*Proof of Proposition* 1.3.14. i) ⇒ ii), iii): Suppose that $S$ is invertible. We use Lemma 1.3.15 to show that $S^{-1}$ is the antipode of $(A, \Delta)^{\mathrm{op}}$ and of $(A, \Delta)^{\mathrm{cop}}$. By Proposition 1.3.12,

$$\sum a_{(2)}S^{-1}(a_{(1)}) = \sum S^{-1}(a_{(1)}S(a_{(2)})) = S^{-1}(\eta(\epsilon(a))) = \eta(\epsilon(a))$$

and similarly $\sum S^{-1}(a_{(2)})a_{(1)} = \eta(\epsilon(a))$ for all $a \in A$.

ii), iii) $\Rightarrow$ i): Let $(A, \Delta)^{\mathrm{op}}$ or $(A, \Delta)^{\mathrm{cop}}$ be a Hopf algebra with antipode $T$. Then by Proposition 1.3.12 and Lemma 1.3.15,

$$\begin{aligned} S(T(a)) &= \sum \epsilon(a_{(2)})S(T(a_{(1)})) = \sum a_{(3)}T(a_{(2)})S(T(a_{(1)})) \\ &= \sum a_{(2)}T(a_{(1)})_{(1)}S(T(a_{(1)})_{(2)}) = \sum a_{(2)}\epsilon(T(a_{(1)})) = a \end{aligned}$$

and similarly $T(S(a)) = a$ for all $a \in A$. □

Note that for every Hopf algebra $(A, \Delta)$, the bialgebra $(A, \Delta)^{\mathrm{op,cop}}$ is a Hopf algebra with the same antipode as $(A, \Delta)$.

**Corollary 1.3.16.** *For every commutative or cocommutative Hopf algebra $(A, \Delta)$, we have $S^2 = \mathrm{id}_A$.*

*Proof.* In both cases, $S^{-1}$ and $S$ are antipodes for $(A, \Delta)^{\mathrm{op}}$ and $(A, \Delta)^{\mathrm{cop}}$. By Remark 1.3.7 iv), $S = S^{-1}$. □

**Proposition 1.3.17.** *Let $(A, \Delta_A)$ and $(B, \Delta_B)$ be Hopf algebras and $F : A \to B$ a unital and counital morphism of bialgebras. Then $F \circ S_A = S_B \circ F$, that is, $F$ is a morphism of Hopf algebras.*

*Proof.* Consider the convolution algebra $\mathrm{Hom}_{\Bbbk}(A, B)$. By Remark 1.3.8 ii), its unit is $\eta_B \circ \epsilon_A = \eta_B \circ \epsilon_B \circ F = F \circ \eta_A \circ \epsilon_A$, and by Remark 1.3.8 iii),

$$\begin{aligned} (S_B \circ F) * F &= (S_B * \mathrm{id}_B) \circ F = \eta_B \circ \epsilon_B \circ F \\ &= F \circ \eta_A \circ \epsilon_A = F \circ (\mathrm{id}_A * S_A) = F * (F \circ S_A). \end{aligned}$$

Therefore $F$ is invertible with respect to the convolution, and its inverse is $S_B \circ F = F \circ S_A$. □

### 1.3.4 Another characterization of Hopf algebras

By Remark 1.3.9, Hopf algebras can be characterized as those bialgebras for which the identity map is invertible with respect to the convolution product. Now, we present another characterization of Hopf algebras which is particularly well suited for generalizations to non-unital algebras and $C^*$-algebras. This characterization was given by Van Daele [174], who used it for the definition of multiplier Hopf algebras (see Chapter 2).

Let $(A, \Delta)$ be a unital bialgebra. Consider the linear maps

$$\begin{aligned} T_1 := (\mathrm{id} \otimes m) \circ (\Delta \otimes \mathrm{id})&: A \otimes A \to A \otimes A, \\ a \otimes b &\mapsto \sum a_{(1)} \otimes a_{(2)}b = \Delta(a)(1_A \otimes b), \end{aligned} \tag{1.11}$$

and

$$T_2 := (m \otimes \mathrm{id}) \circ (\mathrm{id} \otimes \Delta) \colon A \otimes A \to A \otimes A, \\ a \otimes b \mapsto \sum ab_{(1)} \otimes b_{(2)} = (a \otimes 1_A)\Delta(b). \tag{1.12}$$

These maps will play a central rôle in later chapters.

**Theorem 1.3.18.** *A unital bialgebra is a Hopf algebra if and only if the associated maps $T_1$ and $T_2$ are bijective.*

This theorem shows that the existence of a counit and of an antipode can be expressed solely in terms of the bialgebra structure. Before we proceed to the proof, we consider an analogous result concerning semigroups for motivation:

**Proposition 1.3.19.** *Let $\Gamma$ be a semigroup. Denote by $m_\Gamma \colon \Gamma \times \Gamma \to \Gamma$ the semigroup operation, and by $\Delta_\Gamma \colon \Gamma \to \Gamma \times \Gamma$ the diagonal embedding. Then $\Gamma$ is a group if and only if the following two maps are bijective:*

$$\Phi_1 := (\mathrm{id}_\Gamma \times m_\Gamma) \circ (\Delta_\Gamma \times \mathrm{id}_\Gamma) \colon \Gamma \times \Gamma \to \Gamma \times \Gamma, \quad (x, y) \mapsto (x, xy),$$
$$\Phi_2 := (m_\Gamma \times \mathrm{id}_\Gamma) \circ (\mathrm{id}_\Gamma \times \Delta_\Gamma) \colon \Gamma \times \Gamma \to \Gamma \times \Gamma, \quad (x, y) \mapsto (xy, y).$$

*Proof.* If $\Gamma$ is a group, then the maps $\Psi_1, \Psi_2 \colon \Gamma \times \Gamma \to \Gamma \times \Gamma$ given by $\Psi_1(x, y) := (x, x^{-1}y)$ and $\Psi_2(x, y) := (xy^{-1}, y)$ are inverse to $\Phi_1$ and $\Phi_2$, respectively.

Conversely, assume that the maps $\Phi_1$ and $\Phi_2$ are bijective. First, we show that $\Gamma$ has a unit. Fix $x \in \Gamma$. By assumption, the map $l_x \colon \Gamma \to \Gamma$ given by $y \mapsto xy$ is bijective. Hence there exists an element $e \in \Gamma$ such that $x = xe$. Then $zxe = zx$ for all $z \in \Gamma$, and by surjectivity of the map $r_x \colon \Gamma \to \Gamma$, $z \mapsto zx$, we get $y = ye$ for all $y \in \Gamma$. On the other hand, $xy = xey$ for all $y \in \Gamma$, and using injectivity of the map $l_x$, we find $y = ey$ for all $y \in \Gamma$. Thus $e$ is a unit.

Finally, by surjectivity of $l_x$ and $r_x$, the element $x$ has a right and a left inverse, which necessarily coincide. Since $x \in \Gamma$ was arbitrary, $\Gamma$ is a group. □

**Remark 1.3.20.** For a finite semigroup $\Gamma$, the maps $\Phi_1$ and $\Phi_2$ are bijective if and only if they are injective, and this holds if and only if $\Gamma$ has the following *cancellation property:* if $x, y_1, y_2 \in \Gamma$ satisfy $xy_1 = xy_2$, then $y_1 = y_2$, and if $x_1, x_2, y \in \Gamma$ satisfy $x_1 y = x_2 y$, then $x_1 = x_2$.

Now we return to the maps $T_1$ and $T_2$ introduced above and take one step towards the proof of Theorem 1.3.18:

**Lemma 1.3.21.** *Let $(A, \Delta)$ be a Hopf algebra. Then the maps $T_1$ and $T_2$ defined in (1.11) and (1.12), respectively, are bijective.*

*Proof.* Consider the maps

$$R_1 := (\mathrm{id} \otimes m) \circ (\mathrm{id} \otimes S \otimes \mathrm{id}) \circ (\Delta \otimes \mathrm{id})\colon A \otimes A \to A \otimes A,$$
$$a \otimes b \mapsto \sum a_{(1)} \otimes S(a_{(2)})b = ((\mathrm{id} \otimes S)(\Delta(a)))(1 \otimes b),$$

and

$$R_2 := (m \otimes \mathrm{id}) \circ (\mathrm{id} \otimes S \otimes \mathrm{id}) \circ (\mathrm{id} \otimes \Delta)\colon A \otimes A \to A \otimes A,$$
$$a \otimes b \mapsto \sum aS(b_{(1)}) \otimes b_{(2)} = (a \otimes 1)((S \otimes \mathrm{id})(\Delta(b))).$$

We show that $R_1$ is inverse to $T_1$, and a similar calculation shows that $R_2$ is inverse to $T_2$: for all $a, b \in A$,

$$\begin{aligned} R_1(T_1(a \otimes b)) &= R_1\Big(\sum a_{(1)} \otimes a_{(2)}b\Big) \\ &= \sum a_{(1)} \otimes S(a_{(2)})a_{(3)}b = \sum a_{(1)} \otimes \epsilon(a_{(2)})b = a \otimes b \end{aligned}$$

and

$$\begin{aligned} T_1(R_1(a \otimes b)) &= T_1\Big(\sum a_{(1)} \otimes S(a_{(2)})b\Big) \\ &= \sum a_{(1)} \otimes a_{(2)}S(a_{(3)})b = \sum a_{(1)} \otimes \epsilon(a_{(2)})b = a \otimes b. \end{aligned}$$ □

Before we prove the reverse implication, let us collect some useful relations between the maps $T_1$, $T_2$, the multiplication $m$, and the comultiplication $\Delta$:

$$\begin{aligned} (\Delta \otimes \mathrm{id}) \circ T_1 &= (\mathrm{id} \otimes T_1) \circ (\Delta \otimes \mathrm{id})\colon a \otimes b \mapsto \sum a_{(1)} \otimes a_{(2)} \otimes a_{(3)}b, \\ T_1 \circ (\mathrm{id} \otimes m) &= (\mathrm{id} \otimes m) \circ (T_1 \otimes \mathrm{id})\colon a \otimes b \otimes c \mapsto \sum a_{(1)} \otimes a_{(2)}bc, \\ (\mathrm{id} \otimes \Delta) \circ T_2 &= (T_2 \otimes \mathrm{id}) \circ (\mathrm{id} \otimes \Delta)\colon a \otimes b \mapsto \sum ab_{(1)} \otimes b_{(2)} \otimes b_{(3)}, \\ T_2 \circ (m \otimes \mathrm{id}) &= (m \otimes \mathrm{id}) \circ (\mathrm{id} \otimes T_2)\colon a \otimes b \otimes c \mapsto \sum abc_{(1)} \otimes c_{(2)}. \end{aligned} \tag{1.13}$$

The following proposition completes the proof of Theorem 1.3.18:

**Proposition 1.3.22** ([174]). *Let $(A, \Delta)$ be a unital bialgebra. If the maps $T_1$ and $T_2$ defined in* (1.11) *and* (1.12) *are bijective, then $(A, \Delta)$ is a Hopf algebra.*

*Proof.* We need to construct a counit and an antipode for $(A, \Delta)$. Let us start with the counit. The proof of the previous lemma shows that if $(A, \Delta)$ were a Hopf algebra, then we could express the counit $\epsilon$ in terms of the map $T_1$ as follows: $\epsilon(a) = \sum a_{(1)}S(a_{(2)}) = m(T_1^{-1}(a \otimes 1_A))$ for all $a \in A$. So, consider the map

$$E\colon A \to A,\ a \mapsto m(T_1^{-1}(a \otimes 1_A)).$$

We show that the image of $E$ is contained in $\Bbbk \cdot 1_A$. Using the first relation in (1.13), we find that for all $a, b \in A$,

$$\begin{aligned}(\mathrm{id} \otimes E)((a \otimes 1_A)\Delta(b)) &= (a \otimes 1_A) \cdot (\mathrm{id} \otimes m)((\mathrm{id} \otimes T_1^{-1})(\Delta(b) \otimes 1_A)) \\ &= (a \otimes 1_A) \cdot (\mathrm{id} \otimes m)((\Delta \otimes \mathrm{id})(T_1^{-1}(b \otimes 1_A))) \\ &= (a \otimes 1_A) \cdot T_1(T_1^{-1}(b \otimes 1_A)) = ab \otimes 1_A.\end{aligned}$$

Since $T_2$ is surjective, elements of the form $(a \otimes 1_A)\Delta(b)$ span $A \otimes A$. Therefore, the calculation above shows that the image of $E$ is contained in $\Bbbk \cdot 1_A$.

Define $\epsilon\colon A \to \Bbbk$ by $E(a) = \epsilon(a) \cdot 1_A$ for all $a \in A$. We show that $\epsilon$ is a counit. Let $b \in A$. By the calculation above, $(\mathrm{id} \otimes \epsilon)(\Delta(b)) = b$. Using the second equation in (1.13), we find

$$\begin{aligned}(\epsilon \otimes \mathrm{id})(\Delta(b)) &= \sum m(T_1^{-1}(b_{(1)} \otimes 1_A)) \cdot b_{(2)} \\ &= m((\mathrm{id} \otimes m)(T_1^{-1}(b_{(1)} \otimes 1_A) \otimes b_{(2)})) \\ &= m(T_1^{-1}(b_{(1)} \otimes b_{(2)})) = m(b \otimes 1_A) = b.\end{aligned}$$

It remains to show that $\epsilon$ is a homomorphism. The previous results and multiplicativity of $\Delta$ imply

$$\sum \epsilon(a_{(1)})\epsilon(b_{(1)})a_{(2)}b_{(2)}c = abc = \sum \epsilon(a_{(1)}b_{(1)})a_{(2)}b_{(2)}c \quad \text{for all } a, b, c \in A.$$

Because $T_2$ is surjective, we can replace $\sum a_{(1)} \otimes b_{(1)} \otimes a_{(2)}b_{(2)}c$ by $a' \otimes b' \otimes 1_A$, where $a', b' \in A$ are arbitrary. Thus we find $\epsilon(a')\epsilon(b') = \epsilon(a'b')$ for all $a', b' \in A$.

Next, we construct the antipode. The proof of the previous lemma shows that if $(A, \Delta)$ were a Hopf algebra, then we could express the antipode $S$ in terms of $T_1$ and $\epsilon$ as follows: $S(a) = \sum \epsilon(a_{(1)})S(a_{(2)}) = (\epsilon \otimes \mathrm{id})(T_1^{-1}(a \otimes 1_A))$ for all $a \in A$. So, consider the map

$$S\colon A \to A,\ a \mapsto (\epsilon \otimes \mathrm{id})(T_1^{-1}(a \otimes 1_A)).$$

Let $a \in A$. From the relation $(\mathrm{id} \otimes m) \circ (T_1 \otimes \mathrm{id})^{-1} = T_1^{-1} \circ (\mathrm{id} \otimes m)$, we deduce

$$\begin{aligned}\sum S(a_{(1)})a_{(2)} &= \sum (\epsilon \otimes \mathrm{id})(T_1^{-1}(a_{(1)} \otimes 1_A)) \cdot a_{(2)} \\ &= \sum (\epsilon \otimes \mathrm{id})(T_1^{-1}(a_{(1)} \otimes a_{(2)})) = (\epsilon \otimes \mathrm{id})(a \otimes 1_A) = \epsilon(a)1_A.\end{aligned}$$

Since $(\Delta \otimes \mathrm{id}) \circ T_1^{-1} = (\mathrm{id} \otimes T_1)^{-1} \circ (\Delta \otimes \mathrm{id})$ and $(\mathrm{id} \otimes \epsilon) \circ \Delta = \mathrm{id}$,

$$\begin{aligned}\sum a_{(1)}S(a_{(2)}) &= \sum a_{(1)} \cdot (\epsilon \otimes \mathrm{id})(T_1^{-1}(a_{(2)} \otimes 1_A)) \\ &= \sum (m \circ (\mathrm{id} \otimes \epsilon \otimes \mathrm{id}) \circ (\mathrm{id} \otimes T_1^{-1}))(a_{(1)} \otimes a_{(2)} \otimes 1_A) \\ &= (m \circ (\mathrm{id} \otimes \epsilon \otimes \mathrm{id}) \circ (\Delta \otimes \mathrm{id}) \circ T_1^{-1})(a \otimes 1_A) \\ &= m(T_1^{-1}(a \otimes 1_A)) = \epsilon(a)1_A.\end{aligned}$$

□

**Remark 1.3.23.** In the proof of Lemma 1.3.21, we did not use the fact that the counit $\epsilon$ is an algebra homomorphism. By Theorem 1.3.18, this assumption can be dropped from the definition of a Hopf algebra because it is implied by the remaining assumptions.

### 1.3.5 Hopf $*$-algebras

Hopf $*$-algebras are Hopf algebras equipped with a conjugate-linear involution that is compatible with the bialgebra structure in a natural way:

**Definition 1.3.24.** An *involution* on a complex vector space $A$ is a map $*\colon A \to A$, $a \mapsto a^*$, that is conjugate-linear and involutive in the sense that

$$(a+b)^* = a^* + b^*, \quad (\lambda a)^* = \bar{\lambda} a^*, \quad (a^*)^* = a \quad \text{for all } a, b \in A \text{ and } \lambda \in \mathbb{C}.$$

A complex vector space with a fixed involution is also called a $*$-*vector space.*

A linear map $\phi\colon A \to B$ of $*$-vector spaces is $*$-*linear* if $\phi(a^*) = \phi(a)^*$ for all $a \in A$.

A $*$-*algebra* is a complex algebra $A$ equipped with an involution such that $(ab)^* = b^* a^*$ for all $a, b \in A$. A $*$-*coalgebra* is a complex coalgebra $(A, \Delta)$, where $A$ is equipped with an involution such that $\Delta(a^*) = \sum {a_{(1)}}^* \otimes {a_{(2)}}^*$ for all $a \in A$. A $*$-*bialgebra* is a complex bialgebra $(A, \Delta)$, where $A$ is a $*$-algebra and $(A, \Delta)$ a $*$-coalgebra. A $*$-bialgebra that is a Hopf algebra is called a *Hopf* $*$-*algebra.*

A *morphism of* $*$-*algebras/*$*$-*coalgebras/*$*$-*bialgebras/Hopf* $*$-*algebras* is a $*$-linear morphism of the underlying algebras/coalgebras/bialgebras/Hopf algebras. A morphism of $*$-algebras is also called a $*$-*homomorphism.*

An important class of Hopf $*$-algebras – the class of algebraic compact quantum groups – is studied in detail in Chapter 3, and analogues of Hopf $*$-algebras in the setting of $C^*$-algebras and von Neumann algebras are discussed in Part II.

**Remarks 1.3.25.** i) Note the following asymmetry in the definition of $*$-algebras and $*$-coalgebras: for a $*$-algebra $A$, the involution reverses the multiplication and can be considered as a homomorphism $A \to A^{\mathrm{op}}$, whereas for a $*$-coalgebra $(A, \Delta)$, the involution does *not* reverse the comultiplication but is a coalgebra morphism $(A, \Delta) \to (A, \Delta)$.

ii) Given a $*$-coalgebra $(A, \Delta)$ and an element $a \in A$, the expressions $\sum {a^*}_{(1)} \otimes {a^*}_{(2)}$ and $\sum {a_{(1)}}^* \otimes {a_{(2)}}^*$ coincide; hence we shortly write $\sum a^*_{(1)} \otimes a^*_{(2)}$.

**Examples 1.3.26.** i) Let $G$ be a finite group. Then the Hopf algebra $\mathbb{C}(G)$ defined in Example 1.2.4 is a Hopf $*$-algebra with respect to the involution $f \mapsto f^*$ given by $f^*(x) := \overline{f(x)}$ for all $x \in G$ and $f \in \mathbb{C}(G)$.

ii) For a topological group $G$, the same formula as above turns the Hopf algebra $\mathrm{Rep}(G)$ defined in Example 1.2.5 into a Hopf $*$-algebra; note that $\mathrm{Rep}(G)$ is closed under this involution.

iii) Let $G$ be a discrete group. Then the Hopf algebra $\mathbb{C}G$ defined in Example 1.2.8 is a Hopf $*$-algebra with respect to the involution $U_x \mapsto U_{x^{-1}}$, $x \in G$.

iv) The group $\mathrm{SU}_2(\mathbb{C})$ is equal to the subgroup of $\mathrm{SL}_2(\mathbb{C})$ that consists of all matrices $A$ that satisfy $A^* = A^{-1}$. In terms of matrix elements, the last relation reads

$$\begin{pmatrix} \bar{\alpha} & \bar{\gamma} \\ \bar{\beta} & \bar{\delta} \end{pmatrix} = \begin{pmatrix} \alpha & \beta \\ \gamma & \delta \end{pmatrix}^* = \begin{pmatrix} \alpha & \beta \\ \gamma & \delta \end{pmatrix}^{-1} = \begin{pmatrix} \delta & -\beta \\ -\gamma & \alpha \end{pmatrix}.$$

Motivated by this equation, we equip the Hopf algebra $\mathcal{O}(\mathrm{SL}_2(\mathbb{C}))$ defined in Example 1.2.6 with the involution given by

$$u_{11}{}^* := u_{22}, \quad u_{12}{}^* := -u_{21}, \quad u_{21}{}^* := -u_{12}, \quad u_{22}{}^* := u_{11}.$$

It is easy to check that this involution turns $\mathcal{O}(\mathrm{SL}_2(\mathbb{C}))$ into a Hopf $*$-algebra. This Hopf $*$-algebra corresponds to the group $\mathrm{SU}_2(\mathbb{C})$ and will be denoted by $\mathcal{O}(\mathrm{SU}_2(\mathbb{C}))$.

In a Hopf $*$-algebra, the behavior of the multiplication and comultiplication with respect to the involution is prescribed by the definition. For the counit and antipode, we obtain the following relations:

**Proposition 1.3.27.** *The counit of a counital $*$-coalgebra is $*$-linear.*

*Proof.* Let $(A, \Delta)$ be a $*$-coalgebra with counit $\epsilon$. Then the map $\epsilon^* \colon A \to \mathbb{C}$ given by $\epsilon^*(a) = \overline{\epsilon(a^*)}$ for all $a \in A$ is a counit for $A$ as well, since

$$(\mathrm{id} \otimes \epsilon^*)(\Delta(a^*)) = \sum a^*_{(1)}\epsilon^*(a^*_{(2)}) = \sum (a_{(1)}\epsilon(a_{(2)}))^* = a^*$$

and similarly $(\epsilon^* \otimes \mathrm{id})(\Delta(a^*)) = a^*$ for all $a \in A$. By Remark 1.3.2 i), $\epsilon = \epsilon^*$. ☐

**Proposition 1.3.28.** *The antipode of a Hopf $*$-algebra is bijective and satisfies $S \circ * \circ S \circ * = \mathrm{id}$.*

*Proof.* Let $(A, \Delta)$ be a Hopf $*$-algebra. Then the map $S^* \colon A \to A$ given by $S^*(a) = S(a^*)^*$ for all $a \in A$ is an antipode for the bialgebra $(A, \Delta)^{\mathrm{op}}$, since

$$\begin{aligned} \sum a^*_{(2)} S^*(a^*_{(1)}) &= \sum a^*_{(2)} S(a_{(1)})^* \\ &= \sum (S(a_{(1)})a_{(2)})^* = \eta(\epsilon(a))^* = \eta(\epsilon(a^*)) \end{aligned}$$

and likewise $\sum S^*(a^*_{(2)})a^*_{(1)} = \eta(\epsilon(a^*))$ for all $a \in A$. By Proposition 1.3.14, $S^* = S^{-1}$. ☐

**Corollary 1.3.29.** *The antipode of a Hopf $*$-algebra is $*$-linear if and only if it is involutive in the sense that* $S^2 = \mathrm{id}$.

*Proof.* By the previous proposition, the map $S \circ *$ is invertible, and

$$* \circ S = S \circ * \;\Leftrightarrow\; (S \circ *) \circ (* \circ S) = (S \circ *) \circ (S \circ *) \;\Leftrightarrow\; S^2 = \mathrm{id}\,. \qquad \square$$

## 1.4 The duality of Hopf algebras

The concept of a bialgebra and of a Hopf algebra has an intrinsic symmetry that gives rise to a duality which can be considered as an analogue of Pontrjagin duality. In the finite-dimensional case, this duality behaves very nicely. In the infinite-dimensional case, additional concepts and stronger assumptions are needed, and a satisfying duality theory will only be achieved in Chapter 2.

### 1.4.1 The duality of finite-dimensional Hopf algebras

Let us fix some notation. Consider a vector space $V$. We denote by $V' = \mathrm{Hom}_{\Bbbk}(V, \Bbbk)$ the dual space, and by $\iota_V : V \to V''$ the natural embedding given by $(\iota_V(v))(f) := f(v)$ for all $v \in V$, $f \in V'$. Furthermore, we consider $V' \otimes V'$ as a subspace of $(V \otimes V)'$ via the embedding given by $(f \otimes g)(v \otimes w) = f(v)g(w)$ for all $v, w \in V$ and $f, g \in V'$. It is easy to see that this embedding, like the embedding $\iota_V$, is an isomorphism if and only if $V$ has finite dimension. Finally, recall that every linear map of vector spaces $F : V \to W$ induces a dual map $F' : W' \to V'$ by composition, that is, $F'(f) = f \circ F$ for all $f \in W'$.

**Theorem 1.4.1.** i) *Let $(A, \Delta_A)$ be a coalgebra. Then the dual space $A'$ is an algebra with respect to the multiplication*

$$m_{A'} : A' \otimes A' \hookrightarrow (A \otimes A)' \xrightarrow{(\Delta_A)'} A', \; (fg)(a) = (f \otimes g)(\Delta(a)).$$

*The algebra $A'$ is unital if and only if $(A, \Delta_A)$ is counital, and in this case, the unit of $A'$ coincides with the counit of $(A, \Delta_A)$.*

ii) *Let $A$ be a finite-dimensional algebra. Then the dual space $A'$ is a coalgebra with respect to the comultiplication*

$$\Delta_{A'} : A' \xrightarrow{(m_A)'} (A \otimes A)' \cong A' \otimes A', \quad (\Delta_{A'}(f))(a \otimes b) = f(ab).$$

*The coalgebra $(A', \Delta_{A'})$ is counital if and only if $A$ is unital, and in this case, the counit of $(A', \Delta_{A'})$ coincides with evaluation at the unit of $A$.*

iii) *Let $(A, \Delta_A)$ be a finite-dimensional bialgebra. Then $A'$, equipped with the multiplication and comultiplication defined above, is a bialgebra. If $(A, \Delta_A)$ is a*

*Hopf algebra, then* $(A', \Delta_{A'})$ *is a Hopf algebra with antipode* $S_{A'} = (S_A)'$. *The natural isomorphism* $\iota_A \colon A \xrightarrow{\cong} A''$ *of vector spaces is an isomorphism of bialgebras or Hopf algebras, respectively.*

iv) *Let* $(A, \Delta_A)$ *be a Hopf* $*$*-algebra. Then* $A'$ *is a* $*$*-algebra with respect to the involution given by* $f^*(a) := \overline{f(S(a)^*)}$ *for all* $a \in A$, $f \in A'$. *If* $(A, \Delta_A)$ *is a finite-dimensional Hopf* $*$*-algebra, then* $(A', \Delta_{A'})$ *is a Hopf* $*$*-algebra again, and the natural isomorphism* $\iota_A \colon A \xrightarrow{\cong} A''$ *is an isomorphism of Hopf* $*$*-algebras.*

*Proof.* i)–iii) All statements follow easily from the symmetry of the commutative diagrams that express the axioms for the structure maps involved: If we apply the dualization functor $A \mapsto A'$, $F \mapsto F'$, to the commutative diagrams

$$(1.1),\ (1.2),\ (1.5),\ (1.6),\ (1.7),\ (1.8),\ (1.3) \ \text{ for } A$$

and use the isomorphism $(A \otimes A)' \cong A' \otimes A'$ in ii), iii), we find that the diagrams

$$(1.5),\ (1.6),\ (1.1),\ (1.2),\ (1.7),\ (1.8),\ (1.3) \ \text{ for } A'$$

commute. Let us give an example. If we apply the dualization functor to the right square of diagram (1.3), we obtain

$$\begin{array}{ccc} A & \xrightarrow{\Delta_A} & A\otimes A \\ {\scriptstyle \eta_A\circ\epsilon_A}\downarrow & & \downarrow{\scriptstyle \mathrm{id}\otimes S_A} \\ A & \xleftarrow[m_A]{} & A\otimes A \end{array} \quad\mapsto\quad \begin{array}{ccc} A' & \xleftarrow{(\Delta_A)'} & (A\otimes A)' \\ {\scriptstyle (\eta_A\circ\epsilon_A)'}\uparrow & & \uparrow{\scriptstyle (\mathrm{id}\otimes S_A)'} \\ A' & \xrightarrow[(m_A)']{} & (A\otimes A)' \end{array} \quad\equiv\quad \begin{array}{ccc} A' & \xleftarrow{m_{A'}} & A'\otimes A' \\ {\scriptstyle \eta_{A'}\circ\epsilon_{A'}}\uparrow & & \uparrow{\scriptstyle \mathrm{id}\otimes S_{A'}} \\ A' & \xrightarrow[\Delta_{A'}]{} & A'\otimes A', \end{array}$$

which is the right square of diagram (1.3) for $A'$.

iv) Let us show that the involution defined in iv) turns $A'$ into a $*$-algebra. We denote by $* \colon \mathbb{C} \to \mathbb{C}$ the complex conjugation. Then $f^* = * \circ f \circ * \circ S_A$ for all $f \in A'$, and from Propositions 1.3.28 and 1.3.12, we find

$$(f^*)^* = * \circ (* \circ f \circ * \circ S_A) \circ * \circ S_A = f$$

and

$$\begin{aligned} (fg)^* &= * \circ (f \otimes g) \circ \Delta_A \circ * \circ S_A \\ &= ((* \circ g \circ * \circ S_A) \otimes (* \circ f \circ * \circ S_A)) \circ \Delta_A = g^* f^* \end{aligned}$$

for all $f, g \in A'$. Now, assume that $A$ has finite dimension, and let $f \in A'$. By definition and by Proposition 1.3.12,

$$\begin{aligned} \Delta_{A'}(f^*) &= f^* \circ m_A = * \circ f \circ * \circ S_A \circ m_A \\ &= * \circ f \circ m_A \circ ((* \circ S_A) \otimes (* \circ S_A)). \end{aligned}$$

Inserting the relation $f \circ m_A = \Delta_{A'}(f) = \sum f_{(1)} \otimes f_{(2)}$, we find

$$\Delta_{A'}(f^*) = \sum (* \circ f_{(1)} \circ * \circ S_A) \otimes (* \circ f_{(2)} \circ * \circ S_A) = \sum f_{(1)}{}^* \otimes f_{(2)}{}^*.$$

Thus $(A', \Delta_{A'})$ is a $*$-coalgebra. By iii), it is also a Hopf algebra, and hence a Hopf $*$-algebra.

Finally, let us show that the natural isomorphism $\iota_A \colon A \xrightarrow{\cong} A''$ of Hopf algebras is $*$-linear. By definition and by Proposition 1.3.28,

$$\begin{aligned}(\iota(a))^*(f) &= \overline{\iota(a)(S_{A'}(f)^*)} = \overline{(S_{A'}(f)^*)(a)} \\ &= (S_{A'}(f))(S_A(a)^*) = (f \circ S_A \circ * \circ S_A)(a) = f(a^*) \\ &= (\iota(a^*))(f)\end{aligned}$$

for all $a \in A$ and $f \in A'$. □

The following two examples explain the relation between the duality of finite-dimensional Hopf algebras and the Pontrjagin duality of finite abelian groups.

**Example 1.4.2.** Let $G$ be a finite group and consider the associated Hopf algebra $\Bbbk(G)$ defined in Example 1.2.4. Denote by $(\delta_x)_{x \in G}$ the canonical basis of $\Bbbk(G)$, and by $(\varepsilon_x)_{x \in G}$ the dual basis of $\Bbbk(G)'$, determined by $\varepsilon_x(\delta_y) = \delta_{x,y}$ for all $x, y \in G$. We compute the structure maps of the Hopf algebra $\Bbbk(G)'$. To distinguish between the counit, comultiplication, and antipode of $\Bbbk(G)$ and $\Bbbk(G)'$, we index the structure maps of $\Bbbk(G)'$.

- The product of two elements $\varepsilon_x, \varepsilon_y \in \Bbbk(G)'$ is determined by

  $$(\varepsilon_x \cdot \varepsilon_y)(\delta_z) = (\varepsilon_x \otimes \varepsilon_y)(\Delta(\delta_z)) = \sum_{\substack{x',y' \in G \\ x'y'=z}} \varepsilon_x(\delta_{x'}) \cdot \varepsilon_y(\delta_{y'}) = \delta_{xy,z},$$

  whence $\varepsilon_x \varepsilon_y = \varepsilon_{xy}$. If $e \in G$ denotes the unit, then $\varepsilon_e$ is the unit of $\Bbbk(G)'$.
- The coproduct of an element $\varepsilon_z \in \Bbbk(G)'$ is determined by

  $$(\Delta_{\Bbbk(G)'}(\varepsilon_z))(\delta_x \otimes \delta_y) = \varepsilon_z(\delta_x \cdot \delta_y) = \delta_{z,x} \cdot \delta_{z,y},$$

  so $\Delta_{\Bbbk(G)'}(\varepsilon_z) = \varepsilon_z \otimes \varepsilon_z$.
- The counit of $\Bbbk(G)'$ acts by evaluation at the unit $1_{\Bbbk(G)} = \sum_z \delta_z$, whence $\epsilon_{\Bbbk(G)'}(\varepsilon_z) = 1$ for all $z \in G$.
- The antipode $S_{\Bbbk(G)'}$, applied to an element $\varepsilon_x \in \Bbbk(G)'$, acts by

  $$(S_{\Bbbk(G)'}(\varepsilon_x))(\delta_y) = \varepsilon_x(S(\delta_y)) = \varepsilon_x(\delta_{y^{-1}}) = \delta_{x,y^{-1}},$$

  and hence $S_{\Bbbk(G)'}(\varepsilon_x) = \varepsilon_{x^{-1}}$.

Comparing with the definition of the Hopf algebra $\Bbbk G$ in Example 1.2.8, we find that the map $\Bbbk(G)' \to \Bbbk G$ given by $\varepsilon_x \mapsto U_x$ for all $x \in G$ is an isomorphism of Hopf algebras. If $\Bbbk = \mathbb{C}$, then this is also an isomorphism of Hopf $*$-algebras, because ${\varepsilon_x}^* = \varepsilon_{x^{-1}}$ for all $x \in G$:

$$ {\varepsilon_x}^*(\delta_y) = \overline{\varepsilon_x(S(\delta_y))} = \overline{\varepsilon_x(\delta_{y^{-1}})} = \delta_{x,y^{-1}}. $$

**Example 1.4.3.** Let $G$ be a finite abelian group. By the previous example, $\mathbb{C}(G)' \cong \mathbb{C}G$. Moreover, the Fourier transform $\mathcal{F} : \mathbb{C}G \to \mathbb{C}(\widehat{G})$ given by $(\mathcal{F} U_x)(\chi) := \chi(x)$ for all $\chi \in \widehat{G}$ and $x \in G$ is an isomorphism of Hopf algebras: this is just a special case of the isomorphism $\mathbb{C}G \xrightarrow{\cong} \mathrm{Rep}(\widehat{G})$ explained in Example 1.2.8, and since $\widehat{G}$ is finite, $\mathrm{Rep}(\widehat{G}) = \mathbb{C}(\widehat{G})$. Summarizing, we find that the Pontrjagin duality of finite abelian groups and the duality of finite-dimensional Hopf algebras fits into the following scheme:

$$ \begin{array}{ccc} G & \xmapsto[\text{group}]{\text{dual}} & \widehat{G} \\ \vdots & & \vdots \\ \mathbb{C}(G) & \xmapsto[\text{Hopf algebra}]{\text{dual}} & \mathbb{C}(\widehat{G}) \end{array} $$

The duality established in Theorem 1.4.1 iii) does not easily extend to infinite-dimensional bialgebras or Hopf algebras: For an infinite-dimensional algebra $A$, the inclusion $A' \otimes A' \subseteq (A \otimes A)'$ is strict. Therefore, the dual of the multiplication map $(m_A)' : A' \to (A \otimes A)'$ need not define a comultiplication on $A'$ – its image may fail to be contained in the subspace $A' \otimes A' \subset (A \otimes A)'$. This problem can be addressed in several ways. We can

- consider *dual pairings* of $A$ with other Hopf algebras;
- look for a subalgebra $A^\circ \subset A'$ that satisfies $(m_A)'(A^\circ) \subseteq A^\circ \otimes A^\circ$;
- consider non-unital Hopf algebras $(\widehat{A}, \widehat{\Delta})$, for which the image of $\widehat{\Delta}$ is contained in a space of *multipliers* $M(\widehat{A} \otimes \widehat{A})$ that is larger than $\widehat{A} \otimes \widehat{A}$.

In the following sections, we discuss the first and second approach; the third one involves several additional concepts and is presented in Chapter 2.

## 1.4.2 Dual pairings of Hopf algebras

Let $(A, \Delta_A)$ be a finite-dimensional bialgebra. In Theorem 1.4.1, we saw that the dual space $A'$ of $A$ is a bialgebra again. The relations between the structure maps of $(A, \Delta_A)$ and $(A', \Delta_{A'})$ can be expressed conveniently in terms of the natural pairings

$$(\cdot|\cdot)\colon A \times A' \to \Bbbk, \qquad (\cdot|\cdot)\colon (A \otimes A) \times (A' \otimes A') \to \Bbbk,$$
$$(a|f) := f(a), \qquad (a_1 \otimes a_2|f_1 \otimes f_2) := f_1(a_1) \cdot f_2(a_2),$$

as follows: for all $a, a_1, a_2 \in A$ and $f, f_1, f_2 \in A'$,

$$(a|f_1 f_2) = (a|m_{A'}(f_1 \otimes f_2)) = (\Delta_A(a)|f_1 \otimes f_2) = \sum (a_{(1)}|f_1)(a_{(2)}|f_2)$$

and

$$(a_1 a_2|f) = (m_A(a_1 \otimes a_2)|f) = (a_1 \otimes a_2|\Delta_{A'}(f)) = \sum (a_1|f_{(1)})(a_2|f_{(2)}).$$

Furthermore, if $(A, \Delta_A)$ is a Hopf algebra (or Hopf $*$-algebra), then so is $(A', \Delta_{A'})$, and in that case, the unit, counit, antipode (and involution) of $(A, \Delta_A)$ and $(A', \Delta_{A'})$ are related by similar equations. These relations motivate the following definition:

**Definition 1.4.4.** A *dual pairing* between two Hopf algebras (Hopf $*$-algebras) $(A, \Delta_A)$ and $(B, \Delta_B)$ is a bilinear map $(\cdot|\cdot)\colon A \times B \to \Bbbk$, written $(a, b) \mapsto (a|b)$, that satisfies

$$(a|b_1 b_2) = \sum (a_{(1)}|b_1)(a_{(2)}|b_2), \quad (a_1 a_2|b) = \sum (a_1|b_{(1)})(a_2|b_{(2)}),$$

$$(a|1_B) = \epsilon_A(a), \quad (1_A|b) = \epsilon_B(b), \quad (S_A(a)|b) = (a|S_B(b)) \tag{1.14}$$

$$(\text{and} \quad (a|b^*) = \overline{(S_A(a)^*|b)}, \quad (a^*|b) = \overline{(a|S_B(b)^*)}, \quad \text{respectively})$$

for all $a, a_1, a_2 \in A$ and $b, b_1, b_2 \in B$. The dual pairing is called *perfect* or *non-degenerate* if for each non-zero $a_0 \in A$ and each non-zero $b_0 \in B$, there exist $a \in A$ and $b \in B$ such that $(a_0|b) \neq 0$ and $(a|b_0) \neq 0$.

**Remarks 1.4.5.** i) Let $(\cdot|\cdot)\colon A \times B \to \Bbbk$ be a dual pairing of Hopf algebras. Then each $a \in A$ defines a linear map $(a|\cdot)\colon B \to \Bbbk$, $b \mapsto (a|b)$, and the map $A \to B'$ given by $a \mapsto (a|\cdot)$ is a unital homomorphism of algebras. Similarly, we obtain a unital homomorphism $B \to A'$, $b \mapsto (\cdot|b)$. The pairing is perfect if and only if these homomorphisms are injective.

ii) For every finite-dimensional Hopf algebra $(A, \Delta_A)$, the canonical pairing between $A$ and $A'$ is a perfect dual pairing of Hopf algebras. If $(B, \Delta_B)$ is another Hopf algebra and $(\cdot|\cdot)\colon A \times B \to \Bbbk$ is a perfect dual pairing, then $B$ has finite dimension and the homomorphisms $A \to B', a \mapsto (a|\cdot)$, and $B \to A', b \mapsto (\cdot|b)$, are isomorphisms of Hopf algebras.

iii) Let $(A, \Delta_A)$ and $(B, \Delta_B)$ be Hopf $*$-algebras and $(\cdot|\cdot)\colon A \times B \to \Bbbk$ a dual pairing of Hopf algebras. Then $(a|b^*) = \overline{(S_A(a)^*|b)}$ for all $a \in A$ and $b \in B$

if and only if $(a^*|b) = \overline{(a|S_B(b)^*)}$ for all $a \in A$ and $b \in B$. Indeed, if the first condition holds, then by Proposition 1.3.28,

$$(a^*|b) = (S_A(S_A(a)^*)|b) = (S_A(a)^*|S_B(b)) = \overline{(a|S_B(b)^*)}$$

for all $a \in A$, $b \in B$, and the reverse implication follows similarly. Thus, one of these conditions may be omitted in the definition of a dual pairing of Hopf $*$-algebras.

Let us consider several examples of dual pairings.

**Example 1.4.6.** There exists a dual pairing $\mathcal{O}(\mathrm{SL}_n(\mathbb{C})) \times U(\mathfrak{sl}_n(\mathbb{C})) \to \mathbb{C}$, determined by $(u_{ij}|X) := u_{ij}(X)$ for all $i, j = 1, \dots, n$ and $X \in \mathfrak{sl}_n(\mathbb{C})$. Here, $u_{ij}\colon M_n(\mathbb{C}) \to \mathbb{C}$ denotes the $(i, j)$th coordinate function as in Example 1.2.6, and the Lie algebra $\mathfrak{sl}_n(\mathbb{C})$ is identified with the space of $n \times n$-matrices with vanishing trace. A detailed discussion of this dual pairing for $n = 2$ can be found in [79, Section V.7]; a generalization is given in Example 1.4.14.

**Example 1.4.7.** Let $G$ be a compact connected Lie group with Lie algebra $\mathfrak{g}$. There exists a dual pairing between the Hopf algebra of representative functions $\mathrm{Rep}(G)$ and the universal enveloping algebra $U(\mathfrak{g})$, which can be described as follows.

First, observe that the Hopf algebra $\mathrm{Rep}(G)$ is contained in $C^\infty(G)$. Indeed, every continuous group homomorphism of Lie groups is automatically smooth [22, Proposition I.3.12], and therefore every continuous finite-dimensional representation and every representative function of $G$ is smooth.

Next, recall that every $X \in \mathfrak{g}$ determines a left-invariant vector field on $G$ and hence also a first-order differential operator $D_X\colon C^\infty(G) \to C^\infty(G)$. The map $X \mapsto D_X$ extends to an injective unital algebra homomorphism $D\colon U(\mathfrak{g}) \hookrightarrow \mathrm{End}_{\mathbb{C}}(C^\infty(G))$, $\omega \mapsto D_\omega$; explicitly, the differential operator $D_\omega$ associated to an element $\omega = X_1 \dots X_n \in U(\mathfrak{g})$ is given by

$$\begin{aligned}(D_\omega f)(y) &= (D_{X_1} \dots D_{X_n} f)(y) \\ &= \frac{\partial^n}{\partial t_1 \dots \partial t_n}\bigg|_{t_1,\dots,t_n=0} f(y \cdot \exp(t_1 X_1) \dots \exp(t_n X_n)),\end{aligned}$$

see [181, Section 3.4].

Denote by $e \in G$ the unit element. Then the bilinear map

$$(\cdot|\cdot)\colon U(\mathfrak{g}) \times \mathrm{Rep}(G) \to \mathbb{C}, \quad (\omega|f) := (D_\omega f)(e),$$

is a perfect dual pairing of Hopf algebras. Let us prove this assertion.

i) Denote by $1_U$ and $1_{\mathrm{Rep}}$ the units and by $\epsilon_U$ and $\epsilon_{\mathrm{Rep}}$ the counits of $U(\mathfrak{g})$ and $\mathrm{Rep}(G)$, respectively. Since the embedding $D\colon U(\mathfrak{g}) \hookrightarrow \mathrm{End}_{\mathbb{C}}(C^\infty(G))$ is unital, $(1_U|f) = (D_{1_U} f)(e) = f(e) = \epsilon_{\mathrm{Rep}}(f)$ for all $f \in \mathrm{Rep}(G)$. Moreover,

$(D_{X_1\dots X_n})1_{\mathrm{Rep}} = 0$ for all $X_1, \dots, X_n \in \mathfrak{g}$, whence $(\omega|1_{\mathrm{Rep}}) = (D_\omega 1_{\mathrm{Rep}})(e)$ equals $\epsilon_U(\omega)$ for all $\omega \in U(\mathfrak{g})$.

ii) We show that $(\omega_1\omega_2|f) = \sum(\omega_1|f_{(1)})(\omega_2|f_{(2)})$ for all $\omega_1, \omega_2 \in U(\mathfrak{g})$ and $f \in \mathrm{Rep}(G)$. Since $\mathfrak{g}$ and $1_U$ generate $U(\mathfrak{g})$ as an algebra, it suffices to prove

$$(X_1 \dots X_n|f) = \sum(X_1|f_{(1)}) \dots (X_n|f_{(n)}) \tag{1.15}$$

for all $X_1, \dots, X_n \in \mathfrak{g}$ and $f \in \mathrm{Rep}(G)$. By definition,

$$(X_1 \dots X_n|f) = \frac{\partial^n}{\partial t_1 \dots \partial t_n}\bigg|_{t_1,\dots,t_n=0} f(\exp(t_1 X_1) \dots \exp(t_n X_n)).$$

We insert the relation

$$f(x_1 \dots x_n) = (\Delta^{(n-1)}(f))(x_1, \dots, x_n) = \sum f_{(1)}(x_1) \dots f_{(n)}(x_n)$$

with $x_i := \exp(t_i X_i)$ and find that $(X_1 \dots X_n|f)$ is equal to

$$\begin{aligned}\frac{\partial^n}{\partial t_1 \dots \partial t_n}\bigg|_{t_1,\dots,t_n=0} &\sum f_{(1)}(\exp(t_1 X_1)) \dots f_{(n)}(\exp(t_n X_n)) \\ &= \sum(D_{X_1} f_{(1)})(e) \dots (D_{X_n} f_{(n)})(e) = \sum(X_1|f_{(1)}) \dots (X_n|f_{(n)}).\end{aligned}$$

iii) Let us show that $(\omega|gh) = \sum(\omega_{(1)}|g)(\omega_{(2)}|h)$ for all $\omega \in U(\mathfrak{g})$ and $g, h \in \mathrm{Rep}(G)$. The Leibniz rule implies that for every $X \in \mathfrak{g}$,

$$\begin{aligned}(X|gh) &= (D_X(gh))(e) = (D_X g)(e) \cdot h(e) + g(e) \cdot (D_X h)(e) \\ &= (X|g)(1_U|h) + (1_U|g)(X|h).\end{aligned}$$

By definition of the comultiplication on $U(\mathfrak{g})$, $X \otimes 1_U + 1_U \otimes X = \sum X_{(1)} \otimes X_{(2)}$. Hence the equation above can be rewritten as follows:

$$(X|gh) = \sum(X_{(1)}|g)(X_{(2)}|h) \quad \text{for all } X \in \mathfrak{g}.$$

Using relation (1.15) with $f = gh$, or the Leibniz rule for higher order derivatives of the product $gh$, it is easy to see that the equation above remains valid if we replace $X \in \mathfrak{g}$ by an element $\omega = X_1 \dots X_n \in U(\mathfrak{g})$.

iv) For all $X \in \mathfrak{g}$ and $f \in \mathrm{Rep}(G)$,

$$\begin{aligned}(X|S(f)) &= \frac{\partial}{\partial t}\bigg|_{t=0} f(\exp(tX)^{-1}) \\ &= \frac{\partial}{\partial t}\bigg|_{t=0} f(\exp(-tX)) = (-X|f) = (S(X)|f).\end{aligned}$$

Using Proposition 1.3.12 and equation (1.15), it is easy to deduce that $(\omega|S(f)) = (S(\omega)|f)$ for all $\omega \in U(\mathfrak{g})$.

v) Let $\omega \in U(\mathfrak{g})$, and assume that $(\omega|f) = (D_\omega f)(e) = 0$ for all $f \in \mathrm{Rep}(G)$. We show that then $\omega = 0$. Since the operator $D_\omega$ is left-invariant and the space $\mathrm{Rep}(G)$ is closed under left translation, it follows that $D_\omega(\mathrm{Rep}(G)) = 0$. We show that then also $D_\omega(C^\infty(G)) = 0$, and this implies $\omega = 0$. The space $C^\infty(G)$ carries a natural locally convex topology, and $D_\omega$ is continuous with respect to this topology [32, Chapter XVII]. So, it suffices to prove that $\mathrm{Rep}(G)$ is dense in $C^\infty(G)$. The group $G$ acts continuously (in a suitable sense) on $C^\infty(G)$ via left translations, and by a generalization of the Peter–Weyl theorem [22, III, Theorem 5.7], the subspace of $C^\infty(G)$ spanned by finite-dimensional $G$-invariant subspaces is dense. Every such subspace is contained in $\mathrm{Rep}(G)$ and hence $\mathrm{Rep}(G)$ is dense in $C^\infty(G)$.

vi) Let $f \neq 0$ be a representative function on $G$. We show that $(\omega|f) \neq 0$ for some $\omega \in U(\mathfrak{g})$. Denote by $V \subseteq \mathrm{Rep}(G)$ the linear span of all right translates of $f$, and by $\pi\colon G \to \mathrm{Aut}(V)$ the representation given by $\pi(x)g := g(\cdot\, x)$ for all $x \in G$ and $g \in V$. Note that $V$ has finite dimension and that $\pi$ is smooth. Recall that the Lie algebra of $\mathrm{Aut}(V)$ is $\mathrm{End}(V)$. Differentiating $\pi$ at $e \in G$, we obtain a homomorphism of Lie algebras $\mathfrak{g} \to \mathrm{End}(V)$ which extends to a unital algebra homomorphism $D\pi\colon U(\mathfrak{g}) \to \mathrm{End}(V)$. It is easy to check that $D\pi(X)g = D_X g$ for every $X \in \mathfrak{g}$ and $g \in V$. Choose some $x \in G$ such that $f(x) \neq 0$. Since $G$ is compact and connected, there exists an $X \in \mathfrak{g}$ such that $\exp(X) = x$ [22, IV, Theorem 2.2]. For each $n \in \mathbb{N}$, put $\omega_n := \sum_{k=1}^n X^k/k! \in U(\mathfrak{g})$. Then

$$\lim_{n\to\infty} D\pi(\omega_n) = \exp(D\pi(X)) = \pi(\exp(X)) = \pi(x)$$

in $\mathrm{Hom}(V)$ (with respect to every norm topology), and therefore

$$\lim_{n\to\infty} (\omega_n|f) = \lim_{n\to\infty} (D\pi(\omega_n)f)(e) = (\pi(x)f)(e) = f(x) \neq 0.$$

Consequently, $(\omega_n|f) \neq 0$ for some $n \in \mathbb{N}$.

### 1.4.3 The restricted dual of a Hopf algebra

For every unital algebra $A$, there exists a largest subspace $A^\circ \subseteq A'$ for which the map $\Delta_{A'} = (m_A)'\colon A' \to (A \otimes A)'$ defines a comultiplication. Elements of this subspace can be characterized as follows:

**Lemma 1.4.8.** *Let $A$ be a unital algebra and $f \in A'$. Then the following conditions are equivalent:*

i) $\Delta_{A'}(f) \in A' \otimes A'$,

ii) $\ker f$ *contains a left ideal of $A$ that has finite codimension,*

iii) $\ker f$ *contains a right ideal of $A$ that has finite codimension,*

iv) $\ker f$ *contains an ideal of $A$ that has finite codimension.*

*Proof.* i) $\Rightarrow$ ii): Write $\Delta_{A'}(f) = \sum_i g_i \otimes h_i$ with $g_i, h_i \in A'$, where the $g_i$ are linearly independent and the $h_i$ are non-zero. Then $J := \bigcap_i \ker h_i$ is a left ideal of $A$ with finite codimension and contained in $\ker f$:

- $AJ \subseteq J$: if $b \in A$ and $c \in J$, then
$$0 = \sum_i g_i(ab)h_i(c) = f(abc) = \sum_i g_i(a)h_i(bc) \quad \text{for all } a \in A,$$
and the linear independence of the $g_i$ implies $h_i(bc) = 0$ for all $i$;
- $J$ has finite codimension in $A$, because $\ker h_i \subseteq A$ has codimension 1 for each $i$ and the intersection $\bigcap_i \ker h_i$ is finite;
- $f(J) = 0$ since $f(a) = f(1_A \cdot a) = \sum_i g_i(1_A)h_i(a) = 0$ for each $a \in J$.

i) $\Rightarrow$ iii): The proof is similar to the proof of the implication i) $\Rightarrow$ ii).

ii) $\Rightarrow$ iv): Let $J \subseteq A$ be a left ideal of finite codimension that is contained in $\ker f$. Then the map $\pi\colon A \to \mathrm{Hom}_{\Bbbk}(A/J)$ given by $\pi(a)(b + J) := ab + J$ is an algebra homomorphism. Since $A/J$ and hence also $\mathrm{Hom}_{\Bbbk}(A/J)$ have finite dimension, the kernel $I := \ker \pi$ has finite codimension in $A$. Finally, the relation $\pi(I) = 0$ implies $I \subseteq J \subseteq \ker f$.

iii) $\Rightarrow$ iv): Again, the proof is similar to the proof given above.

vi) $\Rightarrow$ i): Let $I \subseteq A$ be an ideal of finite codimension that is contained in $\ker f$. Denote by $\pi\colon A \to A/I$ the quotient map and by $\pi'\colon (A/I)' \to A'$ its transpose. We show that $\Delta_{A'}(f)$ belongs to the space
$$(\pi' \otimes \pi')((A/I)' \otimes (A/I)') \subseteq A' \otimes A'.$$
Since $\Delta_{A'}(f)$ vanishes on $A \otimes I + I \otimes A$, it defines an element
$$g \in ((A \otimes A)/(A \otimes I + I \otimes A))' \cong ((A/I) \otimes (A/I))'.$$
Since $A/I$ has finite dimension, $((A/I) \otimes (A/I))'$ is isomorphic to $(A/I)' \otimes (A/I)'$ and $g$ can be considered as an element of $(A/I)' \otimes (A/I)'$. It is easy to see that $\Delta_{A'}(f) = (\pi' \otimes \pi')(g)$, and this relation completes the proof. □

**Definition 1.4.9.** The *restricted dual* of a unital algebra $A$ is the subspace $A^\circ \subseteq A'$ consisting of all linear maps that satisfy the equivalent conditions given in Lemma 1.4.8.

**Proposition 1.4.10.** i) *Let $A$ be a unital algebra. Then the space $A^\circ$ satisfies $\Delta_{A'}(A^\circ) \subseteq A^\circ \otimes A^\circ$. Put $\Delta_{A^\circ} := \Delta_{A'}|_{A^\circ}$. Then $(A^\circ, \Delta_{A^\circ})$ is a counital coalgebra, where the counit is given by $f \mapsto f(1_A)$.*

ii) *Let* $(A, \Delta_A)$ *be a unital bialgebra. Then* $A^\circ \subseteq A'$ *is a subalgebra and* $(A^\circ, \Delta_{A^\circ})$ *is a counital bialgebra. If* $(A, \Delta_A)$ *is a Hopf algebra, then so is* $(A^\circ, \Delta_{A^\circ})$.

iii) *For every Hopf* $*$*-algebra* $(A, \Delta_A)$*, the formula* $f^*(a) := \overline{f(S(a)^*)}$ *defines an involution on* $A^\circ$ *that turns* $(A^\circ, \Delta_{A^\circ})$ *into a Hopf* $*$*-algebra.*

*Proof.* i) First, we show that $\Delta_{A'}(A^\circ) \subseteq A^\circ \otimes A^\circ$. Let $f \in A^\circ$ and write $\Delta_{A'}(f) = \sum_i g_i \otimes h_i$ with $g_i, h_i \in A'$, where the $h_i$ are linearly independent. By assumption on the $h_i$, we can choose for every $j$ an element $a_j \in A$ such that $h_i(a_j) = \delta_{i,j}$ for all $i, j$. Then

$$g_j(ab) = \sum_i g_i(ab)h_i(a_j) = f(aba_j) = \sum_i g_i(a)h_i(ba_j) \quad \text{for all } a, b \in A,$$

and consequently $\Delta_{A'}(g_j) = \sum_i g_i \otimes h_i(\cdot a_j) \in A' \otimes A'$ and $g_j \in A^\circ$. Therefore, $\Delta_{A'}(f) \in A^\circ \otimes A'$, and a similar argument shows that $\Delta_{A'}(f) \in A' \otimes A^\circ$. We conclude that $\Delta_{A'}(f) \in A^\circ \otimes A^\circ$ for every $f \in A^\circ$, so that $\Delta_{A'}$ restricts to a map $\Delta_{A^\circ} \colon A^\circ \to A^\circ \otimes A^\circ$. This map is coassociative because the multiplication $m_A$ is associative. The assertion concerning the counit is evident.

ii) Suppose that $A$ is a unital bialgebra. Then $A^\circ \subseteq A'$ is a subalgebra because

$$\Delta_{A'}(fg) = \Delta_{A'}(f)\Delta_{A'}(g) \in (A' \otimes A') \cdot (A' \otimes A') \subseteq A' \otimes A' \quad \text{for all } f, g \in A^\circ.$$

The multiplication and comultiplication on $A^\circ$ are compatible because the comultiplication and multiplication on $A$ are compatible.

Assume that $(A, \Delta_A)$ is a Hopf algebra. Denote by $S_{A^\circ} := (S_A)'|_{A^\circ} \colon A^\circ \to A'$ the restriction of the transpose of $S_A$. We claim that $S_{A^\circ}(A^\circ) \subseteq A^\circ$. By Proposition 1.3.12, we have for all $f \in A^\circ$ and $a, b \in A$

$$\begin{aligned}(\Delta_{A'}(f \circ S_A))(a \otimes b) &= f(S_A(ab)) \\ &= f(S_A(b)S_A(a)) \\ &= (\Delta_{A'}(f))(S_A(b) \otimes S_A(a)),\end{aligned}$$

and hence

$$\Delta_{A'}(f \circ S_A) = \sum (f_{(2)} \circ S_A) \otimes (f_{(1)} \circ S_A) \in A' \otimes A'.$$

The claim follows. Now we conclude as in Theorem 1.4.1 that $(A^\circ, \Delta_{A^\circ})$ is a Hopf algebra with antipode $S_{A^\circ}$.

iii) A similar calculation as in ii) shows that for every $f \in A^\circ$, the functional $f^* \in A'$ given by $a \mapsto \overline{f(S(a)^*)}$ belongs to $A^\circ$. The compatibility of this involution with the Hopf algebra structure on $A^\circ$ is verified as in Theorem 1.4.1 iv). □

Dual pairings and restricted duals of Hopf algebras are related to each other as follows:

**Proposition 1.4.11.** i) *For every Hopf algebra $(A, \Delta_A)$, the pairing $A \times A^\circ \to \Bbbk$ given by $(a, f) \mapsto f(a)$ is a dual pairing of Hopf algebras.*

ii) *If $(\cdot|\cdot): A \times B \to \Bbbk$ is a (perfect) dual pairing of Hopf algebras, then the induced maps $A \to B'$, $a \mapsto (a|\cdot)$, and $B \to A'$, $b \mapsto (\cdot|b)$, restrict to (injective) morphisms of Hopf algebras $A \to B^\circ$ and $B \to A^\circ$.*

*Proof.* Statement i) is obvious from the definitions, and statement ii) follows easily from Remark 1.4.5 i) and Lemma 1.4.8. □

**Example 1.4.12.** Let $G$ be a discrete group. We determine $(\Bbbk G)^\circ$.

Evidently, $(\Bbbk G)'$ can be identified with $\Bbbk(G)$: each $\phi \in (\Bbbk G)'$ corresponds to the function $z \mapsto \phi(U_z)$, and each function $f \in \Bbbk(G)$ corresponds to the linear map $U_z \mapsto f(z)$. We claim that this identification induces an isomorphism

$$(\Bbbk G)^\circ \cong \mathrm{Rep}_{\Bbbk}(G)$$

of Hopf algebras, where $\mathrm{Rep}_{\Bbbk}(G)$ denotes the Hopf algebra of representative $\Bbbk$-valued functions on $G$, see the last remarks of Example 1.2.5. In fact, we only prove that the spaces $(\Bbbk G)^\circ$ and $\mathrm{Rep}_{\Bbbk}(G)$ coincide; the compatibility of the structure maps of $(\Bbbk G)^\circ$ and $\mathrm{Rep}_{\Bbbk}(G)$ follows from similar calculations as in Example 1.2.8 and Example 1.4.2.

Assume that $f \in \Bbbk(G) \cong (\Bbbk G)'$ vanishes on an ideal $I \subseteq \Bbbk G$ that has finite codimension. Then left multiplication defines a representation $\pi$ of $G$ on the finite-dimensional vector space $V := (\Bbbk G)/I$, in formulas, $\pi(x)(U_y + I) := U_{xy} + I$. Denote by $\phi \in V'$ the map given by $U_x + I \mapsto f(x)$. Then $f(x) = \phi(U_x + I) = \phi(\pi(x)(U_e + I))$ for all $x \in G$, so $f \in \mathrm{Rep}_{\Bbbk}(G)$.

Conversely, assume that $f \in \mathrm{Rep}_{\Bbbk}(G)$ has the form $f(x) = \phi(\pi(x)v)$, where $\pi$ is a representation of $G$ on some finite-dimensional vector space $V$ and $\phi \in V'$, $v \in V$. Then the functional in $(\Bbbk G)'$ corresponding to $f$ vanishes on the kernel of the representation $\tilde{\pi}: \Bbbk G \to \mathrm{Hom}_{\Bbbk}(V)$, $U_x \mapsto \pi(x)$, which is an ideal in $\Bbbk G$ of codimension less than or equal to $\dim \mathrm{Hom}_{\Bbbk}(V)$.

**Remark 1.4.13.** For every unital algebra $A$, the space $A^\circ$ is the largest subspace of $A'$ on which $\Delta_{A'}$ defines a comultiplication, as can be seen from Lemma 1.4.8 i). Unfortunately, the space $A^\circ$ may be quite small – there exist groups $G$ for which $(\Bbbk G)^\circ \cong \mathrm{Rep}_{\Bbbk}(G) = \Bbbk\epsilon$, see [1, Exercise 2.5] or [18, 2.7].

**Example 1.4.14.** If $G$ is a semisimple connected simply connected affine algebraic group over an algebraically closed field $\Bbbk$ of characteristic 0 and $\mathfrak{g}$ is the Lie algebra of $G$, then there exists a perfect dual pairing $U(\mathfrak{g}) \times \mathcal{O}(G) \to \Bbbk$ that induces an isomorphism $\mathcal{O}(G) \cong U(\mathfrak{g})^\circ$, see [65, Theorem 3.1]. However, $U(\mathfrak{g}) \ncong \mathcal{O}(G)^\circ$ – for any affine algebraic group $G$ over $\Bbbk$,

$$U(\mathfrak{g}) \cong \{\phi \in \mathcal{O}(G)' \mid \phi(\mathfrak{m}_1^n) = 0 \text{ for some } n > 0\},$$

where $\mathfrak{m}_1 = \{f \in \mathcal{O}(G) \mid f(1) = 0\}$ [1, p. 198]. A concrete example of this pairing is given in Example 1.4.6.

# Chapter 2
# Multiplier Hopf algebras and their duality

A multiplier Hopf algebra is a non-unital generalization of a Hopf algebra, where the target of the comultiplication is no longer the twofold tensor product of the underlying algebra, but an enlarged multiplier algebra. A motivation for this generalization, the pertaining definitions, and the main example are presented in Section 2.1.

A remarkable feature of multiplier Hopf algebras is that they admit a nice duality which extends the (incomplete) duality of Hopf algebras presented in Section 1.4. This duality is based on left- and right-invariant linear functionals called integrals, which are analogues of the Haar measures of a group. These integrals and their modular properties are discussed in Section 2.2, and an account of the duality theory is given in Section 2.3.

The theory of multiplier Hopf algebras was developed by Van Daele; all results presented in this chapter are taken from the articles [174], [177].

## 2.1 Definition of multiplier Hopf algebras

Our primary examples of Hopf algebras were algebras of functions on suitable groups, where the comultiplication, counit, and antipode were defined as the transposes of the group multiplication, of the inclusion of the unit, and of the group inversion, respectively, as in equation (1.4). Recall that the functions belonging to such a Hopf algebra had to be chosen carefully and in accordance with the flavor of the groups, as, for example,

- arbitrary functions if the groups are finite,
- representative functions if the groups are compact, or
- polynomial functions if the groups are affine algebraic.

Now assume that $G$ is an infinite discrete group. The discussion in Section 1.2.2 showed that the algebra $\Bbbk(G)$ of all functions on $G$ is too large to carry the comultiplication $\Delta$. A natural replacement would be the subalgebra $\Bbbk_{\text{fin}}(G) \subset \Bbbk(G)$ of functions with finite support. But unfortunately, the image $\Delta(\Bbbk_{\text{fin}}(G)) \subseteq \Bbbk(G \times G)$ is not contained in the subspace $\Bbbk_{\text{fin}}(G) \otimes \Bbbk_{\text{fin}}(G) \cong \Bbbk_{\text{fin}}(G \times G)$, so that $\Delta$ does not restrict to a comultiplication on $\Bbbk_{\text{fin}}(G)$.

To solve this problem, we extend the notion of a morphism of algebras by allowing a morphism from an algebra $A$ to an algebra $B$ to take values in an enlarged multiplier algebra $M(B)$. In the situation above, the multiplier algebra $M(\Bbbk_{\text{fin}}(G) \otimes \Bbbk_{\text{fin}}(G))$ will be equal to $\Bbbk(G \times G)$.

### 2.1.1 Multipliers of algebras

The concept of a multiplier Hopf algebra is based on the notion of a multiplier of an algebra:

**Definition 2.1.1.** Let $A$ be an algebra. A *left / right multiplier of $A$* is a linear map $T: A \to A$ that satisfies $T(ab) = T(a)b \,/\, T(ab) = aT(b)$ for all $a, b \in A$. A *multiplier of $A$* is a pair $(T_l, T_r)$ consisting of a left multiplier $T_l$ and a right multiplier $T_r$ satisfying $bT_l(a) = T_r(b)a$ for all $a, b \in A$. The set of all multipliers of $A$ is denoted by $M(A)$. It is an algebra with respect to the multiplication $(T_l, T_r) \cdot (S_l, S_r) := (T_l \circ S_l, S_r \circ T_r)$, as one can easily check.

**Remark 2.1.2.** For every commutative algebra $A$, a map $T: A \to A$ is a left multiplier if and only if it is a right multiplier.

When working with multipliers, it is convenient to restrict to algebras and homomorphisms that are non-degenerate in the following sense:

**Definition 2.1.3.** i) An algebra $A$ is *non-degenerate* if

(a) for every $a \in A$, $a \neq 0$, we have $Aa \neq 0$ and $aA \neq 0$, and

(b) the linear span of $AA$ is equal to $A$.

ii) Let $A$ and $B$ be non-degenerate algebras. A homomorphism $\phi: A \to M(B)$ is *non-degenerate* if the linear span of $\phi(A)B$ and the linear span of $B\phi(A)$ both are equal to $B$.

We shall frequently use the following properties of multipliers:

**Proposition 2.1.4.** *Let $A$ and $B$ be non-degenerate algebras.*

i) *For every element $a \in A$, the maps $b \mapsto ab$ and $b \mapsto ba$ define a multiplier, and the induced map $A \to M(A)$ embeds $A$ as an ideal in $M(A)$.*

*From now on, we consider $A$ as an ideal of $M(A)$ as in* i).

ii) *$aT = T_r(a)$ and $Ta = T_l(a)$ for all $a \in A$ and $T = (T_l, T_r) \in M(A)$.*

iii) *If $A$ is unital, then $T1_A = T = 1_A T \in A$ for each $T \in M(A)$, and $M(A) = A$.*

iv) *Let $A$ be a $*$-algebra. For every $T \in M(A)$, the formulas $T^*a := (a^*T)^*$ and $aT^* := (Ta^*)^*$, $a \in A$, define a multiplier $T^* \in M(A)$. The involution $T \mapsto T^*$ turns $M(A)$ into a $*$-algebra.*

v) *The tensor product $A \otimes B$ is non-degenerate and one has a natural embedding $M(A) \otimes M(B) \hookrightarrow M(A \otimes B)$.*

vi) *Every non-degenerate homomorphism* $\phi\colon A \to M(B)$ *extends uniquely to a homomorphism* $M(A) \to M(B)$*. If* $A$ *and* $B$ *are* $*$*-algebras and* $\phi$ *is a* $*$*-homomorphism, then the extension is a* $*$*-homomorphism again.*

*Proof.* Statements i)–iv) are easy to check.

v) Consider an element $x \in A \otimes B$. Write $x = \sum_i a_i \otimes b_i$, where the $b_i$ are linearly independent. Assume that $x(c \otimes d) = 0$ for all $c \in A$ and $d \in B$. Then

$$(f \otimes \mathrm{id})(x(c \otimes d)) = \sum_i f(a_i c) b_i d = 0 \quad \text{for all } c \in A,\ d \in B,\ f \in A',$$

and using the fact that $B$ is non-degenerate, we conclude that $\sum_i f(a_i c) b_i = 0$ for all $c \in A$, $d \in B$, $f \in A'$. Since the $b_i$ are linearly independent, we must have $f(a_i c) = 0$ for all $i$ and all $c \in A$, $f \in A'$. But $A$ is non-degenerate, so $a_i = 0$ for all $i$ and hence $x = 0$.

A similar argument shows that $x = 0$ if $(c \otimes d)x = 0$ for all $c \in A$ and $d \in B$. Finally, it is clear that $(A \otimes B)(A \otimes B)$ is linearly dense in $A \otimes B$.

vi) Let $T \in M(A)$. Every extension $\tilde{\phi}\colon M(A) \to M(B)$ of $\phi$ must satisfy

$$\sum_i \tilde{\phi}(T)\phi(a_i)b_i = \sum_i \phi(Ta_i)b_i \quad \text{and} \quad \sum_i b_i\phi(a_i)\tilde{\phi}(T) = \sum_i b_i\phi(a_i T) \tag{2.1}$$

for all $a_i \in A$ and $b_i \in B$. If $\sum_i \phi(a_i)b_i = 0$, then

$$\phi(c)\sum_i \phi(Ta_i)b_i = \sum_i \phi(cTa_i)b_i = \phi(cT)\sum_i \phi(a_i)b_i = 0 \quad \text{for all } c \in A,$$

and since $B$ and $\phi$ are non-degenerate, $\sum_i \phi(Ta_i)b_i = 0$. This and a similar argument show that (2.1) defines a multiplier $\tilde{\phi}(T) \in M(B)$. It is easy to check that the assignment $T \mapsto \tilde{\phi}(T)$ is a homomorphism and that $\tilde{\phi}$ is a $*$-homomorphism if $\phi$ is a $*$-homomorphism. □

### 2.1.2 Multiplier bialgebras

Motivated by the discussion in the introduction to this section, we generalize the notion of a bialgebra as follows:

**Definition 2.1.5.** A *multiplier bialgebra* is a non-degenerate algebra $A$ equipped with a non-degenerate homomorphism $\Delta\colon A \to M(A \otimes A)$ such that

i) the following subsets of $M(A \otimes A)$ are contained in $A \otimes A \subseteq M(A \otimes A)$:

$$\Delta(A)(1 \otimes A), \quad \Delta(A)(A \otimes 1), \quad (A \otimes 1)\Delta(A), \quad (1 \otimes A)\Delta(A);$$

ii) $\Delta$ is coassociative in the sense that the following diagram commutes:

$$\begin{array}{ccc} A & \xrightarrow{\Delta} & M(A \otimes A) \\ {\scriptstyle \Delta}\downarrow & & \downarrow{\scriptstyle \mathrm{id} \otimes \Delta} \\ M(A \otimes A) & \xrightarrow[\Delta \otimes \mathrm{id}]{} & M(A \otimes A \otimes A). \end{array}$$

If $A$ is a $*$-algebra and $\Delta$ a $*$-homomorphism, then $(A, \Delta)$ is called a *multiplier $*$-bialgebra.*

A *morphism* of multiplier ($*$-)bialgebras $(A, \Delta_A)$ and $(B, \Delta_B)$ is a non-degenerate ($*$-)homomorphism $F\colon A \to M(B)$ that satisfies $\Delta_B \circ F = (F \otimes F) \circ \Delta_A$, that is, the following square commutes:

$$\begin{array}{ccc} A & \xrightarrow{F} & M(B) \\ {\scriptstyle \Delta_A}\downarrow & & \downarrow{\scriptstyle \Delta_B} \\ M(A \otimes A) & \xrightarrow[F \otimes F]{} & M(B \otimes B). \end{array}$$

**Remarks 2.1.6.** i) In condition i) above, the symbol 1 denotes the unit in $M(A)$, and the spaces $A \otimes 1$ and $1 \otimes A$ are considered as subsets of $M(A) \otimes M(A) \subseteq M(A \otimes A)$.

ii) In the diagrams above, the homomorphisms $\Delta \otimes \mathrm{id}$, $\mathrm{id} \otimes \Delta\colon A \otimes A \to A \otimes A \otimes A$ and $F \otimes F\colon A \otimes A \to B \otimes B$ have been extended to the respective multiplier algebras.

iii) Given multiplier bialgebras $(A, \Delta_A)$ and $(B, \Delta_B)$, we can construct new multiplier bialgebras $(A, \Delta_A)^{\mathrm{op}}$, $(A, \Delta_A)^{\mathrm{cop}}$, $(A, \Delta_A)^{\mathrm{op,cop}}$ and equip $A \oplus B$ and $A \otimes B$ with the structure of multiplier bialgebras in a similar way as in the case of bialgebras, compare Remark 1.3.7 iii).

The Sweedler notation 1.3.3 can be extended to multiplier bialgebras, but then it has to be used with much more care. Although we shall rarely use it, we explain this notation in detail because it illustrates the difference between multiplier bialgebras and ordinary bialgebras.

**Notation 2.1.7.** Let $(A, \Delta)$ be a multiplier bialgebra and $a \in A$. In general, the multiplier $\Delta(a) \in M(A \otimes A)$ can not be written as a sum $\sum_i a_{1,i} \otimes a_{2,i}$, where $a_{1,i}, a_{2,i} \in M(A)$. Nevertheless, we write this multiplier as a formal sum

$$\Delta(a) =: \sum a_{(1)} \otimes a_{(2)}.$$

As for ordinary bialgebras, we extend this notation to iterated applications of the comultiplication as follows. Since $\Delta$ is non-degenerate, we can define non-degenerate homomorphisms $\Delta^{(n)}\colon A \to M(A^{\otimes n+1})$ by

$$\Delta^{(0)} := \mathrm{id}_A \quad \text{and} \quad \Delta^{(n+1)} := (\Delta^{(n)} \otimes \mathrm{id}_A) \circ \Delta \quad \text{for } n \geq 0.$$

For each $n \in \mathbb{N}$, we write the multiplier $\Delta^{(n)}(a) \in M(A^{\otimes n+1})$ as a formal sum

$$\Delta^{(n)}(a) =: \sum a_{(1)} \otimes \cdots \otimes a_{(n+1)}.$$

We think of the multiplier $\Delta^{(n)}(a)$ as having $n+1$ *legs* which are represented by the symbols $a_{(1)}, \ldots, a_{(n+1)}$, and treat the formal sums introduced above like ordinary sums of elementary tensors in a tensor product of algebras. Thus we write

$$\Delta(a)(b \otimes 1) = \sum a_{(1)}b \otimes a_{(2)}, \quad (b \otimes 1)\Delta(a) = \sum ba_{(1)} \otimes a_{(2)},$$
$$\Delta(a)(1 \otimes b) = \sum a_{(1)} \otimes a_{(2)}b, \quad (1 \otimes b)\Delta(a) = \sum a_{(1)} \otimes ba_{(2)}$$

for all $a, b \in A$. The definition of a multiplier bialgebra implies that these four expressions belong to $A \otimes A$; thus, the formal sums above stand for finite sums of elementary tensors in $A \otimes A$. We say that in the products in the first line above, the first leg $a_{(1)}$ of $\Delta(a)$ is *covered* by $b$, whereas in the products in the second line, the *second leg* $a_{(2)}$ of $\Delta(a)$ is *covered* by $b$.

More generally, every product of $\Delta^{(n)}(a) = \sum a_{(1)} \otimes \cdots \otimes a_{(n+1)}$ with elements of $A$, where at least $n$ legs of $\Delta^{(n)}(a)$ are covered by elements of $A$, belongs to $A^{\otimes n+1}$ and is equal to a finite sum of elementary tensors in $A^{\otimes n+1}$.

A drawback of the Sweedler notation for multiplier bialgebras is that it may be difficult to see whether certain products of formal sums $\sum a_{(1)} \otimes \cdots \otimes a_{(n+1)}$ belong to an ordinary tensor product or not.

**Example 2.1.8.** Let $(A, \Delta)$ be multiplier bialgebra and $a, b, c, d \in A$. The sum

$$\sum b_{(1)}a_{(1)} \otimes ca_{(2)}b_{(2)} \otimes a_{(3)}d \otimes b_{(3)}$$

is a well-defined element of $A \otimes A \otimes A$: First, note that $\sum a_{(1)} \otimes ca_{(2)} \in A \otimes A$. Therefore, $\sum a_{(1)} \otimes ca_{(2)} \otimes a_{(3)}d = \sum a_{(1)} \otimes \Delta(ca_{(2)})(1 \otimes d) \in A \otimes A \otimes A$, and hence the first two legs of $\sum b_{(1)} \otimes b_{(2)} \otimes b_{(3)}$ are covered by elements of $A$.

### 2.1.3 Multiplier Hopf algebras

Multiplier Hopf algebras form a special class of multiplier bialgebras:

**Definition 2.1.9.** A multiplier ($*$-)bialgebra $(A, \Delta)$ is a *multiplier Hopf($*$-)algebra* if the linear maps $T_1, T_2\colon A \otimes A \to A \otimes A$ given by

$$T_1(a \otimes b) = \Delta(a)(1 \otimes b) \quad \text{and} \quad T_2(a \otimes b) = (a \otimes 1)\Delta(b)$$

are bijective. A multiplier Hopf algebra $(A, \Delta)$ is *regular* if the multiplier bialgebras $(A, \Delta)^{\mathrm{op}}$ and $(A, \Delta)^{\mathrm{cop}}$ are multiplier Hopf algebras.

A *morphism of multiplier Hopf* $(*\text{-})$*algebras* is simply a morphism of the underlying multiplier $(*\text{-})$bialgebras.

This definition is similar to the characterization of Hopf algebras among bialgebras given in Section 1.3.4.

Before we consider examples of multiplier Hopf algebras and additional structure maps like the counit and antipode, let us list some easy observations:

**Remarks 2.1.10.** i) Every Hopf algebra is a multiplier Hopf algebra by Theorem 1.3.18. In particular, the notion of a regular Hopf algebra is well defined, and by Proposition 1.3.14, a Hopf algebra is regular if and only if its antipode is bijective. We shall see an analogue of this statement for multiplier Hopf algebras in Proposition 2.1.12 iii).

ii) If $(A, \Delta)$ is a multiplier Hopf algebra, then also $(A, \Delta)^{\mathrm{op,cop}}$ is a multiplier Hopf algebra, and $(A, \Delta)^{\mathrm{op}}$ is a multiplier Hopf algebra if and only if $(A, \Delta)^{\mathrm{cop}}$ is a multiplier Hopf algebra.

iii) Every multiplier Hopf $*$-algebra is regular.

**Example 2.1.11.** Suppose that $G$ is a discrete group. Then $M(\Bbbk_{\mathrm{fin}}(G) \otimes \Bbbk_{\mathrm{fin}}(G)) \cong \Bbbk(G \times G)$, and the algebra $\Bbbk_{\mathrm{fin}}(G)$ equipped with the comultiplication

$$\Delta\colon \Bbbk_{\mathrm{fin}}(G) \to \Bbbk(G \times G) \cong M(\Bbbk_{\mathrm{fin}}(G) \otimes \Bbbk_{\mathrm{fin}}(G)), \quad (\Delta f)(x, y) = f(xy),$$

is a regular multiplier Hopf algebra. Let us explain this in some more detail.

The natural embedding $\Bbbk(G) \otimes \Bbbk(G) \hookrightarrow \Bbbk(G \times G)$ identifies $\Bbbk_{\mathrm{fin}}(G) \otimes \Bbbk_{\mathrm{fin}}(G)$ with $\Bbbk_{\mathrm{fin}}(G \times G)$, and it is easy to see that $M(\Bbbk_{\mathrm{fin}}(G \times G))$ is isomorphic to $\Bbbk(G \times G)$. The isomorphism $M(\Bbbk_{\mathrm{fin}}(G) \otimes \Bbbk_{\mathrm{fin}}(G)) \cong \Bbbk(G \times G)$ thus obtained identifies a multiplier $T$ with the function $f_T$ defined by

$$f_T(x, y) = \lambda \;\Leftrightarrow\; T(\delta_x \otimes \delta_y) = \lambda(\delta_x \otimes \delta_y) \quad \text{for all } x, y \in G.$$

The map $\Delta$ is evidently coassociative, and the maps $T_1$ and $T_2$ considered in Definition 2.1.9 are bijective because they are the transposes of the bijections $\Phi_1, \Phi_2\colon G \times G \to G \times G$ considered in Proposition 1.3.19.

If $\Bbbk = \mathbb{C}$, then $\Bbbk_{\mathrm{fin}}(G)$ is a multiplier Hopf $*$-algebra with respect to pointwise conjugation of functions.

Multiplier Hopf algebras are similar to Hopf algebras in many respects. For example, they also possess a counit and an antipode:

**Proposition 2.1.12.** *Let $(A, \Delta)$ be a multiplier Hopf algebra.*

i) *There exists a unique non-degenerate homomorphism* $\epsilon\colon A \to \Bbbk$, *called the* counit *of* $(A, \Delta)$, *such that the following diagram commutes:*

$$\begin{array}{ccccc} M(A \otimes A) & \xleftarrow{\Delta} & A & \xrightarrow{\Delta} & M(A \otimes A) \\ \Big\downarrow{\scriptstyle \epsilon\otimes\mathrm{id}} & & \Big\downarrow & & \Big\downarrow{\scriptstyle \mathrm{id}\otimes\epsilon} \\ M(\Bbbk \otimes A) & \xrightarrow[\cong]{} & M(A) & \xleftarrow[\cong]{} & M(A \otimes \Bbbk). \end{array} \tag{2.2}$$

*If* $(A, \Delta)$ *is a multiplier Hopf* $*$-*algebra, then* $\epsilon$ *is a* $*$-*homomorphism.*

ii) *There exists a unique linear map* $S\colon A \to M(A)$, *called the* antipode *of* $(A, \Delta)$, *such that*

$$\begin{aligned} m_{M(A)}((S \otimes \mathrm{id}_A)(\Delta(a)(1 \otimes b))) &= \epsilon(a)b, \\ m_{M(A)}((\mathrm{id}_A \otimes S)((a \otimes 1)\Delta(b))) &= a\epsilon(b) \quad \text{for all } a, b \in A. \end{aligned} \tag{2.3}$$

*Here,* $m_{M(A)}$ *denotes the multiplication in* $M(A)$. *Moreover,* $S$ *is a morphism of multiplier Hopf algebras* $(A, \Delta) \to (A, \Delta)^{\mathrm{op,cop}}$.

iii) *A multiplier Hopf algebra* $(A, \Delta)$ *is regular if and only if the antipode* $S$ *is a linear isomorphism* $A \xrightarrow{\cong} A$.

iv) *If* $(A, \Delta)$ *is a multiplier Hopf* $*$-*algebra, then* $S \circ * \circ S \circ * = \mathrm{id}_A$.

v) *Let* $F\colon A \to M(B)$ *be a morphism of multiplier Hopf algebras. Then* $S_B \circ F = F \circ S_A$.

*Proof.* The construction of $\epsilon$ and $S$ proceeds along the same lines as in the proof of Proposition 1.3.22 and differs mainly in more involved notation; for details, see [174, Sections 3–5]. The remaining assertions about $\epsilon$ and $S$ follow similarly as for Hopf algebras; for details, see [174, Proposition 5.2] and [177, Proposition 2.8]. □

**Remark 2.1.13.** In extended Sweedler notation, equation (2.3) takes the form

$$\sum S(a_{(1)})a_{(2)}b = \epsilon(a)b, \qquad \sum ab_{(1)}S(b_{(2)}) = a\epsilon(b) \quad \text{for all } a, b \in A.$$

**Theorem 2.1.14.** *A multiplier bialgebra* $(A, \Delta)$ *is a regular multiplier Hopf algebra if and only if there exist a homomorphism* $\epsilon\colon A \to \Bbbk$ *and a bijective linear map* $S\colon A \to A$ *such that diagram* (2.2) *commutes and equation* (2.3) *holds.*

*Proof.* This follows from the previous proposition and a similar argument as in Lemma 1.3.21, see also [177, Proposition 2.9]. □

## 2.2 Integrals and their modular properties

The concept of an integral on a multiplier Hopf algebra is the natural analogue of the Haar measure of a locally compact group and fundamental to the duality theory of multiplier Hopf algebras presented in the next section. In contrast to locally compact groups, which always have an invariant measure, multiplier Hopf algebras need not possess integrals. Uniqueness, however, holds under very mild assumptions.

This section is organized as follows. First, we introduce the concept of an integral, give some examples, and list some easy properties that can be found in most books on Hopf algebras, as, for example, in [1], [80], [111], [145]. Next, we show that integrals are unique and faithful, closely following the original article [177] of Van Daele. Like the Haar measure of a locally compact group, every integral on a multiplier Hopf algebra enjoys several modular properties. These are studied in the last sections; again, we follow [177].

### 2.2.1 The concept of an integral

To motivate the definition of left and right integrals, let us reformulate the concept of a Haar measure in terms of Hopf algebras.

Let $G$ be a locally compact group with left Haar measure $\lambda$ and let $A \subseteq \mathbb{C}(G)$ be some Hopf algebra of functions on $G$ with comultiplication, counit, and antipode as in equation (1.4) on page 6. If the functions in $A$ are integrable, that is, if $A$ is contained in $L^1(G, \lambda)$, the Haar measure $\lambda$ defines a linear map

$$\phi\colon A \to \mathbb{C}, \quad f \mapsto \int_G f(y)d\lambda(y). \tag{2.4}$$

Left-invariance of $\lambda$ amounts to the fact that for each function $f \in A$, the function $F$ on $G$ defined by

$$F(x) := \int_G f(xy)d\lambda(y) \quad \text{for all } x \in G$$

satisfies $F(x) = \phi(f)$ for all $x \in G$. We replace the multiplication of $G$ by the comultiplication of $A$, using the relation

$$f(xy) = (\Delta(f))(x,y) = \sum f_{(1)}(x) f_{(2)}(y) \quad \text{for all } x, y \in G,$$

and obtain

$$F = \sum \Big( f_{(1)} \int_G f_{(2)}(y)d\lambda(y) \Big) = (\mathrm{id} \otimes \phi)(\Delta(f)).$$

Thus, the invariance condition $F(x) = \phi(f)$, $x \in G$, takes the form

$$(\mathrm{id} \otimes \phi)(\Delta(f)) = \phi(f)1_A.$$

Now, let $(A, \Delta)$ be an arbitrary multiplier Hopf algebra. Given a linear map $\phi\colon A \to \Bbbk$ and an element $a \in A$, we define $(\mathrm{id} \otimes \phi)(\Delta(a)) \in M(A)$ by

$$((\mathrm{id} \otimes \phi)(\Delta(a)))b := (\mathrm{id} \otimes \phi)(\Delta(a)(b \otimes 1)) = \sum a_{(1)} b \phi(a_{(2)}),$$
$$b((\mathrm{id} \otimes \phi)(\Delta(a))) := (\mathrm{id} \otimes \phi)((b \otimes 1)\Delta(a)) = \sum b a_{(1)} \phi(a_{(2)}).$$

Similarly we define $(\phi \otimes \mathrm{id})(\Delta(a)) \in M(A)$.

**Definition 2.2.1.** Let $(A, \Delta)$ be a multiplier Hopf algebra. A linear map $\phi\colon A \to \Bbbk$ is

- *left-invariant* if $(\mathrm{id} \otimes \phi)(\Delta(a)) = \phi(a)1_{M(A)}$ for all $a \in A$, and
- *right-invariant* if $(\phi \otimes \mathrm{id})(\Delta(a)) = \phi(a)1_{M(A)}$ for all $a \in A$.

If $\phi$ is non-zero and left- or right-invariant, we call it a *left* or *right integral on* $(A, \Delta)$, respectively. A map that is a left and a right integral is briefly called an *integral*.

We call $(A, \Delta)$ a *multiplier Hopf algebra with integrals* if there exist a left and a right integral on $(A, \Delta)$, and *unimodular* if every left integral on $(A, \Delta)$ is a right integral and vice versa.

Recall that a linear map $\phi\colon A \to \Bbbk$ on an algebra $A$ is called

- *faithful* if $\phi(aA) \neq 0$ and $\phi(Aa) \neq 0$ for every non-zero $a \in A$,
- *positive* if $A$ is a $*$-algebra and $\phi(a^*a) \geq 0$ for all $a \in A$, and
- *normalized* if $A$ is unital and $\phi(1_A) = 1$.

We call a positive normalized functional on a unital $*$-algebra a *state*.

**Remarks 2.2.2.** i) Let $(A, \Delta)$ be a multiplier Hopf algebra. It is easy to see that a functional $\phi \in A'$ is left-invariant if and only if for all $f \in A'$ and $a, b \in A$,

$$(f \otimes \phi)(\Delta(a)(b \otimes 1)) = f(b)\phi(a), \quad (f \otimes \phi)((b \otimes 1)\Delta(a)) = f(b)\phi(a).$$

Likewise, $\phi$ is right-invariant if and only if for all $f \in A'$ and $a, b \in A$,

$$(\phi \otimes f)(\Delta(a)(1 \otimes b)) = \phi(a)f(b), \quad (\phi \otimes f)((1 \otimes b)\Delta(a)) = \phi(a)f(b).$$

ii) For a Hopf algebra $(A, \Delta)$, the invariance of linear maps can be characterized in terms of the convolution product defined in Section 1.3.2: Let $\phi \in A'$. Then

$$\begin{aligned} \phi \text{ is left-invariant } &\Leftrightarrow\ f * \phi = f(1_A)\phi \text{ for all } f \in A', \\ \phi \text{ is right-invariant } &\Leftrightarrow\ \phi * f = f(1_A)\phi \text{ for all } f \in A'. \end{aligned}$$

This follows from the relations

$$f((\mathrm{id} \otimes \phi)(\Delta(a))) = (f * \phi)(a) \quad \text{and} \quad f((\phi \otimes \mathrm{id})(\Delta(a))) = (\phi * f)(a),$$

which hold for all $a \in A$.

iii) If a Hopf algebra has a left integral, then its antipode is bijective [29, Theorem 5.2.3, Corollary 5.4.6], [145, Corollary 5.1.7].

iv) Let $(A, \Delta)$ be a multiplier Hopf $*$-algebra and $\phi$ a non-zero positive functional on $A$.

(a) $\phi$ is $*$-linear: Since $A$ is non-degenerate, it suffices to show that $\phi(a^*b) = \overline{\phi(b^*a)}$ for all $a, b \in A$. Now the inequalities $\phi((a+b)^*(a+b)) \geq 0$, $\phi(a^*a) \geq 0$, $\phi(b^*b) \geq 0$ imply $\Im\phi(a^*b) + \Im\phi(b^*a) = 0$, and replacing $b$ by $ib$, we find $\Re\phi(a^*b) - \Re\phi(b^*a) = 0$.

(b) If $A$ is unital, then there exists a real number $r > 0$ such that $r\phi$ is normalized: The Cauchy–Schwarz inequality implies $|\phi(a)|^2 \leq \phi(1)\phi(a^*a)$ for all $a \in A$, whence $\phi(1) > 0$.

(c) $\phi$ is faithful if and only if $\phi(a^*a) > 0$ for all non-zero $a \in A$: The Cauchy–Schwarz inequality implies that for each $a \in A$, the functional $\phi(a\,\cdot)$ or $\phi(\cdot\, a)$ vanishes on $A$ if and only if $\phi(aa^*) = 0$ or $\phi(a^*a) = 0$, respectively.

**Examples 2.2.3.** Let us consider the Hopf algebras introduced in Section 1.2.2:

i) Let $G$ be a compact group with normalized Haar measure $\lambda$. Then the map $\phi\colon \mathrm{Rep}(G) \to \mathbb{C}$ given by $f \mapsto \int_G f d\lambda$ is a positive faithful normalized integral.

ii) Let $G$ be a discrete group. Then the map $\phi\colon \Bbbk_{\mathrm{fin}}(G) \to \Bbbk$ given by $\phi(\delta_x) = 1$ for all $x \in G$ is a faithful integral; it corresponds to integration with respect to the counting measure on $G$. For $\Bbbk = \mathbb{C}$, this integral is positive.

iii) Let $G$ be a discrete group. Then the map $\phi\colon \Bbbk G \to \Bbbk$ given by $\phi(U_e) = 1$ and $\phi(U_x) = 0$ for $x \in G, x \neq e$, is a faithful normalized integral. For $\Bbbk = \mathbb{C}$, this integral is positive.

iv) Consider the additive group $\Bbbk^n$, where $n \in \mathbb{N}$ and $\Bbbk \in \{\mathbb{R}, \mathbb{C}\}$. The Hopf algebra $\mathcal{O}(\Bbbk^n)$ does not contain any non-zero function that is integrable with respect to the Haar measure on $\mathbb{k}^n$, and therefore formula (2.4) does not define an integral on $\mathcal{O}(\mathbb{k}^n)$. In fact, the Hopf algebra $\mathcal{O}(\mathbb{k}^n)$ does not possess any left or right integral. We prove this assertion for $n = 1$, the general case can be treated similarly. Assume that $\phi\colon \mathcal{O}(\mathbb{k}) \to \mathbb{k}$ is left-invariant. For every $m \in \mathbb{N}$,

$$\Delta(X^m) = \sum_{k=0}^{m} \binom{m}{k} X^{m-k} \otimes X^k$$

and

$$1_{\mathcal{O}(\mathbb{k})} \cdot \phi(X^m) = (\mathrm{id} \otimes \phi)(\Delta(X^m)) = \sum_{k=0}^{m} \binom{m}{k} X^{m-k} \phi(X^k).$$

Thus $\phi(X^k) = 0$ for all $k < m$ and $m \in \mathbb{N}$, that is, $\phi = 0$. A similar argument shows that every right-invariant functional vanishes.

v) Consider the multiplicative group $\mathbb{k}^n_\times$, where $n \in \mathbb{N}$ and $\mathbb{k} \in \{\mathbb{R}, \mathbb{C}\}$. Again, the Haar measure on $\mathbb{k}^n_\times$ does not yield a left or right integral on the Hopf algebra $\mathcal{O}(\mathbb{k}^n_\times)$ because the only integrable function in $\mathcal{O}(\mathbb{k}^n_\times)$ is 0.

Nevertheless, the Hopf algebra $\mathcal{O}(\mathbb{k}^n_\times)$ does possess a normalized integral $\phi$ for every field $\mathbb{k}$, given by $1_{\mathcal{O}(\mathbb{k}^n_\times)} \mapsto 1$ and $X_1^{k_1} \dots X_n^{k_n} \mapsto 0$ for all $(k_1, \dots, k_n) \in \mathbb{Z}^n \setminus \{0\}$. With respect to the natural isomorphism $\mathcal{O}(\mathbb{k}^n_\times) \cong \mathbb{k}\mathbb{Z}^n$ (see the end of Example 1.2.8), this map corresponds to the integral described in ii). For $\mathbb{k} = \mathbb{C}$, we have an isomorphism $\mathcal{O}(\mathbb{C}^n_\times) \cong \mathrm{Rep}(\mathbb{T}^n)$ (see the end of Example 1.2.8), and the map $\phi$ corresponds to integration over $\mathbb{T}^n$ with respect to the Haar measure as in i).

The following result is fundamental:

**Proposition 2.2.4.** *Every left integral and every right integral on a regular multiplier Hopf algebra is faithful.*

*Proof.* To simplify the presentation, we restrict ourselves to the case of a Hopf algebra; see [177, Proposition 3.4] for the general case. So, let $\phi$ be a left integral on a regular Hopf algebra $(A, \Delta)$ and $a \in A$. Assume that $\phi(Aa) = 0$. Then

$$0 = c\phi(ba) = (\mathrm{id} \otimes \phi)((c \otimes 1)\Delta(b)\Delta(a)) \quad \text{for all } c, b \in A.$$

Since the map $T_2\colon A \otimes A \to A \otimes A$, $c \otimes d \to (c \otimes 1)\Delta(d)$, is surjective, we can replace $(c \otimes 1)\Delta(b)$ by an arbitrary element $d \otimes e \in A \otimes A$, and find $0 = \sum da_{(1)}\phi(ea_{(2)})$ for all $d, e \in A$. We put $d = 1_A$, apply $\Delta$, and get

$$0 = \sum a_{(1)} \otimes a_{(2)}\phi(ea_{(3)}) \quad \text{for all } e \in A.$$

Write $\sum a_{(1)} \otimes a_{(2)} \otimes a_{(3)} = \sum_i a_{1,i} \otimes a_{2,i} \otimes a_{3,i}$, choosing the elements $a_{2,i}$ linearly independent. Then $0 = a_{1,i}\phi(ea_{3,i})$ for all $e \in A$ and all $i$. Replacing $e$ by $fS(a_{2,i})$ for each $i$, where $f \in A$ is fixed but arbitrary, and summing again, we obtain

$$0 = \sum_i a_{1,i}\phi(fS(a_{2,i})a_{3,i}) = \sum a_{(1)}\phi(f\epsilon(a_{(2)})) = a\phi(f) \quad \text{for all } f \in A.$$

This shows $a = 0$. A modification of this argument shows that $\phi(aA) = 0$ implies $a = 0$. The proof of the assertion concerning right integrals is similar. □

**Corollary 2.2.5.** *Every left or right integral $\phi$ on a multiplier Hopf $*$-algebra $(A, \Delta)$ satisfies $\phi(a^*a) > 0$ for all $a \in A$, $a \neq 0$.*

*Proof.* Combine Proposition 2.2.4 with Remark 2.2.2 iv) (c). □

Before we consider the question of uniqueness and existence of left and right integrals, let us clarify the relation between these two notions. Recall that for a locally compact group, the inversion of the group interchanges the left with the right Haar measure, and that for a compact group, the left and the right Haar measure coincide. Similar results hold for multiplier Hopf algebras:

**Proposition 2.2.6.** i) *Let $(A, \Delta)$ be a regular multiplier Hopf algebra with a left/right integral $\phi$. Then $\phi \circ S$ and $\phi \circ S^{-1}$ are right/left integrals on $(A, \Delta)$.*

ii) *On every Hopf algebra, there exists at most one normalized left/right integral $\phi$, and this $\phi$ is simultaneously a right/left integral and satisfies $\phi = \phi \circ S = \phi \circ S^{-1}$.*

*Proof.* i) Let $\phi$ be a left integral on $(A, \Delta)$. By Proposition 2.1.12 iii), the antipode $S$ is a linear isomorphism of $A$, whence $\phi \circ S$ is non-zero. Furthermore,

$$\begin{aligned} S((\phi \circ S \otimes \mathrm{id})(\Delta(a)(1 \otimes b))) &= (\phi \otimes \mathrm{id})((S \otimes S)(\Delta(a)(1 \otimes b))) \\ &= (\mathrm{id} \otimes \phi)((S(b) \otimes 1)\Delta(S(a))) = S(b)\phi(S(a)) \end{aligned}$$

for all $a, b \in A$. An application of $S^{-1}$ shows $(\phi \circ S \otimes \mathrm{id})(\Delta(a)) = \phi(S(a))$, so $\phi \circ S$ is right-invariant. The proofs of the other assertions are similar.

ii) Let $(A, \Delta)$ be a Hopf algebra with a normalized left integral $\phi$. By Remark 2.2.2 iii) and Proposition 2.1.12 iii), $(A, \Delta)$ is regular, so $\psi := \phi \circ S$ is a right integral. For every normalized left integral $\tilde{\phi}$, we have $\tilde{\phi} = \psi$ because

$$\tilde{\phi}(a) = \psi(1_A\tilde{\phi}(a)) = (\psi \otimes \tilde{\phi})(\Delta(a)) = \tilde{\phi}(\psi(a)1_A) = \psi(a), \quad a \in A.$$

The proofs of the other assertions are similar. □

For left and for right integrals on multiplier Hopf $*$-algebras, it is natural to demand positivity. Unfortunately, the correspondence between left and right integrals established above need not preserve positivity. Nevertheless, the following result holds:

**Proposition 2.2.7** ([94, Theorem 9.9]). *A multiplier Hopf $*$-algebra has a positive left integral if and only if it has a positive right integral.*

*Comments on the proof.* This result seems to be accessible only via operator theory and a fair amount of work; therefore we do not include the proof. Roughly, one proceeds as follows. Given a multiplier Hopf $*$-algebra $(A, \Delta)$ with positive left integral $\phi$, one constructs a GNS-representation $(H, \Lambda, \pi)$ for $(A, \phi)$ and considers the unbounded conjugate-linear map $G$ on $H$ given by $\Lambda(a) \mapsto \Lambda(S(a)^*)$. Let $G = IM^{1/2}$ be the polar decomposition of $G$. Then $I$ is an antiunitary involutive map and one can show that for each $a \in A$, the operator $I\pi(a^*)I$ is contained in $\pi(A)$. Therefore, one can define a map $R\colon A \to A$, called the *unitary antipode of* $(A, \Delta)$, by the formula $\pi(R(a)) = I\pi(a^*)I$. Finally, the map $\phi \circ R$ turns out to be a positive right-invariant functional on $(A, \Delta)$.

**Definition 2.2.8.** An *algebraic quantum group* is a multiplier Hopf $*$-algebra with a positive left integral and a positive right integral.

### 2.2.2 Existence and uniqueness

The existence of integrals on multiplier Hopf algebras can be shown under special assumptions only, for example

- every finite-dimensional Hopf algebra has a left integral and a right integral [177, Proposition 5.1];
- a Hopf algebra has a normalized integral if and only if it is *cosemisimple*, which, roughly, means that the dual algebra $A'$ is semisimple [29, Exercise 5.5.9], [80, Section 11.2], [145, Chapter XIV].

A comprehensive characterization of all Hopf algebras that possess left integrals can be found in [29, Section 5.3]. Hopf $*$-algebras with positive integrals, also called algebraic compact quantum groups, are studied in Chapter 3, and further examples of such Hopf $*$-algebras are given in Chapter 6.

In contrast to existence, uniqueness of integrals already holds for all regular multiplier Hopf algebras. This class includes all multiplier Hopf $*$-algebras and all Hopf algebras that have a left integral, see Remark 2.2.2 iii) and Proposition 2.1.12 iii). The first main step towards the proof of uniqueness is the following result which will be of individual interest later on.

**Proposition 2.2.9.** *Let $(A, \Delta)$ be a regular multiplier Hopf algebra with a left integral $\phi$ and a right integral $\psi$. Then*

$$\{\psi(\cdot a) \mid a \in A\} = \{\phi(\cdot a) \mid a \in A\} \quad \textit{and} \quad \{\psi(a \cdot) \mid a \in A\} = \{\phi(a \cdot) \mid a \in A\}.$$

*In particular, these spaces do not depend on the choice of $\phi$ and $\psi$.*

*Proof.* We show that for each $a \in A$ there exists some $c \in A$ such that $\phi(\cdot a) = \psi(\cdot c)$; the rest of the proof is similar and uses the regularity of $(A, \Delta)$. So, let $a \in A$ and choose $b \in A$ such that $\psi(b) = 1$. We apply Remark 2.2.2 i) to $\phi$ and $f := \psi$, and find

$$\phi(xa) = \psi(b)\phi(xa) = (\psi \otimes \phi)(\Delta(xa)(b \otimes 1)) \quad \text{for all } x \in A.$$

Since the map $T_1 : A \otimes A \to A \otimes A$, $c \otimes d \mapsto \Delta(c)(1 \otimes d)$, is surjective, we can write $\Delta(a)(b \otimes 1) = \sum_i \Delta(c_i)(1 \otimes d_i)$ with $c_i, d_i \in A$. Using Remark 2.2.2 i) again, but this time for $\psi$ and $f := \phi$, we find

$$\begin{aligned}\phi(xa) &= \sum_i (\psi \otimes \phi)(\Delta(x)\Delta(c_i)(1 \otimes d_i)) \\ &= \sum_i \phi(\psi(xc_i)d_i) = \sum_i \psi(xc_i\phi(d_i))\end{aligned}$$

for all $x \in A$. Thus $\phi(\cdot a) = \psi(\cdot c)$, where $c := \sum_i c_i\phi(d_i)$. □

**Theorem 2.2.10.** *Let $(A, \Delta)$ be a regular multiplier Hopf algebra with integrals. Then the space of all left-invariant functionals and the space of all right-invariant functionals on $A$ both have dimension* 1.

*Proof.* Let $\phi_1$ and $\phi_2$ be two left integrals and $\psi$ a right integral on $(A, \Delta)$. Choose $a, b \in A$ such that $\psi(ab) = 1$. Let $x \in A$. Since the map $T_2 : A \otimes A \to A \otimes A$, $c \otimes y \to (c \otimes 1)\Delta(y)$, is surjective, we can choose $c_i, y_i \in A$ such that $\sum_i (c_i \otimes 1)\Delta(y_i) = (1 \otimes x)\Delta(a)$. Then

$$\begin{aligned}\phi_1(x) = \psi(ab)\phi_1(x) &= (\psi \otimes \phi_1)((1 \otimes x)\Delta(a)\Delta(b)) \\ &= (\psi \otimes \phi_1)\Big(\sum_i (c_i \otimes 1)\Delta(y_i)\Delta(b)\Big) \qquad (2.5) \\ &= \sum_i \psi(c_i)\phi_1(y_i b).\end{aligned}$$

By Proposition 2.2.9, there exists some $d \in A$ such that $\phi_1(\cdot b) = \phi_2(\cdot d)$. Reversing the transformations in (2.5), we obtain

$$\phi_1(x) = \sum_i \psi(c_i)\phi_2(y_i d) = \psi(ad)\phi_2(x) \quad \text{for all } x \in A.$$ □

**Remark 2.2.11.** Let $(A, \Delta)$ be a multiplier Hopf $*$-algebra with a positive left integral $\phi$. Then there exists a number $z \in \mathbb{C}$ with $|z| = 1$ such that the right integral $z\phi \circ S$ is positive. This follows immediately from Propositions 2.2.6, 2.2.7 and Theorem 2.2.10.

### 2.2.3 The modular element of an integral

An integral on a multiplier Hopf algebra has similar modular properties like the Haar measure of a locally compact group: every left integral $\phi$ is related to the associated right integral $\phi \circ S$ via a modular multiplier which is an analogue of the modular function of a locally compact group. Let us recall this modular function. Consider a locally compact group $G$ with left Haar measure $\lambda$ and denote by $i : G \to G$ the inversion $x \mapsto x^{-1}$. Then the image $\lambda^{-1} := i_*(\lambda)$ of the left Haar measure under $i$ is a right-invariant measure, and there exists a strictly positive continuous function $\delta_G$ on $G$ – called the *modular function of* $G$ – such that

$$\int_G f(y) d\lambda(y) = \int_G f(y^{-1})\delta_G(y^{-1}) d\lambda(y) = \int_G f(z)\delta_G(z) d\lambda^{-1}(z) \tag{2.6}$$

for all positive measurable functions $f$ on $G$. The function $\delta_G$ is a homomorphism from $G$ to the multiplicative group of positive real numbers, that is,

$$\delta_G(xy) = \delta_G(x)\delta_G(y), \quad \delta_G(e) = 1, \quad \delta_G(x^{-1}) = \delta_G(x)^{-1} \quad \text{for all } x, y \in G; \tag{2.7}$$

here, $e \in G$ denotes the unit as usual.

Let us translate equation (2.6) into the language of integrals on Hopf algebras. Suppose that $A \subseteq L^1(G)$ is a multiplier Hopf algebra with structure maps as in equation (1.4) on page 6 and that $\phi$ is the left integral on $(A, \Delta)$ given by integration with respect to $\lambda$ as in equation (2.4) on page 47. Then equation (2.6) takes the form

$$\phi(f) = (\phi \circ S)(f\delta_G) \quad \text{for all } f \in A.$$

Here, we assume for the moment that $\delta_G$ belongs to $M(A)$. The relations in (2.7) are equivalent to the relations $\Delta(\delta_G) = \delta_G \otimes \delta_G$, $\epsilon(\delta_G) = 1$, and $S(\delta_G) = \delta_G^{-1}$.

Every multiplier Hopf algebra with integrals has a modular element $\delta$ that behaves very much like $\delta_G$. There is, however, one small change:

*The conventional choice of $\delta$ corresponds to $\delta_G^{-1}$ instead of $\delta_G$.*

The first step towards the construction of $\delta$ is the following lemma, which is also of individual interest:

**Lemma 2.2.12.** *Let $(A, \Delta)$ be a regular multiplier Hopf algebra with a left integral $\phi$ and a right integral $\psi$. Then for all $a, b, x \in A$,*

$$(\psi \otimes S)((x \otimes 1)\Delta(a)) = (\psi \otimes \mathrm{id})(\Delta(x)(a \otimes 1))$$

*and*

$$(S \otimes \phi)(\Delta(b)(1 \otimes x)) = (\mathrm{id} \otimes \phi)((1 \otimes b)\Delta(x)).$$

In Sweedler notation, these formulas can be rewritten as follows:

$$\sum \psi(xa_{(1)})S(a_{(2)}) = \sum \psi(x_{(1)}a)x_{(2)},$$
$$\sum S(b_{(1)})\phi(b_{(2)}x) = \sum x_{(1)}\phi(bx_{(2)}).$$

*Proof.* We only prove the first equation, the second one follows similarly. To simplify notation, we assume that $(A, \Delta)$ is a Hopf algebra; for the general case, see [177, Proof of Proposition 3.11]. Let $x, a \in A$. Since $\psi$ is right-invariant,

$$\begin{aligned}\sum \psi(xa_{(1)})S(a_{(2)}) &= \sum \psi(x_{(1)}a_{(1)})x_{(2)}a_{(2)}S(a_{(3)}) \\ &= \sum \psi(x_{(1)}a_{(1)})x_{(2)}\epsilon(a_{(2)}) = \sum \psi(x_{(1)}a)x_{(2)}. \qquad \square\end{aligned}$$

**Proposition 2.2.13.** *Let $(A, \Delta)$ be a regular multiplier Hopf algebra with a left integral $\phi$.*

i) *There exists an invertible multiplier $\delta \in M(A)$ such that*

$$(\phi \otimes \mathrm{id})(\Delta(a)) = \phi(a)\delta \quad \text{for all } a \in A. \tag{2.8}$$

ii) *The extensions of $\Delta, \epsilon$, and $S$ to $M(A)$ act on $\delta$ as follows:*

$$\Delta(\delta) = \delta \otimes \delta, \quad \epsilon(\delta) = 1, \quad S(\delta) = \delta^{-1}. \tag{2.9}$$

iii) *$\phi(S(a)) = \phi(a\delta)$ for all $a \in A$.*

*Proof.* To simplify notation, we assume that $(A, \Delta)$ is a Hopf algebra; the general case is treated similarly, see [177, Propositions 3.8–3.10].

i) It suffices to show that the map $(\phi \otimes \mathrm{id}) \circ \Delta$ factorizes through $\phi$. It follows from Remark 2.2.2 ii) that for every $\omega \in A'$, the convolution product $\phi * \omega$ is left-invariant, and by Theorem 2.2.10, this product vanishes on $\ker \phi$. Now $(\phi \otimes \mathrm{id})(\Delta(\ker \phi)) = 0$ results from the equation

$$\omega((\phi \otimes \mathrm{id})(\Delta(\ker \phi))) = (\phi * \omega)(\ker \phi) = 0 \quad \text{for all } \omega \in A'.$$

ii) We apply $\Delta$ and $\epsilon$ to the equation $\phi(a)\delta = \sum \phi(a_{(1)})a_{(2)}$ and obtain

$$\Delta(\phi(a)\delta) = \sum \phi(a_{(1)})a_{(2)} \otimes a_{(3)} = \sum \phi(a_{(1)})\delta \otimes a_{(2)} = \delta \otimes \phi(a)\delta,$$
$$\epsilon(\phi(a)\delta) = \sum \phi(a_{(1)})\epsilon(a_{(2)}) = \phi(a)$$

for all $a \in A$. Consequently, $\Delta(\delta) = \delta \otimes \delta$ and $\epsilon(\delta) = 1$. Furthermore,

$$S(\delta)\delta = m((S \otimes \mathrm{id})(\Delta(\delta))) = \epsilon(\delta)1_A = 1_A$$

and similarly $\delta S(\delta) = 1_A$; in particular, $\delta$ is invertible.

iii) By right-invariance of $\phi \circ S$ and by the second equation in Lemma 2.2.12,

$$\phi(S(a))\phi(b) = \sum \phi(S(a_{(1)}))\phi(a_{(2)}b) = \sum \phi(b_{(1)})\phi(ab_{(2)}) = \phi(b)\phi(a\delta)$$

for all $a, b \in A$. □

**Remarks 2.2.14.** Let $(A, \Delta)$ and $\phi$, $\delta$ be as above.

i) If $\psi$ is a right integral on $(A, \Delta)$, then $(\mathrm{id} \otimes \psi)(\Delta(a)) = \psi(a)\delta^{-1}$ for all $a \in A$. This follows easily from equations (2.8), (2.9) and the fact that $\psi$ is a multiple of $\phi \circ S$.

ii) $\phi(S^2(a)) = \phi(S(a)\delta) = \phi(S(\delta^{-1}a)) = \phi(\delta^{-1}a\delta)$ for all $a \in A$.

### 2.2.4 The modular automorphism of an integral

In general, an integral $\phi$ on a multiplier Hopf algebra $(A, \Delta)$ need not be tracial in the sense that $\phi(ab) = \phi(ba)$ for all $a, b \in A$. However, there exists some control on the non-commutativity of $A$ with respect to $\phi$, namely, there exists an automorphism $\sigma$ of $A$ that satisfies $\phi(ab) = \phi(b\sigma(a))$ for all $a, b \in A$. This modular automorphism is an analogue of the modular automorphism of a weight on a von Neumann algebra, see Section 8.1.3.

The construction of this modular automorphism proceeds in several steps.

**Lemma 2.2.15.** *For every Hopf algebra $(A, \Delta)$, the maps $R, T: A \otimes A \to A \otimes A$ given by $T(a \otimes b) := \sum a_{(1)} \otimes bS^2(a_{(2)})$ and $R(a \otimes b) := \sum a_{(1)} \otimes bS(a_{(2)})$ are inverse to each other.*

*Proof.* By Proposition 1.3.12,

$$\begin{aligned} T(R(a \otimes b)) &= \sum T(a_{(1)} \otimes bS(a_{(2)})) = \sum a_{(1)} \otimes bS(a_{(3)})S^2(a_{(2)}) \\ &= \sum a_{(1)} \otimes bS(S(a_{(2)})a_{(3)}) = \sum a_{(1)} \otimes b\epsilon(a_{(2)}) = a \otimes b \end{aligned}$$

and similarly $R(T(a \otimes b)) = a \otimes b$ for all $a, b \in A$. □

**Proposition 2.2.16.** *Let $\phi$ be a left integral on a regular multiplier Hopf algebra $(A, \Delta)$. Then the subspaces $\{\phi(\cdot a) \mid a \in A\}$ and $\{\phi(b \cdot) \mid b \in A\}$ of $A'$ coincide.*

*Proof.* To simplify notation, we assume that $(A, \Delta)$ is a Hopf algebra; the general case is proved similarly, see [177, Proposition 3.11]. We show that for every $a \in A$, there exists a $b \in A$ such that $\phi(\cdot\, a) = \phi(b\, \cdot)$, the converse follows from a similar argument. Let $\psi$ be a right integral on $(A, \Delta)$. By Proposition 2.2.9, there exists an $a' \in A$ such that $\phi(\cdot\, a) = \psi(\cdot\, a')$. Two applications of Lemma 2.2.12 show that for all $c, d, x \in A$,

$$\sum \psi(xc_{(1)})\phi(dS(c_{(2)})) = \sum \psi(x_{(1)}c)\phi(dx_{(2)}) = \sum \psi(S(d_{(1)})c)\phi(d_{(2)}x).$$

Since the map $R$ defined in the previous lemma is surjective, the composition $(\mathrm{id} \otimes \phi) \circ R\colon A \otimes A \to A$ is surjective as well. Hence we can write $a'$ in the form

$$a' = \sum_i (\mathrm{id} \otimes \phi)(R(c_i \otimes d_i)) = \sum_i c_{i(1)}\phi(d_i S(c_{i(2)})) \quad \text{for some } c_i, d_i \in A.$$

The equation above, applied to $c := c_i$ and $d := d_i$ for each $i$ individually, shows that $b := \sum_i \psi(S(d_{i(1)})c)d_{i(2)}$ satisfies $\psi(\cdot\, a') = \phi(b\, \cdot)$. □

Let $\phi$ be a left integral on a regular multiplier Hopf algebra $(A, \Delta)$. Since the map $\phi \circ S^2$ is left-invariant, there exists a unique scalar $\tau \in \Bbbk$ such that $\phi \circ S^2 = \tau\phi$, see Theorem 2.2.10. This element $\tau$ figures in the next theorem:

**Theorem 2.2.17.** *Let $(A, \Delta)$ be a regular multiplier Hopf algebra with a left integral $\phi$. Then there exists an automorphism $\sigma$ of $A$ such that*

$$\phi(ax) = \phi(x\sigma(a)) \quad \textit{for all } a, x \in A.$$

*Furthermore,*

$$\phi \circ \sigma = \phi, \quad \Delta \circ \sigma = (S^2 \otimes \sigma) \circ \Delta, \quad \sigma(\delta) = \tau^{-1}\delta.$$

*Proof.* By Proposition 2.2.16, there exists for every $a \in A$ an element $\sigma(a) \in A$ such that $\phi(ax) = \phi(x\sigma(a))$ for all $x \in A$, and this element is unique because $\phi$ is faithful (Proposition 2.2.4). The assignment $\sigma\colon A \to A$, $a \mapsto \sigma(a)$, is injective because $\phi$ is faithful, surjective by Proposition 2.2.16, and a homomorphism because

$$\phi(x\sigma(ab)) = \phi(abx) = \phi(bx\sigma(a)) = \phi(x\sigma(a)\sigma(b)) \quad \text{for all } a, b, x \in A.$$

Dropping $x$ in the relation above, we obtain $\phi(\sigma(c)) = \phi(c)$ for all $c \in \operatorname{span} A^2 = A$.

Let us prove that $\Delta \circ \sigma = (S^2 \otimes \sigma) \circ \Delta$. Since functionals of the form $\omega := \phi(b\, \cdot)$, where $b \in A$, separate the elements of $A$ (Proposition 2.2.4), it is enough to show that for all such functionals $\omega$ and for all $a \in A$,

$$(\mathrm{id} \otimes \omega)(\Delta(\sigma(a))) = (\mathrm{id} \otimes \omega)((S^2 \otimes \sigma)(\Delta(a))),$$

that is,

$$(\mathrm{id}\otimes\phi)((1\otimes b)\Delta(\sigma(a))) = (\mathrm{id}\otimes\phi)((1\otimes b)((S^2\otimes\sigma)(\Delta(a)))).$$

Two applications of Lemma 2.2.12 show that the left-hand side is equal to

$$\begin{aligned}(S\otimes\phi)(\Delta(b)(1\otimes\sigma(a))) &= (S\otimes\phi)((1\otimes a)\Delta(b))\\ &= (S^2\otimes\phi)(\Delta(a)(1\otimes b)) = (\mathrm{id}\otimes\phi)((1\otimes b)((S^2\otimes\sigma)(\Delta(a)))).\end{aligned}$$

Finally, by definition of $\tau$ and by Remark 2.2.14 ii),

$$\tau\phi(a) = \phi(S^2(a)) = \phi(\delta^{-1}a\delta) = \phi(a\delta\sigma(\delta^{-1}))$$

for all $a \in A$. Therefore, $\delta\sigma(\delta^{-1}) = \tau$ and $\tau^{-1}\delta = \sigma(\delta)$. □

**Remarks 2.2.18.** i) For every right integral $\psi$ on a regular multiplier Hopf algebra $(A, \Delta)$, there exists a modular automorphism $\sigma'$ of $A$ such that $\psi(ax) = \psi(x\sigma'(a))$ for all $a, x \in A$, and this $\sigma'$ satisfies

$$\sigma' = S^{-1}\circ\sigma^{-1}\circ S = \delta\sigma(\cdot)\delta^{-1} \quad\text{and}\quad \Delta\circ\sigma' = (\sigma'\otimes S^{-2})\circ\Delta.$$

This follows easily from the fact that $\psi$ is a multiple of $\phi\circ S = \phi(\cdot\,\delta)$.

ii) For a multiplier Hopf $*$-algebra, the modular automorphism of a left or right integral need not be a $*$-homomorphism.

## 2.3 Duality

Let $(A, \Delta)$ be a regular multiplier Hopf algebra with integrals. Then one can associate to $(A, \Delta)$ a dual regular multiplier Hopf algebra $(\widehat{A}, \widehat{\Delta})$ with integrals and show that the bidual $(\widehat{\widehat{A}}, \widehat{\widehat{\Delta}})$ is naturally isomorphic to $(A, \Delta)$. This duality extends to algebraic quantum groups and constitutes an algebraic analogue of Pontrjagin duality, compare also with the discussion in Example 1.4.3. These results and constructions are due to Van Daele; we follow his article [177].

### 2.3.1 The duality of regular multiplier Hopf algebras

Let $(A, \Delta)$ be a regular multiplier Hopf algebra with a left integral $\phi$ and a right integral $\psi$. Put

$$\widehat{A} := \{\phi(\cdot\, a) \mid a \in A\} \subseteq A'.$$

By Theorem 2.2.17 and Proposition 2.2.9,

$$\widehat{A} = \{\phi(a\,\cdot) \mid a \in A\} = \{\psi(a\,\cdot) \mid a \in A\} = \{\psi(\cdot\, a) \mid a \in A\}.$$

In particular, $\widehat{A}$ does not depend on the choice of $\phi$; of course, this follows from Theorem 2.2.10 as well.

We equip $\widehat{A}$ with a multiplication and a comultiplication that are dual to the comultiplication and multiplication, respectively, of $(A, \Delta)$. Let us begin with the multiplication. Given elements $\omega_1 = \phi(\cdot a_1)$ and $\omega_2 = \phi(\cdot a_2)$ of $\widehat{A}$, we can define a linear map $\omega_1\omega_2 \colon A \to \Bbbk$ by the formula

$$(\omega_1\omega_2)(x) := (\omega_1 \otimes \omega_2)(\Delta(x)) = (\phi \otimes \phi)(\Delta(x)(a_1 \otimes a_2)). \tag{2.10}$$

**Lemma 2.3.1.** *Equation* (2.10) *defines an associative and non-degenerate multiplication on* $\widehat{A}$.

*Proof.* Let $\omega_1 = \phi(\cdot a_1)$ and $\omega_2 = \phi(\cdot a_2)$ as above. We have to show that the map $\omega_1\omega_2$ belongs to $\widehat{A}$. Since $(A, \Delta)$ is regular, we can write $a_1 \otimes a_2 = \sum_i \Delta(c_i)(d_i \otimes 1)$ with suitably chosen $c_i, d_i \in A$. Then

$$(\omega_1\omega_2)(x) = \sum_i (\phi \otimes \phi)(\Delta(xc_i)(d_i \otimes 1)) = \sum_i \phi(xc_i)\phi(d_i) = \phi(xb)$$

for all $x \in A$, where $b = \sum_i c_i\phi(d_i)$. Hence $\omega_1\omega_2 \in \widehat{A}$. Moreover, using regularity of $(A, \Delta)$, we see that $\widehat{A}\widehat{A}$ spans $\widehat{A}$.

Associativity of the multiplication follows easily from coassociativity of $\Delta$. Let us show that the multiplication is non-degenerate. If $\omega_1\omega_2 = 0$ for all $\omega_2$, then

$$0 = (\omega_1\omega_2)(b) = (\phi \otimes \phi)(\Delta(b)(a_1 \otimes a_2)) \quad \text{for all } b, a_2 \in A.$$

Since the map $T_1 \colon A \otimes A \to A \otimes A$, $b \otimes a_2 \mapsto \Delta(b)(1 \otimes a_2)$, is surjective, we can replace $\Delta(b)(1 \otimes a_2)$ by arbitrary elements $c \otimes d \in A \otimes A$. Then we obtain $\phi(ca_1)\phi(d) = 0$ for all $c, d \in A$, and by Proposition 2.2.4, also $a_1 = 0$ and $\omega_1 = 0$. □

Next, we turn to the comultiplication $\widehat{\Delta} \colon \widehat{A} \to M(\widehat{A} \otimes \widehat{A})$. The obvious formula $(\widehat{\Delta}(\omega))(x \otimes y) := \omega(xy)$ only defines a map $\widehat{\Delta} \colon \widehat{A} \to (A \otimes A)'$, and it is not immediately clear that the image of this map can be identified with a subspace of $M(\widehat{A} \otimes \widehat{A})$. Therefore we take another approach:

**Lemma 2.3.2.** *There exists a homomorphism* $\widehat{\Delta} \colon \widehat{A} \to M(\widehat{A} \otimes \widehat{A})$ *such that*

i) $(\omega_1 \otimes 1)\widehat{\Delta}(\omega_2) \in \widehat{A} \otimes \widehat{A}$ *and* $\widehat{\Delta}(\omega_2)(1 \otimes \omega_1) \in \widehat{A} \otimes \widehat{A}$ *for all* $\omega_1, \omega_2 \in \widehat{A}$,

ii) *for all* $x, y \in A$,

$$((\omega_1 \otimes 1)\widehat{\Delta}(\omega_2))(x \otimes y) = (\omega_1 \otimes \omega_2)(\Delta(x)(1 \otimes y)), \tag{2.11}$$

$$(\widehat{\Delta}(\omega_2)(1 \otimes \omega_1))(x \otimes y) = (\omega_2 \otimes \omega_1)((x \otimes 1)\Delta(y)). \tag{2.12}$$

Before we give the proof, let us briefly comment on equation (2.11) and (2.12). If the linear map $\epsilon \in A'$ belonged to $\widehat{A}$, then it would be the unit, and inserting $\epsilon$ for $\omega_1$ in (2.11) and (2.12) would yield

$$(\widehat{\Delta}(\omega_2))(x \otimes y) = \left\{ \begin{matrix} (\epsilon \otimes \omega_2)(\Delta(x)(1 \otimes y)) \\ (\omega_2 \otimes \epsilon)((x \otimes 1)\Delta(y)) \end{matrix} \right\} = \omega_2(xy).$$

*Proof.* The proof uses the same techniques as the proofs given in the previous section; therefore we only indicate the main steps. One shows that:

- for all $\omega_1, \omega_2 \in \widehat{A}$, equations (2.11) and (2.12) define elements of $\widehat{A} \otimes \widehat{A}$;
- for each $\omega_2 \in \widehat{A}$, the assignments $\omega_1 \otimes 1 \mapsto (\omega_1 \otimes 1)\widehat{\Delta}(\omega_2)$ and $1 \otimes \omega_1 \mapsto \widehat{\Delta}(\omega_2)(1 \otimes \omega_1)$ extend to a right multiplier $\widehat{\Delta}(\omega_2)_r$ and a left multiplier $\widehat{\Delta}(\omega_2)_l$ on $\widehat{A} \otimes \widehat{A}$;
- $\widehat{\Delta}(\omega_2)_r$ and $\widehat{\Delta}(\omega_2)_l$ form a multiplier $\widehat{\Delta}(\omega_2) \in M(\widehat{A} \otimes \widehat{A})$ because
$$(\omega_1 \otimes 1) \cdot (\widehat{\Delta}(\omega_2)_l(1 \otimes \omega_3)) = ((\omega_1 \otimes 1)\widehat{\Delta}(\omega_2)_r) \cdot (1 \otimes \omega_3)$$
for all $\omega_1, \omega_3 \in \widehat{A}$;
- the map $\widehat{\Delta}\colon \widehat{A} \to M(\widehat{A} \otimes \widehat{A})$ thus obtained is an algebra homomorphism.

The proofs of the individual steps are lengthy but do not require any substantial new idea. For the first step, see the proof of Proposition 2.3.5; for further details, see [177, Section 4]. □

**Proposition 2.3.3.** *The pair $(\widehat{A}, \widehat{\Delta})$ is a regular multiplier Hopf algebra. Its antipode $\widehat{S}$ and counit $\hat{\epsilon}$ are given by $(\widehat{S}(\omega))(a) = \omega(S(a))$ and*

$$\hat{\epsilon}(\phi(\cdot a)) = \phi(a) = \hat{\epsilon}(\phi(a \cdot)), \quad \hat{\epsilon}(\psi(\cdot a)) = \psi(a) = \hat{\epsilon}(\psi(a \cdot))$$

*for all $a \in A$.*

*Proof.* Again, the proof uses the same techniques as the proofs of the preceding results; for details, see [177, Section 4]. □

**Remark 2.3.4.** If $(A, \Delta)$ is a Hopf algebra, then the counit $\hat{\epsilon}$ on the dual multiplier Hopf algebra $(\widehat{A}, \widehat{\Delta})$ is given by $\hat{\epsilon}(\omega) = \omega(1_A)$ for all $\omega \in \widehat{A}$. This follows immediately from Proposition 2.3.3.

Next, we identify integrals on the multiplier Hopf algebra $(\widehat{A}, \widehat{\Delta})$. We reserve the symbol $\hat{\phi}$ for left-invariant and the symbol $\hat{\psi}$ for right-invariant functionals.

**Proposition 2.3.5.** *The maps $\hat{\psi}\colon \widehat{A} \to \Bbbk$ and $\hat{\phi}\colon \widehat{A} \to \Bbbk$ given by*

$$\hat{\psi}(\omega) := \epsilon(a) \ \textit{for}\ \omega = \phi(\cdot a), \quad \hat{\phi}(\omega) := \epsilon(a) \ \textit{for}\ \omega = \psi(a \cdot)$$

*are a right and a left integral on $(\widehat{A}, \widehat{\Delta})$, respectively.*

*Proof.* We only prove left-invariance of $\hat{\phi}$, right-invariance of $\hat{\psi}$ follows similarly. Consider elements $\omega_1 = \psi(a_1 \cdot)$ and $\omega_2 = \psi(a_2 \cdot)$ of $\widehat{A}$. First, we compute the product $(\omega_1 \otimes 1)\widehat{\Delta}(\omega_2)$. For all $x, y \in A$,

$$((\omega_1 \otimes 1)\widehat{\Delta}(\omega_2))(x \otimes y) = (\psi \otimes \psi)((a_1 \otimes a_2)\Delta(x)(1 \otimes y)).$$

Since $(A, \Delta)$ is regular, we can write $a_1 \otimes a_2 = \sum_i (1 \otimes c_i)\Delta(b_i)$ with suitably chosen $b_i, c_i \in A$. Then

$$\begin{aligned}((\omega_1 \otimes 1)\widehat{\Delta}(\omega_2))(x \otimes y) &= \sum_i (\psi \otimes \psi)((1 \otimes c_i)\Delta(b_i x)(1 \otimes y)) \\ &= \sum_i \psi(b_i x)\psi(c_i y).\end{aligned}$$

Hence $(\omega_1 \otimes 1)\widehat{\Delta}(\omega_2) = \sum_i \psi(b_i \cdot) \otimes \psi(c_i \cdot)$, and by definition of $\hat{\phi}$,

$$(\mathrm{id} \otimes \hat{\phi})((\omega_1 \otimes 1)\widehat{\Delta}(\omega_2)) = \sum_i \psi(b_i \cdot)\epsilon(c_i).$$

Inserting the relation

$$\sum_i b_i \epsilon(c_i) = \sum_i (\mathrm{id} \otimes \epsilon)((1 \otimes c_i)\Delta(b_i)) = (\mathrm{id} \otimes \epsilon)(a_1 \otimes a_2) = a_1 \epsilon(a_{(2)}),$$

we find $(\mathrm{id} \otimes \hat{\phi})((\omega_1 \otimes 1)\widehat{\Delta}(\omega_2)) = \psi(a_1 \cdot)\epsilon(a_2) = \omega_1 \hat{\phi}(\omega_2)$. Consequently, $(\mathrm{id} \otimes \hat{\phi})(\widehat{\Delta}(\omega_2)) = 1_{M(\widehat{A})} \cdot \hat{\phi}(\omega_2)$. Since $\omega_2 \in \widehat{A}$ was arbitrary, $\hat{\phi}$ is left-invariant. □

Summarizing, we obtain the following duality result:

**Theorem 2.3.6.** *If $(A, \Delta)$ is a regular multiplier Hopf algebra with integrals, then the dual $(\widehat{A}, \widehat{\Delta})$ is a regular multiplier Hopf algebra with integrals.* □

**Example 2.3.7.** Let $G$ be a discrete group and consider the multiplier Hopf algebra $\Bbbk_{\mathrm{fin}}(G)$. Recall that the counting measure on $G$ yields an integral $\phi$ on $\Bbbk_{\mathrm{fin}}(G)$ (Example 2.2.3 ii)). Similar calculations as in Example 1.4.2 show that the dual algebra $\widehat{\Bbbk_{\mathrm{fin}}(G)}$ is unital and that the linear map

$$\widehat{\Bbbk_{\mathrm{fin}}(G)} \to \Bbbk G, \quad \phi(\cdot\, \delta_x) \mapsto U_x,$$

is an isomorphism of Hopf algebras. The dual integral $\hat{\phi} = \hat{\psi}$ is given by $\hat{\psi}(\phi(\cdot\, \delta_x)) = \epsilon(\delta_x) = \delta_{x,e}$ for all $x \in G$ and corresponds to the integral on $\Bbbk G$ defined in Example 2.2.3 iii).

Likewise, $\widehat{\Bbbk G} \cong \Bbbk_{\mathrm{fin}}(G)$ as multiplier Hopf algebras, and again, one can easily verify that the integral on $\Bbbk_{\mathrm{fin}}(G)$ corresponds to the dual integral on $\widehat{\Bbbk G}$ defined in Proposition 2.3.5.

The pattern observed in the previous example is typical: The bidual $(\hat{\hat{A}}, \hat{\hat{\Delta}})$ of a regular multiplier Hopf algebra $(A, \Delta)$ with integrals is naturally isomorphic to $(A, \Delta)$. The main step towards the proof of this assertion is the following lemma.

**Lemma 2.3.8.** *Let $\omega_1, \omega_2 \in \hat{A}$, where $\omega_2 = \phi(\cdot a_2)$ for some $a_2 \in A$. Then $\hat{\psi}(\omega_1\omega_2) = \omega_1(S^{-1}(a_2))$.*

*Proof.* To simplify notation, we assume that $(A, \Delta)$ is a Hopf algebra; the general case is proved similarly, see [177, Lemma 4.11]. Let $\omega_1 = \phi(\cdot a_1) \in \hat{A}$, where $a_1 \in A$, and choose $c_i, d_i \in A$ such that

$$a_1 \otimes a_2 = \sum_i \Delta(c_i)(d_i \otimes 1) = \sum_i \sum c_{i(1)} d_i \otimes c_{i(2)}. \tag{2.13}$$

By the proof of Lemma 2.3.1, $\omega_1\omega_2 = \phi(\cdot b)$, where $b := \sum_i c_i \phi(d_i)$, and by the definition of $\hat{\psi}$ (Proposition 2.3.5),

$$\hat{\psi}(\omega_1\omega_2) = \epsilon(b) = \sum_i \epsilon(c_i)\phi(d_i) = \sum_i \phi(\epsilon(c_i)d_i).$$

Now we insert the relation $\eta(\epsilon(c_i)) = \sum S^{-1}(c_{i(2)})c_{i(1)}$ (see Proposition 1.3.14) in the right-hand side, use equation (2.13), and obtain

$$\hat{\psi}(\omega_1\omega_2) = \sum_i \sum \phi(S^{-1}(c_{i(2)})c_{i(1)}d_i) = \phi(S^{-1}(a_2)a_1) = \omega_1(S^{-1}(a_2)). \quad \square$$

**Theorem 2.3.9.** *Let $(A, \Delta)$ be a regular multiplier Hopf algebra with integrals. Then the map $\iota\colon A \to (\hat{A})'$ given by $(\iota(a))(\omega) := \omega(a)$ for all $a \in A$, $\omega \in \hat{A}$ takes values in $\hat{\hat{A}}$ and defines an isomorphism of multiplier Hopf algebras $(A, \Delta) \cong (\hat{\hat{A}}, \hat{\hat{\Delta}})$.*

*Proof.* First, we show that the map $\iota$ is injective. Assume that $\iota(a) = 0$ for some $a \in A$. Then $\phi(ab) = (\iota(a))(\phi(\cdot b)) = 0$ for all $b \in A$, and since $\phi$ is faithful (Proposition 2.2.4), $a = 0$.

Next, we show that the image of $\iota$ is $\hat{\hat{A}}$. Every element of $\hat{\hat{A}}$ is of the form $\hat{\psi}(\cdot\,\omega)$, where $\omega \in \hat{A}$. Furthermore, since the antipode $S$ is bijective, every $\omega \in \hat{A}$ can be written in the form $\omega = \phi(\cdot\, S(a))$ for some $a \in A$. If $\omega$ is written in that form, then $\hat{\psi}(\cdot\,\omega) = \iota(a)$: by the previous lemma,

$$\hat{\psi}(\omega_1\omega) = \omega_1(S^{-1}(S(a))) = \omega_1(a) = (\iota(a))(\omega_1) \quad \text{for all } \omega_1 \in \hat{A}.$$

Thus the map $\iota$ restricts to an isomorphism of vector spaces $A \cong \hat{\hat{A}}$. Straightforward but lengthy calculations show that this isomorphism is compatible with the multiplication and comultiplication on $A$ and $\hat{\hat{A}}$, see [177, Theorem 4.12]. $\square$

## 2.3.2 The duality of algebraic quantum groups

The duality of regular multiplier Hopf algebras with integrals extends to a duality of algebraic quantum groups as follows.

**Lemma 2.3.10.** *Let $(A, \Delta)$ be a multiplier Hopf $*$-algebra. Then the formula $\omega^*(a) := \overline{\omega(S(a)^*)}$, where $a \in A$ and $\omega \in \widehat{A}$, defines an involution on $\widehat{A}$ that turns $(\widehat{A}, \widehat{\Delta})$ into a multiplier Hopf $*$-algebra.*

*Proof.* This follows from similar calculations as in the proof of Theorem 1.4.1. □

The following result is an analogue of the Plancherel/Parseval identity:

**Theorem 2.3.11.** *Let $(A, \Delta)$ be a multiplier Hopf $*$-algebra with a positive left integral $\phi$, and denote by $\hat{\psi}$ the right integral on $(\widehat{A}, \widehat{\Delta})$ defined in Proposition 2.3.5. Then*

$$\hat{\psi}(\phi(\cdot\, a_1)^* \phi(\cdot\, a_2)) = \phi(a_1^* a_2) \quad \text{for all } a_1, a_2 \in A. \tag{2.14}$$

*In particular, $\hat{\psi}$ is positive. Likewise, if $\psi$ is a positive right integral on $(A, \Delta)$, then the left integral $\hat{\phi}$ on $(\widehat{A}, \widehat{\Delta})$ defined in Proposition 2.3.5 is positive.*

*Proof.* We only prove equation (2.14). Given $a_1, a_2 \in A$, we apply Lemma 2.3.8 to $\omega_1 = \phi(\cdot\, a_1)^*$ and $\omega_2 = \phi(\cdot\, a_2)$, insert the definition of $\omega_1^*$, and find

$$\begin{aligned} \hat{\psi}(\phi(\cdot\, a_1)^* \phi(\cdot\, a_2)) &= \hat{\psi}(\omega_1^* \omega_2) = \omega_1^*(S^{-1}(a_2)) \\ &= \overline{\omega_1(S(S^{-1}(a_2))^*)} = \overline{\omega_1(a_2^*)} = \overline{\phi(a_2^* a_1)} = \phi(a_1^* a_2). \end{aligned}$$ □

**Remark 2.3.12.** The classical Parseval/Plancherel identity has the form

$$\int_G \overline{f(x)} g(x) dx = \int_{\widehat{G}} \overline{(\mathcal{F} f)(\chi)} (\mathcal{F} g)(\chi) d\chi \quad \text{for all } f, g \in L^2(G),$$

where $G$ is a locally compact abelian group, $\widehat{G}$ its dual group, $\mathcal{F} : L^2(G) \to L^2(\widehat{G})$ the Fourier transformation, and the integrals are taken with respect to Haar measures with appropriate scaling.

Summarizing, we obtain the following duality theorem for algebraic quantum groups:

**Theorem 2.3.13.** *Let $(A, \Delta)$ be an algebraic quantum group.*

i) *The dual $(\widehat{A}, \widehat{\Delta})$ is an algebraic quantum group.*

ii) *The map $\iota\colon A \to (\widehat{A})'$ given by $(\iota(a))(\omega) := \omega(a)$ for all $a \in A$, $\omega \in \widehat{A}$ takes values in $\widehat{\widehat{A}}$ and defines an isomorphism of algebraic quantum groups $(A, \Delta) \cong (\widehat{\widehat{A}}, \widehat{\widehat{\Delta}})$.*

*Proof.* i) This follows directly from Theorem 2.3.6, Lemma 2.3.10, and Theorem 2.3.11.

ii) The map $\iota$ is an isomorphism of multiplier Hopf algebras by Theorem 2.3.9, and $*$-linear by a similar calculation as in the proof of Theorem 1.4.1. □

# Chapter 3
# Algebraic compact quantum groups

Algebraic compact quantum groups form a class of quantum groups that is particularly well understood. They can be approached from several directions:

First, their theory can be formulated either in terms of Hopf $*$-algebras or in terms of $C^*$-bialgebras. One can pass back and forth between these two levels, which provide two descriptions of the same underlying objects. In this chapter, we focus on the algebraic setting. In Chapter 5, we switch to the setting of $C^*$-algebras and explain in detail the close relation between algebraic and $C^*$-algebraic compact quantum groups.

Second, algebraic compact quantum groups can be characterized either by the existence of a positive integral or in terms of their corepresentation theory. Corepresentations of Hopf algebras are analogues of group representations, and from a categorical perspective, the corepresentation theory of compact quantum groups is very similar to the representation theory of compact groups. If a positive integral exists, then it can be used to prove many properties of the corepresentations in a similar way as the Haar measure is used in the theory of group representations. Conversely, if the corepresentations of a Hopf $*$-algebra satisfy certain natural properties, then the existence of a positive integral can be deduced.

Our starting point is the following definition:

**Definition.** An *algebraic compact quantum group* or *compact algebraic quantum group* is a unital algebraic quantum group, that is, a Hopf $*$-algebra with a positive integral. The unique normalized positive integral on an algebraic compact quantum group is called its *Haar state* and denoted by $h$.

A *morphism* of algebraic compact quantum groups is simply a morphism of the underlying Hopf $*$-algebras (see Definition 1.3.24).

**Remark.** Uniqueness of a normalized positive integral follows from iv) (b) in Remark 2.2.2 and Proposition 2.2.6.

In the literature, algebraic compact quantum groups are sometimes also called *compact quantum group algebras* or simply *compact quantum groups*. In an early article [40], Effros and Ruan called them *discrete quantum groups*, but this terminology has become standard for the dual of a compact quantum group, see Section 3.3.

**Examples.** We already met the following examples of algebraic compact quantum groups:

- the Hopf $*$-algebra $\mathbb{C}(G)$ of functions on a finite group $G$ (see Examples 1.2.4, 1.3.26 i), 2.2.3 i));
- more generally, the Hopf $*$-algebra $\mathrm{Rep}(G)$ of representative functions on a compact group $G$ (see Examples 1.2.5, 1.3.26 ii), 2.2.3 i));
- the group algebra $\mathbb{C}G$ of a discrete group $G$ (see Examples 1.2.8, 1.3.26 iii), 2.2.3 iii)).

A detailed discussion of further important examples can be found in Chapter 6.

Our study of algebraic compact quantum groups proceeds as follows. First, we introduce the concept of a corepresentation and reformulate it in several equivalent ways (Section 3.1). Then, we develop the theory of corepresentations (Section 3.2) and characterize algebraic compact quantum groups in terms of their corepresentations. Finally, we dicuss the duals of compact quantum groups, namely, discrete algebraic quantum groups (Section 3.3).

Much of the theory developed in this chapter carries over to ordinary Hopf algebras with normalized integrals – these are precisely the *cosemisimple* Hopf algebras [29], [80], [145]. However, our primary interest lies on Hopf $*$-algebras and their $C^*$-algebraic variants. Most of the results presented in the first two sections can also be found in [80, Section 11]. The original reference are the two articles [193] and [202] by Woronowicz, but also the article [40] by Effros and Ruan.

## 3.1 Corepresentations of Hopf $*$-algebras

The concept of an algebra, dualized, leads to the concept of a coalgebra. Similarly, the notion of a module over an algebra, dualized, leads to the notion of a comodule over a coalgebra. We focus on particular comodules over Hopf $*$-algebras, called corepresentations, that are relevant for our discussion of algebraic compact quantum groups in the next section. Most of the following definitions, constructions, and results apply in wider generality.

### 3.1.1 Definition and examples

Throughout this section, let $(A, \Delta)$ be a Hopf $*$-algebra. In the next definition, we use the following construction: Given a vector space $V$ equipped with a Hermitian inner product $\langle \cdot | \cdot \rangle_V$, we define an $A$-valued sesquilinear inner product on $V \otimes A$ by the formula

$$\langle v \otimes a | w \otimes b \rangle_A := \langle v | w \rangle_V a^* b \quad \text{for all } v, w \in V,\ a, b \in A.$$

**Definition 3.1.1.** A *corepresentation* of a Hopf $*$-algebra $(A, \Delta)$ on a complex vector space $V$ is a linear map $\delta\colon V \to V \otimes A$ that satisfies

$$(\delta \otimes \mathrm{id}_A) \circ \delta = (\mathrm{id}_V \otimes \Delta) \circ \delta \quad \text{and} \quad (\mathrm{id}_V \otimes \epsilon) \circ \delta = \mathrm{id}_V .$$

Let $\delta$ and $\tilde{\delta}$ be corepresentations of $(A, \Delta)$ on vector spaces $V$ and $\tilde{V}$, respectively. We call

- a subspace $W \subseteq V$ *invariant* with respect to $\delta$ if $\delta(W) \subseteq W \otimes A$;
- $\delta$ *irreducible* if $V$ contains no non-trivial invariant subspace;
- $\delta$ *unitary* if $\langle \delta(v)|\delta(w)\rangle_A = \langle v|w\rangle_V \cdot 1_A$ for all $v, w \in V$;
- a linear map $T\colon V \to \tilde{V}$ an *intertwiner from $\delta$ to $\tilde{\delta}$* if $\tilde{\delta} \circ T = (T \otimes \mathrm{id}_A) \circ \delta$;
- $\delta$ and $\tilde{\delta}$ *equivalent*, written $\delta \simeq \tilde{\delta}$, if $\delta$ and $\tilde{\delta}$ admit an invertible intertwiner;
- $\mathcal{C}(\delta) := \mathrm{span}\{(f \otimes \mathrm{id})(\delta(v)) \mid v \in V,\ f \in V'\} \subseteq A$ the *space of matrix elements* of $\delta$.

We denote the space of all intertwiners from $\delta$ to $\tilde{\delta}$ by $\mathrm{Hom}(\delta, \tilde{\delta})$.

**Remark 3.1.2.** If $\delta$ and $\tilde{\delta}$ are equivalent corepresentations, then $\mathcal{C}(\delta) = \mathcal{C}(\tilde{\delta})$, and $\delta$ is irreducible if and only if $\tilde{\delta}$ is irreducible. The proof is straightforward.

We extend the Sweedler notation (see Notation 1.3.3) to corepresentations as follows:

**Notation 3.1.3.** Let $\delta$ be a corepresentation on a vector space $V$. Given $v \in V$, we can write $\delta(v) = \sum_i v_{0,i} \otimes v_{1,i}$, where $v_{0,i} \in V$ and $v_{1,i} \in A$. To simplify notation, we suppress the summation index $i$ and write

$$\delta(v) = \sum_i v_{0,i} \otimes v_{1,i} =: \sum v_{(0)} \otimes v_{(1)}.$$

Define $\delta^{(n)}\colon V \to V \otimes A^{\otimes n}$ for each $n \in \mathbb{N}$ by

$$\delta^{(0)} := \mathrm{id}_V \quad \text{and} \quad \delta^{(n+1)} := (\delta^{(n)} \otimes \mathrm{id}_A) \circ \delta \quad \text{for } n \geq 0.$$

Note that $\delta^{(n+1)} = (\mathrm{id}_V \otimes \Delta^{(n)}) \circ \delta$ for all $n \in \mathbb{N}$. We write

$$\delta^{(n)}(v) =: \sum v_{(0)} \otimes v_{(1)} \otimes \cdots \otimes v_{(n)}.$$

**Example 3.1.4.** Let $\delta$ be a corepresentation on a Hermitian vector space $V$ and denote by $\langle \cdot|\cdot\rangle_A$ the $A$-valued inner product on $V \otimes A$ defined above. Then $\langle \delta(v)|\delta(w)\rangle_A = \sum \langle v_{(0)}|w_{(0)}\rangle_V \cdot v_{(1)}{}^* w_{(1)}$ for all $v, w \in V$.

The following example shows how group representations can be expressed in terms of corepresentations.

**Example 3.1.5.** Let $G$ be a compact group and consider the Hopf $*$-algebra $\mathrm{Rep}(G)$ of representative functions on $G$ (see Examples 1.2.5, 1.3.26 ii)). For every finite-dimensional vector space $V$, continuous representations of $G$ on $V$ correspond bijectively with corepresentations of $\mathrm{Rep}(G)$ on $V$:

Let $\pi$ be a continuous representation of $G$ on $V$. Then for every $v \in V$, the function $\delta(v)\colon G \to V$ given by $x \mapsto \pi(x)v$ belongs to $C(G;V)$. We identify $C(G;V)$ with $V \otimes C(G)$ and regard $\delta(v)$ as an element of $V \otimes C(G)$. Then $\delta(v)$ is contained in the subspace $V \otimes \mathrm{Rep}(G) \subseteq V \otimes C(G)$ and the map $v \mapsto \delta(v)$ is a corepresentation. Let us prove the first claim. Choose a basis $(v_i)_i$ of $V$, and denote by $(\phi_i)_i$ the dual basis of $V'$. Then $\delta(v) = \sum_i v_i \otimes f_i$ for some $f_i \in C(G)$, and

$$\sum_i f_i(x)v_i = (\mathrm{id}_V \otimes \mathrm{ev}_x)(\delta(v)) = \pi(x)v = \sum_i \phi_i(\pi(x)v)v_i \quad \text{for all } x \in G,$$

were $\mathrm{ev}_x\colon C(G) \to \mathbb{C}$ denotes the evaluation at $x \in G$ as usual. Thus $\delta(v)$ is contained in $V \otimes \mathrm{Rep}(G)$. The map $\delta\colon v \mapsto \delta(v)$ is a corepresentation since for all $x, y \in G$, the composition

$$\begin{aligned}(\mathrm{id}_V \otimes \mathrm{ev}_x \otimes \mathrm{ev}_y) \circ (\delta \otimes \mathrm{id}_A) \circ \delta &= (\mathrm{id}_V \otimes \mathrm{ev}_x) \circ \delta \circ (\mathrm{id}_V \otimes \mathrm{ev}_y) \circ \delta \\ &= \pi(x)\pi(y)\end{aligned}$$

is equal to

$$\begin{aligned}(\mathrm{id}_V \otimes \mathrm{ev}_x \otimes \mathrm{ev}_y) \circ (\mathrm{id}_V \otimes \Delta) \circ \delta &= (\mathrm{id}_V \otimes (\mathrm{ev}_x \otimes \mathrm{ev}_y)\Delta) \circ \delta \\ &= (\mathrm{id}_V \otimes \mathrm{ev}_{xy}) \circ \delta = \pi(xy).\end{aligned}$$

Conversely, let $\delta$ be a corepresentation of $\mathrm{Rep}(G)$ on $V$. Then the map $\pi\colon G \to \mathrm{Hom}(V)$ given by $\pi(x)v := (\mathrm{id} \otimes \mathrm{ev}_x)(\delta(v))$ is a representation of $G$ – this follows from the two equations displayed above and the relation $\pi(e) = (\mathrm{id}_V \otimes \mathrm{ev}_e) \circ \delta = (\mathrm{id}_V \otimes \epsilon) \circ \delta = \mathrm{id}_V$, where $e$ denotes the unit of $G$.

The matrix elements of a representation $\pi$ appear in the associated corepresentation $\delta$ as follows: If we identify $\mathrm{Hom}(V)$ with $M_n(\mathbb{C})$ via the basis $(v_i)_i$ and consider $\pi\colon G \to \mathrm{Hom}(V)$ as an element of $C(G; M_n(\mathbb{C})) \cong M_n(C(G))$, then $\delta(v_j) = \sum_i v_i \otimes f_{ij}$ for all $i, j$ if and only if $\pi$ corresponds to the matrix $(f_{ij})_{i,j}$.

Finally, if $V$ is Hermitian, then $\pi$ is unitary if and only if $\delta$ is unitary: for all $v, w \in V$, the element $\langle \delta(v)|\delta(w)\rangle_{\mathrm{Rep}(G)} \in \mathrm{Rep}(G)$ corresponds to the function $x \mapsto \langle \pi(x)v|\pi(x)w\rangle$.

**Example 3.1.6.** The comultiplication $\Delta$ is a corepresentation of the Hopf $*$-algebra $(A, \Delta)$ on the vector space $A$. This corepresentation is called the *regular corepresentation* of $(A, \Delta)$. If $(A, \Delta)$ has a positive integral $h$, then the regular corepresentation can be equipped with a unitary structure as follows: The map $(a, b) \mapsto h(a^*b)$ defines a Hermitian inner product on $A$ because $h$ is positive and faithful, and for all $a, b \in A$,

$$\langle \Delta(a)|\Delta(b)\rangle_A = (\mathrm{id} \otimes h)(\Delta(a^*)\Delta(b)) = h(a^*b)1_A = \langle a|b\rangle 1_A$$

because $h$ is left-invariant.

The space of matrix elements $\mathcal{C}(\Delta)$ of the regular corepresentation $\Delta$ is equal to $A$ because every element $a \in A$ can be written as $a = (\epsilon \otimes \mathrm{id})(\Delta(a))$.

Every non-zero corepresentation $\delta$ on some vector space $V$ admits a non-zero intertwiner to the regular corepresentation. Indeed, for every $f \in V'$, the map $T_f : V \to A$ given by $T_f(v) := (f \otimes \mathrm{id})(\delta(v))$ is an intertwiner from $\delta$ to $\Delta$ because the following diagram commutes:

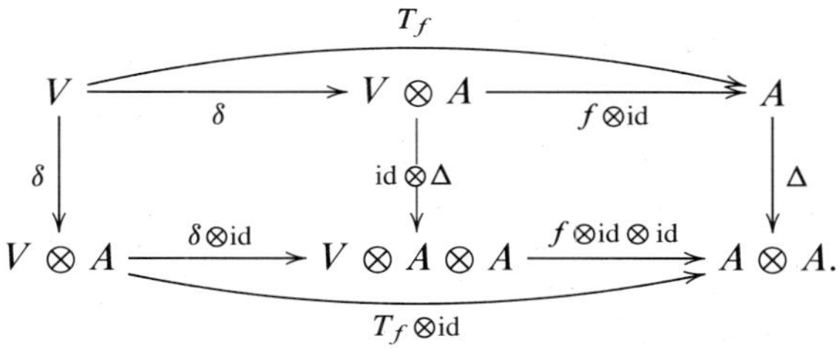


Note that $T_f \neq 0$ for some $f \in V'$ because $\delta(V) \neq 0$.

### 3.1.2 Reformulation of the concept of a corepresentation

In Example 3.1.5, we observed a correspondence between corepresentations of a Hopf $*$-algebra $\mathrm{Rep}(G)$ and representations of the group $G$. In general, corepresentations of a Hopf $*$-algebra $(A, \Delta)$ can be described in many equivalent ways, in particular, also in terms of representations of the dual $*$-algebra $A'$. These different descriptions are the topic of this subsection.

Let us fix some notation. Recall that $A'$ is a $*$-algebra with respect to the operations

$$(fg)(a) = (f \otimes g)(\Delta(a)), \quad f^*(a) = \overline{f(S(a)^*)},$$

where $f, g \in A'$ and $a \in A$ (Theorem 1.4.1). Denote by $\Sigma : A \otimes A \to A \otimes A$ the flip $a \otimes b \mapsto b \otimes a$. Given a vector space $V$ and an element $X \in \mathrm{Hom}(V) \otimes A$, we define $X_{[12]}, X_{[13]} \in \mathrm{Hom}(V) \otimes A \otimes A$ by $X_{[12]} := X \otimes 1_A$ and $X_{[13]} := (\mathrm{id} \otimes \Sigma)(X_{[12]})$. In statement i) of the following proposition, we use the bilinear map

$$(\mathrm{Hom}(V) \otimes A) \times (V \otimes A) \to V \otimes A, \quad (f \otimes a)(v \otimes b) := f(v) \otimes ab.$$

If we embed $A$ in $\mathrm{Hom}(A)$ via left multiplication, we obtain an embedding $\mathrm{Hom}(V) \otimes A \hookrightarrow \mathrm{Hom}(V \otimes A)$, and the map above corresponds to the evaluation map $\mathrm{Hom}(V \otimes A) \times (V \otimes A) \to V \otimes A$.

**Proposition 3.1.7.** *Let $V$ be a vector space with basis $(v_i)_i$ and let $\delta\colon V \to V \otimes A$ be a linear map. Define*

- *a map $\pi\colon A' \to \mathrm{Hom}(V)$ by $\pi(f)v := (\mathrm{id}_V \otimes f)(\delta(v))$;*
- *a family $a = (a_{ij})_{i,j}$ of elements of $A$ by $\delta(v_j) = \sum_i v_i \otimes a_{ij}$; note that for each $j$, there exist only finitely many $i$ such that $a_{ij} \neq 0$;*
- *if $V$ has finite dimension, an element $X \in \mathrm{Hom}(V) \otimes A$ by $X := \sum_{i,j} e_{ij} \otimes a_{ij}$, where $e_{ij} \in \mathrm{Hom}(V)$ denotes the map $v_k \mapsto v_i \cdot \delta_{j,k}$.*

*In the following statements, conditions involving $X$ are to be taken into account only if $V$ has finite dimension. Qualifiers like "for all $i, j$" are omitted, and the sums formed in* (a3), (b3), (d3), (e3), *and* (e4) *are finite.*

i) $\pi(f) = (\mathrm{id} \otimes f)(X)$ *for all* $f \in A'$ *and* $\delta(v) = X(v \otimes 1_A)$ *for all* $v \in V$.

ii) *The following conditions are equivalent:*

(a1) $(\delta \otimes \mathrm{id}_A) \circ \delta = (\mathrm{id}_V \otimes \Delta) \circ \delta$;

(a2) $\pi$ *is an algebra homomorphism;*

(a3) $\Delta(a_{ij}) = \sum_k a_{ik} \otimes a_{kj}$;

(a4) $(\mathrm{id} \otimes \Delta)(X) = X_{[12]}X_{[13]}$.

*Assume that* (a1)–(a4) *hold.*

iii) *The following conditions are equivalent:*

(b1) $\delta$ *is injective;*

(b2) $\pi$ *is non-degenerate: if $\pi(A')v = 0$, then $v = 0$;*

(b3) *$a$ is invertible: there is $b_{ij} \in A$ such that $\sum_k b_{ik}a_{kj} = \delta_{i,j} = \sum_k a_{ik}b_{kj}$;*

(b4) *as* (b3), *but with* $b_{ij} = S(a_{ij})$;

(b5) *$X$ is invertible;*

(b6) $X^{-1} = (\mathrm{id} \otimes S)(X)$;

(c1) $(\mathrm{id}_V \otimes \epsilon) \circ \delta = \mathrm{id}_V$;

(c2) $\pi(\epsilon) = \mathrm{id}_V$;

(c3) $\epsilon(a_{ij}) = \delta_{i,j}$;

(c4) $(\mathrm{id} \otimes \epsilon)(X) = \mathrm{id}_V$.

*Assume that* (b1)–(c4) *hold, so that $\delta$ is a corepresentation.*

iv) *Let $\tilde{\delta}$ be a corepresentation on a vector space $\tilde{V}$ with basis $(\tilde{v}_k)_k$ and let $T\colon V \to \tilde{V}$ be a linear map. Define a family of complex numbers $(t_{kj})_{k,j}$ by $Tv_j = \sum_k \tilde{v}_k t_{kj}$ for all $j$, and $\tilde{\pi}$, $\tilde{a} = (\tilde{a}_{kl})_{k,l}$, $\tilde{X}$ similarly as $\pi$, $a$, $X$. Then the following conditions are equivalent:*

(d1) *$T$ intertwines $\delta$ and $\tilde{\delta}$;*

(d2) *$T$ intertwines $\pi$ and $\tilde{\pi}$;*

(d3) $\sum_i t_{ki}a_{ij} = \sum_l \tilde{a}_{kl}t_{lj}$;

(d4) $(T \otimes 1_A)X = \tilde{X}(T \otimes 1_A)$.

v) *If $V$ is Hermitian and $(v_i)_i$ orthonormal, then the following conditions are equivalent:*

(e1) $\delta$ *is unitary;*

(e2) $\pi$ *is a $*$-homomorphism;*

(e3) $\sum_k a_{ki}{}^* a_{kj} = \delta_{i,j} = \sum_k a_{ik} a_{jk}{}^*$;

(e4) $\sum_k a_{ki}{}^* a_{kj} = \delta_{i,j}$;

(e5) $S(a_{ij}) = a_{ji}{}^*$;

(e6) $X^* X = \mathrm{id}_V \otimes 1_A = X X^*$.

vi) *Let $W \subseteq V$ be a subspace and $e \in \mathrm{Hom}(V)$ an idempotent such that $eV = W$. Then the following conditions are equivalent:*

(f1) $\delta(W) \subseteq W \otimes A$;

(f2) $\pi(A')W \subseteq W$;

(f3) $(e \otimes \mathrm{id}_A) \circ \delta \circ e = \delta \circ e$;

(f4) $e\pi(f)e = \pi(f)e$ *for all* $f \in A'$;

(f5) $(e \otimes 1)X(e \otimes 1) = X(e \otimes 1)$.

vii) $\mathcal{C}(\delta) = \mathrm{span}\{a_{ij} \mid i, j\} = \{(\omega \otimes \mathrm{id})(X) \mid \omega \in \mathrm{Hom}(V)'\} = \mathrm{span}\{\omega(\pi(f)) \mid \omega \in \mathrm{Hom}(V)', f \in A'\}$.

*Proof.* In the following calculations, all sums taken over infinite index sets have only finitely many non-zero summands.

i) For all $f \in A'$ and all $j$,

$$\pi(f)v_j = (\mathrm{id} \otimes f)(\delta(v_j)) = \sum_i v_i f(a_{ij}) \quad \text{and} \quad (\mathrm{id} \otimes f)(X) = \sum_{i,j} e_{ij} f(a_{ij}).$$

The first assertion follows, and the second one follows from the relation

$$X(v_j \otimes 1_A) = \sum_i v_i \otimes a_{ij} = \delta(v_j) \quad \text{for all } j.$$

We now give the main steps of the proofs of the assertions ii)–vi).

(a1) $\Leftrightarrow$ (a2): For all $f, g \in A'$ and $v \in V$,

$$\begin{aligned}\pi(f)\pi(g)v = \pi(fg)v &\Leftrightarrow (\pi(f) \otimes g)(\delta(v)) = (\mathrm{id} \otimes fg)(\delta(v)) \\ &\Leftrightarrow (\mathrm{id} \otimes f \otimes g)((\delta \otimes \mathrm{id})(\delta(v))) = (\mathrm{id} \otimes f \otimes g)((\mathrm{id} \otimes \Delta)(\delta(v))).\end{aligned}$$

(a1) $\Leftrightarrow$ (a3): Compare coefficients in

$$\begin{aligned}(\mathrm{id} \otimes \Delta)(\delta(v_j)) &= \sum_i v_i \otimes \Delta(a_{ij}), \\ (\delta \otimes \mathrm{id})(\delta(v_j)) &= \sum_k \delta(v_k) \otimes a_{kj} = \sum_{i,k} v_i \otimes a_{ik} \otimes a_{kj}.\end{aligned}$$

(a3) $\Leftrightarrow$ (a4): Compare coefficients in

$$(\mathrm{id} \otimes \Delta)(X) = \sum_{i,j} e_{ij} \otimes \Delta(a_{ij}),$$
$$X_{[12]}X_{[13]} = \sum_{i,j,k,l} e_{ik}e_{lj} \otimes a_{ik} \otimes a_{lj} = \sum_{i,j,k} e_{ij} \otimes a_{ik} \otimes a_{kj}.$$

(c1) $\Leftrightarrow$ (c2): Evident from the definition.
(c1) $\Leftrightarrow$ (c3): Consider the coefficients in $(\mathrm{id}_V \otimes \epsilon)(\delta(v_j)) = \sum_i v_i \epsilon(a_{ij})$.
(c2) $\Leftrightarrow$ (c4): Immediate from i).
(b1) $\Leftrightarrow$ (b2): For every $v \in V$,

$$0 = \delta(v) \;\Leftrightarrow\; 0 = (\mathrm{id} \otimes f)(\delta(v)) = \pi(f)v \text{ for all } f \in A'.$$

(b2) $\Leftrightarrow$ (c2): This follows from the fact that $\epsilon$ is the unit of $A'$.
(b3) $\Rightarrow$ (b1): Consider an element $v = \sum_i \lambda_i v_i \in V$. If $0 = \delta(v)$, then

$$0 = \delta(v) = \sum_i \lambda_i \delta(v_i) = \sum_{k,i} v_k \otimes \lambda_i a_{ki},$$

so $\sum_i \lambda_i a_{ki} = 0$ for all $k$, and by invertibility of $(a_{ij})_{i,j}$ also $\lambda_i = 0$ for all $i$.
(b4) $\Rightarrow$ (b3): Trivial.
(c3) $\Leftrightarrow$ (b4): The axioms for the antipode and (a3) imply

$$\sum_k a_{ik} S(a_{kj}) = \epsilon(a_{ij}) = \sum_k S(a_{ik}) a_{kj}.$$

(b3) $\Leftrightarrow$ (b5) and (b4) $\Leftrightarrow$ (b6): Evidently, $(b_{ij})_{i,j}$ is inverse to $(a_{ij})_{i,j}$ if and only if $Y := \sum_{i,j} e_{ij} \otimes b_{ij}$ is inverse to $X$:

$$XY = \sum_{i,j,k} e_{ij} \otimes a_{ik} b_{kj} \quad \text{and} \quad YX = \sum_{i,j,k} e_{ij} \otimes b_{ik} a_{kj}.$$

(d1) $\Leftrightarrow$ (d2): Condition (d1) holds if and only if for all $f \in A'$ and $v \in V$,

$$(\mathrm{id}_V \otimes f)(\tilde{\delta}(Tv)) = (\mathrm{id}_V \otimes f)((T \otimes \mathrm{id}_A)(\delta(v))) = T((\mathrm{id}_V \otimes f)(\delta(v))).$$

Here, the left- and right-hand side are equal to $\tilde{\pi}(f)Tv$ and $T\pi(f)v$, respectively.
(d1) $\Leftrightarrow$ (d3): Compare coefficients in

$$\tilde{\delta}(T(v_j)) = \sum_l \tilde{\delta}(\tilde{v}_l) t_{lj} = \sum_{k,l} \tilde{v}_k \otimes \tilde{a}_{kl} t_{lj},$$
$$(T \otimes \mathrm{id})(\delta(v_j)) = \sum_i T v_i \otimes a_{ij} = \sum_{k,i} \tilde{v}_k \otimes t_{ki} a_{ij}.$$

(d2) $\Leftrightarrow$ (d4): By i), we have for all $f \in A'$

$$\tilde{\pi}(f)T = T\pi(f) \;\Leftrightarrow\; (\mathrm{id} \otimes f)(\widetilde{X}(T \otimes 1_A)) = (\mathrm{id} \otimes f)((T \otimes 1_A)X).$$

(e1) $\Leftrightarrow$ (e4): This follows from the calculation

$$\langle \delta(v_i)|\delta(v_j)\rangle_A = \sum_{k,l} \langle v_l|v_k\rangle_V \, a_{li}{}^* a_{kj} = \sum_k a_{ki}{}^* a_{kj}.$$

(e2) $\Leftrightarrow$ (e5): By definition of the involution on $A'$, we have for all $f \in A'$

$$\begin{aligned} \langle v_i|\pi(f)v_j\rangle_V &= \langle \pi(f^*)v_i|v_j\rangle_V \\ \Leftrightarrow\; \sum_k \langle v_i|v_k\rangle_V f(a_{kj}) &= \sum_l \overline{f^*(a_{li})}\langle v_l|v_j\rangle_V \\ \Leftrightarrow\; f(a_{ij}) &= \overline{f^*(a_{ji})} \\ \Leftrightarrow\; f(a_{ij}) &= f(S(a_{ji})^*). \end{aligned}$$

(e3) $\Rightarrow$ (e4): Trivial. (e4) $\Rightarrow$ (e3): This follows from the invertibility of $a$, see (b3).

(e3) $\Leftrightarrow$ (e5): By (b4), the family $a^* = ((a_{ji})^*)_{i,j}$ is inverse to $a$ if and only if $S(a_{ij}) = (a_{ji})^*$ for all $i, j$.

(e3) $\Leftrightarrow$ (e6): This follows from a similar calculation as in (b3) $\Leftrightarrow$ (b5).

(f1) $\Leftrightarrow$ (f2): $\delta(W) \subseteq W \otimes A$ if and only if for all $f \in A'$,

$$(\mathrm{id} \otimes f)(\delta(W)) \subseteq W, \quad \text{that is,} \quad \pi(f)W \subseteq W.$$

(f1) $\Leftrightarrow$ (f3) and (f2) $\Leftrightarrow$ (f4): Evident.

(f4) $\Leftrightarrow$ (f5): Immediate from i).

vii) These relations follow easily from the definition. □

The preceding proposition motivates the following definitions:

**Definition 3.1.8.** Let $I$ be some index set. A family $(a_{ij})_{i,j \in I}$ of elements of $A$ is called a *corepresentation matrix* if $\{i \in I \mid a_{ij} \neq 0\}$ is finite for each $j$ and

$$\Delta(a_{ij}) = \sum_k a_{ik} \otimes a_{kj} \quad \text{and} \quad \epsilon(a_{ij}) = \delta_{i,j} \quad \text{for all } i, j \in I.$$

Let $a = (a_{ij})_{i,j}$ and $\tilde{a} = (\tilde{a}_{kl})_{k,l}$ be corepresentation matrices. We call

- the matrix $a$ *unitary* if $\sum_k a_{ki}^* a_{kj} = \delta_{i,j}$ for all $i, j \in I$;
- a family $(t_{ki})_{k,i}$ of complex numbers an *intertwiner* from $a$ to $\tilde{a}$ if the set $\{k \mid t_{ki} \neq 0\}$ is finite for each $i$ and $\sum_i t_{ki} a_{ij} = \sum_l \tilde{a}_{kl} t_{lj}$ for all $k, j$;
- $a$ and $\tilde{a}$ *equivalent*, written $a \simeq \tilde{a}$, if they admit an invertible intertwiner;

- $\mathcal{C}(a) = \mathrm{span}\{a_{ij} \mid i, j \in I\}$ the space of *matrix elements* of $a$.

We denote the space of all intertwiners from $a$ to $\tilde{a}$ by $\mathrm{Hom}(a, \tilde{a})$.

**Definition 3.1.9.** Let $V$ be a finite-dimensional vector space. An operator $X \in \mathrm{Hom}(V) \otimes A$ is called a *corepresentation operator* if

$$(\mathrm{id} \otimes \Delta)(X) = X_{[12]}X_{[13]} \quad \text{and} \quad (\mathrm{id} \otimes \epsilon)(X) = \mathrm{id}_V .$$

Let $X$ and $\widetilde{X}$ be corepresentation operators on finite-dimensional vector spaces $V$ and $\widetilde{V}$, respectively. We call

- a subspace $W \subseteq V$ *invariant* if $X(e \otimes 1_A) = (e \otimes 1_A)X(e \otimes 1_A)$ for every idempotent $e \in \mathrm{Hom}(V, W)$;
- $X$ irreducible if $V$ contains no non-trivial invariant subspace;
- $X$ *unitary* if $V$ is Hermitian and $XX^* = \mathrm{id}_V \otimes 1_A = X^*X$;
- a linear map $T: V \to \widetilde{V}$ an *intertwiner* from $X$ to $\widetilde{X}$ if $(T \otimes 1_A)X = \widetilde{X}(T \otimes 1_A)$;
- $X$ and $\widetilde{X}$ equivalent, written $X \simeq \widetilde{X}$, if they admit an invertible intertwiner;
- $\mathcal{C}(X) = \{(\omega \otimes \mathrm{id})(X) \mid \omega \in \mathrm{Hom}(V)'\}$ the space of *matrix elements* of $X$.

We denote the space of all intertwiners from $X$ to $\widetilde{X}$ by $\mathrm{Hom}(X, \widetilde{X})$.

Let us summarize the main results of Proposition 3.1.7:

i) For every vector space $V$ with a fixed basis $(v_i)_{i \in I}$, we have a bijective correspondence between corepresentations $\delta$ on $V$ and corepresentation matrices $(a_{ij})_{i,j \in I}$, prescribed by the relation

$$\delta(v_j) = \sum_i v_i \otimes a_{ij} \quad \text{for all } j \in I.$$

ii) For every finite-dimensional vector space $V$, we have a bijective correspondence between corepresentations $\delta$ and corepresentation operators $X$ on $V$, prescribed by the relation

$$X(v \otimes 1_A) = \delta(v) \quad \text{for all } v \in V.$$

iii) These correspondences preserve unitarity, intertwiners, spaces of matrix elements, and in case ii) also invariant subspaces and irreducibility.

From now on, we freely identify corepresentations, corepresentation matrices, and corepresentation operators by means of the bijective correspondences i) and ii). We call a corepresentation matrix *irreducible* if it corresponds to an irreducible corepresentation.

**Remarks 3.1.10.** i) Let $X$ be a corepresentation operator on a finite-dimensional Hermitian vector space $V$. If we identify elements of $A$ with left multiplication operators in $\mathrm{Hom}(A)$ and consider $X$ as an element of $\mathrm{Hom}(V \otimes A)$, we can express the space of matrix elements $\mathcal{C}(X)$ in terms of maps of the form

$$|w\rangle \otimes 1 \colon A \to V \otimes A,\ a \mapsto w \otimes a, \qquad \langle v| \otimes 1 \colon V \otimes A \to A,\ u \otimes a \mapsto \langle v|u\rangle a,$$

where $v, w \in V$, as follows: $\mathcal{C}(X) = \mathrm{span}\{(\langle v| \otimes 1)X(|w\rangle \otimes 1) \mid v, w \in V\}$.

ii) If $\mathcal{C}$ is the space of matrix elements of some corepresentation, then $\Delta(\mathcal{C}) \subseteq \mathcal{C} \otimes \mathcal{C}$; if the corepresentation is unitary, then $S(\mathcal{C}) = \mathcal{C}^*$. This follows from i) and Proposition 3.1.7 (a3), (e5).

iii) If $\delta$ is a corepresentation on some vector space $V$ and $\pi \colon A' \to \mathrm{Hom}(V)$ is the associated representation defined in Proposition 3.1.7, then

$$\ker \pi = \{f \in A' \mid f(a) = 0 \text{ for all } a \in \mathcal{C}(\delta)\}.$$

Indeed, $\pi(f) = 0$ if and only if for all $v \in V$ and $\phi \in V'$,

$$0 = \phi(\pi(f)v) = (\phi \otimes f)(\delta(v)) = f((\phi \otimes \mathrm{id})\delta(v)).$$

iv) Let $V$ and $W$ be finite-dimensional vector spaces. If $X \in \mathrm{Hom}(V) \otimes A$ is a corepresentation operator and $T \in \mathrm{Hom}(V, W)$ is invertible, then the element $(T \otimes 1)X(T \otimes 1)^{-1} \in \mathrm{Hom}(W) \otimes A$ is a corepresentation operator as well.

### 3.1.3 Construction of new corepresentations

Corepresentations admit several standard constructions – direct sum, tensor product, and conjugation – which turn the category of corepresentations of a Hopf $*$-algebra into a monoidal category [79], [104]. We explain these constructions and describe them in terms of corepresentations, corepresentation matrices, and corepresentation operators. For most of the statements, the proofs are straightforward and therefore will be omitted.

**Direct sum.** Let $(\delta_\alpha)_\alpha$ be a family of corepresentations on vector spaces $(V_\alpha)_\alpha$. Then the map

$$\boxplus_\alpha \delta_\alpha \colon \bigoplus_\alpha V \xrightarrow{\oplus_\alpha \delta_\alpha} \bigoplus_\alpha (V_\alpha \otimes A) \cong \Big(\bigoplus_\alpha V_\alpha\Big) \otimes A$$

is a corepresentation again, called the *direct sum* of $(\delta_\alpha)_\alpha$. This is the unique corepresentation on $\bigoplus_\alpha V_\alpha$ for which the inclusion $\iota_\beta \colon V_\beta \hookrightarrow \bigoplus_\alpha V_\alpha$ is an intertwiner for each $\beta$.

The direct sum construction is functorial: for every second family of corepresentations $(\tilde{\delta}_\alpha)_\alpha$, there exists a map

$$\prod_\alpha \operatorname{Hom}(\delta_\alpha, \tilde{\delta}_\alpha) \to \operatorname{Hom}\Big(\boxplus_\alpha \delta_\alpha, \boxplus_\alpha \tilde{\delta}_\alpha\Big), \quad (T_\alpha)_\alpha \mapsto \boxplus_\alpha T_\alpha := \bigoplus_\alpha T_\alpha.$$

If each $\delta_\alpha$ is unitary, then $\boxplus_\alpha \delta_\alpha$ is unitary with respect to the canonical Hermitian structure on $\bigoplus_\alpha V_\alpha$ (note that we take algebraic direct sums).

Clearly,

$$\mathcal{C}\Big(\boxplus_\alpha \delta_\alpha\Big) = \sum_\alpha \mathcal{C}(\delta_\alpha).$$

For each $\alpha$, denote by $\pi_\alpha \colon A' \to \operatorname{Hom}(V_\alpha)$ the representation associated to $\delta_\alpha$. Then the representation $\pi \colon A' \to \operatorname{Hom}(\bigoplus_\alpha V_\alpha)$ associated to $\boxplus_\alpha \delta_\alpha$ is given by

$$\pi(f)(v_\alpha)_\alpha = (\pi_\alpha(f)v_\alpha)_\alpha \quad \text{for all } f \in A' \text{ and } (v_\alpha)_\alpha \in \bigoplus_\alpha V_\alpha.$$

For each $\alpha$, let $I_\alpha$ be some index set and $u^\alpha = (u^\alpha_{ij})_{i,j\in I_\alpha}$ an $I_\alpha \times I_\alpha$-corepresentation matrix. Put $I := \coprod_\alpha I_\alpha$ (disjoint union). Then the $I \times I$-matrix

$$\boxplus_\alpha u^\alpha := (u_{ij})_{i,j\in I}, \quad \text{where } u_{ij} := \begin{cases} u^\alpha_{ij} & i,j \in I_\alpha, \\ 0 & \text{otherwise,} \end{cases}$$

is a corepresentation matrix, called the direct sum of $(u^\alpha)_\alpha$. If each $u^\alpha$ is the corepresentation matrix corresponding to $\delta_\alpha$ and some basis $(e^\alpha_i)_i$ of $V_\alpha$, then $\boxplus_\alpha u^\alpha$ is the corepresentation matrix corresponding to $\boxplus_\alpha \delta_\alpha$ and the basis $(e^\alpha_i)_{i\in I}$ of $\bigoplus_\alpha V_\alpha$.

Assume that $\bigoplus_\alpha V_\alpha$ has finite dimension and that $(X_\alpha)_\alpha$ is a family of corepresentation operators on the spaces $(V_\alpha)_\alpha$. Then there exists a unique corepresentation operator

$$\boxplus_\alpha X_\alpha \in \operatorname{Hom}\Big(\bigoplus_\alpha V_\alpha\Big) \otimes A$$

such that for each $\beta$, the inclusion $\iota_\beta \colon V_\beta \hookrightarrow \bigoplus_\alpha V_\alpha$ is an intertwiner from $X_\beta$ to $\boxplus_\alpha X_\alpha$. This operator $\boxplus_\alpha X_\alpha$ is the image of the family $(X_\alpha)_\alpha$ under the canonical map

$$\prod_\alpha (\operatorname{Hom}(V_\alpha) \otimes A) \cong \Big(\prod_\alpha \operatorname{Hom}(V_\alpha)\Big) \otimes A \hookrightarrow \operatorname{Hom}\Big(\bigoplus_\alpha V_\alpha\Big) \otimes A.$$

If each $X_\alpha$ is the corepresentation operator corresponding to $\delta_\alpha$, then $\boxplus_\alpha X_\alpha$ is the corepresentation operator corresponding to $\boxplus_\alpha \delta_\alpha$.

**Tensor product.** Let $\delta_V$ and $\delta_W$ be corepresentations on vector spaces $V$ and $W$, respectively. Then the map

$$\delta_V \boxtimes \delta_W : V \otimes W \mapsto (V \otimes W) \otimes A, \quad v \otimes w \mapsto \sum v_{(0)} \otimes w_{(0)} \otimes v_{(1)} w_{(1)},$$

is a corepresentation again, called the *tensor product* of $\delta_V$ and $\delta_W$.

The tensor product construction is functorial: for every second pair of corepresentations $\delta_{\tilde{V}}$ and $\delta_{\tilde{W}}$, there exists a map

$$\begin{aligned} \mathrm{Hom}(\delta_V, \delta_{\tilde{V}}) \times \mathrm{Hom}(\delta_W, \delta_{\tilde{W}}) &\to \mathrm{Hom}(\delta_V \boxtimes \delta_W, \delta_{\tilde{V}} \boxtimes \delta_{\tilde{W}}), \\ (S, T) &\mapsto S \boxtimes T := S \otimes T. \end{aligned}$$

If $\delta_V$ and $\delta_W$ are unitary, then $\delta_V \boxtimes \delta_W$ is unitary with respect to the canonical Hermitian structure on $V \otimes W$. Indeed, for all $v, \tilde{v} \in V$ and $w, \tilde{w} \in W$,

$$\begin{aligned} &\big\langle (\delta_V \boxtimes \delta_W)(v \otimes w) \big| (\delta_V \boxtimes \delta_W)(\tilde{v} \otimes \tilde{w}) \big\rangle_A \\ &\quad = \sum \langle v_{(0)} | \tilde{v}_{(0)} \rangle_V \langle w_{(0)} | \tilde{w}_{(0)} \rangle_W w_{(1)}^* v_{(1)}^* \tilde{v}_{(1)} \tilde{w}_{(1)} \\ &\quad = \langle v | \tilde{v} \rangle_V \langle w | \tilde{w} \rangle_W 1_A = \langle v \otimes w | \tilde{v} \otimes \tilde{w} \rangle_{(V \otimes W)}. \end{aligned}$$

Clearly,

$$\mathcal{C}(\delta_V \boxtimes \delta_W) = \mathrm{span}\, \mathcal{C}(\delta_V) \mathcal{C}(\delta_W).$$

Let $I$ and $J$ be two index sets and $v = (v_{ij})_{i,j \in I}$ and $w = (w_{k,l})_{k,l \in J}$ two corepresentation matrices. Then the $(I \times J) \times (I \times J)$-matrix

$$v \boxtimes w = (v_{ij} w_{kl})_{(i,k),(j,l)}$$

is a corepresentation matrix again. If $v$, $w$ are the corepresentation matrices corresponding to $\delta_V$, $\delta_W$ and some bases $(e_i)_{i \in I}$, $(f_k)_{k \in J}$ of $V$, $W$, respectively, then $v \boxtimes w$ is the corepresentation matrix corresponding to $\delta_V \boxtimes \delta_W$ and the basis $(e_i \otimes f_k)_{(i,k)}$ of $V \otimes W$.

Assume that $V$ and $W$ have finite dimension and that $X$ and $Y$ are corepresentation operators on $V$ and $W$, respectively. Then the operator

$$X \boxtimes Y := X_{[13]} Y_{[23]} \in \mathrm{Hom}(V) \otimes \mathrm{Hom}(W) \otimes A \subseteq \mathrm{Hom}(V \otimes W) \otimes A$$

is a corepresentation operator again. Here we have $Y_{[23]} = \mathrm{id}_V \otimes Y$ and $X_{[13]} = (\Sigma \otimes \mathrm{id}_A)(\mathrm{id}_W \otimes X)$, where $\Sigma \colon \mathrm{Hom}(W) \otimes \mathrm{Hom}(V) \to \mathrm{Hom}(V) \otimes \mathrm{Hom}(W)$ denotes the flip. If $X$ and $Y$ are the corepresentation operators corresponding to $\delta_V$ and $\delta_W$, then $X \boxtimes Y$ is the corepresentation operator corresponding to $\delta_V \boxtimes \delta_W$.

The tensor product is associative in a natural sense. If the Hopf $*$-algebra $(A, \Delta)$ is commutative, then the tensor product is commutative – in that case, the natural isomorphism $V \otimes W \xrightarrow{\cong} W \otimes V$ intertwines $\delta_V \boxtimes \delta_W$ and $\delta_W \boxtimes \delta_V$, and $\delta_V \boxtimes \delta_W \simeq \delta_W \boxtimes \delta_V$. If $(A, \Delta)$ is not commutative, this relation need not hold. However, there exist interesting examples of non-commutative Hopf $*$-algebras for which the bifunctors

$$(\delta_V, \delta_W) \mapsto \delta_V \boxtimes \delta_W \quad \text{and} \quad (\delta_V, \delta_W) \mapsto \delta_W \boxtimes \delta_V$$

are naturally equivalent. A natural equivalence between these bifunctors is called a *braiding* if it satisfies some additional coherence conditions, and Hopf $*$-algebras that possess a braiding are called *braided.* An interesting discussion of braided Hopf $*$-algebras and their relation to knot invariants is given in [79].

The tensor product is distributive: for each pair of families of corepresentations $(\delta_\alpha)_\alpha$ and $(\tilde{\delta}_\beta)_\beta$, there exists a natural isomorphism

$$\Big(\boxplus_\alpha \delta_\alpha\Big) \boxtimes \Big(\boxplus_\beta \tilde{\delta}_\beta\Big) \simeq \boxplus_{\alpha,\beta} (\delta_\alpha \boxtimes \tilde{\delta}_\beta).$$

**Conjugation.** Let $\delta_V$ be a corepresentation on a vector space $V$. Denote by $\overline{V}$ the conjugate vector space of $V$ and by $v \mapsto \bar{v}$ the canonical conjugate-linear isomorphism. Then the map

$$\overline{\delta_V} : \overline{V} \to \overline{V} \otimes A, \quad \bar{v} \mapsto \sum \overline{v_{(0)}} \otimes v_{(1)}^*,$$

is a corepresentation again, called the *conjugate* of $\delta_V$.

For every corepresentation $\delta_W$ on a vector space $W$, there exists a conjugate-linear map

$$\mathrm{Hom}(\delta_V, \delta_W) \to \mathrm{Hom}(\overline{\delta_V}, \overline{\delta_W}), \quad T \mapsto \overline{T},$$

where $\overline{T}\bar{v} = \overline{Tv}$ for each $T \in \mathrm{Hom}(V, W)$ and $v \in V$. Moreover, a subspace $\overline{V_0} \subseteq \overline{V}$ is invariant for $\overline{\delta_V}$ if and only if $V_0 \subseteq V$ is invariant for $\delta_V$; in particular, $\overline{\delta_V}$ is irreducible if and only if $\delta_V$ is irreducible.

If $\delta_V$ is unitary, then $\overline{\delta_V}$ need not be unitary with respect to the natural Hermitian structure on $\overline{V}$.

Evidently,

$$\mathcal{C}(\overline{\delta_V}) = \mathcal{C}(\delta_V)^*.$$

Let $I$ be an index set and $v = (v_{ij})_{i,j \in I}$ a corepresentation matrix. Then

$$\bar{v} := (v_{ij}{}^*)_{i,j \in I}$$

is a corepresentation matrix again. If $v$ corresponds to $\delta_V$ and some basis $(e_i)_i$ of $V$, then $\bar{v}$ corresponds to $\overline{\delta_V}$ and the basis $(\overline{e_i})_i$ of $\overline{V}$. If $v$ is unitary, then $\bar{v} = (S(v_{ji}))_{i,j \in I}$ by Proposition 3.1.7 v).

If $V$ has finite dimension and $X$ is a corepresentation operator on $V$, then the image of $X$ under the conjugate-linear map

$$\mathrm{Hom}(V) \otimes A \to \mathrm{Hom}(\overline{V}) \otimes A, \quad T \otimes a \mapsto \overline{T} \otimes a^*,$$

is a corepresentation operator again, which we denote by $\overline{X}$. If $X$ corresponds to $\delta_V$, then $\overline{X}$ corresponds to $\overline{\delta_V}$, as one can easily check.

Conjugation is compatible with direct sums and tensor products in the following sense: For every family of corepresentations $(\delta_\alpha)_\alpha$ and for each pair of corepresentations $\delta_V, \delta_W$, there exist natural isomorphisms

$$\overline{\Big(\boxplus_\alpha \delta_\alpha\Big)} \simeq \boxplus_\alpha \overline{\delta_\alpha}, \qquad \overline{(\delta_V \boxtimes \delta_W)} \simeq \overline{\delta_W} \boxtimes \overline{\delta_V}.$$

The following result is called *Frobenius reciprocity*:

**Proposition 3.1.11.** *Let $\delta_U$, $\delta_V$, $\delta_W$ be corepresentations on spaces $U$, $V$, $W$, where $\delta_U$ is unitary and finite-dimensional. Then the natural isomorphisms*

$$\phi_U \colon \mathrm{Hom}(\bar{U} \otimes V, W) \xrightarrow{\cong} \mathrm{Hom}(V, U \otimes W),$$
$$\psi_U \colon \mathrm{Hom}(V \otimes U, W) \xrightarrow{\cong} \mathrm{Hom}(V, W \otimes \bar{U})$$

*restrict to isomorphisms*

$$\mathrm{Hom}(\overline{\delta_U} \boxtimes \delta_V, \delta_W) \xrightarrow{\cong} \mathrm{Hom}(\delta_V, \delta_U \boxtimes \delta_W),$$
$$\mathrm{Hom}(\delta_V \boxtimes \delta_U, \delta_W) \xrightarrow{\cong} \mathrm{Hom}(\delta_V, \delta_W \boxtimes \overline{\delta_U}).$$

*Proof.* We only prove the assertion concerning $\phi_U$; for $\psi_U$, the proof is similar. Let $u$, $v$, $w$ be the corepresentation matrices corresponding to $\delta_U, \delta_V, \delta_W$ and bases $(e_i)_i$, $(f_k)_k$, $(g_r)_r$ of $U$, $V$, $W$, respectively, where $(e_i)_i$ is orthonormal. With respect to the bases $(\overline{e_i} \otimes f_k)_{(i,k)}$, $(g_r)_r$, $(f_k)_k$, $(e_i \otimes g_r)_{(i,r)}$ of $\bar{U} \otimes V$, $W$, $V$, $U \otimes W$, respectively, the map $\phi_U$ is given by

$$(t_{r(i,k)})_{r,(i,k)} \mapsto (t_{r(i,k)})_{(i,r),k}.$$

Now $(t_{r(i,k)})_{r,(i,k)} \in \mathrm{Hom}(\bar{u} \boxtimes v, w) \Leftrightarrow (t_{r(i,k)})_{(i,r),k} \in \mathrm{Hom}(v, u \boxtimes w)$ because

$$\sum_s w_{rs} t_{s(j,l)} \text{ equals } \sum_{i,k} t_{r(i,k)} (\bar{u} \boxtimes v)_{(i,k)(j,l)} = \sum_{i,k} t_{r(i,k)} u^*_{ij} v_{kl} \quad \text{for all } r, j, l$$

if and only if

$$\sum_{s,j} u_{ij} w_{rs} t_{s(j,l)} = \sum_{s,j} (u \boxtimes w)_{(i,r)(j,s)} t_{s(j,l)} \text{ equals } \sum_k t_{r(i,k)} v_{kl} \quad \text{for all } r, i, l.$$

□

## 3.2 Corepresentation theory and structure theory

The corepresentations of an algebraic compact quantum group enjoy the following nice properties:

1. Every corepresentation is equivalent to a direct sum of irreducible finite-dimensional unitary corepresentations (Theorem 3.2.1).
2. Two irreducible corepresentations either admit no non-zero intertwiner or they are equivalent and the space of intertwiners has dimension one (Proposition 3.2.2).
3. The quantum group is spanned by the matrix elements of irreducible corepresentations and these matrix elements satisfy certain orthogonality relations with respect to the Haar state (Proposition 3.2.6 and 3.2.9).
4. The modular properties of the Haar state can be described in terms of a one-parameter family of characters on the quantum group which can be constructed out of certain intertwiners of irreducible corepresentations (Theorem 3.2.19).
5. Those Hopf $*$-algebras that are algebraic compact quantum groups can be characterized in terms of corepresentations (Theorem 3.2.12).

All of these results are proved in the following sections. For later applications, we state some results in a slightly wider generality.

### 3.2.1 Decomposition into irreducible corepresentations

In this subsection, we show that the category of corepresentations of an algebraic compact quantum group can be described in terms of its irreducible corepresentations. Throughout this subsection, $(A, \Delta)$ is a Hopf $*$-algebra.

The main step is the following theorem:

**Theorem 3.2.1.** *Let $\delta$ be a corepresentation of $(A, \Delta)$ on a vector space $V$.*

i) *If $(A, \Delta)$ is an algebraic compact quantum group, then $\delta$ is equivalent to a unitary corepresentation.*

ii) *If $\delta$ is unitary and $W \subseteq V$ is an invariant subspace, then the orthogonal complement $W^{\perp} \subseteq V$ is invariant again.*

iii) *Every element $v \in V$ is contained in some finite-dimensional invariant subspace of $V$. In particular, $V$ has finite dimension if $\delta$ is irreducible.*

iv) *If for every finite-dimensional invariant subspace $W \subseteq V$, the restriction $\delta|_W$ is equivalent to a unitary corepresentation, in particular, if $(A, \Delta)$ is an algebraic compact quantum group, then $\delta$ is equivalent to a direct sum of finite-dimensional irreducible unitary corepresentations.*

*Proof.* i) Choose some Hermitian inner product $\langle\cdot|\cdot\rangle_V$ on $V$. Since the Haar state $h$ of $(A, \Delta)$ is positive and faithful, the formula $(a, b) \mapsto h(a^*b)$ defines a Hermitian inner product on $A$. By a standard argument, the formula $\langle v \otimes a|w \otimes b\rangle :=$ $\langle v|w\rangle_V h(a^*b)$ defines a Hermitian inner product on $V \otimes A$. Because $\delta$ is injective, we can define a second Hermitian inner product $\langle\cdot|\cdot\rangle'_V$ on $V$ by the formula

$$\langle v|w\rangle'_V := \langle \delta(v)|\delta(w)\rangle = \sum \langle v_{(0)}|w_{(0)}\rangle_V \cdot h(v^*_{(1)}w_{(1)}) \quad \text{for all } v, w \in V.$$

The associated $A$-valued inner product $\langle\cdot|\cdot\rangle'_A$ on $V \otimes A$ is given by

$$\langle \delta(v)|\delta(w)\rangle'_A = \sum \langle v_{(0)}|w_{(0)}\rangle'_V v^*_{(1)}w_{(1)} = \sum \langle v_{(0)}|w_{(0)}\rangle_V h(v^*_{(1)}w_{(1)})v^*_{(2)}w_{(2)}$$

for all $v, w \in V$. Using the fact that $h$ is right-invariant, we find

$$\langle \delta(v)|\delta(w)\rangle'_A = \sum \langle v_{(0)}|w_{(0)}\rangle_V h(v^*_{(1)}w_{(1)})1_A = \langle v|w\rangle'_V 1_A.$$

Therefore, $\delta$ is unitary with respect to the inner product $\langle\cdot|\cdot\rangle'_V$.

ii) By Proposition 3.1.7 vi), $W \subseteq V$ is invariant for $\delta$ if and only if it is invariant for the associated $*$-representation $\pi\colon A' \to \operatorname{Hom}(V)$. Now the claim follows from a standard argument.

iii) Denote by $\pi\colon A' \to \operatorname{Hom}(V)$ the representation associated to $\delta$ in Proposition 3.1.7. Since $\pi(\epsilon) = \mathrm{id}_V$, the subspace $\pi(A')v \subseteq V$ contains $v$. Evidently, $\pi(A')v$ is invariant for $\pi$, so by Proposition 3.1.7 vi) also for $\delta$. If $\delta(v) = \sum_i v_i \otimes a_i$, where $v_i \in V$ and $a_i \in A$, then $\pi(A')v$ is contained in the linear span of the $v_i$. Therefore, $\pi(A')v$ has finite dimension.

iv) This follows from ii) and iii) by an application the Lemma of Zorn. □

The next step is an analogue of the Lemma of Schur [22, Chapter II, Lemma 2.1], [62, Chapter VII, Theorem 27.9] for intertwiners of corepresentations:

**Proposition 3.2.2.** *Let $\delta_V$ and $\delta_W$ be corepresentations on vector spaces $V$ and $W$, respectively.*

i) *For every $T \in \operatorname{Hom}(\delta_V, \delta_W)$, the subspaces $\ker T \subseteq V$ and $\operatorname{Im} T \subseteq W$ are invariant.*

ii) *Assume that $\delta_V$ and $\delta_W$ are irreducible. Then either $\dim \operatorname{Hom}(\delta_V, \delta_W) = 1$ and $\delta_V \simeq \delta_W$, or $\operatorname{Hom}(\delta_V, \delta_W) = 0$.*

*Proof.* i) Let $T \in \operatorname{Hom}(\delta_V, \delta_W)$ and denote by $\pi_V$ and $\pi_W$ the representations of $A'$ associated to $\delta_V$ and $\delta_W$, respectively. Then $\pi_W(A')T = T\pi_V(A')$ by Proposition 3.1.7 iv), whence $\pi_W(A') \operatorname{Im} T \subseteq \operatorname{Im} T$ and $\pi_V(A') \ker T \subseteq \ker T$. Now the claim follows from Proposition 3.1.7 vi).

ii) By i), every non-zero intertwiner from $\delta_V$ to $\delta_W$ is bijective. If $S$ and $T$ are non-zero intertwiners, then $\det(S - \lambda T)$, considered as a polynomial in $\lambda$, must vanish at some $\lambda_0 \in \mathbb{C}$. Since $S - \lambda_0 T$ is not bijective, we must have $S - \lambda_0 T = 0$. Note that the determinant $\det(S - \lambda T)$ is well defined because the dimension of $V$ and $W$ is finite by Theorem 3.2.1 iii). □

**Proposition 3.2.3.** *Two irreducible unitary corepresentations are equivalent if and only if they admit a unitary intertwiner.*

*Proof.* Let $\delta$ be an irreducible corepresentation on a vector space $V$. We only need to show that up to multiplication by a constant factor, there exists only one Hermitian inner product on $V$ with respect to which $\delta$ is unitary. Given such an inner product $\langle \cdot | \cdot \rangle_V$, denote by $F : \overline{V} \otimes V \to \mathbb{C}$ the linear map given by $\bar{v} \otimes w \mapsto \langle v|w\rangle_V$, and by $\tau : \mathbb{C} \to \mathbb{C} \otimes A, 1 \mapsto 1 \otimes 1_A$, the trivial corepresentation. Then for all $v, w \in V$,

$$\langle v|w\rangle_V 1_A = F(\bar{v} \otimes w)1_A = \tau(F(\bar{v} \otimes w))$$

and

$$\langle \delta(v)|\delta(w)\rangle_A = \sum F(\overline{v_{(0)}} \otimes w_{(0)}) \otimes v_{(1)}^* w_{(1)} = (F \otimes \mathrm{id})((\bar{\delta} \boxtimes \delta)(\bar{v} \otimes w)).$$

Consequently, $\delta$ is unitary with respect to $\langle \cdot | \cdot \rangle_V$ if and only if $F \in \mathrm{Hom}(\bar{\delta} \boxtimes \delta, \tau)$. But by Frobenius reciprocity (Proposition 3.1.11), $\dim \mathrm{Hom}(\bar{\delta} \boxtimes \delta, \tau) = \dim \mathrm{Hom}(\delta, \delta) = 1$, and the claim follows. □

Using the preceding results, we can describe the category of corepresentations of an algebraic compact quantum group as follows:

**Corollary 3.2.4.** *Let $(A, \Delta)$ be an algebraic compact quantum group and $(\delta_\alpha)_{\alpha \in I}$ a maximal family of pairwise inequivalent irreducible corepresentations of $(A, \Delta)$.*

i) *Every corepresentation of $(A, \Delta)$ is equivalent to a direct sum $\boxplus_k \delta_{\alpha_k}$, where $(\alpha_k)_k$ is some family of elements of $I$.*

ii) *Let $(\alpha_k)_k$ and $(\beta_l)_l$ be families of elements of $I$. Then*

$$\mathrm{Hom}\Big(\boxplus_k \delta_{\alpha_k}, \boxplus_l \delta_{\beta_l}\Big) \cong \prod_k \mathrm{Hom}\Big(\delta_{\alpha_k}, \boxplus_l \delta_{\beta_l}\Big) \cong \prod_k \bigoplus_{\substack{l \text{ s.t.} \\ \alpha_k = \beta_l}} \mathbb{C}. \qquad \square$$

This corollary shows that the corepresentation category of an algebraic compact quantum group, regarded as an ordinary category, has a very simple structure. The interesting and non-trivial information is contained in the monoidal structure of this category, that is, in the tensor product. Note that the conjugation of corepresentations can be reconstructed from the monoidal structure: If $\delta_V$

and $\delta_W$ are irreducible finite-dimensional corepresentations, then $\delta_W \simeq \overline{\delta_V}$ if and only if $\dim \operatorname{Hom}(\tau, \delta_V \boxtimes \delta_W) \neq 0$, where $\tau$ denotes the trivial corepresentation. This follows from Frobenius reciprocity (Proposition 3.1.11) and Schur's Lemma (Proposition 3.2.2 ii)).

### 3.2.2 Schur's orthogonality relations

The matrix elements of irreducible corepresentations satisfy an analogue of Schur's orthogonality relations known from the representation theory of compact groups [22, Chapter II, Theorem 4.5, 4.6], [62, Chapter VII, Theorem 27.19]. To prove these relations, we shall use the following lemma:

**Lemma 3.2.5.** *Let $(A, \Delta)$ be a Hopf $*$-algebra with a normalized integral $h$, let $\delta_V$ and $\delta_W$ be corepresentations on finite-dimensional vector spaces $V$ and $W$, respectively, and let $R \in \operatorname{Hom}(V, W)$. Denote by $X$ and $Y$ the corepresentation operators corresponding to $\delta_V$ and $\delta_W$, respectively, and define $S, T \in \operatorname{Hom}(V, W)$ by*

$$S := (\mathrm{id} \otimes h)(Y^{-1}(R \otimes 1)X) \quad \text{and} \quad T := (\mathrm{id} \otimes h)(Y(R \otimes 1)X^{-1}).$$

*Then $S, T \in \operatorname{Hom}(\delta_V, \delta_W)$. If $R \in \operatorname{Hom}(\delta_V, \delta_W)$, then $S = T = R$.*

*Proof.* Since $Y$ and $X$ are corepresentation operators, the composition

$$\begin{aligned} Y^{-1}(S \otimes 1)X &= Y^{-1} \cdot (\mathrm{id} \otimes h)(Y^{-1}(R \otimes 1)X) \cdot X \\ &= (\mathrm{id} \otimes h \otimes \mathrm{id})(Y^{-1}_{[13]} Y^{-1}_{[12]} (R \otimes 1 \otimes 1) X_{[12]} X_{[13]}) \end{aligned}$$

can be rewritten in the form

$$(\mathrm{id} \otimes h \otimes \mathrm{id})((\mathrm{id} \otimes \Delta)(Y^{-1}) \cdot (R \otimes 1 \otimes 1) \cdot (\mathrm{id} \otimes \Delta)(X)).$$

By right-invariance of $h$, this composition is equal to

$$(\mathrm{id} \otimes h)(Y^{-1}(R \otimes 1)X) \otimes 1 = S \otimes 1.$$

Thus $(S \otimes 1)X = Y(S \otimes 1)$. A similar calculation shows that $(T \otimes 1)X = Y(T \otimes 1)$; here, one has to use left-invariance of $h$. Proposition 3.1.7 iv) implies $S, T \in \operatorname{Hom}(\delta_V, \delta_W)$. If $R \in \operatorname{Hom}(\delta_V, \delta_W)$, then $Y^{-1}(R \otimes 1)X = R \otimes 1 = Y(R \otimes 1)X^{-1}$ by Proposition 3.1.7 iv), and hence $S = R = T$. □

Combining Schur's Lemma 3.2.2 with Lemma 3.2.5, we obtain the following orthogonality relations:

**Proposition 3.2.6.** *Let $(A, \Delta)$ be a Hopf $*$-algebra with a normalized integral $h$, and let $\delta_V$ and $\delta_W$ be inequivalent irreducible corepresentations of $(A, \Delta)$ on vector spaces $V$ and $W$, respectively. Then for all $a \in \mathcal{C}(\delta_V)$ and $b \in \mathcal{C}(\delta_W)$,*

$$h(S(b)a) = 0 = h(bS(a)). \tag{3.1}$$

*If $\delta_V$ and $\delta_W$ are unitary, then $h(b^*a) = 0 = h(ba^*)$ for all $a \in \mathcal{C}(\delta_V)$ and $b \in \mathcal{C}(\delta_W)$.*

*Proof.* By Theorem 3.2.1 iii), $V$ and $W$ have finite dimension. Let $X$ and $Y$ be the corepresentation operators associated to $\delta_V$ and $\delta_W$, respectively, and let $a = (\langle\eta'| \otimes \mathrm{id})X(|\eta\rangle \otimes \mathrm{id}) \in \mathcal{C}(X)$ and $b = (\langle\xi'| \otimes \mathrm{id})Y(|\xi\rangle \otimes \mathrm{id}) \in \mathcal{C}(Y)$, where $\eta, \eta' \in V$ and $\xi, \xi' \in W$ (see Remark 3.1.10 i)). Then by Proposition 3.1.7 iii),

$$\begin{aligned} h(S(b)a) &= \langle\xi'|S\eta\rangle, \quad \text{where } S = (\mathrm{id} \otimes h)(Y^{-1}(|\xi\rangle\langle\eta'| \otimes \mathrm{id})X), \\ h(bS(a)) &= \langle\xi'|T\eta\rangle, \quad \text{where } T = (\mathrm{id} \otimes h)(Y(|\xi\rangle\langle\eta'| \otimes \mathrm{id})X^{-1}), \end{aligned}$$

and by Proposition 3.2.2 and Lemma 3.2.5, $S, T \in \mathrm{Hom}(\delta_V, \delta_W) = \{0\}$. Thus we have proved (3.1). The second assertion follows from (3.1) and Remark 3.1.10 ii). □

**Corollary 3.2.7.** *Let $(A, \Delta)$ be a Hopf $*$-algebra with a normalized integral $h$. Then $h(\mathcal{C}(\delta)) = 0$ for every non-trivial irreducible corepresentation $\delta$.* □

Combining the preceding results with the fact that every integral is faithful (Proposition 2.2.4), we find:

**Corollary 3.2.8.** *Let $(A, \Delta)$ be a Hopf $*$-algebra with a normalized integral $h$. Then for each pair of corepresentations $\delta_V$ and $\delta_W$,*

$$\begin{aligned} \mathrm{Hom}(\delta_V, \delta_W) = 0 &\Leftrightarrow h(\mathcal{C}(\delta_V)S(\mathcal{C}(\delta_W))) = 0 \\ &\Leftrightarrow h(S(\mathcal{C}(\delta_V))\mathcal{C}(\delta_W)) = 0 \Leftrightarrow \mathrm{Hom}(\delta_W, \delta_V) = 0. \end{aligned}$$ □

The next result describes expressions of the form $h(b^*a)$, where $b$ and $a$ are matrix elements of the same irreducible corepresentation:

**Proposition 3.2.9.** *Let $(A, \Delta)$ be a Hopf $*$-algebra with a normalized integral $h$, and let $a \in M_n(A)$ be an irreducible unitary corepresentation matrix, where $n \in \mathbb{N}$, such that the conjugate $\bar{a}$ is equivalent to a unitary corepresentation matrix.*

i) *The matrix $a^t := (a_{ji})_{i,j} \in M_n(A)$ is invertible and its inverse $a^{-t} := (a^t)^{-1}$ is a corepresentation matrix.*

ii) *There exists a unique intertwiner $\widetilde{F} \in \mathrm{Hom}(\bar{a}, a^{-t})$ such that $\mathrm{Tr}\,\widetilde{F} = \mathrm{Tr}(\widetilde{F}^{-1}) > 0$, and this $\widetilde{F}$ is invertible and positive definite.*

iii) *For all* $i, j, k, l$,

$$h(S(a_{ji})a_{kl}) = h(a_{ij}{}^* a_{kl}) = \frac{\delta_{j,l}}{\operatorname{Tr}(\widetilde{F}^{-1})}(\widetilde{F}^{-1})_{ik},$$
$$h(a_{ij} S(a_{lk})) = h(a_{ij} a_{kl}{}^*) = \frac{\delta_{i,k}}{\operatorname{Tr}\widetilde{F}}\widetilde{F}_{jl}.$$

iv) *The elements* $(a_{ij})_{i,j}$ *are linearly independent.*

*Proof.* i) By assumption, there exists an invertible $T \in M_n(\mathbb{C})$ such that $b := T\bar{a}T^{-1}$ is unitary. Then $a^t = \bar{a}^* = T^* b^* (T^{-1})^*$ is invertible and

$$a^{-t} = (\bar{a}^*)^{-1} = T^* b (T^*)^{-1} = T^* T \bar{a} (T^* T)^{-1}. \tag{3.2}$$

Remark 3.1.10 iv) implies that $a^{-t}$ is a corepresentation matrix.

ii) Uniqueness of $\widetilde{F}$ follows from the fact that $\bar{a}$ is irreducible (see Section 3.1.3) and from Proposition 3.2.2 ii). We prove existence. Let $T$ be as in i). Since $T$ is invertible, $T^*T$ is positive definite. Choose $\lambda$ such that $\widetilde{F} := \lambda T^* T$ satisfies $\operatorname{Tr}\widetilde{F} = \operatorname{Tr}\widetilde{F}^{-1} > 0$. By equation (3.2), $\widetilde{F} \in \operatorname{Hom}(\bar{a}, a^{-t})$.

iii) First, note that $h(S(a_{ji})a_{kl}) = h(a_{ij}{}^* a_{kl})$ and $h(a_{ij} S(a_{lk})) = h(a_{ij} a_{kl}{}^*)$ by Proposition 3.1.7 v).

Denote by $(e_i)_i$ the standard basis of $\mathbb{C}^n$, and by $Y := \sum_{i,j} |e_i\rangle\langle e_j| \otimes a_{ij}$ and $\bar{Y} := \sum_{i,j} |e_i\rangle\langle e_j| \otimes a_{ij}{}^*$ the corepresentation operators associated to $a$ and $\bar{a}$, respectively. Note that the operator $(\bar{Y}^*)^{-1}$ corresponds to the matrix $a^{-t}$. Straightforward calculations show that

$$h(a_{ij}{}^* a_{kl}) = \begin{cases} \langle e_j | S_{ik} e_l \rangle, & \text{where } S_{ik} := (\mathrm{id} \otimes h)(Y^*(|e_i\rangle\langle e_k| \otimes 1)Y), \\ \langle e_i | T_{jl} e_k \rangle, & \text{where } T_{jl} := (\mathrm{id} \otimes h)(\bar{Y}(|e_j\rangle\langle e_l| \otimes 1)\bar{Y}^*). \end{cases} \tag{3.3}$$

By Lemma 3.2.5, $S_{ik} \in \operatorname{Hom}(a, a)$ and $T_{jl} \in \operatorname{Hom}(a^{-t}, \bar{a})$, and by Proposition 3.2.2 and ii), $S_{ik} = \lambda_{ik}\,\mathrm{id}$ and $T_{jl} = \mu_{jl}\widetilde{F}^{-1}$ for some $\lambda_{ik}, \mu_{jl} \in \mathbb{C}$. Hence

$$\mu_{jl}(\widetilde{F}^{-1})_{ik} = \langle e_i | T_{jl} e_k \rangle = h(a_{ij}{}^* a_{kl}) = \langle e_j | S_{ik} e_l \rangle = \lambda_{ik}\delta_{j,l}$$

for all $i, j, k, l$. This equation shows that $\mu_{jl} = 0$ for $j \neq l$ and that $\mu := \mu_{jj}$ does not depend on $j$. We can read off $\mu = 1/\operatorname{Tr}(\widetilde{F}^{-1})$ from the equation

$$n\mu \operatorname{Tr}(\widetilde{F}^{-1}) = \sum_{i,j} \langle e_i | T_{jj} e_i \rangle = \sum_{i,j} \langle e_j | S_{ii} e_j \rangle = \sum_i \operatorname{Tr} S_{ii} = (\operatorname{Tr} \otimes h)(Y^* Y) = n.$$

Thus, we have proved the first equation in assertion iii). The proof of the second one is similar.

iv) Assume that $0 = \sum_{k,l} a_{kl}\lambda_{kl}$ for some $\lambda_{kl} \in \mathbb{C}$. Then for all $i, j$,

$$0 = \sum_{k,l} h(a_{ij}{}^* a_{kl})\lambda_{kl} = \frac{1}{\mathrm{Tr}(\widetilde{F}^{-1})} \sum_k (\widetilde{F}^{-1})_{ik}\lambda_{kj}.$$

Since $\widetilde{F}^{-1}$ is invertible, we must have $\lambda_{kj} = 0$ for all $k, j$. □

**Remark 3.2.10.** If $(A, \Delta)$ is an algebraic compact quantum group and $a \in M_n(A)$ is a corepresentation matrix, then $\bar{a}$ is equivalent to a unitary corepresentation matrix by Theorem 3.2.1 i) and Proposition 3.1.7 v).

For later use, we note the following result:

**Proposition 3.2.11.** *Let $\delta$ be an irreducible corepresentation of a Hopf $*$-algebra $(A, \Delta)$. Then the corepresentation $\Delta|_{\mathcal{C}(\delta)}$ is equivalent to $\delta^{\boxplus n}$ for some $n \in \mathbb{N}$.*

*Proof.* Denote by $V$ the underlying vector space of $\delta$ and let $(f_i)_i$ be a basis of $V'$. Then the maps $T_i : V \to A$ given by $v \mapsto (f_i \otimes \mathrm{id})(\delta(v))$ intertwine $\delta$ and the regular corepresentation $\Delta$ (see Example 3.1.6), and $\sum_i \mathrm{Im}\, T_i = \mathcal{C}(\delta)$. If $T_i$ is not 0, then it is injective and its image is invariant by Proposition 3.2.2. Hence, $\Delta|_{\mathrm{Im}\, T_i}$ is equivalent to $\delta$, in particular, it is irreducible. Since $\mathrm{Im}\, T_i \cap (\sum_{j \neq i} \mathrm{Im}\, T_j)$ is an invariant subspace of $\Delta|_{\mathrm{Im}\, T_i}$, the intersection is either equal to 0 or to $\mathrm{Im}\, T_i$. Now an easy inductive argument shows that $\mathcal{C}(\delta)$ is equal to the direct sum $\bigoplus_{j \in J} \mathrm{Im}\, T_j$ for some subset of indices $J$. □

### 3.2.3 Characterization of compact quantum groups

Algebraic compact quantum groups can be characterized in terms of their corepresentations as follows:

**Theorem 3.2.12.** *The following conditions on a Hopf $*$-algebra $(A, \Delta)$ are equivalent:*

i) *$(A, \Delta)$ has a positive integral.*

ii) *Every finite-dimensional corepresentation of $(A, \Delta)$ is equivalent to a unitary corepresentation.*

iii) *$A$ is spanned by the matrix elements of its irreducible finite-dimensional unitary corepresentations.*

iv) *As an algebra, $A$ is generated by the matrix elements of all finite-dimensional corepresentations that are equivalent to unitary corepresentations.*

v) *Let $(u^\alpha)_\alpha$ be a maximal family of pairwise inequivalent irreducible unitary corepresentation matrices, where $u^\alpha = (u^\alpha_{ij})_{i,j}$. Then $(u^\alpha_{ij})_{\alpha,i,j}$ is a basis of $A$.*

*Proof.* i) $\Rightarrow$ ii): This is Theorem 3.2.1 i).

ii) $\Rightarrow$ iii): By Theorem 3.2.1 iv), there exists a family $(\delta_\alpha)_\alpha$ of finite-dimensional irreducible unitary corepresentations such that $\Delta \simeq \boxplus_\alpha \delta_\alpha$, and $A = \mathcal{C}(\Delta) = \sum_\alpha \mathcal{C}(\delta_\alpha)$.

iii) $\Rightarrow$ iv): Obvious.

iv) $\Rightarrow$ iii): Let us call a corepresentation unitarizable if it is equivalent to a unitary corepresentation. If $\delta_V$ and $\delta_W$ are unitarizable finite-dimensional corepresentations, then also $\delta_V \boxplus \delta_W$ and $\delta_V \boxtimes \delta_W$ are unitarizable, and $\mathcal{C}(\delta_V)+\mathcal{C}(\delta_W) = \mathcal{C}(\delta_V \boxplus \delta_W)$ and $\mathcal{C}(\delta_V)\mathcal{C}(\delta_W) \subseteq \mathcal{C}(\delta_V \boxtimes \delta_W)$. Hence, the algebra generated by the matrix elements of finite-dimensional unitarizable corepresentations is equal to the vector space spanned by the matrix elements of such corepresentations. By Theorem 3.2.1 iv), every such corepresentation is equivalent to a direct sum of irreducible unitary corepresentations.

iii) $\Rightarrow$ v): Let the family $(u^\alpha)_\alpha$ be given. From Proposition 3.2.11 and Proposition 3.2.2, one easily deduces $\mathcal{C}(u^\alpha) \cap \sum_{\beta\neq\alpha} \mathcal{C}(u^\beta) = 0$ for all $\alpha$.

We show that $(A, \Delta)$ has a normalized integral $h$. Observe that the unit $1_A$ is the matrix element of the trivial corepresentation $\mathbb{C} \to \mathbb{C} \otimes A$, $\lambda \mapsto \lambda \otimes 1_A$. Therefore we can define a linear map $h\colon A \to \mathbb{C}$ by $1_A \mapsto 1$ and $u^\alpha_{ij} \mapsto 0$ if $u^\alpha$ does not correspond to the trivial corepresentation. Since $\Delta(u^\alpha_{ij}) \in \operatorname{span}\{u^\alpha_{kl} \otimes u^\alpha_{mn} \mid k,l,m,n\}$ for every $\alpha$, the map $h$ is a normalized integral.

Finally, we show that for each $\alpha$, the irreducible corepresentation matrix $\overline{u^\alpha} = (u^{\alpha\,*}_{ij})_{i,j}$ is equivalent to a unitary one, and then the claim follows from Proposition 3.2.9 iv). By Corollary 3.2.8 and Proposition 3.2.2, it suffices to show that $h(S(\mathcal{C}(\overline{u^\alpha}))\mathcal{C}(u^\beta)) \neq 0$ for some $\beta$. But this follows from the assumption that $A = \sum_\beta \mathcal{C}(u^\beta)$ and the fact that $h$ is faithful (Proposition 2.2.4).

v) $\Rightarrow$ i): The proof of the implication iii) $\Rightarrow$ v) shows that $(A, \Delta)$ has a normalized integral $h$ and that for each $\alpha$, the corepresentation matrix $\overline{u^\alpha}$ is equivalent to a unitary one. We prove that $h$ is positive. For each $\alpha$, denote by $\widetilde{F}_\alpha \in \operatorname{Hom}(\overline{u^\alpha}, (u^\alpha)^{-t})$ the intertwiner constructed in Proposition 3.2.9. Consider an element $a = \sum_{\alpha,i,j} \lambda^\alpha_{ij} u^\alpha_{ij} \in A$. By Proposition 3.2.6 and 3.2.9,

$$h(a^*a) = \sum_{\alpha,i,j,\beta,k,l} \overline{\lambda^\alpha_{ij}}\, h(u^{\alpha\,*}_{ij}\, u^\beta_{kl})\, \lambda^\beta_{kl} = \sum_{\alpha,i,j,k} \frac{1}{\operatorname{Tr}(\widetilde{F}_\alpha^{-1})} \overline{\lambda^\alpha_{ij}} (\widetilde{F}_\alpha^{-1})_{ik} \lambda^\alpha_{kj}.$$

Since each $\widetilde{F}_\alpha$ is positive definite, so is each $\widetilde{F}_\alpha^{-1}$. Hence $h(a^*a) > 0$. □

### 3.2.4 Characters of corepresentations

Let $(A, \Delta)$ be an algebraic compact quantum group with Haar state $h$. An important tool in the study of corepresentations of $(A, \Delta)$ are the associated characters:

**Definition 3.2.13.** The *character* of a corepresentation matrix $a \in M_n(A)$ is the element $\chi(a) := \sum_{i=1}^n a_{ii} \in A$.

Recall that the equivalence classes of finite-dimensional corepresentations of $(A, \Delta)$ form a semiring with respect to the direct sum and the tensor product (see Section 3.1.3). Frequently, it is more convenient to work with rings instead of semirings. Forming the characters of corepresentations, we can embed the semiring of (equivalence classes of) finite-dimensional corepresentations into a subring of $A$:

**Proposition 3.2.14.** *Let $a \in M_m(A)$ and $b \in M_n(A)$ be corepresentation matrices of $(A, \Delta)$. Then*

$$\begin{gathered}\chi(a \boxplus b) = \chi(a) + \chi(b), \quad \chi(a \boxtimes b) = \chi(a)\chi(b), \\ \chi(\bar{a}) = \chi(a)^* = S(\chi(a)), \quad \epsilon(\chi(a)) = m, \\ m = n \text{ and } a \simeq b \Rightarrow \chi(a) = \chi(b).\end{gathered}$$

*Proof.* All assertions follow easily from the definitions, so we only prove $\chi(\bar{a}) = \chi(a)^* = S(\chi(a))$ and $\epsilon(\chi(a)) = m$. We may assume that $a$ is unitary, and then $\epsilon(\chi(a)) = \sum_{i=1}^m \epsilon(a_{ii}) = m$ and $S(\chi(a)) = \sum_{i=1}^m S(a_{ii}) = \sum_{i=1}^m a_{ii}^* = \chi(\bar{a})$ by Proposition 3.1.7 iii), v). □

The following results show that the map $a \mapsto \chi(a)$ is an embedding:

**Proposition 3.2.15.** *If $a, b$ are irreducible corepresentation matrices, then*

$$h(\chi(a)^*\chi(b)) = h(\chi(a)\chi(b)^*) = \begin{cases} 1, & a \simeq b, \\ 0, & \text{otherwise.} \end{cases}$$

*Proof.* This follows directly from Proposition 3.2.6 and 3.2.9. □

**Corollary 3.2.16.** *Let $a$ be a corepresentation matrix of $(A, \Delta)$, and let $(u^\alpha)_\alpha$ be a maximal family of pairwise inequivalent irreducible corepresentation matrices of $(A, \Delta)$. For each $\alpha$, put $n_\alpha := h(\chi(u^\alpha)^*\chi(a))$. Then*

$$\begin{gathered}a \simeq \boxplus_\alpha (u^\alpha)^{\boxplus n_\alpha}, \quad \chi(a) = \sum_\alpha n_\alpha \chi(u^\alpha), \\ \dim \operatorname{Hom}(a, a) = \sum_\alpha n_\alpha^2 = h(\chi(a)^*\chi(a)).\end{gathered}$$

*In particular, for every other corepresentation matrix $b$ of $(A, \Delta)$, we have $a \simeq b$ if and only if $\chi(a) = \chi(b)$.* □

### 3.2.5 Modular properties of the Haar state

Let $(A, \Delta)$ be an algebraic compact quantum group. The modular properties of the Haar state of $(A, \Delta)$ can conveniently be expressed in terms of a one-parameter family $(f_z)_{z\in\mathbb{C}}$ of characters. These characters are related to certain intertwiners which are constructed in the following proposition.

Given a matrix $c = (c_{ij})_{i,j} \in M_n(A)$, we put $S(c) := (S(c_{ij}))_{i,j}$.

**Proposition 3.2.17.** *Let $a \in M_n(A)$ be an irreducible unitary corepresentation matrix, where $n \in \mathbb{N}$. Then $S^2(a)$ is an irreducible corepresentation matrix and there exists a unique invertible intertwiner $F \in \mathrm{Hom}(a, S^2(a))$ such that $\mathrm{Tr}\, F = \mathrm{Tr}\, F^{-1} > 0$. This $F$ is positive definite, and the intertwiner $\widetilde{F} \in \mathrm{Hom}(\bar{a}, a^{-t})$ of Proposition 3.2.9 is equal to $\overline{F} = F^t$. In particular, $\mathrm{Tr}\, F = \mathrm{Tr}\, \widetilde{F}$.*

*Proof.* By Theorem 3.2.1 i), there exists an invertible matrix $T \in M_n(\mathbb{C})$ such that $b := T\bar{a}T^{-1}$ is unitary. By Proposition 3.1.7 v),

$$\bar{a} = S(a)^t \quad \text{and} \quad \overline{T}a\overline{T}^{-1} = \bar{b} = S(b)^t = T^{-t}S(\bar{a})^tT^t = T^{-t}S^2(a)T^t.$$

Here, $T^{-t} := (T^t)^{-1} = (T^{-1})^t$. As in Proposition 3.2.9 i), ii), we conclude that $S^2(a)$ is an irreducible corepresentation matrix and that $F$ is unique and equal to $\lambda T^t\overline{T} = \lambda\overline{T}^*\overline{T}$ for some $\lambda > 0$. The proof of Proposition 3.2.9 i), ii) shows that $\widetilde{F} = \tilde{\lambda} \cdot T^*T$ for some $\tilde{\lambda} > 0$, so $\widetilde{F} = \overline{F}\tilde{\lambda}/\lambda$. Since $\mathrm{Tr}\, F = \mathrm{Tr}\, F^{-1}$ and $\mathrm{Tr}\, \widetilde{F} = \mathrm{Tr}\, \widetilde{F}^{-1}$, we must have $\tilde{\lambda}/\lambda = 1$. Finally, $\mathrm{Tr}\, F = \mathrm{Tr}\, \overline{F} = \mathrm{Tr}\, \widetilde{F}$ because $F$ is positive. □

Now we turn to the one-parameter family $(f_z)_{z\in\mathbb{C}}$. We will need the following concept from function theory: An entire (that is, holomorphic on $\mathbb{C}$) function $g$ is *of exponential growth on the right half-plane* if there exist $C > 0$ and $d \in \mathbb{R}$ such that $|g(z)| \leq Ce^{d\,\mathrm{Re}(z)}$ for all $z \in \mathbb{C}$ with $\mathrm{Re}(z) > 0$.

**Lemma 3.2.18.** *If $g_1$ and $g_2$ are entire functions of exponential growth on the right half-plane such that $g_1(n) = g_2(n)$ for all $n \in \mathbb{N}$, then $g_1 = g_2$.*

*Proof.* This is a generalization of Carlson's Theorem [64, Theorem 11.3.3], see [192, p. 228]. □

**Theorem 3.2.19.** *Let $(A, \Delta)$ be an algebraic compact quantum group. Then there exists a family $(f_z)_{z\in\mathbb{C}}$ of characters on $A$ such that for all $z, z' \in \mathbb{C}$ and all $a, b \in A$, the following conditions are satisfied:*

i) *The function $z'' \mapsto f_{z''}(a)$ is entire and of exponential growth on the right half-plane.*

ii) $f_0 = \epsilon$ *and* $f_z * f_{z'} = f_{z+z'}$.

iii) $f_z(1_A) = 1$, $f_z(S(a)) = f_{-z}(a)$, *and* $f_z(a^*) = \overline{f_{-\bar{z}}(a)}$.

iv) $S^2(a) = f_{-1} * a * f_1$.

v) $h(ab) = h(b(f_1 * a * f_1))$ *for all* $a, b \in A$.

*The family* $(f_z)_{z\in\mathbb{C}}$ *is uniquely determined by the conditions* i), ii), v).

*Proof.* We define the family $(f_z)_z$ as follows. Let $(u^\alpha)_\alpha$ be a family of irreducible unitary corepresentation matrices as in Theorem 3.2.12 v). For each $\alpha$, denote by $F_\alpha \in \mathrm{Hom}(u^\alpha, S^2(u^\alpha))$ the intertwiner constructed in Proposition 3.2.17 and put $f_z(u^\alpha_{ij}) := (F^z_\alpha)_{ij}$ for all $z \in \mathbb{C}$ and all $i, j$.

We prove the assertions in the order i), ii), iv), v), vi), iii), where "vi)" denotes the claim that each $f_z$ is a character and that the family $(f_z)_z$ is uniquely determined by condition v).

i) This is an immediate consequence of the definition of the family $(f_z)_z$.

ii) By construction, $f_0(u^\alpha_{ij}) = \delta_{i,j} = \epsilon(u^\alpha_{ij})$ and

$$\begin{aligned}(f_z * f_{z'})(u^\alpha_{ij}) &= (f_z \otimes f_{z'})(\Delta(u^\alpha_{ij})) = \sum_k f_z(u^\alpha_{ik}) f_{z'}(u^\alpha_{kj}) \\ &= \sum_k (F^z_\alpha)_{ik}(F^{z'}_\alpha)_{kj} = (F^{z+z'}_\alpha)_{ij} = f_{z+z'}(u^\alpha_{ij}).\end{aligned}$$

iv) The relation $S^2(u^\alpha) = F_\alpha u^\alpha F^{-1}_\alpha$ implies

$$S^2(u^\alpha_{ij}) = \sum_{k,l} (F_\alpha)_{ik} u^\alpha_{kl} (F^{-1}_\alpha)_{lj} = (f_1 \otimes \mathrm{id} \otimes f_{-1})(\Delta^{(2)}(u^\alpha_{ij})) = f_{-1} * u^\alpha_{ij} * f_1.$$

v) By Proposition 3.2.6, we may assume $a \in \mathcal{C}(u^\alpha)$ and $b \in \mathcal{C}(u^\alpha)^*$ for some $\alpha$. Using Proposition 3.2.9 iii) and the relation $\widetilde{F} = F^t$ (Proposition 3.2.17), we find

$$\begin{aligned}h((u^\alpha_{kl})^*(f_1 * u^\alpha_{ij} * f_1)) &= \sum_{m,n} h((u^\alpha_{kl})^*(F_\alpha)_{im} u^\alpha_{mn} (F_\alpha)_{nj}) \\ &= \sum_{m,n} (F_\alpha)_{im} \frac{\delta_{l,n}}{\mathrm{Tr}\, F_\alpha} (F^{-1}_\alpha)_{mk} (F_\alpha)_{nj} \\ &= \frac{\delta_{i,k}}{\mathrm{Tr}\, F_\alpha} (F_\alpha)_{lj} = h(u^\alpha_{ij} u^{\alpha}_{kl}{}^*).\end{aligned}$$

vi) First, we show that $f_z$ is a character for each $z$. By v), the modular automorphism $\sigma$ of $h$ (Proposition 2.2.17) is given by $a \mapsto f_1 * a * f_1$. Assertion ii) implies $f_2(a) = (f_1 * \epsilon * f_1)(a) = \epsilon(f_1 * a * f_1)$ for all $a \in A$, that is, $f_2 = \epsilon \circ \sigma$. Thus $f_2$ is a character. Since the convolution of characters is a character again,

$f_{2k} = f_2 * \cdots * f_2$ ($k$ times) is a character for all $k \in \mathbb{N}$. Now an application of Lemma 3.2.18 shows that $f_z$ is a character for each $z \in \mathbb{C}$.

Next, we prove that the family $(f_z)_z$ is uniquely determined by conditions i), ii), v). By Proposition 2.2.17, condition v) uniquely determines the automorphism $a \mapsto f_1 * a * f_1$, and therefore also the character $f_2$. Next, $f_{2k}$ is uniquely determined for all $k \in \mathbb{N}$ by condition ii), and Lemma 3.2.18 implies that the family $(f_z)_z$ is uniquely determined by the family $(f_{2k})_{k \in \mathbb{N}}$.

iii) By construction, $f_z(1_A) = 1$ for each $z \in \mathbb{C}$, and by ii) and vi),

$$\begin{aligned} f_z(S(a)) &= \sum f_z(S(a_{(1)})) \cdot f_0(a_{(2)}) \\ &= \sum f_z(S(a_{(1)})) \cdot f_z(a_{(2)}) f_{-z}(a_{(3)}) \\ &= \sum f_z(S(a_{(1)})a_{(2)}) f_{-z}(a_{(3)}) = f_z(1_A) f_{-z}(a). \end{aligned}$$

Finally, by Proposition 3.1.7 v),

$$f_z((u_{ij}^\alpha)^*) = f_z(S(u_{ji}^\alpha)) = f_{-z}(u_{ji}^\alpha) = (F_\alpha^{-z})_{ji} = \overline{(F_\alpha^{-\bar{z}})_{ij}} = \overline{f_{-\bar{z}}(u_{ij}^\alpha)}. \quad \square$$

For each $z \in \mathbb{C}$, define $\rho_{z,z'} \colon A \to A$ by $a \mapsto f_z * a * f_{z'}$. The preceding theorem implies:

**Corollary 3.2.20.** *For all $z, z' \in \mathbb{C}$, the map $\rho_{z,z'}$ is an algebra automorphism of $A$, and for all $w, w' \in \mathbb{C}$,*

$$\begin{aligned} \rho_{0,0} &= \mathrm{id}, & \rho_{z,z'} \circ \rho_{w,w'} &= \rho_{z+w,z'+w'}, \\ h \circ \rho_{z,z'} &= h, & \rho_{z,z'} \circ * &= * \circ \rho_{-\bar{z},-\bar{z}'}, \\ \rho_{z,z'} \circ S &= S \circ \rho_{-z',-z}, & \Delta \circ \rho_{z,z'} &= (\rho_{w,z'} \otimes \rho_{z,-w}) \circ \Delta, \\ S^{-1} &= \rho_{1,-1} \circ S. \end{aligned} \quad \square$$

## 3.3 Discrete algebraic quantum groups

A classical result says that the Pontrjagin dual of a compact abelian group is discrete and that the Pontrjagin dual of a discrete abelian group is compact. Similarly, the dual of an algebraic compact quantum group is a discrete algebraic quantum group and vice versa. Like the compact ones, discrete algebraic quantum groups can be characterized in several equivalent ways. They were studied by Van Daele in [176] and [177, Section 5], and by Effros and Ruan in [40]; beware that Effros and Ruan use non-standard terminology.

We take the following definition as our starting point:

**Definition 3.3.1.** An algebraic quantum group is *discrete* if, as a $*$-algebra, it is isomorphic to an algebraic direct sum of matrix algebras, where a matrix algebra means a $*$-algebra of the form $M_n(\mathbb{C})$ for some $n \in N$.

**Theorem 3.3.2.** *Let $(A, \Delta)$ be an algebraic quantum group and $(\widehat{A}, \widehat{\Delta})$ its dual (see Section 2.3). Then the following conditions are equivalent:*

i) *$(A, \Delta)$ is compact.*

ii) *There exists an element $h \in \widehat{A}$ such that $\omega h = \hat{\epsilon}(\omega)h = h\omega$ for all $\omega \in \widehat{A}$.*

iii) *$(\widehat{A}, \widehat{\Delta})$ is discrete.*

*Proof.* i) $\Rightarrow$ ii): Denote by $h$ the Haar state of $(A, \Delta)$. Since $A$ is unital, $h = h(\cdot\, 1_A)$ belongs to $\widehat{A}$, and by Remark 2.2.2 ii) and Remark 2.3.4,

$$\omega h = \omega * h = \omega(1_A)h = \hat{\epsilon}(\omega)h = h * \omega = h\omega \quad \text{for all } \omega \in \widehat{A}.$$

ii) $\Rightarrow$ i): Let $h \in \widehat{A}$ be such that $\omega h = \hat{\epsilon}(\omega)h = h\omega$ for all $\omega \in \widehat{A}$. We show that $h$ is a left integral on $(A, \Delta)$. Let $\phi$ be some left integral on $A$, let $a \in A$, and put $T := (\mathrm{id} \otimes h)(\Delta(a)) \in M(A)$. Then for all $b \in A$,

$$\phi(Tb) = (\phi \otimes h)(\Delta(a)(b \otimes 1)) = (\phi(\cdot\, b)h)(a) = \hat{\epsilon}(\phi(\cdot\, b))h(a) = \phi(b)h(a).$$

Since $\phi$ is faithful (Proposition 2.2.4), $T = 1_{M(A)}h(a)$. The claim follows.

By definition of $\widehat{A}$, we can write $h = h(\cdot\, e_1) = h(e_2\, \cdot)$ with some $e_1, e_2 \in A$. Then $h(e_2ab) = h(ab) = h(abe_1)$ for all $a, b \in A$, and since $h$ is faithful (Proposition 2.2.4), it follows that $e_2a = a$ and $b = be_1$ for all $a, b \in A$. Thus $e_1 = e_2$ is a unit of $A$.

iii) $\Rightarrow$ ii): Assume that $\widehat{A} \cong \bigoplus_i M_{n_i}(\mathbb{C})$ as $*$-algebras, where $(n_i)_i$ is some family of natural numbers. Since the counit $\hat{\epsilon}\colon \widehat{A} \to \mathbb{C}$ is a non-zero homomorphism and each summand $M_{n_i}(\mathbb{C})$ is a simple algebra, there exists precisely one index $i_0$ such that $\hat{\epsilon}$ corresponds to the projection $(x_i)_i \mapsto x_{i_0}$, and $n_{i_0} = 1$. Now the element $h \in \widehat{A}$ that corresponds to the family $(x_i)_i$ given by $x_{i_0} = 1$ and $x_i = 0$ for $i \neq i_0$ obviously satisfies $\omega h = \hat{\epsilon}(\omega)h = h\omega$ for all $\omega \in \widehat{A}$.

i) $\Rightarrow$ iii): By Theorem 3.2.12, there exists a family of irreducible unitary matrix corepresentations $u^\alpha$ of $(A, \Delta)$ such that $(u^\alpha_{ij})_{\alpha,i,j}$ is a basis of $A$. For each $\alpha$, denote by $\widetilde{F}_\alpha \in \mathrm{Hom}(\overline{u^\alpha}, (u^\alpha)^{-t})$ the intertwiner constructed in Proposition 3.2.9 and put

$$\omega^\alpha_{ij} := (\mathrm{Tr}\, \widetilde{F}_\alpha) \sum_k (\widetilde{F}_\alpha)_{ik} h((u^\alpha_{kj})^* \cdot) \in \widehat{A} \quad \text{for all } i, j.$$

Since each $\widetilde{F}_\alpha$ is invertible, the family $(\omega^\alpha_{ij})_{\alpha,i,j}$ forms a basis of $\widehat{A}$. By Proposition 3.2.9 iii),

$$\omega^\alpha_{ij}(u^\beta_{kl}) = \delta_{\alpha,\beta} \cdot \delta_{i,k} \cdot \delta_{j,l} \quad \text{for all } \alpha, \beta, i, j, k, l.$$

From this equation and the relations $\Delta(u^\alpha_{ij}) = \sum_k u^\alpha_{ik} \otimes u^\alpha_{kj}$ and $S(u^\alpha_{ij})^* = u^\alpha_{ji}$ (see Proposition 3.1.7), it is easy to deduce that

$$\omega^\alpha_{ij}\omega^\beta_{kl} = \delta_{\alpha,\beta} \cdot \delta_{j,k} \cdot \omega^\alpha_{il} \quad \text{and} \quad (\omega^\alpha_{ij})^* = \omega^\alpha_{ji} \quad \text{for all } \alpha, \beta, i, j, k, l.$$

Thus $\hat{A}$ is isomorphic to an algebraic direct sum of matrix algebras. □

**Remark 3.3.3.** The preceding result can be strengthened as follows. Call a multiplier Hopf $*$-algebra $(A, \Delta)$ *discrete* if, as a $*$-algebra, it is isomorphic to an algebraic direct sum of matrix algebras. Van Daele [176] showed that every discrete multiplier Hopf $*$-algebra has a positive left integral and a positive right integral, and therefore is a discrete algebraic quantum group.

# Part II

# Quantum groups in the setting of $C^*$-algebras and von Neumann algebras

# Chapter 4
# First definitions and examples

In Part I of this book, we regarded Hopf algebras as purely algebraic objects. This point of view is adequate for the study of quantum analogues of discrete, compact, or affine algebraic groups. In Part II, we want to study quantum analogues of general locally compact groups; therefore we need to consider topological variants of Hopf algebras. More precisely, the topological aspects of locally compact groups and of their quantum analogues will be covered by a theory based on $C^*$-algebras, and the measurable aspects will be covered by a theory based on von Neumann algebras. Throughout Part II of this book, we shall focus on the setting of $C^*$-algebras.

To find a definition of a Hopf $C^*$-algebra or Hopf–von Neumann algebra that assumes few axioms but covers many examples is not an easy task. Some of the related difficulties and the main existing approaches are summarized in Chapter 4. The classical examples of Hopf $C^*$-algebras and Hopf–von Neumann algebras associated to locally compact groups are also discussed in this chapter.

Particularly accessible and well understood is the theory of $C^*$-algebraic compact quantum groups developed by Woronowicz [193], [202]. Much of the corepresentation theory of algebraic compact quantum groups carries over to the $C^*$-algebraic setting, and the two classes of quantum groups are closely related. These topics form the contents of Chapter 5 .

General locally compact quantum groups were introduced and studied by Vaes and Kustermans [88], [91], [158], and by Masuda, Nakagami, and Woronowicz [110]. The theories developed by these two groups are very satisfying in terms of the results that can be proved and the examples that are covered; however, the details of the theories are highly intricate. We outline the approach of Vaes and Kustermans in Chapter 8, but focus on motivation and give no proofs.

Fundamental to almost all approaches to Hopf algebras and generalized Pontrjagin duality in the setting of $C^*$-algebras or von Neumann algebras is the concept of a multiplicative unitary. Examples of such unitaries were used in various proofs and constructions for a long time till Baaj and Skandalis put them center-stage, formulated an abstract definition, and gave a comprehensive treatment [7]. We discuss multiplicative unitaries in Chapter 7; they will reappear in Chapters 8, 9, 10, and 11.

## 4.1 $C^*$-bialgebras and von Neumann bialgebras

**Notation.** A short summary on $C^*$-algebras and von Neumann algebras as well as standard references can be found in the appendix. We shall use the following notation.

Given a $C^*$-algebra $A$, we denote by $M(A)$ the multiplier algebra of $A$; this is a $C^*$-algebra again. If $\phi\colon A \to M(B)$ is a $*$-homomorphism of $C^*$-algebras that is non-degenerate in the sense that $\overline{\operatorname{span}}\, \phi(A)B = B$, we denote the unique extension to a $*$-homomorphism $M(A) \to M(B)$ by $\phi$ again. Given $C^*$-algebras $A_1, A_2$, we denote by $A_1 \otimes A_2$ the minimal tensor product of $A_1$ and $A_2$.

Given a von Neumann algebra $M$, we denote by $M_*$ its predual which is the space of all normal linear functionals on $M$. The von Neumann-algebraic tensor product of von Neumann algebras $M_1$ and $M_2$ will be denoted by $M_1 \bar{\otimes} M_2$.

From now on, we denote the purely algebraic tensor product of vector spaces by "$\odot$" instead of "$\otimes$".

***$C^*$*-bialgebras and von Neumann bialgebras.** The concept of a comultiplication or a bialgebra carries over to the setting of $C^*$-algebras and von Neumann algebras easily:

**Definition 4.1.1.** A *$C^*$-bialgebra* is a $C^*$-algebra $A$ equipped with a non-degenerate $*$-homomorphism $\Delta\colon A \to M(A \otimes A)$ called the *comultiplication* such that

i) $\Delta$ is *coassociative* in the sense that $(\Delta \otimes \mathrm{id}_A) \circ \Delta = (\mathrm{id}_A \otimes \Delta) \circ \Delta$,

ii) $\Delta(A)(1 \otimes A)$ and $\Delta(A)(A \otimes 1)$ are contained in $A \otimes A$.

A $C^*$-bialgebra $(A, \Delta)$ is called *bisimplifiable* if each of the sets $\Delta(A)(1 \otimes A)$ and $\Delta(A)(A \otimes 1)$ is linearly dense in $A \otimes A$.

A *morphism of $C^*$-bialgebras* $(A, \Delta_A)$ and $(B, \Delta_B)$ is a non-degenerate $*$-homomorphism $F\colon A \to M(B)$ that satisfies $\Delta_B \circ F = (F \otimes F) \circ \Delta_A$.

**Remarks 4.1.2.** i) In the definition above, the non-degenerate homomorphisms $\Delta \otimes \mathrm{id}_A, \mathrm{id}_A \otimes \Delta\colon A \otimes A \to A \otimes A \otimes A$ and $F \otimes F\colon A \otimes A \to B \otimes B$ have been extended to the multiplier $C^*$-algebra $M(A \otimes A)$.

ii) Given $C^*$-bialgebras $(A, \Delta_A)$ and $(B, \Delta_B)$, one can construct new $C^*$-bialgebras $(A, \Delta_A)^{\mathrm{op}}$, $(A, \Delta_A)^{\mathrm{cop}}$, $(A, \Delta_A)^{\mathrm{op,cop}}$, and $(A \oplus B, \Delta_{A \oplus B})$, $(A \otimes B, \Delta_{A \otimes B})$ in much the same way as in the case of bialgebras and multiplier bialgebras, see Remarks 1.3.7 iii).

**Definition 4.1.3.** A *von Neumann bialgebra* is a von Neumann algebra $M$ equipped with a normal unital $*$-homomorphism $\Delta\colon M \to M \bar{\otimes} M$ called the *comultiplication* such that $(\Delta \bar{\otimes} \mathrm{id}_M) \circ \Delta = (\mathrm{id}_M \bar{\otimes} \Delta) \circ \Delta$.

A *morphism* of von Neumann bialgebras $(M, \Delta_M)$ and $(N, \Delta_N)$ is a normal $*$-homomorphism $F\colon M \to N$ that satisfies $\Delta_N \circ F = (F \bar{\otimes} F) \circ \Delta_M$.

**Remark 4.1.4.** Given von Neumann-bialgebras $(M, \Delta_M)$ and $(N, \Delta_N)$, one can construct new $C^*$-bialgebras $(M, \Delta_M)^{\mathrm{op}}$, $(M, \Delta_M)^{\mathrm{cop}}$, $(M, \Delta_M)^{\mathrm{op,cop}}$, and $(M \oplus N, \Delta_{M \oplus N})$, $(M \otimes N, \Delta_{M \otimes N})$ in much the same way as in the case of bialgebras and multiplier bialgebras, see Remarks 1.3.7 iii).

**Convolution.** Let $(A, \Delta)$ be a $C^*$-bialgebra. Then we can define a convolution product on $A'$, very much as in the case of bialgebras (see Section 1.3.2): Given $f, g \in A'$, the functional $f \otimes g$ on $A \otimes A'$ extends uniquely to a strictly continuous functional on $M(A \otimes A)$ (Corollary 12.1.2), and the *convolution product* of $f$ and $g$ is the functional

$$f * g := (f \otimes g) \circ \Delta \in A'.$$

The convolution product is associative: given $f, g, h \in A'$, the products $(f * g) * h$ and $f * (g * h)$ both are equal to the composition of the strictly continuous functional $f \otimes g \otimes h$ on $M(A \otimes A \otimes A)$ with the map $(\mathrm{id} \otimes \Delta) \circ \Delta = (\Delta \otimes \mathrm{id}) \circ \Delta$.

For each $f \in A'$, we can define slice maps $\mathrm{id} \otimes f, f \otimes \mathrm{id} \colon M(A \otimes A) \to M(A)$ that are norm-continuous and strictly continuous on bounded subsets (see Section 12.4). We put

$$f * a := (\mathrm{id} \otimes f)(\Delta(a)), \quad a * f := (f \otimes \mathrm{id})(\Delta(a))$$

for all $f \in A'$, $a \in A$. The multipliers $f * a$ and $a * f$ belong to $A$ by condition ii) of Definition 4.1.1 and Proposition 12.4.3. Similarly as in the case of bialgebras, the maps $(f, a) \mapsto f * a$ and $(a, f) \mapsto a * f$ turn $A$ into a (Banach) bimodule over $A'$.

Similarly, if $(M, \Delta)$ is a von Neumann bialgebra, then the space of normal linear functionals $M_*$ carries an associative convolution product, given by

$$f * g := (f \mathbin{\bar{\otimes}} g) \circ \Delta \in M_*,$$

and $M$ carries a natural structure of a bimodule over $M_*$.

**From bialgebras to Hopf algebras.** If we want to define analogues of Hopf algebras in the setting of $C^*$-algebras or von Neumann algebras, and extend Pontrjagin duality to such generalized Hopf algebras, the following problems arise:

- In many examples, the counit and the antipode are unbounded and densely defined only. In those cases, it is difficult to make sense of the axioms $(\epsilon \otimes \mathrm{id}) \circ \Delta = \mathrm{id} = (\mathrm{id} \otimes \epsilon) \circ \Delta$ and $m \circ (S \otimes \mathrm{id}) \circ \Delta = \eta \circ \epsilon = m \circ (\mathrm{id} \otimes S) \circ \Delta$ that characterize Hopf algebras – how can the maps $\epsilon \otimes \mathrm{id}$, $\mathrm{id} \otimes \epsilon$ and $S \otimes \mathrm{id}$, $\mathrm{id} \otimes S$, which are unbounded and densely defined, be extended to the image of $\Delta$?

- Further problems arise from the axiom $m \circ (S \otimes \mathrm{id}) \circ \Delta = \eta \circ \epsilon = m \circ (\mathrm{id} \otimes S) \circ \Delta$, because the multiplication map $m \colon A \odot A \to A$ need not extend to the minimal $C^*$-tensor product $A \otimes A$.

- To generalize Pontrjagin duality, we need an analogue of the Haar measure of a locally compact group. But in contrast to the situation of locally compact groups, it seems to be very difficult to deduce the existence of an invariant integral on a $C^*$-bialgebra or von Neumann bialgebra from a reasonable set of axioms.

So, at the moment, we have to leave the answer to the question "What is a Hopf $C^*$-algebra or a Hopf–von Neumann algebra?" open. A satisfying answer that was proposed by Vaes and Kustermans will be presented in Chapter 8.

## 4.2 Bialgebras associated to groups

To every locally compact group, one can associate several $C^*$-bialgebras and von Neumann bialgebras:

**Example 4.2.1.** Let $G$ be a locally compact group. Then the $C^*$-algebra $C_0(G)$, equipped with the comultiplication

$$\Delta\colon C_0(G) \to C_b(G \times G) \cong M(C_0(G) \otimes C_0(G)), \quad (\Delta f)(x, y) := f(xy),$$

is a $C^*$-bialgebra. It is bisimplifiable, as can be seen from the equations

$$(\Delta(f)(1 \otimes g))(x, y) = f(xy)g(y), \quad (\Delta(f)(g \otimes 1))(x, y) = f(xy)g(x),$$

where $f, g \in C_0(G)$ and $x, y \in G$, and from the fact that the maps $G \times G \to G \times G$ given by $(x, y) \mapsto (xy, y)$ and $(x, y) \mapsto (x, xy)$, respectively, are homeomorphisms.

The $C^*$-bialgebra $C_0(G)$ has a well behaved counit and antipode: the $*$-homomorphism $\epsilon\colon C_0(G) \to \mathbb{C}$ given by $\epsilon(f) := f(e)$, where $e$ denotes the unit of $G$, and the $*$-homomorphism $S\colon C_0(G) \to C_0(G)$ given by $(Sf)(x) := f(x^{-1})$ for all $x \in G$ and $f \in C_0(G)$ satisfy the counit identity and the antipode identity known from Hopf algebras.

**Example 4.2.2.** To every locally compact group $G$, one can associate a full/universal group $C^*$-bialgebra $C^*(G)$ and a reduced group $C^*$-bialgebra $C^*_r(G)$ as follows.

Let us begin with the full group $C^*$-algebra $C^*(G)$. Denote by $\lambda$ the left Haar measure and by $\delta$ the modular function of $G$. Then the space $L^1(G, \lambda)$ is a Banach $*$-algebra with respect to the multiplication and involution given by

$$(f * g)(x) = \int_G f(y)g(y^{-1}x)d\lambda(y), \quad f^*(x) = \overline{f(x^{-1})}\delta(x)^{-1}. \tag{4.1}$$

This Banach $*$-algebra has a bounded approximate unit. Indeed, a routine argument shows that if $(V_\nu)_\nu$ is a neighborhood basis of $e \in G$, and if $(f_\nu)_\nu$ is a net of non-negative measurable functions on $G$ such that $\operatorname{supp} f_\nu \subseteq V_\nu$ and $\int_G f_\nu d\lambda = 1$ for all $\nu$, then $(f_\nu)_\nu$ is a bounded approximate unit.

The group $C^*$-algebra $C^*(G)$ is the enveloping $C^*$-algebra of $L^1(G,\lambda)$, that is, the completion of $L^1(G,\lambda)$ with respect to the norm

$$\|f\| := \sup\{\|\rho(f)\| \mid \rho \text{ is a } *\text{-representation of } L^1(G,\lambda) \text{ on some Hilbert space } H\}.$$

If $G$ is discrete, then for each $x \in G$, the element $U_x \in L^1(G,\lambda)$ given by $y \mapsto \delta_{x,y}$ is a unitary element of $C^*(G)$, and the family $(U_x)_{x\in G}$ is linearly dense in $C^*(G)$. In the general case, the map

$$U_x \colon L^1(G,\lambda) \to L^1(G,\lambda), \quad (U_x f)(y) = f(x^{-1}y),$$

is a multiplier for each $x \in G$; if $G$ is discrete, we simply recover the family $(U_x)_{x\in G}$ defined before. For each $x \in G$, the multiplier $U_x \in M(L^1(G,\lambda))$ extends to a multiplier $U_x \in M(C^*(G))$ (see Corollary 12.5.2). Evidently,

$$U_x U_y = U_{xy} \quad \text{and} \quad (U_x)^* = U_{x^{-1}} \quad \text{for all } x, y \in G.$$

One can show that the linear span of the family $(U_x)_x$ is strictly dense in $M(C^*(G))$.

Using the relation $fg = \int_G f(y) U_y g \, d\lambda(y)$, which holds for all $f, g \in L^1(G,\lambda)$, one can show that

$$f = \int_G f(x) U_x d\lambda(x) \quad \text{for every } f \in L^1(G,\lambda). \tag{4.2}$$

Here, the integral of a (suitable) function $h \colon G \to E^{**}$, where $E$ is a Banach space and $E^{**}$ its bidual, is defined to be the element of $E^{**}$ given by $\phi \mapsto \int_G \phi(h(x)) d\lambda(x)$.

The $C^*$-algebra $C^*(G)$ has the following universal property. Recall that a strongly continuous unitary representation of $G$ on a Hilbert space $H$ is a homomorphism $\pi$ from $G$ to the group of unitary operators on $H$ such that for every $\xi \in H$, the map $G \to H$ given by $x \mapsto \pi(x)\xi$ is continuous. For every such representation $\pi$, there exists a unique non-degenerate $*$-homomorphism $C^*(G) \to \mathcal{L}(H)$ whose extension to $M(C^*(G))$ maps $U_x$ to $\pi(x)$ for every $x \in G$. Indeed, the map $\tilde{\pi} \colon L^1(G,\lambda) \to \mathcal{L}(H)$ given by $\tilde{\pi}(f)\xi := \int_G f(x)\pi(x)\xi d\lambda(x)$ defines a $*$-representation of $L^1(G,\lambda)$ and hence a representation of $C^*(G)$ that has the desired property.

The universal property of $C^*(G)$ can be used to construct non-degenerate $*$-homomorphisms

$$\Delta \colon C^*(G) \to M(C^*(G) \otimes C^*(G)), \quad \epsilon \colon C^*(G) \to \mathbb{C}, \quad S \colon C^*(G) \to C^*(G)^{\mathrm{op}},$$

such that the extensions of these maps to $M(C^*(G))$ act as follows:

$$\Delta(U_x) = U_x \otimes U_x, \quad \epsilon(U_x) = 1, \quad S(U_x) = U_{x^{-1}}.$$

Equation (4.2) shows that for each $f \in L^1(G,\lambda) \subseteq C^*(G)$,

$$\Delta(f) = \int_G (U_x \otimes U_x) f(x) d\lambda(x),$$
$$\epsilon(f) = \int_G f(x) d\lambda(x),$$

and

$$S(f) = \int_G f(x) U_{x^{-1}} d\lambda(x)$$
$$= \int_G f(x^{-1}) U_x d\lambda(x^{-1}) = \int_G f(x^{-1})\delta(x)^{-1} U_x d\lambda(x),$$

that is, $(S(f))(x) = f(x^{-1})\delta(x)^{-1}$ for all $x \in G$.

It is easy to check that the $C^*$-bialgebra $(C^*(G), \Delta)$ is bisimplifiable.

Let us turn to the reduced group $C^*$-algebra $C_r^*(G)$. By definition, this is the $C^*$-subalgebra of $\mathcal{L}(L^2(G,\lambda))$ generated by the left regular representation $L\colon L^1(G,\lambda) \to \mathcal{L}(L^2(G,\lambda))$,

$$(L(f)\xi)(x) = \int_G f(y)\xi(y^{-1}x) d\lambda(y) \quad \text{for } f \in L^1(G,\lambda),\ \xi \in L^2(G,\lambda),\ x \in G.$$

The representation $L$ extends to a representation of $C^*(G)$, and this extension yields a surjective $*$-homomorphism $L\colon C^*(G) \to C_r^*(G)$. We put

$$L_x := L(U_x) \in M(C_r^*(G)) \subseteq \mathcal{L}(L^2(G,\lambda)) \quad \text{for each } x \in G.$$

The quotient map $L\colon C^*(G) \to C_r^*(G)$ is an isomorphism if and only if $G$ is amenable, that is, if and only if there exists a state $\phi$ on $L^\infty(G,\lambda)$ which is left-invariant in the sense that $\phi(f) = \phi(f(x\,\cdot))$ for all $f \in L^\infty(G)$, $x \in G$ [119, Theorem 4.21], [121, Theorem 7.3.9]; here, $f(x\,\cdot)$ denotes the function $y \mapsto f(xy)$.

There exists a $*$-homomorphism $\Delta_r\colon C_r^*(G) \to M(C_r^*(G) \otimes C_r^*(G))$ such that $\Delta_r(L_x) = L_x \otimes L_x$ for all $x \in G$, and the following diagram commutes:

$$\begin{array}{ccc} C^*(G) & \xrightarrow{\Delta} & M(C^*(G) \otimes C^*(G)) \\ {\scriptstyle L}\downarrow & & \downarrow{\scriptstyle L\otimes L} \\ C_r^*(G) & \xrightarrow[\Delta_r]{} & M(C_r^*(G) \otimes C_r^*(G)). \end{array}$$

The construction of $\Delta_r$ involves a multiplicative unitary and is given in Example 7.2.13. Like $(C^*(G), \Delta)$, also $(C^*_r(G), \Delta_r)$ is a bisimplifiable $C^*$-bialgebra.

The counit $\epsilon\colon C^*(G) \to \mathbb{C}$ factorizes to a $*$-homomorphism $\epsilon_r\colon C^*_r(G) \to \mathbb{C}$ if and only if the group $G$ is amenable. The "if" part is obvious, for the "only if" part, see [119, Proof of Theorem 4.21].

Finally, the antipode $S\colon C^*(G) \to C^*(G)^{\mathrm{op}}$ factorizes to a $*$-homomorphism $S_r\colon C^*_r(G) \to C^*_r(G)^{\mathrm{op}}$, that is, there exists a commutative diagram

$$\begin{array}{ccc} C^*(G) & \xrightarrow{S} & C^*(G)^{\mathrm{op}} \\ {\scriptstyle L}\downarrow & & \downarrow{\scriptstyle L^{\mathrm{op}}} \\ C^*_r(G) & \xrightarrow[S_r]{} & C^*_r(G)^{\mathrm{op}}. \end{array}$$

This can be seen as follows: The formula $(I\xi)(x) := \overline{\xi(x)}$, where $\xi \in L^2(G,\lambda)$ and $x \in G$, defines a conjugate-linear isometric map $I$ on $L^2(G,\lambda)$, and a short calculation shows that $S_r(L(f)) = IL(f)^*I$ for all $f \in L^1(G,\lambda)$.

Every continuous homomorphism of locally compact groups $\phi\colon G \to H$ induces a morphism of $C^*$-bialgebras $\phi_*\colon C^*(G) \to M(C^*(H))$ by the formula $U_x \mapsto U_{\phi(x)}$, that is, $f \mapsto \int_G U_{\phi(x)} f(x) d\lambda(x)$. This follows easily from the universal property of $C^*(G)$. However, there need not exist a $*$-homomorphism $\phi_{*,r}\colon C^*_r(G) \to M(C^*_r(H))$ that makes the following diagram commute:

$$\begin{array}{ccc} C^*(G) & \xrightarrow{\phi_*} & M(C^*(H)) \\ {\scriptstyle L_G}\downarrow & & \downarrow{\scriptstyle L_H} \\ C^*_r(G) & \xrightarrow[\phi_{*,r}]{} & M(C^*_r(H)); \end{array}$$

here, $L_G$ and $L_H$ denote the respective quotient maps.

In the abelian case, the $C^*$-bialgebras introduced above are related via Pontrjagin duality as follows (compare also with Example 1.4.3). Let $G$ be a locally compact abelian group $G$. Then $G$ is amenable [119, Proposition 0.15], whence $C^*_r(G) \cong C^*(G)$. For each $x \in G$, denote by $\mathrm{ev}_x\colon \widehat{G} \to \mathbb{C}$ the function given by $\chi \mapsto \chi(x)$. Note that $\mathrm{ev}_x$ belongs to $C_b(\widehat{G}) \cong M(C_0(\widehat{G}))$.

**Proposition 4.2.3.** *Let $G$ be a locally compact abelian group with dual group $\widehat{G}$. There exists an isomorphism of $C^*$-bialgebras $\Phi\colon C^*(G) \xrightarrow{\cong} C_0(\widehat{G})$ such that $\Phi(U_x) = \mathrm{ev}_x$ for all $x \in G$.*

*Proof.* By the universal property of $C^*(G)$, there exists a $*$-homomorphism $\Phi\colon C^*(G) \to C_0(\widehat{G})$ whose extension to $M(C^*(G))$ maps $U_x$ to $\mathrm{ev}_x$ for each $x \in G$. The universal property of the $C^*$-algebra $C^*(G)$ furthermore implies that its spectrum, that is, its space of characters or continuous one-dimensional representations, can be identified with the space of characters of $G$, that is, with $\widehat{G}$. Since $\Phi$ is induced by this identification, $\Phi$ is an isomorphism. The calculation

$$\begin{aligned}((\Phi \otimes \Phi)(\Delta(U_x)))(\chi, \chi') &= (\Phi(U_x))(\chi) \cdot (\Phi(U_x))(\chi') \\ &= \chi(x)\chi'(x) = (\chi\chi')(x),\end{aligned}$$

$$(\Delta(\Phi(U_x)))(\chi, \chi') = (\Phi(U_x))(\chi\chi') = (\chi\chi')(x), \quad \chi, \chi' \in \widehat{G},\ x \in G,$$

shows that $\Delta \circ \Phi = (\Phi \otimes \Phi) \circ \Delta$. □

Let us briefly consider the setting of von Neumann algebras:

**Example 4.2.4.** Let $G$ be a locally compact group with left Haar measure $\lambda$. Then $L^\infty(G, \lambda)$, equipped with the comultiplication given by $(\Delta f)(x, y) = f(xy)$ for all $x, y \in G$ and $f \in L^\infty(G, \lambda)$, is a von Neumann bialgebra.

The group von Neumann algebra $L(G) \subseteq \mathcal{L}(L^2(G, \lambda))$ is the von Neumann algebra generated by the unitaries $L_x \in \mathcal{L}(L^2(G, \lambda))$ defined in Example 4.2.2. There exists a normal unital $*$-homomorphism $L(G) \to L(G) \mathbin{\bar{\otimes}} L(G)$ such that $L_x \mapsto L_x \otimes L_x$ for all $x \in G$, and this $*$-homomorphism turns $L(G)$ into a von Neumann bialgebra. The proof of this assertion proceeds via a multiplicative unitary, see Example 7.2.13

## 4.3 Approaches to quantum groups in the setting of von Neumann algebras and $C^*$-algebras

### The three guises of a quantum group

In the setting of $C^*$-algebras and von Neumann algebras, a quantum group usually appears in several guises: as a full/universal $C^*$-bialgebra, as a reduced $C^*$-bialgebra, and as a von Neumann bialgebra. The first two $C^*$-bialgebras, however, may coincide. Furthermore, a quantum group may be of an algebraic origin in the sense that it can be described by a (multiplier) Hopf $*$-algebra.

This multitude of bialgebras associated to one (quantum) group appeared already in Example 4.2.2 and 4.2.4: Every locally compact group $G$ gives rise to the $C^*$-bialgebra $C_0(G)$ and the von Neumann bialgebra $L^\infty(G)$, and, dually, to the full group $C^*$-algebra $C^*(G)$, the reduced group $C^*$-algebra $C_r^*(G)$, and the group von Neumann algebra $L(G)$. If the group $G$ is compact (or discrete), then the multiplier Hopf $*$-algebras $\mathrm{Rep}(G)$ and $\widehat{\mathrm{Rep}(G)}$ (or $C_c(G)$ and $\mathbb{C}G$, respectively), provide further descriptions of the same underlying quantum group and its dual.

The universal $C^*$-bialgebra, the reduced $C^*$-bialgebra, and the von Neumann bialgebra (as well as the multiplier Hopf $*$-algebra, if present) of a quantum group provide equivalent views on one and the same underlying object, and one can pass back and forth between these different points of view. The distinction between the reduced and the universal $C^*$-bialgebra amounts to the choice whether a quantum group is studied in terms of its regular representation or in terms of all of its representations. Naturally, the "reduced theory" is closer to the von Neumann algebraic setting, and in this book, we focus on that "reduced theory".

## Existing approaches

In the setting of $C^*$-algebras and von Neumann algebras, several approaches to quantum groups of varying levels of generality and technical complexity have been developed:

**Kac algebras.** The first satisfactory extension of Pontrjagin duality to all locally compact groups was given in the framework of Kac algebras developed by Enock and Schwartz [47] and by Kac and Vainerman [167], [168] in the seventies. This theory was formulated in the setting of von Neumann algebras; later, it was extended to the setting of $C^*$-algebras by Vallin and Enock [49], [170]. The existence of a Haar weight – the analogue of a Haar measure – is postulated as an axiom.

After the development of the theory, many examples of quantum groups were found that fit into this framework [31], [48], [77], [78], [96], [106], [130], [132], [165], [173], [208].

A severe limitation of the theory of Kac algebras is that the antipode is assumed to be bounded and to commute with the involution.

**Compact quantum groups.** The theory of compact quantum groups developed by Woronowicz [193], [194], [202] is perhaps most easily accessible and closest to the purely algebraic setting. For $C^*$-algebraic compact quantum groups, the Haar weight is bounded, that is, a state, and its existence can be deduced from a few natural axioms. Inside every $C^*$-algebraic compact quantum group, one can identify a unique dense algebraic compact quantum group, and one can pass back and forth between the algebraic and the $C^*$-algebraic level. We discuss $C^*$-algebraic compact quantum groups in Chapter 5.

The first example of a compact quantum group – the famous quantum group $\mathrm{SU}_\mu(2)$ introduced by Woronowicz (see Section 6.2) – showed that the antipode of a compact quantum group need not be bounded and need not commute with the involution. In particular $\mathrm{SU}_\mu(2)$ is not a Kac algebra.

**Multiplicative unitaries.** A fundamental tool for the study of quantum groups in the setting of $C^*$-algebras and von Neumann algebras, in particular in relation with Pontrjagin duality, are multiplicative unitaries. Their theory was developed by Baaj and Skandalis [7]; an important contribution was made by Woronowicz [201]. Roughly, every Hopf $*$-algebra, $C^*$-bialgebra, and von Neumann bialgebra equipped with a positive integral or a Haar weight gives rise to a multiplicative unitary; conversely, under certain regularity assumptions, a multiplicative unitary gives rise to a pair of $C^*$-bialgebras and von Neumann bialgebras. Both transitions, applied subsequently, can be used to construct the Pontrjagin dual of a suitable $C^*$-bialgebra or von Neumann bialgebra. Moreover, multiplicative unitaries facilitate the transition between the different guises of a quantum group.

**Locally compact quantum groups / weighted Hopf $C^*$-algebras.** The most comprehensive and at the same time technically most demanding approaches are the theory of locally compact quantum groups developed by Vaes and Kustermans [91], [93] and the theory of weighted Hopf $C^*$-algebras developed by Masuda, Nakagami and Woronowicz [110]. In both theories, the existence of a left Haar weight is assumed. Additionally, the first theory assumes the existence of a right Haar weight, whereas the second theory assumes the existence of an antipode. The theory of locally compact quantum groups involves fewer axioms; here, the antipode, like the counit, is constructed out of the Haar weight. Both theories include a nice generalization of Pontrjagin duality and make substantial use of multiplicative unitaries. We give a survey on locally compact quantum groups in Chapter 8.

**Hopf $C^*$-algebras.** Vaes and Van Daele propose a definition of a Hopf $C^*$-algebra [164] which is based on the characterization of Hopf algebras given in Theorem 1.3.18. Out of the comultiplication, Vaes and Van Daele construct a counit and an antipode on a subset of the Hopf $C^*$-algebra which is dense in all known examples.

# Chapter 5
# $C^*$-algebraic compact quantum groups

$C^*$-algebraic compact quantum groups are particularly well understood:

1. They can be defined in terms of a few simple axioms.
2. From these axioms one can deduce the existence of a Haar state which is the analogue of the Haar measure of a group and of the Haar state of an algebraic compact quantum group (Section 5.1).
3. Using the Haar state, one can show that their corepresentation theory is essentially the same as the representation theory of compact groups and the corepresentation theory of algebraic compact quantum groups (Sections 5.2 and 5.3).
4. Like algebraic compact quantum groups, they can be characterized in terms of their corepresentations (Section 5.3).
5. Using the corepresentation theory, one can identify a unique dense algebraic compact quantum group inside every $C^*$-algebraic compact quantum group. This algebraic compact quantum group answers most questions concerning the counit, the antipode, and the modular properties of the Haar state (Section 5.4, but see also Example 8.1.22 and 8.3.7).
6. Every algebraic compact quantum group can be completed so that one obtains a $C^*$-algebraic compact quantum group. In general, several completions exist, but one can always identify a minimal and a maximal one (Section 5.4).

The original references for $C^*$-algebraic compact quantum groups are the articles [193], [202] by Woronowicz; detailed accounts can also be found in [80], [105].

Throughout this section, we use the notation and definitions introduced in Section 4.1 and the background on $C^*$-algebras summarized in Section 12.1.

## 5.1 Definition and examples

The definition of a $C^*$-algebraic compact quantum group is very brief:

**Definition 5.1.1.** A *$C^*$-algebraic compact quantum group* is a unital bisimplifiable $C^*$-bialgebra.

A *morphism* of $C^*$-algebraic compact quantum groups is just a morphism of the underlying $C^*$-bialgebras (see Definition 4.1.1).

More explicitly, a $C^*$-algebraic compact quantum group is a unital $C^*$-algebra $A$ equipped with a unital $*$-homomorphism $\Delta\colon A \to A \otimes A$ such that

i) $\Delta$ is coassociative in the sense that $(\Delta \otimes \mathrm{id}_A) \circ \Delta = (\mathrm{id}_A \otimes \Delta) \circ \Delta$, and

ii) each of the sets $\Delta(A)(1_A \otimes A)$ and $\Delta(A)(A \otimes 1_A)$ is linearly dense in $A \otimes A$.

If $(A, \Delta)$ is a $C^*$-algebraic compact quantum group and the comultiplication $\Delta$ is understood, we freely speak of $A$ itself as a $C^*$-algebraic compact quantum group.

**Examples 5.1.2.** i) For every compact group $G$, the $C^*$-bialgebra $C(G)$ introduced in Example 4.2.1 is a $C^*$-algebraic compact quantum group.

ii) For every discrete group $G$, the $C^*$-bialgebras $C^*(G)$ and $C^*_r(G)$ introduced in Example 4.2.2 are $C^*$-algebraic compact quantum groups.

Every commutative $C^*$-algebraic compact quantum group is of the same form as in Example 5.1.2 i):

**Proposition 5.1.3.** *Let $(A, \Delta_A)$ be a $C^*$-algebraic compact quantum group. Then there exists a compact group $G$ and an isomorphism $\Phi\colon A \xrightarrow{\cong} C(G)$ of $C^*$-bialgebras.*

*Proof.* By the Gelfand theorem, there exist a compact space $G$, a continuous map $m\colon G \times G \to G$, and an isomorphism $\Phi\colon A \xrightarrow{\cong} C(G)$ such that the map $\Delta_G := (\Phi \otimes \Phi)\Delta_A\Phi^{-1}$ is given by $(\Delta_G(f))(x, y) = f(m(x, y))$ (here, we identify $C(G) \otimes C(G)$ with $C(G \times G)$).

Since $\Delta_A$ is coassociative, the map $m$ is associative and turns $G$ into a commutative semigroup. The map $G \times G \to G \times G$ given by $(x, y) \mapsto (m(x, y), y)$ is injective because $\Delta_G(C(G))(1 \otimes C(G))$ is linearly dense in $C(G \times G)$ and

$$(f \otimes g)(m(x, y), y) = \Delta_G(f)(x, y)g(y) = (\Delta_G(f)(1 \otimes g))(x, y)$$

for all $x, y \in G$ and $f, g \in C(G)$. Likewise, the map $G \times G \to G \times G$ given by $(x, y) \mapsto (x, m(x, y))$ is injective. Therefore, $G$ is a compact semigroup with the cancellation property. A classical result says that every such semigroup is a group, see [63, Theorem II.9.16] or [105, Proposition 3.2]. □

For a $C^*$-algebraic compact quantum group, the analogue of the Haar measure of a compact group is an invariant state. The precise definition is as follows; for motivation, see Section 2.2.

**Definition 5.1.4.** Let $(A, \Delta)$ be a $C^*$-algebraic compact quantum group. A state $h$ on $A$ is *left-invariant* if $(\mathrm{id}_A \otimes h)(\Delta(a)) = 1_A h(a)$ for all $a \in A$, and *right-invariant* if $(h \otimes \mathrm{id}_A)(\Delta(a)) = 1_A h(a)$ for all $a \in A$.

**Remark 5.1.5.** Let $(A, \Delta)$ be a $C^*$-algebraic compact quantum group. Then a state $h$ on $A$ is left-invariant / right-invariant if and only if $g*h = g(1_A)h$ / $h*g = hg(1_A)$ for each $g \in A'$. This follows from the relations $g \circ (\mathrm{id}_A \otimes h) \circ \Delta = g * h$ and $g \circ (h \otimes \mathrm{id}_A) \circ \Delta = h * g$.

The following result is crucial for everything that follows:

**Theorem 5.1.6.** *Every $C^*$-algebraic compact quantum group has a left-invariant and right-invariant state, and every state that is left- or right-invariant coincides with this state.*

This result was first proved by Woronowicz [193, Theorem 4.2], [202, Theorem 1.3] under an additional separability assumption which was later removed by Van Daele [175]. Note that in the purely algebraic setting, an analogous statement does not hold.

The proof depends on two lemmas. Let $(A, \Delta)$ be a $C^*$-algebraic compact quantum group.

**Lemma 5.1.7.** *For every state $\rho$ on $A$, there exists a state $h$ on $A$ such that $h * \rho = \rho * h = h$.*

*Proof.* Let $\rho$ be a state on $A$. For each $k \in \mathbb{N}$, denote by $\rho^{*k}$ the $k$-fold convolution product of $\rho$. Since the unit ball of $A'$ is weak-$*$-compact, the sequence $(h_n)_n$ given by $h_n := (\rho + \rho^{*2} + \dots + \rho^{*(n-1)} + \rho^{*n})/n$ has a weak-$*$-accumulation point, $h$, say. It is easy to see that for all $n \in \mathbb{N}$,

$$\rho * h_n = h_n * \rho = h_n + \frac{1}{n}(\rho^{*(n+1)} - \rho).$$

Consequently, $\rho * h = h * \rho = h$. □

**Lemma 5.1.8.** *Let $h$ and $\rho$ be states on $A$ such that $h * \rho = \rho * h = h$. If $\omega \in A'$ satisfies $0 \le \omega \le \rho$, then $h * \omega = \omega * h = \omega(1_A)h$.*

*Proof.* Assume that $\omega \in A'$ satisfies $0 \le \omega \le \rho$. We show that $\omega * h = \omega(1_A)h$, and a similar argument shows that $h * \omega = \omega(1_A)h$.

Denote by $L_{h\otimes\rho} := \{q \in A \otimes A \mid (h \otimes \rho)(q^*q) = 0\}$ the left ideal related to the state $h \otimes \rho$, and define $L_{h\otimes\omega}$ similarly. Then $L_{h\otimes\rho} \subseteq L_{h\otimes\omega}$ because $\omega \le \rho$, and $L_{h\otimes\omega} \subseteq \ker(h \otimes \omega)$ by the Cauchy–Schwarz inequality. Define $\Psi_L \colon A \to A$ by

$$\Psi_L(a) := h * a - 1_A h(a) \quad \text{for all } a \in A.$$

We shall show that $(\mathrm{id} \otimes \Psi_L)(\Delta(A)) \subseteq L_{h\otimes\rho}$. Combining the relations $1_A \otimes A \subseteq [(A \otimes 1_A)\Delta(A)]$ and $(A \otimes 1_A)L_{h\otimes\rho} \subseteq L_{h\otimes\rho} \subseteq \ker(h \otimes \omega)$, we then find

$$\begin{aligned} 1_A \otimes \Psi_L(A) &\subseteq [(\mathrm{id} \otimes \Psi_L)((A \otimes 1_A)\Delta(A))] \\ &= [(A \otimes 1_A)(\mathrm{id} \otimes \Psi_L)(\Delta(A))] \subseteq [(A \otimes 1_A)L_{h\otimes\rho}] \subseteq \ker(h \otimes \omega), \end{aligned}$$

and this relation implies $\omega * h = \omega(1_A)h$ because for all $a \in A$,

$$0 = (h \otimes \omega)(1_A \otimes \Psi_L(a)) = \omega(\Psi_L(a))$$
$$= \omega(h * a) - \omega(1_A)h(a) = (\omega * h)(a) - \omega(1_A)h(a).$$

So, let us prove that $(\mathrm{id} \otimes \Psi_L)(\Delta(A)) \subseteq L_{h\otimes\rho}$. Given $c \in A$, put $d := h * c$. Then

$$\begin{aligned} q &:= (\mathrm{id} \otimes \Psi_L)(\Delta(c)) \\ &= (\mathrm{id} \otimes \mathrm{id} \otimes h)(\Delta^{(2)}(c)) - (\mathrm{id} \otimes h)(\Delta(c)) \otimes 1_A = \Delta(d) - d \otimes 1_A \end{aligned}$$

and $(h \otimes \rho)(q^*q) = X - Y - Y^* + Z$, where

$$\begin{aligned} X &= (h \otimes \rho)(\Delta(d^*d)) = (h * \rho)(d^*d), \\ Y &= (h \otimes \rho)((d^* \otimes 1_A)\Delta(d)) = h(d^*(\rho * d)) = h(d^*(\rho * h * c)) = h(d^*d), \\ Z &= (h \otimes \rho)(d^*d \otimes 1_A) = h(d^*d). \end{aligned}$$

The assumption implies $X = Z = Y = Y^*$, so $(h \otimes \rho)(q^*q) = 0$ and $q \in L_{h\otimes\rho}$. The proof is finished. □

*Proof of Theorem* 5.1.6. For each positive functional $\omega$ on $A$, define $K_\omega$ to be the set of states $h$ on $A$ that satisfy $\omega * h = h * \omega = \omega(1_A)h$. Then $K_\omega \subset A'$ is compact with respect to the weak-$*$-topology, and non-empty by Lemma 5.1.7.

By Lemma 5.1.8, $K_{\omega_1+\omega_2} \subseteq K_{\omega_1} \cap K_{\omega_2}$ for all positive functionals $\omega_1, \omega_2$. By compactness, the intersection of all $K_\omega$ is non-empty and contains a left- and right-invariant state $h$.

Finally, if $h'$ is another left- or right-invariant state, then $h' = h * h' = h$ or $h' = h' * h = h$, respectively (see Remark 5.1.5). □

**Definition 5.1.9.** The unique left- and right-invariant state of a $C^*$-algebraic compact quantum group is called its *Haar state* and denoted by $h$. A $C^*$-algebraic compact quantum group is *reduced* if its Haar state is faithful.

For $C^*$-algebraic compact quantum groups associated to groups, the Haar states are easily identified:

**Examples 5.1.10.** i) Let $G$ be a compact group. Then the Haar state of the $C^*$-algebraic compact quantum group $C(G)$ is given by $f \mapsto \int_G f d\lambda$, where $\lambda$ denotes the normalized Haar measure of $G$. Evidently, this Haar state is faithful.

ii) Let $G$ be a discrete group. Then the Haar states of the $C^*$-algebraic compact quantum groups $C^*(G)$ and $C^*_r(G)$ are given by $U_x \mapsto \delta_{x,e}$ and $L_x \mapsto \delta_{x,e}$ for all $x \in G$, respectively, where $e \in G$ denotes the unit, and $(U_x)_x$ and $(L_x)_x$ denote the canonical generators as in Example 4.2.2. Note that these Haar states are tracial in the sense that $h(ab) = h(ba)$ for all $a, b \in C^*_{(r)}(G)$.

Let us show that the Haar state of $C_r^*(G)$ is faithful. In terms of the standard basis $(\varepsilon_x)_{x\in G}$ of $l^2(G)$, it is given by $T \mapsto \langle\varepsilon_e|T\varepsilon_e\rangle$. The map $C_r^*(G) \to l^2(G)$ given by $T \mapsto T\varepsilon_e$ is injective because $T\varepsilon_y = TR_{y^{-1}}\varepsilon_e = R_{y^{-1}}T\varepsilon_e$ for all $y \in G$, where $R_{y^{-1}} \in \mathcal{L}(l^2(G))$ denotes the right shift $\varepsilon_x \mapsto \varepsilon_{xy}$. Therefore, $\langle\varepsilon_e|T^*T\varepsilon_e\rangle = \langle T\varepsilon_e|T\varepsilon_e\rangle \neq 0$ whenever $T \neq 0$.

Since the Haar state of $C^*(G)$ factorizes through the quotient map $L : C^*(G) \to C_r^*(G)$, it is faithful if and only if this quotient map is faithful, that is, an isomorphism. This happens if and only if $G$ is amenable [119, Theorem 4.21], [121, Theorem 7.3.9].

## 5.2 Corepresentations of $C^*$-bialgebras

Like corepresentations of Hopf $*$-algebras, corepresentations of $C^*$-bialgebras can be described in several equivalent ways:

- Unitary corepresentations can elegantly be described in the language of $C^*$-modules. To simplify the presentation, we explain this approach for $C^*$-algebraic compact quantum groups only instead of general $C^*$-bialgebras.
- Non-unitary corepresentations are more conveniently studied in terms of corepresentation operators or corepresentation matrices. We explain the connection to the unitary corepresentations mentioned above but consider general $C^*$-bialgebras.

As in the setting of Hopf $*$-algebras, one can associate to every corepresentation of a $C^*$-bialgebra a representation of the dual algebra, construct new corepresentations out of given ones, and associate to every $C^*$-algebraic compact quantum group a particular regular corepresentation. We present the pertaining definitions and constructions, frequently referring to Section 3.1 for motivation.

We shall use the notation and definitions introduced in Section 4.1, the slice maps discussed in Section 12.4, and the background on $C^*$-modules summarized in Section 12.2. As before, we use the symbol "$\odot$" to denote algebraic tensor products and the symbol "$\otimes$" to denote minimal tensor products of $C^*$-algebras and completed internal or external tensor products of $C^*$-modules.

### 5.2.1 Unitary corepresentations of $C^*$-algebraic compact quantum groups

Recall that a unitary corepresentation of an algebraic compact quantum group $(A_0, \Delta_0)$ on a finite-dimensional Hilbert space $H_0$ is a linear map $\delta_0 \colon H_0 \to H_0 \odot A_0$ that satisfies

$$(\delta_0 \odot \mathrm{id}) \circ \delta_0 = (\mathrm{id} \odot \Delta_0) \circ \delta_0$$

and

$$\langle \delta_0(\eta) | \delta_0(\xi) \rangle_{A_0} = \langle \eta | \xi \rangle \cdot 1_{A_0} \quad \text{for all } \eta, \xi \in H_0.$$

Here, the $A_0$-valued inner product $\langle \cdot | \cdot \rangle_{A_0}$ on $H_0 \odot A_0$ was given by $\langle \eta \odot b | \xi \odot a \rangle_{A_0} = \langle \eta | \xi \rangle b^* a$ for all $\xi, \eta \in H_0$ and $a, b \in A_0$.

This definition can be adapted to the present setting as follows. Let $(A, \Delta)$ be a $C^*$-algebraic compact quantum group and $H$ a Hilbert space.

- We consider $H$ and $A$ as $C^*$-modules over $\mathbb{C}$ and $A$, respectively, and form the tensor product of $C^*$-modules $H \otimes A$ (see Section 12.2). This is a $C^*$-module over $A$. It is a completion of $H \odot A$; its structure maps are given by

  $$\langle \eta \otimes b | \xi \otimes a \rangle = \langle \eta | \xi \rangle b^* a, \quad (\eta \otimes b) a = \eta \otimes ba$$

  for all $\eta, \xi \in H$, $a, b \in A$. Similarly, we define a $C^*$-module $H \otimes A \otimes A$ over $A \otimes A$.

- By definition, the map $\mathrm{id}_H \odot \Delta \colon H \odot A \to H \odot (A \otimes A)$ satisfies

  $$\langle (\mathrm{id}_H \odot \Delta)(y) | (\mathrm{id}_H \odot \Delta)(x) \rangle = \Delta(\langle y | x \rangle) \quad \text{for all } x, y \in H \odot A$$

  and extends to an isometric linear map $\mathrm{id}_H \otimes \Delta \colon H \otimes A \to H \otimes (A \otimes A)$.

- If $\delta \colon H \to H \otimes A$ is a linear map that satisfies $\langle \delta(\eta) | \delta(\xi) \rangle = \langle \eta | \xi \rangle \cdot 1_A$ for all $\eta, \xi \in H$, then the map $\delta \odot \mathrm{id}_A \colon H \odot A \to (H \otimes A) \odot A$ satisfies

  $$\langle (\delta \odot \mathrm{id}_A)(y) | (\delta \odot \mathrm{id}_A)(x) \rangle = 1_A \otimes \langle y | x \rangle \quad \text{for all } x, y \in H \odot A$$

  and extends to an isometric linear map $\delta \otimes \mathrm{id}_A \colon H \otimes A \to H \otimes A \otimes A$.

**Definition 5.2.1.** A *unitary corepresentation* of a $C^*$-algebraic compact quantum group $(A, \Delta)$ on a Hilbert space $H$ is a linear map $\delta \colon H \to H \otimes A$ that satisfies the following conditions:

i) $\langle \delta(\eta) | \delta(\xi) \rangle = \langle \eta | \xi \rangle \cdot 1_A$ for all $\eta, \xi \in H$;

ii) the set $\delta(H)A$ is linearly dense in $H \otimes A$;

iii) $(\mathrm{id}_H \otimes \Delta) \circ \delta = (\delta \otimes \mathrm{id}_A) \circ \delta$, that is, the following diagram commutes:

$$\begin{array}{ccc} H & \xrightarrow{\ \delta\ } & H \otimes A \\ {\scriptstyle \delta}\downarrow & & \downarrow{\scriptstyle \delta \otimes \mathrm{id}} \\ H \otimes A & \xrightarrow[\mathrm{id} \otimes \Delta]{} & H \otimes A \otimes A. \end{array}$$

One can extend this definition to the case of a non-unital $C^*$-bialgebra $(A, \Delta)$ by replacing the $C^*$-module $H \otimes A$ by the multiplier $C^*$-module $M(H \otimes A) = \mathcal{L}(A, H \otimes A)$ and considering linear maps $\delta\colon H \to M(H \otimes A)$ that satisfy obvious analogues of the conditions i)–iii) above. Moreover, one can consider intertwiners, invariant subspaces, and irreducibility for unitary corepresentations, very much like in the setting of Hopf $*$-algebras. However, it turns out to be more convenient to work with corepresentation operators instead of unitary corepresentations.

### 5.2.2 Corepresentation operators of $C^*$-bialgebras

Before we can define corepresentation operators of a general $C^*$-bialgebra, we need to collect some preliminaries and fix some notation. Let $(A, \Delta)$ be a $C^*$-bialgebra and $H$ a Hilbert space.

- The $C^*$-algebra $\mathcal{K}(H) \otimes A$ embeds naturally in $\mathcal{L}_A(H \otimes A)$, and the image of this embedding is $\mathcal{K}_A(H \otimes A)$ [69, Lemma 1.2.7]. Thus

$$\mathcal{L}_A(H \otimes A) \cong M(\mathcal{K}_A(H \otimes A)) \cong M(\mathcal{K}(H) \otimes A). \tag{5.1}$$

  From now on, we shall use these identifications without further notice.

- Since $\Delta$ is non-degenerate, the $*$-homomorphism

$$\operatorname{id} \otimes \Delta\colon \mathcal{K}(H) \otimes A \to \mathcal{K}(H) \otimes M(A \otimes A) \hookrightarrow M(\mathcal{K}(H) \otimes A \otimes A) \tag{5.2}$$

  is non-degenerate and extends to the multiplier algebra (5.1).

- If $(A, \Delta)$ is a $C^*$-algebraic compact quantum group, then the $*$-homomorphism (5.2) can be expressed in terms of the map $\operatorname{id}_H \otimes \Delta\colon H \otimes A \to H \otimes A \otimes A$: with respect to the natural isomorphisms $M(\mathcal{K}(H) \otimes A) \cong \mathcal{L}_A(H \otimes A)$ and $M(\mathcal{K}(H) \otimes A \otimes A) \cong \mathcal{L}_{A \otimes A}(H \otimes A \otimes A)$,

$$((\operatorname{id} \otimes \Delta)(X))(\eta \otimes 1_{(A \otimes A)}) = (\operatorname{id}_H \otimes \Delta)(X(\eta \otimes 1_A)) \tag{5.3}$$

  for all $\eta \in H$ and $X \in \mathcal{L}_A(H \otimes A)$.

- Given $X \in M(\mathcal{K}(H) \otimes A)$, we define $X_{[12]}, X_{[13]} \in M(\mathcal{K}(H) \otimes A \otimes A)$ by

$$X_{[12]} := X \otimes 1 \quad \text{and} \quad X_{[13]} := (\operatorname{id} \otimes \Sigma)(X_{[12]}),$$

  where $\Sigma\colon A \otimes A \to A \otimes A$ denotes the flip as usual. This is a particular example of the leg notation that will be used extensively in Chapter 7.

**Proposition 5.2.2.** *Let $(A, \Delta)$ be a $C^*$-algebraic compact quantum group and let $H$ be a Hilbert space.*

i) *Let $\delta$ be a unitary corepresentation of $(A, \Delta)$ on $H$. Then the map*

$$X: H \odot A \to H \otimes A, \quad \eta \odot a \mapsto \delta(\eta)a,$$

*extends to a unitary operator $X \in \mathcal{L}_A(H \otimes A)$, and $X_{[12]}X_{[13]} = (\mathrm{id} \otimes \Delta)(X)$.*

ii) *If a unitary $X \in \mathcal{L}_A(H \otimes A)$ satisfies $X_{[12]}X_{[13]} = (\mathrm{id} \otimes \Delta)(X)$, the map*

$$\delta: H \to H \otimes A, \quad \eta \mapsto X(\eta \otimes 1_A),$$

*is a unitary corepresentation.*

*Proof.* i) For all elementary tensors $y = \eta \otimes b$ and $x = \xi \otimes a$ in $H \odot A$,

$$\langle X(y)|X(x)\rangle = \langle \delta(\eta)b|\delta(\xi)a\rangle = b^*\langle \delta(\eta)|\delta(\xi)\rangle a = b^*\langle \eta|\xi\rangle 1_A a = \langle y|x\rangle.$$

Thus $X$ extends to an isometric linear map $H \otimes A \to H \otimes A$. Since $X(H \odot A) = \operatorname{span} \delta(H)A$ is dense in $H \otimes A$, the extension is unitary. Finally, the relation $(\delta \otimes \mathrm{id}_A) \circ \delta = (\mathrm{id}_H \otimes \Delta) \circ \delta$ implies

$$\begin{aligned} X_{[12]}X_{[13]}(\eta \otimes 1_{(A \otimes A)}) &= (\delta \otimes \mathrm{id}_A)(\delta(\eta)) \\ &= (\mathrm{id}_H \otimes \Delta)(\delta(\eta)) \\ &= (\mathrm{id}_H \otimes \Delta)(X(\eta \otimes 1_A)) \\ &= ((\mathrm{id} \otimes \Delta)(X))(\eta \otimes 1_{(A \otimes A)}) \end{aligned} \tag{5.4}$$

for all $\eta \in H$, whence $X_{[12]}X_{[13]} = (\mathrm{id} \otimes \Delta)(X)$.

ii) Since $X$ is unitary,

$$\langle \delta(\eta)|\delta(\xi)\rangle = \langle X(\eta \otimes 1_A)|X(\xi \otimes 1_A)\rangle = \langle \eta \otimes 1_A|\xi \otimes 1_A\rangle = \langle \eta|\xi\rangle \cdot 1_A$$

for all $\eta, \xi \in H$, and

$$[\delta(H)A] = [(X(H \otimes 1_A))A] = X(H \otimes A) = H \otimes A.$$

Thus the map $\delta$ satisfies conditions i) and ii) of Definition 5.2.1. A similar calculation like (5.4) shows that $\delta$ also satisfies condition iii). □

Propositions 5.2.2 and 3.1.7 motivate the following definition:

**Definition 5.2.3.** Let $(A, \Delta)$ be a $C^*$-bialgebra. A *(unitary) corepresentation operator* of $(A, \Delta)$ on a Hilbert space $H$ is an invertible (unitary) operator $X \in M(\mathcal{K}(H) \otimes A) \cong \mathcal{L}_A(H \otimes A)$ that satisfies $X_{[12]}X_{[13]} = (\mathrm{id} \otimes \Delta)(X)$. If $H$ has finite dimension, we call $X$ *finite-dimensional*.

Let $X$ and $Y$ be corepresentation operators on Hilbert spaces $H$ and $K$, respectively.

An *intertwiner* from $X$ to $Y$ is an operator $T \in \mathcal{L}(H, K)$ that satisfies the equation $Y(T \otimes \mathrm{id}_A) = (T \otimes \mathrm{id}_A)X$. We denote the space of all intertwiners from $X$ to $Y$ by $\mathrm{Hom}(X, Y)$. We call $X$ and $Y$ *equivalent*, written $X \simeq Y$, if $X$ and $Y$ admit an invertible intertwiner.

A subspace $L \subseteq H$ is *invariant* for $X$ if the orthogonal projection $p_L : H \to L \subseteq H$ satisfies $X(p_L \otimes 1_A) = (p_L \otimes 1_A)X(p_L \otimes 1_A)$. The corepresentation operator $X$ is *irreducible* if there exists no invariant subspace besides 0 and $H$.

The *space of matrix elements* of $X$ is the space

$$\mathcal{C}(X) := \overline{\mathrm{span}}\{(\omega_{\eta,\xi} \otimes \mathrm{id})(X) \mid \eta, \xi \in H\} \subseteq M(A),$$

where $\omega_{\eta,\xi} : \mathcal{K}(H) \to \mathbb{C}$ is given by $T \mapsto \langle \eta | T\xi \rangle$ and the slice map $\omega_{\eta,\xi} \otimes \mathrm{id}$ is defined as in Section 12.4.

**Remarks 5.2.4.** i) For every $C^*$-algebraic compact quantum group $(A, \Delta)$ and every Hilbert space $H$, there exists a bijective correspondence between unitary corepresentations and unitary corepresentation operators of $(A, \Delta)$ on $H$, see Proposition 5.2.2.

ii) The equation $Y(T \otimes 1_A) = (T \otimes 1_A)X$ figuring in the definition of an intertwiner should be considered as an equation in $\mathcal{L}_A(H \otimes A, K \otimes A)$ or in $M(\mathcal{K}(H \oplus K) \otimes A)$.

iii) If $H$ is a finite-dimensional Hilbert space and $A$ is unital, then we have $M(\mathcal{K}(H) \otimes A) = \mathcal{K}(H) \otimes A = \mathcal{L}(H) \otimes A$. Generally, the space $\mathcal{L}(H) \otimes M(A) \cong M(\mathcal{K}(H)) \otimes M(A)$ embeds in $M(\mathcal{K}(H) \otimes A)$ as a strict subspace. For an illustration of the difference between corepresentation operators in $M(\mathcal{K}(H)) \otimes M(A)$ and corepresentation operators in $M(\mathcal{K}(H) \otimes A)$, see the end of Example 5.2.5.

Corepresentations are related to group representations as follows:

**Example 5.2.5.** Let $G$ be a locally compact group. Then corepresentation operators of the $C^*$-bialgebra $C_0(G)$ (see Example 4.2.1) on a Hilbert space $H$ correspond bijectively with strictly continuous representations of $G$ on $H$:

The $C^*$-algebra $M(\mathcal{K}(H) \otimes C_0(G))$ can be identified with the $C^*$-algebra of norm-bounded and strictly continuous functions $G \to M(\mathcal{K}(H)) \cong \mathcal{L}(H)$ [2, Corollary 3.4]. Explicitly, each multiplier $X \in M(\mathcal{K}(H) \otimes C_0(G))$ corresponds to the function

$$\pi : G \to M(\mathcal{K}(H)) \cong \mathcal{L}(H), \quad x \mapsto (\mathrm{id} \otimes \mathrm{ev}_x)(X),$$

where $\mathrm{ev}_x : C_0(G) \to \mathbb{C}$ denotes evaluation at $x \in G$ as usual. The relations

$$\pi(x)\pi(y) = (\mathrm{id} \otimes \mathrm{ev}_x)(X) \cdot (\mathrm{id} \otimes \mathrm{ev}_y)(X) = (\mathrm{id} \otimes \mathrm{ev}_x \otimes \mathrm{ev}_y)(X_{[12]}X_{[13]}),$$
$$\pi(xy) = (\mathrm{id} \otimes \mathrm{ev}_{xy})(X) = (\mathrm{id} \otimes \mathrm{ev}_x \otimes \mathrm{ev}_y)((\mathrm{id} \otimes \Delta)(X)), \quad x, y \in G,$$

show that $X$ is a corepresentation operator if and only if $\pi$ is a representation. Evidently, $X$ is unitary if and only if $\pi$ is unitary. Furthermore, a subspace $K \subseteq H$ is invariant for $X$ if and only if it is invariant for $\pi$; in particular, $X$ is irreducible if and only if $\pi$ is irreducible.

Note that every strictly continuous representation is strongly continuous. Conversely, it is easy to see that every strongly continuous unitary representation is strictly continuous.

Finally, let $G$ be compact and let $X \in M(\mathcal{K}(H) \otimes C(G))$ be a corepresentation operator. Then $X$ is contained in $\mathcal{L}(H) \otimes C(G)$ if and only if the associated representation $\pi$ is norm-continuous, because $\mathcal{L}(H) \otimes C(G)$ corresponds to the $C^*$-algebra of norm-continuous functions $G \to \mathcal{L}(H)$.

### 5.2.3 Constructions related to corepresentation operators

In this section, we introduce several constructions related to corepresentation operators that are analogues of the purely algebraic constructions for corepresentations of Hopf $*$-algebras discussed in Section 3.1.2 and 3.1.3.

Throughout this section, let $(A, \Delta)$ be a $C^*$-bialgebra.

**Corepresentation matrices.** Let $X$ be a corepresentation operator of $(A, \Delta)$ on an $n$-dimensional Hilbert space $H$ with orthonormal basis $(e_i)_{i=1}^n$, where $n \in \mathbb{N}$. Note that then $X \in M(\mathcal{K}(H) \otimes A) = \mathcal{K}(H) \otimes M(A)$. The choice of the basis $(e_i)_i$ defines a natural identification $H \cong \mathbb{C}^n$ and an isomorphism

$$\mathcal{K}(H) \otimes M(A) \cong \mathcal{K}(\mathbb{C}^n) \otimes M(A) = M_n(\mathbb{C}) \otimes M(A) \cong M_n(M(A)). \tag{5.5}$$

With respect to this isomorphism, $X$ corresponds to the matrix

$$a = (a_{ij})_{i,j} \in M_n(M(A)), \quad \text{where } a_{ij} = (\omega_{e_i,e_j} \otimes \mathrm{id})(X) \text{ for all } i, j.$$

As in the setting of Hopf $*$-algebras (Section 3.1), we find

$$\Delta(a_{ij}) = \sum_k a_{ik} \otimes a_{kj} \quad \text{for all } i, j, \tag{5.6}$$

$$\mathcal{C}(X) = \overline{\operatorname{span}}\,\{a_{ij} \mid i, j\}, \tag{5.7}$$

and $X$ is unitary if and only if the matrix $a$ is unitary, that is, if

$$\sum_k a_{ki}{}^* a_{kj} = \delta_{i,j} \cdot 1_A = \sum_k a_{ik} a_{jk}{}^* \quad \text{for all } i, j. \tag{5.8}$$

**Definition 5.2.6.** A *corepresentation matrix* of $(A, \Delta)$ is an invertible matrix $a \in M_n(M(A))$, where $n \in \mathbb{N}$, that satisfies equation (5.6). The matrix $a$ is *irreducible* if the corepresentation operator $\sum_{i,j} e_{ij} \otimes a_{ij}$ on $\mathbb{C}^n$ is irreducible, where $(e_{ij})_{i,j}$ denotes the standard matrix units.

**Representation of the dual algebra.** To every corepresentation operator of $(A, \Delta)$, one can associate a representation of the dual algebra $A'$. The definition of this representation involves slice maps, which are discussed in Section 12.4.

**Proposition 5.2.7.** i) *Let $X$ be a corepresentation operator of $(A, \Delta)$ on a Hilbert space $H$. Then the map*

$$\pi_X : A' \to M(\mathcal{K}(H)) \cong \mathcal{L}(H), \quad f \mapsto (\mathrm{id} \otimes f)(X), \tag{5.9}$$

*is a homomorphism, that is, $\pi_X(f * g) = \pi_X(f)\pi_X(g)$ for all $f, g \in A'$.*

ii) *Let $X$, $H$ and $\pi_X$ be as above and let $K$ be a subspace of $H$. Then $K$ is invariant for $X$ if and only if it is invariant for $\pi_X$.*

iii) *Let $X$ and $Y$ be corepresentation operators of $(A, \Delta)$ on Hilbert spaces $H$ and $K$, respectively, and let $T \in \mathcal{L}(H, K)$. Then $T \in \mathrm{Hom}(X, Y)$ if and only if $T\pi_X(f) = \pi_Y(f)T$ for all $f \in A'$.*

*Proof.* i) The proof consists of a calculation with slice maps; comments on such calculations can be found in Section 12.4. For all $f, g \in A'$,

$$\begin{aligned}(\mathrm{id} \otimes f)(X) \cdot (\mathrm{id} \otimes g)(X) &= (\mathrm{id} \otimes f \otimes g)(X_{[12]}X_{[13]}) \\ &= (\mathrm{id} \otimes f \otimes g)((\mathrm{id} \otimes \Delta)(X)) = (\mathrm{id} \otimes fg)(X).\end{aligned}$$

ii), iii) The proofs are similar to the proofs of the equivalences (f4)⇔(f5) and (d2)⇔(d4) in Proposition 3.1.7. □

In general, the algebra $A'$ is not equipped with a natural involution: the formula $f^*(a) := \overline{f(S(a)^*)}$ known from the setting of Hopf $*$-algebras makes no sense if $(A, \Delta)$ has no antipode $S$. Hence we can not ask whether the representation $\pi_X$ associated to a corepresentation operator $X$ is a $*$-homomorphism. But if $(A, \Delta)$ is a $C^*$-algebraic compact quantum group and $X$ is unitary, then $\pi(A')$ contains a useful non-degenerate $C^*$-subalgebra:

**Proposition 5.2.8.** *Let $(A, \Delta)$ be a $C^*$-algebraic compact quantum group with Haar state $h$, and let $X$ be a corepresentation operator of $(A, \Delta)$ on a Hilbert space $H$. Put $A'_h := \{h(\cdot a) \mid a \in A\} \subseteq A'$.*

i) *The space $C_X := [\pi_X(A'_h)] \subseteq \mathcal{L}(H)$ is a non-degenerate $C^*$-algebra and $X \in M(C_X \otimes A)$.*

ii) *A subspace $K \subseteq H$ is invariant for $X$ if and only if it is invariant for $C_X$.*

iii) *If $p \in \mathcal{L}(H)$ is a projection and $X(p \otimes 1_A) = (p \otimes 1_A)X(p \otimes 1_A)$, then $(p \otimes 1_A)X = X(p \otimes 1_A)$.*

*Proof.* i) Writing out the definition, we find

$$C_X = \overline{\operatorname{span}}\{(\mathrm{id} \otimes h)(X(\mathrm{id}_H \otimes a)) \mid a \in A\}.$$

First, we show that $[C_X^* C_X] = C_X$; this implies that $C_X$ is a $C^*$-algebra. By left-invariance of $h$,

$$C_X \otimes 1_A = [(\mathrm{id} \otimes \mathrm{id} \otimes h)((\mathrm{id} \otimes \Delta)(X(\mathrm{id}_H \otimes A)))],$$

and hence

$$\begin{aligned}[C_X^* C_X] &= [(\mathrm{id} \otimes h)((\mathrm{id}_H \otimes A^*)X^*(C_X \otimes 1_A))] \\ &= [(\mathrm{id} \otimes h \otimes h)((\mathrm{id}_H \otimes A^* \otimes 1_A)\,(X^* \otimes 1_A)\,((\mathrm{id} \otimes \Delta)(X))\,(\mathrm{id}_H \otimes \Delta(A)))].\end{aligned}$$

We insert the relation $(X^* \otimes 1_A)((\mathrm{id} \otimes \Delta)(X)) = X_{[12]}^* X_{[12]} X_{[13]} = X_{[13]}$, move $\mathrm{id}_H \otimes A^* \otimes 1_A$ to the right of $X_{[13]}$, and find

$$[C_X^* C_X] = [(\mathrm{id} \otimes h \otimes h)(X_{[13]}\,(\mathrm{id}_H \otimes A^* \otimes 1_A)\,(\mathrm{id}_H \otimes \Delta(A)))].$$

Since $[(A^* \otimes 1_A)\Delta(A)] = A \otimes A$,

$$[C_X^* C_X] = [(\mathrm{id} \otimes h \otimes h)(X_{[13]}(\mathrm{id}_H \otimes A \otimes A))] = [(\mathrm{id} \otimes h)(X(\mathrm{id}_H \otimes A))] = C_X.$$

Put $\mathcal{K} := \mathcal{K}(H)$. Since $\mathcal{K} \otimes A \subseteq M(\mathcal{K} \otimes A)$ is an ideal and $X \in M(\mathcal{K} \otimes A)$ is unitary, $X(\mathcal{K} \otimes A) = \mathcal{K} \otimes A$ and

$$[C_X \mathcal{K}] = [(\mathrm{id} \otimes h)(X(\mathcal{K} \otimes A))] = [(\mathrm{id} \otimes h)(\mathcal{K} \otimes A)] = \mathcal{K}.$$

Consequently, $C_X$ is non-degenerate.

Let us prove that $X$ belongs to $M(C_X \otimes A)$. By definition of $C_X$,

$$[X(C_X \otimes A)] = [(\mathrm{id} \otimes \mathrm{id} \otimes h)(X_{[12]} X_{[13]}(\mathrm{id}_H \otimes A \otimes A))].$$

We replace $X_{[12]} X_{[13]}$ by $(\mathrm{id} \otimes \Delta)(X)$ and $A \otimes A$ by $[\Delta(A)(A \otimes 1_A)]$, and find

$$[X(C_X \otimes A)] = [(\mathrm{id} \otimes \mathrm{id} \otimes h)((\mathrm{id} \otimes \Delta)(X(\mathrm{id}_H \otimes A))) \cdot (\mathrm{id}_H \otimes A)].$$

Since $h$ is left-invariant, this is equal to

$$[((\mathrm{id} \otimes h)(X(\mathrm{id}_H \otimes A)) \otimes 1_A) \cdot (\mathrm{id}_H \otimes A)] = C_X \otimes A.$$

Similarly, one shows that $(C_X \otimes A)X \subseteq C_X \otimes A$. Thus $X \in M(C_X \otimes A)$.

ii), iii) Let $K \subseteq H$ be some subspace, and denote by $p$ the orthogonal projection onto $K$. If $(p \otimes 1_A)X(p \otimes 1_A) = X(p \otimes 1_A)$, then $pcp = cp$ for all $c \in C_X$, and since $C_X$ is self-adjoint, also $pc = cp$ for all $c \in C_X$. Conversely, if $cp = pc$ for all $c \in C_X$, then also $X(p \otimes 1_A) = (p \otimes 1_A)X$ because $X \in M(C_X \otimes A)$. □

**The regular corepresentation.** To every $C^*$-algebraic compact quantum group, one can associate a regular corepresentation. This corepresentation plays an important rôle in the characterization of compact quantum groups.

To construct this corepresentation, we need to fix some notation and collect some preliminaries. Let $(A, \Delta)$ be a $C^*$-algebraic compact quantum group with Haar state $h$ and associated GNS-representation $(H_h, \Lambda_h, \pi_h)$ (see Section 12.1). Then the map $\Lambda_h \odot \mathrm{id}_A \colon A \odot A \to H_h \odot A$ extends to a continuous linear map $\Lambda_h \otimes \mathrm{id}_A \colon A \otimes A \to H_h \otimes A$ such that

$$(\Lambda_h \otimes \mathrm{id}_A)(x) = (\pi_h \otimes \mathrm{id}_A)(x)\,(\Lambda_h(1_A) \otimes 1_A),$$

and

$$\langle(\Lambda_h \otimes \mathrm{id}_A)(x)|(\Lambda_h \otimes \mathrm{id}_A)(y)\rangle = (h \otimes \mathrm{id}_A)(x^* y) \quad \text{for all } x, y \in A \otimes A.$$

**Theorem 5.2.9.** *Let $(A, \Delta)$ be a $C^*$-algebraic compact quantum group with Haar state $h$ and associated GNS-representation $(H_h, \Lambda_h, \pi_h)$.*

i) *The map $\delta_h \colon H_h \to H_h \otimes A$ given by $\Lambda_h(a) \mapsto (\Lambda_h \otimes \mathrm{id}_A)(\Delta(a))$ is a unitary corepresentation.*

*Denote by $X_h \in M(\mathcal{K}(H_h) \otimes A)$ the corepresentation operator corresponding to $\delta_h$ as in Proposition 5.2.2.*

ii) $X_h(\pi_h(a) \otimes 1_A) = ((\pi_h \otimes \mathrm{id}_A)(\Delta(a)))X_h$ *for all $a \in A$.*

iii) $\mathcal{C}(X_h) = A$.

*Proof.* i) The map $\delta_h \colon \Lambda_h(A) \to H_h \otimes A$ given by $\Lambda_h(a) \mapsto (\Lambda_h \otimes \mathrm{id})(\Delta(a))$ extends to $H_h$ and satisfies condition 5.2.1 i) because $h$ is right-invariant:

$$\langle \delta_h(\Lambda_h(b))|\delta_h(\Lambda_h(a))\rangle = (h \otimes \mathrm{id})(\Delta(b^*)\Delta(a)) = h(b^*a)1_A \quad \text{for all } a, b \in A.$$

The map $\delta_h$ satisfies condition 5.2.1 ii) because $\Delta(A)(1 \otimes A)$ is dense in $A \otimes A$:

$$\begin{aligned}[\delta_h(\Lambda_h(A))A] &= [(\Lambda_h \otimes \mathrm{id})(\Delta(A)) \cdot A] \\ &= [(\Lambda_h \otimes \mathrm{id})(\Delta(A)(1 \otimes A))] = [\Lambda_h(A) \otimes A];\end{aligned}$$

and condition 5.2.1 iii) because $\Delta$ is coassociative:

$$\begin{aligned}(\mathrm{id}_{H_h} \otimes \Delta) \circ \delta_h \circ \Lambda_h &= (\Lambda_h \otimes \mathrm{id} \otimes \mathrm{id}) \circ (\mathrm{id} \otimes \Delta) \circ \Delta \\ &= (\Lambda_h \otimes \mathrm{id} \otimes \mathrm{id}) \circ (\Delta \otimes \mathrm{id}) \circ \Delta = (\delta_h \otimes \mathrm{id}) \circ \delta_h \circ \Lambda_h.\end{aligned}$$

ii) By definition of $X_h$, we have for all $a, b \in A$

$$\begin{aligned}X_h(\pi_h(a)\Lambda_h(b) \otimes 1_A) = \delta_h(\Lambda_h(ab)) &= ((\pi_h \otimes \mathrm{id})(\Delta(a)))\delta_h(\Lambda_h(b)) \\ &= ((\pi_h \otimes \mathrm{id})(\Delta(a)))X_h(\Lambda_h(b) \otimes 1_A).\end{aligned}$$

iii) Since $X_h(\Lambda_h(A) \otimes 1_A) = \delta_h(\Lambda_h(A))$ and $A$ is bisimplifiable,

$$\mathcal{C}(X_h) = [(h \otimes \mathrm{id})((A^* \otimes 1)\Delta(A))] = [(h \otimes \mathrm{id})(A \otimes A)] = A. \qquad \square$$

**Definition 5.2.10.** The *regular corepresentation/regular corepresentation operator* of a $C^*$-algebraic compact quantum group $(A, \Delta)$ is the corepresentation $\delta_h$/the corepresentation operator $X_h$ defined in Theorem 5.2.9.

**Construction of new corepresentation operators.** The corepresentations of every $C^*$-bialgebra admit a direct sum and a tensor product which turn the category of all corepresentations into a monoidal/tensor category [79], [104] and, more precisely, into a concrete monoidal/tensor $W^*$-category [195] and strict monoidal/tensor $C^*$-category [102]. Corepresentations of $C^*$-algebraic compact quantum groups additionally admit a conjugation. All these constructions are similar to the corresponding constructions for corepresentations of Hopf $*$-algebras (see Section 3.1.3); therefore we only give a brief summary.

**Direct sum.** Let $(X_\alpha)_\alpha$ be a family of corepresentation operators on Hilbert spaces $(H_\alpha)_\alpha$. Then there exists a unique corepresentation operator

$$\boxplus_\alpha X_\alpha \in M\big(\mathcal{K}\big(\bigoplus_\alpha H_\alpha\big) \otimes A\big),$$

called the *direct sum* of $(X_\alpha)_\alpha$, such that for each $\beta$, the natural inclusion $\iota_\beta : H_\beta \hookrightarrow \bigoplus_\alpha H_\alpha$ is an intertwiner from $X_\beta$ to $\boxplus_\alpha X_\alpha$.

The direct sum construction is functorial: For every second family of corepresentation operators $(Y_\alpha)_\alpha$ on Hilbert spaces $(K_\alpha)_\alpha$, there exists a natural map

$$\prod_\alpha \operatorname{Hom}(X_\alpha, Y_\alpha) \to \operatorname{Hom}\big(\boxplus_\alpha X_\alpha, \boxplus_\alpha Y_\alpha\big), \quad (T_\alpha)_\alpha \mapsto \bigoplus_\alpha T_\alpha.$$

Clearly, $\boxplus_\alpha X_\alpha$ is unitary if each $X_\alpha$ is unitary, and

$$\mathcal{C}\big(\boxplus_\alpha X_\alpha\big) = \overline{\operatorname{span}} \bigcup_\alpha \mathcal{C}(X_\alpha).$$

**Tensor product.** Let $X$ and $Y$ be corepresentation operators on Hilbert spaces $H$ and $K$, respectively. Then the operator

$$X \boxtimes Y := X_{[13]} Y_{[23]} \in M(\mathcal{K}(H) \otimes \mathcal{K}(K) \otimes A) = M(\mathcal{K}(H \otimes K) \otimes A)$$

is a corepresentation operator again, called the *tensor product* of $X$ and $Y$. Here, $Y_{[23]} = \mathrm{id}_H \otimes Y$ and $X_{[13]} = (\Sigma \otimes \mathrm{id}_A)(\mathrm{id}_K \otimes X)$, where $\Sigma \colon \mathcal{K}(K) \otimes \mathcal{K}(H) \to \mathcal{K}(H) \otimes \mathcal{K}(K)$ denotes the flip as usual.

The tensor product construction is functorial: for every second pair of corepresentation operators $\widetilde{X}$ and $\widetilde{Y}$, there exists a map

$$\operatorname{Hom}(X, \widetilde{X}) \times \operatorname{Hom}(Y, \widetilde{Y}) \to \operatorname{Hom}(X \boxtimes Y, \widetilde{X} \boxtimes \widetilde{Y}), \quad (S, T) \mapsto S \otimes T.$$

It is easy to see that $X \boxtimes Y$ is unitary if $X$ and $Y$ are, and that

$$\mathcal{C}(X \boxtimes Y) = \overline{\operatorname{span}}\, \mathcal{C}(X)\mathcal{C}(Y).$$

If $A$ is commutative, then the natural isomorphism $H \otimes K \cong K \otimes H$ intertwines $X \boxtimes Y$ and $Y \boxtimes X$; in particular,

$$X \boxtimes Y \simeq Y \boxtimes X.$$

If $A$ is not commutative, this relation need not hold. However, there may exist a *braiding* for $(A, \Delta)$, that is, a natural equivalence between the bifunctors $(X, Y) \mapsto X \boxtimes Y$ and $(X, Y) \mapsto Y \boxtimes X$ that satisfies some additional coherence properties (see [79]).

**Conjugation.** Let $X$ be a corepresentation operator on a finite-dimensional Hilbert space $H$. Denote by $\bar{H}$ the conjugate Hilbert space of $H$, and by $\zeta \mapsto \bar{\zeta}$ the canonical conjugate-linear isomorphism $H \xrightarrow{\cong} \bar{H}$; thus, $\lambda\bar{\eta} = \overline{\bar{\lambda}\eta}$ and $\langle\bar{\eta}|\bar{\xi}\rangle = \langle\xi|\eta\rangle$ for all $\eta, \xi \in H$ and $\lambda \in \mathbb{C}$. For each $T \in \mathcal{L}(H)$, the map $\bar{\xi} \mapsto \overline{T\xi}$ defines an operator $\bar{T} \in \mathcal{L}(\bar{H})$, and the map $T \mapsto \bar{T}$ is a conjugate-linear $*$-isomorphism $\mathcal{L}(H) \to \mathcal{L}(\bar{H})$.

Denote by $j : \mathcal{L}(H) \otimes M(A) \to \mathcal{L}(\bar{H}) \otimes M(A)$ the map given by $T \otimes S \mapsto \bar{T} \otimes S^*$, and consider the operator $\bar{X} := j(X) \in \mathcal{L}(\bar{H}) \otimes M(A)$. If

$$X = \sum_{i,j} |e_i\rangle\langle e_j| \otimes a_{ij},$$

where $(e_i)_i$ is some orthonormal basis of $H$, then

$$\bar{X} = \sum_{i,j} |\overline{e_i}\rangle\langle\overline{e_j}| \otimes a_{ij}{}^*.$$

Since $\Delta$ is a $*$-homomorphism, $\bar{X}$ satisfies $\bar{X}_{[12]}\bar{X}_{[13]} = (\mathrm{id} \otimes \Delta)(\bar{X})$. It is not immediately clear whether the operator $\bar{X}$ is invertible. We shall prove that if $(A, \Delta)$ is a $C^*$-algebraic compact quantum group, then $\bar{X}$ is invertible and hence a corepresentation operator (see Corollary 5.3.10).

Of course, we can still define $\mathcal{C}(\bar{X})$ as before, and then $\mathcal{C}(\bar{X}) = \mathcal{C}(X)^*$.

## 5.3 Corepresentation theory and structure theory

The corepresentation theory of $C^*$-algebraic compact quantum groups is very similar to the representation theory of compact groups and to the corepresentation theory of algebraic compact quantum groups:

1. Every corepresentation operator is equivalent to a direct sum of irreducible finite-dimensional unitary corepresentation operators (Theorem 5.3.3).
2. Two irreducible corepresentations either do not admit a non-zero intertwiner, or they are equivalent and the space of intertwiners has dimension one (Proposition 5.3.4).
3. The subspace spanned by the matrix elements of irreducible corepresentations is dense in the quantum group, and these matrix elements satisfy certain orthogonality relations with respect to the Haar state (Propositions 5.3.7, 5.3.8).
4. Moreover, $C^*$-algebraic compact quantum groups can be characterized in terms of corepresentations (Theorem 5.3.11).

Several of these results follow from similar arguments as in the setting of algebraic compact quantum groups (see Section 3.2), but sometimes, we have to use new methods and ideas.

Throughout this section, let $(A, \Delta)$ be a $C^*$-algebraic compact quantum group with Haar state $h$.

### 5.3.1 Decomposition into irreducible corepresentations

In this subsection, we show that corepresentation operators and intertwiners between them can be described in terms of irreducible corepresentation operators.

As a tool, we first construct intertwiners between corepresentation operators by an averaging procedure (cf. Lemma 3.2.5). Let $X$ and $Y$ be corepresentation operators on Hilbert spaces $H$ and $K$, respectively, and let $R \in \mathcal{L}(H, K)$. We identify $M(\mathcal{K}(H) \otimes A)$, $M(\mathcal{K}(K) \otimes A)$, and $\mathcal{L}(H, K) \otimes A$ with subspaces of $M(\mathcal{K}(H \oplus K) \otimes A)$, and use the slice map

$$\mathrm{id} \otimes h \colon M(\mathcal{K}(H \oplus K) \otimes A) \to M(\mathcal{K}(H \oplus K)) \cong \mathcal{L}(H \oplus K)$$

(see Section 12.4) to define operators $S, T \in \mathcal{L}(H, K)$ as follows:

$$S := (\mathrm{id} \otimes h)(Y^{-1}(R \otimes 1_A)X), \quad T := (\mathrm{id} \otimes h)(Y(R \otimes 1_A)X^{-1}). \tag{5.10}$$

**Lemma 5.3.1.** *Let $X$ and $Y$ be corepresentations operators on Hilbert spaces $H$ and $K$, respectively, let $R \in \mathcal{L}(H, K)$, and define $S, T$ as above. Then $S, T \in$ $\mathrm{Hom}(X, Y)$. If $R \in \mathrm{Hom}(X, Y)$, then $S = R = T$. If $R$ is compact, so are $S$ and $T$.*

*Proof.* The first statements follows from similar calculations as in Lemma 3.2.5. If $R$ is compact, then $Y^{-1}(R \otimes 1_A)X$ and $Y(R \otimes 1_A)X^{-1}$ belong to $\mathcal{K}(H, K) \otimes A \subseteq \mathcal{K}(H \oplus K) \otimes A$, so that $S$ and $T$ are compact as well. □

**Remark 5.3.2.** We can equip $\mathcal{L}(H, K)$ with a strict topology by identifying it with a subspace of $M(\mathcal{K}(H \oplus K))$. Then the assignments $R \mapsto S$ and $R \mapsto T$ defined in (5.10), considered as maps $\mathcal{L}(H, K) \to \mathcal{L}(H, K)$, are strictly continuous on bounded subsets: The assignment $R \mapsto R \otimes 1_A$ is strictly continuous on bounded subsets, and left and right multiplication by $Y$, $Y^{-1}$, $X$, or $X^{-1}$, respectively, as well as the slice map $\mathrm{id} \otimes h$ are strictly continuous.

Every corepresentation operator is a direct sum of finite-dimensional irreducible ones (cf. Theorem 3.2.1):

**Theorem 5.3.3.** *Let $X$ be a corepresentation operator on a Hilbert space $H$.*

i) *$X$ is equivalent to a unitary corepresentation operator.*

ii) *If $X$ is unitary, then the orthogonal complement of every invariant subspace of $H$ is invariant again.*

iii) *If $H$ is non-zero, then it contains a non-zero invariant subspace of finite dimension. In particular, $X$ is finite-dimensional if it is irreducible.*

iv) *$X$ is equivalent to a direct sum of finite-dimensional irreducible unitary corepresentation operators.*

*Proof.* i) Put $T := (\mathrm{id} \otimes h)(X^*X) \in \mathcal{L}(H)$. Then $T$ is invertible: $X^*X > \epsilon \cdot 1$ for some $\epsilon > 0$ because $X$ and hence also $X^*X$ are invertible, and then $T > \epsilon \cdot 1$ because $\mathrm{id} \otimes h$ is positive. Consider the operator

$$Y := (T^{1/2} \otimes 1_A)X(T^{-1/2} \otimes 1_A).$$

This is a corepresentation operator which is equivalent to $X$. Furthermore,

$$\begin{aligned} X^*(T \otimes 1_A)X &= (\mathrm{id} \otimes h \otimes \mathrm{id})(X^*_{[13]}X^*_{[12]}X_{[12]}X_{[13]}) \\ &= (\mathrm{id} \otimes h \otimes \mathrm{id})((\mathrm{id} \otimes \Delta)(X^*X)) \\ &= (\mathrm{id} \otimes h)(X^*X) \otimes 1_A = T \otimes 1_A, \end{aligned}$$

so $Y^*Y = (T^{-1/2} \otimes \mathrm{id}_A)X^*(T \otimes \mathrm{id}_A)X(T^{-1/2} \otimes \mathrm{id}_A) = 1$, that is, $Y$ is unitary.

ii) This claim follows from the fact that a subspace is invariant if and only if it is invariant for the $C^*$-subalgebra of $\mathcal{L}(H)$ defined in Proposition 5.2.8.

iii) By i), we may assume that $X$ is unitary. Let $(u_\alpha)_\alpha$ be an approximate unit for $\mathcal{K}(H)$. Since conjugation by $X$ is an automorphism of $\mathcal{K}(H) \otimes A$, the elements $X(u_\alpha \otimes 1_A)X^*$ form an approximate unit for $\mathcal{K}(H) \otimes A$. By Lemma 5.3.1, the elements $(\mathrm{id} \otimes h)(X(u_\alpha \otimes 1_A)X^*)$ belong to the $C^*$-algebra

$$D := \mathcal{K}(H) \cap \mathrm{Hom}(X, X),$$

and since $\mathrm{id} \otimes h$ is continuous on $\mathcal{K}(H) \otimes A$, these elements form an approximate unit for $\mathcal{K}(H)$. Therefore, $D \subseteq \mathcal{K}(H)$ is non-degenerate and contains a non-zero projection $P$. Now $PH \subseteq H$ is a finite-dimensional invariant subspace.

iv) This follows from i)–iii) by a straightforward application of Zorn's Lemma. Alternatively, one could choose a maximal family $(p_\alpha)_\alpha$ of pairwise orthogonal minimal projections in the $C^*$-algebra $D$; then $H \cong \bigoplus_\alpha p_\alpha H$ since $D$ is non-degenerate, and the restriction of $X$ to $p_\alpha H$ is irreducible for each $\alpha$ because $p_\alpha H$ is a minimal invariant subspace. □

We have the following analogue of Schur's Lemma (cf. Proposition 3.2.2):

**Proposition 5.3.4.** *Let $X$ and $Y$ be corepresentation operators on Hilbert spaces $H$ and $K$, respectively.*

i) *For every $T \in \mathrm{Hom}(X, Y)$, the subspaces $\ker T \subseteq H$ and $\mathrm{Im}\, T \subseteq K$ are invariant.*

ii) *If $X$ and $Y$ are irreducible, then either $\mathrm{Hom}(X, Y)$ has dimension 1 and $X \simeq Y$, or $\mathrm{Hom}(X, Y) = 0$.*

*Proof.* The proof is essentially the same as the proof of Proposition 3.2.2, the only difference is that for i), we use the representations $\pi_X$ and $\pi_Y$ associated to $X$ and $Y$, respectively (see Proposition 5.2.7): If $T \in \mathrm{Hom}(X, Y)$, then $T\pi_X(A') = \pi_Y(A')T$, whence $\ker T \subseteq H$ is invariant for $\pi_X(A')$ and $X$, and $\mathrm{Im}\, T \subseteq K$ is invariant for $\pi_Y(A')$ and $Y$. □

As in the setting of algebraic compact quantum groups, the preceding results show that the category of all corepresentation operators of $(A, \Delta)$, regarded as an ordinary category, has a very simple structure (cf. Corollary 3.2.4):

**Corollary 5.3.5.** *Let $(X_\alpha)_{\alpha \in I}$ be a maximal family of pairwise inequivalent irreducible corepresentations operators of $(A, \Delta)$.*

i) *Every corepresentation operator of $(A, \Delta)$ is equivalent to a direct sum $\boxplus_k X_{\alpha_k}$, where $(\alpha_k)_k$ is some family of elements of $I$.*

ii) *Let $(\alpha_k)_k$ and $(\beta_l)_l$ be families of elements of $I$. Then*

$$\mathrm{Hom}\Big(\boxplus_k X_{\alpha_k}, \boxplus_l X_{\beta_l}\Big) \cong \prod_k \mathrm{Hom}\Big(X_{\alpha_k}, \boxplus_l X_{\beta_l}\Big) \cong \prod_k \bigoplus_{\substack{l \text{ s.t.} \\ \alpha_k = \beta_l}} \mathbb{C},$$

*where $\prod$ and $\bigoplus$ denote the direct product and the $l^2$-sum of Banach spaces, respectively.* □

The interesting and non-trivial information of the category of all corepresentation operators is contained in the monoidal structure, that is, in the tensor product.

Finally, we show that every irreducible corepresentation is contained in the regular corepresentation:

**Proposition 5.3.6.** *Let $X$ be an irreducible corepresentation operator. Then there exists an injective intertwiner $T \in \operatorname{Hom}(X, X_h)$, where $X_h$ denotes the regular corepresentation operator (see Theorem 5.2.9).*

*Proof.* Denote by $H$ the underlying Hilbert space of $X$, and by $(a_{ij})_{i,j}$ the corepresentation matrix of $X$ with respect to some orthonormal basis $(e_i)_i$. For a fixed $i$, consider the map $T_i : H \to H_h$ given by $e_j \mapsto \Lambda_h(a_{ij})$ for all $j$. The calculations

$$X_h(T_i \otimes 1_A)(e_j \otimes 1_A) = X_h(\Lambda_h(a_{ij}) \otimes 1_A) = \sum_k \Lambda_h(a_{ik}) \otimes a_{kj},$$

$$(T_i \otimes 1_A)X(e_j \otimes 1_A) = \sum_k (T_i \otimes 1_A)(e_k \otimes a_{kj}) = \sum_k \Lambda_h(a_{ik}) \otimes a_{kj}$$

show that $T_i \in \operatorname{Hom}(X, X_h)$. By Theorem 5.3.3, we may assume that $X$ is unitary. Then $\sum_i a_{ij}{}^* a_{ij} = 1_A$ for each $j$, and hence there exist $i_0$ and $j$ such that $h(a_{i_0 j}{}^* a_{i_0 j}) \neq 0$, that is, $T_{i_0} e_j = \Lambda_h(a_{i_0 j}) \neq 0$. In particular, $T_{i_0}$ is non-zero. By Proposition 5.3.4, $T_{i_0}$ must be injective. □

## 5.3.2 Schur's orthogonality relations

In this subsection, we study the matrix elements of irreducible corepresentation operators and prove an analogue of Schur's orthogonality relations known from the representation theory of compact groups [22, Chapter II, Theorems 4.5, 4.6], [62, Chapter VII, Theorem 27.19], and from the corepresentation theory of algebraic compact quantum groups (cf. Section 3.2.2).

**Proposition 5.3.7.** *Let $X$ and $Y$ be inequivalent irreducible corepresentation operators. Then $h(b^*a) = 0 = h(ba^*)$ for all $a \in \mathcal{C}(X)$ and $b \in \mathcal{C}(Y)$.*

*Proof.* The proof of Proposition 3.2.6 carries over without modifications. □

To obtain a better understanding of expressions of the form $h(b^*a)$, where $b$ and $a$ are matrix elements of the same irreducible corepresentation operator, we need to consider particular intertwiners (cf. Proposition 3.2.9).

**Proposition 5.3.8.** *Let $Y = \sum_{i,j} e_{ij} \otimes a_{ij}$ be an irreducible unitary corepresentation operator on $\mathbb{C}^n$, where $n \in \mathbb{N}$ and $(e_{ij})_{i,j} \in M_n(\mathbb{C})$ denote the standard matrix units.*

i) *$\overline{Y}$ and $\overline{Y}^{-*} := (\overline{Y}^*)^{-1}$ are irreducible corepresentation operators on $\overline{\mathbb{C}^n}$.*

ii) *There exists a unique intertwiner* $\widetilde{F} \in \mathrm{Hom}(\overline{Y}, \overline{Y}^{-*})$ *that satisfies* $\mathrm{Tr}\,\widetilde{F} = \mathrm{Tr}\,\widetilde{F}^{-1} > 0$, *and this operator is positive and invertible.*

iii) *For all* $i, j, k, l$,

$$h(a_{ij}{}^* a_{kl}) = \frac{\delta_{j,l}}{\mathrm{Tr}(\widetilde{F}^{-1})}(\widetilde{F}^{-1})_{ik} \quad \text{and} \quad h(a_{ij} a_{kl}{}^*) = \frac{\delta_{i,k}}{\mathrm{Tr}\,\widetilde{F}} \widetilde{F}_{jl},$$

*where the matrix representations of* $\widetilde{F}$ *and* $\widetilde{F}^{-1}$ *are taken with respect to the standard basis of* $\overline{\mathbb{C}^n}$.

iv) *The elements* $(a_{ij})_{i,j}$ *are linearly independent.*

*Proof.* i) First, we modify the proof of Proposition 5.3.6 and show that $\overline{Y}$ is equivalent to the restriction of the regular corepresentation operator $X_h$ (see Theorem 5.2.9) to an invariant subspace. This implies that $\overline{Y}$ is a corepresentation operator.

Denote by $(e_i)_i$ the standard units of $\mathbb{C}^n$. Fix some $i$, and consider the map

$$T : \overline{\mathbb{C}^n} \to H_h, \quad \overline{e_j} \mapsto \Lambda_h(a_{ij}{}^*).$$

Equation (5.6) on page 116 shows that $\Delta(a_{ij}{}^*) = \sum_{ij} a_{ik}{}^* \otimes a_{kj}{}^*$ for all $i, j$, and as in the proof of Proposition 5.3.6, we deduce

$$X_h(T \otimes 1_A) = (T \otimes 1_A)\overline{Y}. \tag{5.11}$$

Since $Y$ is unitary, $\sum_j a_{ij} a_{ij}{}^* = 1_A$, and hence $h(a_{ij_0} a_{ij_0}{}^*) \neq 0$ for some $j_0$. So, $Te_{j_0} = \Lambda_h(a_{ij_0}{}^*) \neq 0$ and $T \neq 0$. We show that $T$ is injective and that its image is invariant. Denote by $\pi_{X_h}$ and $\pi_Y$ the representations associated to $X_h$ and $Y$, respectively (see Proposition 5.2.7), and by $\pi_{\overline{Y}} : A' \to \mathcal{L}(\overline{\mathbb{C}^n})$ the map given by $f \mapsto (\mathrm{id} \otimes f)(\overline{Y})$. For each $f \in A'$, define $\overline{f} \in A'$ by $a \mapsto \overline{f(a^*)}$. Then $\pi_{\overline{Y}}(f) = \sum_{i,j} \overline{e_{ij}} f(a_{ij}{}^*) = \sum_{i,j} \overline{e_{ij} \overline{f}(a_{ij})} = \overline{\pi_Y(\overline{f})}$ for all $f \in A'$, and combining this relation with equation (5.11), we find

$$\pi_{X_h}(A')T = T\pi_{\overline{Y}}(A') = T\overline{\pi_Y(A')}.$$

Using Proposition 5.2.7, we conclude that $\mathrm{Im}\,T \subseteq H_h$ is invariant for $X_h$ and $\overline{\ker T} \subseteq \mathbb{C}^n$ is invariant for $Y$. Since $Y$ is irreducible and $T \neq 0$, the map $T$ must be injective. Thus, $\overline{Y}$ is equivalent to a restriction of $X_h$ to the invariant subspace $\mathrm{Im}\,T$; in particular, it is a corepresentation operator.

An easy calculation shows that $\overline{Y}^{-*}$ is a corepresentation operator:

$$\overline{Y}^{-*}_{[12]}\overline{Y}^{-*}_{[13]} = (\overline{Y}_{[12]}\overline{Y}_{[13]})^{-*} = ((\mathrm{id} \otimes \Delta)(\overline{Y}))^{-*} = (\mathrm{id} \otimes \Delta)(\overline{Y}^{-*}).$$

The fact that $\overline{Y}^{-*}$ is irreducible will follow from statement ii).

ii) Put $\widetilde{G} := (\mathrm{id} \otimes h)(\overline{Y}^* \overline{Y}) \in \mathcal{L}(\overline{\mathbb{C}^n})$. Evidently, $\widetilde{G}$ is positive, and as in Proposition 3.2.1, we conclude: $\overline{Y}$ is invertible $\Rightarrow \overline{Y}^* \overline{Y}$ is invertible $\Rightarrow \overline{Y}^* \overline{Y} > \epsilon 1$ for some $\epsilon > 0 \Rightarrow \widetilde{G} > \epsilon 1 \Rightarrow \widetilde{G}$ is invertible. By Lemma 5.3.1, $\widetilde{G} \in \mathrm{Hom}(\overline{Y}, \overline{Y}^{-*})$; in particular, $\overline{Y} \simeq \overline{Y}^{-*}$. Now $\widetilde{F}$ is some uniquely determined multiple of $\widetilde{G}$.

iii), iv) This follows as in Proposition 3.2.9. □

Propositions 5.3.4, 5.3.7 and 5.3.8 imply:

**Corollary 5.3.9.** *For all corepresentation operators $X$ and $Y$,*

$$\begin{aligned}\mathrm{Hom}(X, Y) = 0 &\Leftrightarrow h(\mathcal{C}(Y)^* \mathcal{C}(X)) = 0 \\ &\Leftrightarrow h(\mathcal{C}(Y) \mathcal{C}(X)^*) = 0 \Leftrightarrow \mathrm{Hom}(Y, X) = 0.\end{aligned}$$ □

Recall that we associated to every corepresentation operator $X$ on a Hilbert space $H$ a conjugate operator $\overline{X} \in M(\mathcal{K}(\overline{H}) \otimes A)$. Now we can prove:

**Corollary 5.3.10.** *The conjugate of a corepresentation operator is a corepresentation operator again.*

*Proof.* Every corepresentation operator is equivalent to a direct sum of irreducible ones, and for every irreducible corepresentation operator, the conjugate is a corepresentation operator by Proposition 5.3.8. □

### 5.3.3 Characterization of $C^*$-algebraic compact quantum groups

Like algebraic compact quantum groups, $C^*$-algebraic compact quantum groups can be characterized in terms of their corepresentations:

**Theorem 5.3.11.** *The following conditions on a unital $C^*$-bialgebra $(A, \Delta)$ are equivalent:*

i) *$(A, \Delta)$ is a $C^*$-algebraic compact quantum group.*
ii) *The subspace of $A$ spanned by the elements of all irreducible unitary corepresentation matrices $a \in M_n(A)$ ($n \in \mathbb{N}$) whose conjugate $\bar{a} = (a_{ij}{}^*)_{i,j}$ is invertible, is dense in $A$.*
iii) *The $*$-subalgebra of $A$ generated by the elements of all corepresentation matrices $a \in M_n(A)$ ($n \in \mathbb{N}$) whose conjugate $\bar{a}$ is invertible, is dense in $A$.*
iv) *The subalgebra of $A$ generated by the elements of all corepresentation matrices is dense in $A$.*

*Proof.* i) $\Rightarrow$ ii): By Theorem 5.3.3 iv), the regular corepresentation operator is equivalent to a direct sum of finite-dimensional irreducible unitary corepresentation operators, and by Theorem 5.2.9 iii), its space of matrix elements is equal to $A$.

Therefore the matrix elements of finite-dimensional irreducible unitary corepresentations span a dense subspace of $A$. Furthermore, if $a \in M_n(A)$ is a unitary irreducible corepresentation matrix, then the matrix $\bar{a}$ is invertible by Proposition 3.2.9 i).

ii) $\Rightarrow$ iii): Trivial.

iii) $\Rightarrow$ iv): If $a \in M_n(A)$ is a corepresentation matrix and $\bar{a}$ is invertible, then also $\bar{a}$ is a corepresentation matrix.

iv) $\Rightarrow$ i): We have to show that $\Delta(A)(1_A \otimes A)$ and $\Delta(A)(A \otimes 1_A)$ are linearly dense in $A \otimes A$. If $a \in M_n(A)$ is a corepresentation matrix with inverse $b$, then

$$\sum_k \Delta(a_{ik})(1_A \otimes b_{kj}) = \sum_{k,l} a_{il} \otimes a_{lk} b_{kj} = \sum_l a_{il} \otimes \delta_{l,j} 1_A = a_{ij} \otimes 1_A$$

for all $i, j$. Moreover, if

$$c \otimes 1_A = \sum_i \Delta(c_{1,i})(1_A \otimes c_{2,i}) \quad \text{and} \quad d \otimes 1_A = \sum_j \Delta(d_{1,j})(1_A \otimes d_{2,j}),$$

for some $c_{1,i}, c_{2,i}, d_{1,j}, d_{2,j} \in A$, then

$$\begin{aligned} cd \otimes 1_A &= \sum_i \Delta(c_{1,i})(1_A \otimes c_{2,i})(d \otimes 1_A) \\ &= \sum_i \Delta(c_{1,i})(d \otimes 1_A)(1_A \otimes c_{2,i}) = \sum_{i,j} \Delta(c_{1,i} d_{1,j})(1_A \otimes d_{2,j} c_{2,i}). \end{aligned}$$

Now, the assumption and the equations above imply that $\Delta(A)(1_A \otimes A)$ is linearly dense in $A \otimes A$. A similar argument shows that $\Delta(A)(A \otimes 1_A)$ is linearly dense in $A \otimes A$. □

## 5.4 The relation to algebraic compact quantum groups

$C^*$-algebraic and algebraic compact quantum groups are closely related. In this section, we explain how one can associate to every $C^*$-algebraic compact quantum group an algebraic one, and how one can associate to an algebraic compact quantum group a maximal and a minimal $C^*$-algebraic one.

**Notation.** We shall abbreviate the phrase "compact quantum group" by the acronym "CQG" whenever it seems convenient.

### 5.4.1 From $C^*$-algebraic to algebraic CQGs

For every $C^*$-algebraic compact quantum group, the subspace spanned by the matrix elements of finite-dimensional corepresentation operators forms an algebraic compact quantum group:

**Theorem 5.4.1.** *Let $(A, \Delta)$ be a $C^*$-algebraic compact quantum group. Denote by $A_0 \subseteq A$ the subspace spanned by the matrix elements of all finite-dimensional corepresentation operators.*

i) *$A_0$ is dense in $A$, the image $\Delta(A_0)$ is contained in the algebraic tensor product $A_0 \odot A_0$, and $(A_0, \Delta_0)$ is an algebraic compact quantum group, where $\Delta_0 := \Delta|_{A_0}$.*

ii) *If $(\tilde{A}_0, \tilde{\Delta}_0)$ is an algebraic compact quantum group such that $\tilde{A}_0 \subseteq A$ is a dense $*$-subalgebra and $\tilde{\Delta}_0 = \Delta|_{\tilde{A}_0}$, then $(\tilde{A}_0, \tilde{\Delta}_0) = (A_0, \Delta_0)$.*

*We call $(A, \Delta)_0 := (A_0, \Delta_0)$ the* algebraic CQG associated to $(A, \Delta)$.

iii) *Every morphism of $C^*$-algebraic CQGs $(A, \Delta_A) \to (B, \Delta_B)$ restricts to a morphism $(A, \Delta_A)_0 \to (B, \Delta_B)_0$ of algebraic CQGs.*

*Proof.* i) $A_0$ is dense in $A$ by Theorem 5.3.11 ii), and a $*$-algebra because

$$\mathcal{C}(X) + \mathcal{C}(Y) = \mathcal{C}(X \boxplus Y), \quad \mathcal{C}(X)\mathcal{C}(Y) \subseteq \mathcal{C}(X \boxtimes Y), \quad \mathcal{C}(X)^* = \mathcal{C}(\bar{X})$$

for all finite-dimensional corepresentation operators $X$ and $Y$. The unit $1_A$ belongs to $A_0$ because it is the matrix element of the trivial corepresentation. Equations (5.6) and (5.7) on page 116 show that $\Delta(A_0)$ is contained in $A_0 \odot A_0$. Thus $(A_0, \Delta_0)$ is a bialgebra.

Let us show that $(A_0, \Delta_0)$ is a Hopf algebra. Choose a maximal family $(u^\alpha)_\alpha$ of pairwise inequivalent irreducible unitary corepresentation matrices of $(A, \Delta)$. By Theorem 5.3.3 iv), Proposition 5.3.7, and Proposition 5.3.8 iii), iv), the family $(u^\alpha_{ij})_{\alpha,i,j}$ is a basis of $A_0$. Consider the linear maps

$$\epsilon_0 \colon A_0 \to \mathbb{C},\ u^\alpha_{ij} \mapsto \delta_{i,j}, \quad \text{and} \quad S_0 \colon A_0 \to A_0,\ u^\alpha_{ij} \mapsto (u^\alpha_{ji})^*.$$

Equations (5.6) and (5.8) imply that $\epsilon_0$ and $S_0$ satisfy the axioms for the counit and for the antipode of a Hopf algebra. Therefore $(A_0, \Delta_0)$ is a Hopf $*$-algebra. It is an algebraic compact quantum group because the restriction of $h$ to $A_0$ is a positive integral.

ii) By Theorem 3.2.12 iii), $\tilde{A}_0$ is spanned by the components of its finite-dimensional corepresentation matrices; since these matrices are also corepresentation matrices of $(A, \Delta)$, the space $\tilde{A}_0$ is contained in $A_0$. If there exists an irreducible unitary corepresentation matrix $(a_{ij})_{i,j}$ of $A$ whose components are not contained in $\tilde{A}_0$, we obtain a contradiction to the assumption that $\tilde{A}_0$ is dense in $A$ from Proposition 5.3.7 and 5.3.8 iii).

iii) Let $(B, \Delta_B)$ be another $C^*$-algebraic compact quantum group and let $\pi \colon A \to B$ be a morphism of $C^*$-bialgebras. Since $\pi$ is non-degenerate, it is unital. For every corepresentation matrix $(a_{ij})_{i,j} \in M_n(A)$, the image under $\pi$,

$(\pi(a_{ij}))_{i,j} \in M_n(B)$, is a corepresentation matrix as well. Therefore, $\pi(A_0)$ is contained in $B_0$, that is, $\pi$ restricts to a unital morphism of $*$-bialgebras $\pi_0 \colon A_0 \to B_0$. From the construction of the counit and antipode in i), it is easy to see that $\pi_0$ preserves the counit and the antipode. □

**Example 5.4.2.** i) For every compact group $G$, the $C^*$-bialgebra $C(G)$ introduced in Example 4.2.1 is a $C^*$-algebraic CQG, and the associated algebraic CQG is the Hopf $*$-algebra of representative functions $\mathrm{Rep}(G)$ introduced in Example 1.2.5.

ii) For every discrete group $G$, the $C^*$-bialgebras $C^*(G)$ and $C^*_r(G)$ introduced in Example 4.2.2 are $C^*$-algebraic CQGs, and the associated algebraic CQG is both times the group Hopf $*$-algebra $\mathbb{C}G$ introduced in Example 1.2.8.

The preceding theorem shows that every $C^*$-algebraic compact quantum group $(A, \Delta)$ is the completion of some unique algebraic compact quantum group $(A_0, \Delta_0)$ which has the same category of finite-dimensional corepresentations. This result provides a partial answer to the question for the antipode and for the counit of $(A, \Delta)$: at least, both maps are well defined on $A_0$. If $(A, \Delta)$ is reduced, then the antipode of $(A_0, \Delta_0)$ can be extended to a closed (possibly unbounded) linear map on $A$ and can be described in terms of a unitary antipode and a scaling group, see Example 8.3.7. Moreover, in that case, the modular automorphism of the Haar state of $(A_0, \Delta_0)$ (Theorem 2.2.17) extends to a modular automorphism of the Haar state of $(A, \Delta)$, see Example 8.1.22.

## 5.4.2 From algebraic to $C^*$-algebraic CQGs

An algebraic compact quantum group can have several different completions that are $C^*$-algebraic compact quantum groups, see Example 5.4.2 ii). However, there always exist a maximal and a minimal one. First, we describe the maximal completion:

**Theorem 5.4.3.** *Let $(A, \Delta)$ be an algebraic compact quantum group. Then*

$$a \mapsto \|a\|_u := \sup\{p(a) \mid p \text{ is a } C^*\text{-seminorm on } A\}$$

*is a $C^*$-norm on $A$. Denote by $A_u$ the corresponding completion of $A$.*

i) *$A_u$ is a unital $C^*$-algebra, and the comultiplication $\Delta$ extends to a $*$-homomorphism $\Delta_u \colon A_u \to A_u \otimes A_u$ that turns $A_u$ into a $C^*$-algebraic CQG. The counit $\epsilon$ of $A$ extends to a character $\epsilon_u$ on $A_u$ that satisfies $(\epsilon_u \otimes \mathrm{id}) \circ \Delta_u = \mathrm{id} = (\mathrm{id} \otimes \epsilon_u) \circ \Delta_u$.*

ii) *$(A, \Delta)$ is the algebraic compact quantum group associated to $(A_u, \Delta_u)$.*

*We call $(A, \Delta)_u := (A_u, \Delta_u)$ the* universal $C^*$-algebraic CQG *of $(A, \Delta)$.*

iii) *Every morphism of algebraic CQGs* $(A, \Delta_A) \to (B, \Delta_B)$ *extends uniquely to a morphism of* $C^*$*-algebraic CQGs* $(A, \Delta_A)_u \to (B, \Delta_B)_u$.

*Proof.* First, we show that $\|a\|_u$ is finite for each $a \in A$. Let $p$ be a $C^*$-seminorm on $A$. Then for each unitary corepresentation matrix $(a_{ij})_{i,j} \in M_n(A)$ and all $i, j$, we have $p(a_{ij})^2 = p(a_{ij}{}^* a_{ij}) \leq 1$ because $\sum_i a_{ij}{}^* a_{ij} = 1_A$. Since $A$ is spanned by the elements of unitary corepresentation matrices (Theorem 3.2.12 iii)), it follows that $\|a\|_u$ is finite for each $a \in A$.

Let us show that the seminorm $\| \cdot \|_u$ is a norm. Denote by $h$ the Haar state of $(A, \Delta)$, and by $H$ the completion of $A$ with respect to the inner product given by $\langle b|c\rangle := h(b^*c)$ for all $b, c \in A$. Let $(a_{ij})_{i,j}$ be a corepresentation matrix of $(A, \Delta)$, and fix $i, j$. Since $h$ is positive and $\sum_k a_{kj}{}^* a_{kj} = 1_A$,

$$\|a_{ij}b\|_H^2 = h(b^* a_{ij}{}^* a_{ij} b) \leq h(b^* 1_A b) = \|b\|_H^2 \quad \text{for all } b \in A.$$

Therefore the map $b \mapsto a_{ij}b$ extends to a bounded operator $\pi(a_{ij}) \in \mathcal{L}(H)$. Since $A$ is spanned by the elements of unitary corepresentation matrices, $\pi$ extends to a $*$-homomorphism $\pi\colon A \to \mathcal{L}(H)$, where $\pi(a)b = ab$ for all $a, b \in A$. Since $h$ is faithful (Proposition 2.2.4), $\|a\|_u \geq \|\pi(a)\| \geq \|\pi(a)1_A\| = h(a^*a) > 0$ for all non-zero $a \in A$.

i) The maps $a \mapsto \|\Delta(a)\|_{(A_u \otimes A_u)}$ and $a \mapsto |\epsilon(a)|$ are $C^*$-seminorms on $A$. Therefore they are dominated by $\| \cdot \|_u$, and $\Delta$ and $\epsilon$ extend to $*$-homomorphisms $\Delta_u$ and $\epsilon_u$ as claimed. By Theorem 3.2.12 iii) and 5.3.11 ii), $(A_u, \Delta_u)$ is a $C^*$-algebraic compact quantum group. Finally, the equation concerning $\epsilon_u$ follows from the corresponding equation for $\epsilon$ and the density of $A$ in $A_u$.

ii) This follows immediately from Theorem 5.4.1 ii).

iii) Let $(B, \Delta_B)$ be an algebraic compact quantum group and let $\pi\colon A \to B$ be a morphism of Hopf $*$-algebras. By definition of the norm $\| \cdot \|_u$ on $A_u$, the $*$-homomorphism $A \xrightarrow{\pi} B \to B_u$ extends to a $*$-homomorphism $\pi_u\colon A_u \to B_u$. Since $\pi$ preserves the unit and the comultiplication, so does $\pi_u$. □

**Remark 5.4.4.** The antipode of $(A, \Delta)$ need not be bounded with respect to the norm $\| \cdot \|_u$ and need not extend to $A_u$.

Now we turn to the minimal completion:

**Theorem 5.4.5.** *Let* $(A, \Delta)$ *be an algebraic compact quantum group. Denote by* $(H, \Lambda, \pi)$ *the GNS-representation for the Haar state* $h_u$ *of* $(A_u, \Delta_u) = (A, \Delta)_u$, *and put* $A_r := \pi(A_u)$.

i) $\pi$ *restricts to an embedding* $A \hookrightarrow A_r$, *and the comultiplication* $\Delta$ *extends to a* $*$*-homomorphism* $\Delta_r\colon A_r \to A_r \otimes A_r$ *that turns* $(A_r, \Delta_r)$ *into a reduced* $C^*$*-algebraic compact quantum group.*

ii) $(A, \Delta)$ *is the algebraic compact quantum group associated to* $(A_r, \Delta_r)$.

*We call* $(A, \Delta)_r := (A_r, \Delta_r)$ *the* reduced $C^*$-algebraic CQG *of* $(A, \Delta)$.

*Proof.* i) The restriction of $h_u$ to $A$ is a positive normalized integral, and hence equal to the Haar state $h$ of $(A, \Delta)$ (Proposition 2.2.6 ii)). Since $h$ is faithful (Proposition 2.2.4), $\|\pi(a)\|^2 \geq \|\pi(a)\Lambda(1_A)\|^2 = h(a^*a) > 0$ for all $a \in A$.

To show that $\Delta$ extends to a $*$-homomorphism $\Delta_r \colon A_r \to A_r \otimes A_r$, we introduce an auxiliary multiplicative unitary: The map

$$\begin{aligned} V_0 \colon \Lambda(A) \odot \Lambda(A) &\to \Lambda(A) \odot \Lambda(A), \\ \Lambda(b) \odot \Lambda(c) &\mapsto (\Lambda \odot \Lambda)(\Delta(c)(b \odot 1)), \end{aligned}$$

is bijective because $A$ is a Hopf $*$-algebra, and isometric since

$$\begin{aligned} &\langle V_0(\Lambda(b) \odot \Lambda(c)) | V_0(\Lambda(b') \odot \Lambda(c'))\rangle \\ &\quad = (h \odot h)((b^* \odot 1)\Delta(c^*c')(b' \odot 1)) \\ &\quad = h(b^*b')h(c^*c') = \langle \Lambda(b) \odot \Lambda(c) | \Lambda(b') \odot \Lambda(c')\rangle \end{aligned}$$

for all $b, c, b', c' \in A$. Therefore, $V_0$ extends to a unitary $V$ on $H \otimes H$. Now

$$V(1 \otimes \pi(a)) = (\pi \otimes \pi)(\Delta(a))V \quad \text{for all } a \in A;$$

indeed, the operators on the left-hand side and on the right-hand side are given by $\Lambda(b) \odot \Lambda(c) \mapsto (\Lambda \odot \Lambda)(\Delta(ac)(b \odot 1))$ for all $b, c \in A$. Therefore the $*$-homomorphism

$$\Delta_r \colon \mathcal{L}(H) \to \mathcal{L}(H \otimes H), \quad T \mapsto V(1 \otimes T)V^*,$$

extends $\Delta$ as desired. By Theorems 3.2.12 iii) and 5.3.11 ii), $(A_r, \Delta_r)$ is a $C^*$-algebraic compact quantum group.

Finally, let us show that the Haar state $h_r$ on $A_r$ is faithful. Note that the restriction of $h_r$ to $A$ is equal to $h$. Denote by $\sigma$ the modular automorphism of $h$ (see Theorem 2.2.17 or 3.2.19 v)). Let $x \in A_r$, $x \neq 0$. We have to show that $h_r(x^*x) > 0$. Since $\Lambda(A)$ is dense in $H$, we find $b, c \in A$ such that $\langle \Lambda(c) | x\Lambda(b)\rangle > 0$. By the Cauchy–Schwarz inequality, there exists a constant $C_{b,c} > 0$ such that

$$\begin{aligned} |\langle \Lambda(c) | \pi(a)\Lambda(b)\rangle| &= |h(c^*ab)| = |h(ab\sigma(c^*))| \\ &\leq C_{b,c} h(a^*a) = C_{b,c} h_r(\pi(a^*a)) \end{aligned}$$

for all $a \in A$. But then also $0 < |\langle \Lambda(c) | x\Lambda(b)\rangle| \leq C_{b,c} h_r(x^*x)$, so $0 < h_r(x^*x)$.

ii) This follows immediately from Theorem 5.4.1 ii). □

**Example 5.4.6.** i) Let $G$ be a discrete group. Then the group algebra $\mathbb{C}G$ carries the structure of an algebraic CQG (see p. 66), and the associated universal and the associated reduced $C^*$-algebraic CQG are the $C^*$-bialgebras $C^*(G)$ and $C_r^*(G)$, respectively, introduced in Example 4.2.2.

ii) Let $G$ be a compact group. Then the Hopf $*$-algebra of representative functions $\mathrm{Rep}(G)$ carries the structure of an algebraic CQG (see p. 66), and the associated universal and the associated reduced $C^*$-algebraic CQG are both equal to the $C^*$-bialgebra $C(G)$ introduced in Example 4.2.1.

**Remark 5.4.7.** The counit $\epsilon$ of $(A, \Delta)$ need not be bounded with respect to the norm of $A_r$ and need not extend to $A_r$: If $G$ is a discrete group and $A = \mathbb{C}G$, then $A_r = C_r^*(G)$, and $\epsilon$ extends to $A_r$ if and only if $G$ is amenable. In particular, this example shows that the assignment $A \mapsto A_r$ does not extend to a functor, because $\mathbb{C}$ can be considered as an algebraic CQG and $\epsilon$ as a morphism of algebraic CQGs.

Using the preceding constructions, we can associate to every $C^*$-algebraic CQG $(A, \Delta)$ the following three CQGs:

- an algebraic CQG $(A_0, \Delta_0)$,
- the universal $C^*$-algebraic CQG $(A_u, \Delta_u)$ of $(A_0, \Delta_0)$,
- the reduced $C^*$-algebraic CQG $(A_r, \Delta_r)$ of $(A_0, \Delta_0)$.

We call $(A, \Delta)_u := (A_u, \Delta_u)$ and $(A, \Delta)_r := (A_r, \Delta_r)$ the *universal* and the *reduced* CQG associated to $(A, \Delta)$.

**Proposition 5.4.8.** *Let $(A, \Delta)$ be a $C^*$-algebraic CQG. Denote by $\iota_u$, $\iota$, $\iota_r$ the canonical embeddings of $A_0$ in $A_u$, $A$, $A_r$, respectively. Then there exists a commutative diagram*

$$\begin{array}{ccccc} & \overset{\iota_u}{\swarrow} & A_0 & \overset{\iota_r}{\searrow} & \\ & & \downarrow \iota & & \\ A_u & \twoheadrightarrow & A & \twoheadrightarrow & A_r, \end{array}$$

*where the horizontal maps are surjective morphisms of $C^*$-bialgebras.*

*Proof.* Denote by $\|\cdot\|_u$, $\|\cdot\|$ and $\|\cdot\|_r$ the norms on $A_u$, $A$ and $A_r$, respectively. Evidently, it suffices to show that

$$\|\iota_r(a)\|_r \leq \|\iota(a)\| \leq \|\iota_u(a)\|_u \quad \text{for all } a \in A_0.$$

The inequality on the right-hand side follows immediately from the definition of $\|\cdot\|_u$. To prove the inequality on the left-hand side, let us use the notation of

Theorem 5.4.5. Denote by $h$ and $h_0$ the Haar states of $A$ and $A_0$, respectively. For all $a, b \in A_0$,

$$\iota(b)^*\iota(a^*a)\iota(b) \leq \iota(b)^*\iota(b)\|\iota(a)\|^2 \quad \text{in } A$$

and, since $h_0 = h \circ \iota$ by Proposition 2.2.6,

$$\begin{aligned}\|\iota_r(a)\Lambda(b)\|^2 &= |h_0(b^*a^*ab)| = |h(\iota(b^*)\iota(a^*a)\iota(b))| \\ &\leq |h_0(b^*b)| \cdot \|\iota(a)\|^2 = \|\Lambda(b)\|^2 \cdot \|\iota(a)\|^2.\end{aligned}$$

Therefore, $\|\iota_r(a)\| \leq \|\iota(a)\|$ for all $a \in A_0$. □

The previous result motivates the following terminology:

**Definition 5.4.9.** We call a $C^*$-algebraic CQG $(A, \Delta)$ *universal* if the natural surjection $A_u \to A$ is an isomorphism.

We call a CQG $(A, \Delta)$ *amenable* if the associated universal and reduced $C^*$-algebraic CQGs coincide.

Let us summarize the contents of Theorems 5.4.1, 5.4.3 and 5.4.5 in categorical terms. Evidently, the class of all algebraic compact quantum groups forms a category with respect to morphisms of Hopf $*$-algebras, and the class of all universal $C^*$-algebraic compact quantum groups forms a category with respect to morphisms of $C^*$-bialgebras.

**Corollary 5.4.10.** i) *The assignment $(A_0, \Delta_0) \mapsto (A_u, \Delta_u)$ defines a natural equivalence between the category of algebraic CQGs and the category of universal $C^*$-algebraic CQGs.*

ii) *The assignment $(A_0, \Delta_0) \mapsto (A_r, \Delta_r)$ defines a bijective correspondence (up to isomorphism) between all algebraic CQGs and all reduced $C^*$-algebraic CQGs.* □

# Chapter 6
# Examples of compact quantum groups

The general theory of compact quantum groups is complemented by a rich supply of examples. The most important sources of examples are the following:

**$q$-deformations of compact semisimple Lie groups.** In [37] and [70], Drinfeld and Jimbo associated to every semisimple complex Lie algebra $\mathfrak{g}$ a Hopf algebra $U_q(\mathfrak{g})$ which is a deformation of the universal enveloping algebra $U(\mathfrak{g})$ depending on a formal parameter $q$. These $q$-deformations attracted much attention [23], [24], [68], [79], [80], [84], [103], [140], but were studied from a different perspective than that of compact quantum groups. The first contact point for the two approaches was the compact quantum group $\mathrm{SU}_\mu(2)$, which was introduced independently by Soibelman and Vaksman [169] and by Woronowicz [194]; see also [133]. Later, Levendorski and Soibelman [100] and Rosso [134] constructed many more examples of compact quantum groups out of the $q$-deformations of Drinfeld and Jimbo; see also [84].

**Universal/free compact quantum groups.** Another important class of examples are the free unitary and the free orthogonal quantum groups, which were introduced by Wang [185], [186] and generalized by Wang and Van Daele [179], and the quantum permutation groups introduced by Wang [189]. A detailed study of these quantum groups was given by Banica [12], [13], [14].

**General constructions.** There exist several general constructions that produce new compact quantum groups out of given ones, for example

- the free product, tensor product, and crossed product constructions of Wang [186], [187],
- the deformation of compact groups by toral subgroups and its generalization to compact quantum groups, which were given by Rieffel [131] and Wang [188], respectively.

The examples of the first and second kind listed above belong to the class of compact matrix quantum groups. We begin with a review of this class (Section 6.1) and thereafter discuss the following examples: the compact quantum group $\mathrm{SU}_\mu(2)$ of Woronowicz (Section 6.2); free products and tensor products of compact quantum groups (Section 6.3); and the free unitary and the free orthogonal quantum group (Section 6.4).

Throughout this chapter, we use the results on compact quantum groups that were obtained in Chapters 3 and 5; in particular the correspondence between algebraic and $C^*$-algebraic compact quantum groups described in Section 5.4 and the correspondence between corepresentations, corepresentation operators, and corepresentation matrices described in Sections 3.1 and 5.2.

## 6.1 Compact matrix quantum groups

Compact matrix quantum groups are analogues of compact Lie groups. They include many examples of compact quantum groups like the $q$-deformations of compact semisimple Lie groups and the universal compact quantum groups mentioned above. Their theory was developed by Woronowicz [193]; it preceded the theory of general compact quantum groups.

Let us fix some notation. Throughout this section, we treat the algebraic and the $C^*$-algebraic setting in parallel and use the symbol "$\otimes$" to denote the algebraic tensor product of $*$-algebras and the minimal tensor product of $C^*$-algebras. It will always be clear from the context which tensor product we refer to. Given a $*$-algebra $A$ and a matrix $u = (u_{ij})_{i,j} \in M_n(A)$, we put $\bar{u} := (u_{ij}{}^*)_{i,j}$ and $u^t = (u_{ji})_{i,j}$; thus, $\bar{u}^* = u^t$.

Roughly, a compact matrix quantum group is a compact quantum group that is generated by the entries of one fundamental corepresentation matrix:

**Definition 6.1.1.** A *$C^*$-algebraic compact matrix quantum group* is a unital $C^*$-algebra $A$ equipped with a $*$-homomorphism $\Delta\colon A \to A \otimes A$ and a unitary $u \in M_n(A)$, where $n \in \mathbb{N}$, such that

i) $\Delta(u_{ij}) = \sum_k u_{ik} \otimes u_{kj}$ for all $i, j$;

ii) $\bar{u}$ is an invertible matrix;

iii) the elements $u_{ij}$ $(1 \leq i, j \leq n)$ generate $A$ (as a $C^*$-algebra).

An *algebraic compact matrix quantum group* is a Hopf $*$-algebra $(A, \Delta)$ together with a unitary $u \in M_n(A)$, where $n \in \mathbb{N}$, such that

i) $u$ is a corepresentation matrix;

ii) $\bar{u}$ is equivalent to a unitary corepresentation matrix;

iii) the elements $u_{ij}$ $(1 \leq i, j \leq n)$ generate $A$ (as a $*$-algebra).

In both cases, $u$ is called the *fundamental corepresentation* (*matrix*) of $(A, \Delta, u)$.

**Remarks 6.1.2.** i) In the definition above, $\Delta$ is uniquely determined by $A$ and $u$ because of conditions i) and iii). Therefore one can also refer to the pair $(A, u)$ as a compact matrix quantum group.

ii) Let $(A, \Delta, u)$ be an algebraic compact matrix quantum group. Then the counit and the antipode of the Hopf $*$-algebra $(A, \Delta)$ are determined by $\epsilon(u_{ij}) = \delta_{i,j}$ and $S(u_{ij}) = u^*_{ji}$ for all $i, j$; see Proposition 3.1.7 iii), v).

**Definition 6.1.3.** We call two compact matrix quantum groups $(A, \Delta_A, u)$ and $(B, \Delta_B, v)$ *similar* and write $(A, \Delta_A, u) \simeq (B, \Delta_B, v)$ if there exist a $*$-isomorphism $f : A \to B$, an $n \in \mathbb{N}$, and a matrix $T \in \mathrm{GL}_n(\mathbb{C})$ such that $u \in M_n(A)$, $v \in M_n(B)$, and $v = T(f(u_{ij}))_{i,j}T^{-1}$.

**Notation.** We abbreviate the phrase "compact matrix quantum group" by the acronym "CMQG" whenever it seems convenient, and omit the qualifiers "algebraic" and "$C^*$-algebraic" when we want to refer to both cases.

The terminology introduced in Definition 6.1.1 is justified because of the following result:

**Proposition 6.1.4.** i) *If $(A, \Delta, u)$ is a CMQG, then $(A, \Delta)$ is a CQG and $u, \bar{u}$ are corepresentation matrices of $(A, \Delta)$.*

*Let $(A, \Delta_A, u)$ and $(B, \Delta_B, v)$ be compact matrix quantum groups, where $u \in M_n(A)$ and $v \in M_n(B)$ for some $n \in \mathbb{N}$.*

ii) *Let $f : A \to B$ be a $*$-homomorphism such that $f(u_{ij}) = v_{ij}$ for all $i, j$. Then $f$ is a morphism of compact quantum groups.*

iii) *If $(A, \Delta_A, u) \simeq (B, \Delta_B, v)$ as CMQGs, then $(A, \Delta_A) \cong (B, \Delta_B)$ as CQGs.*

*Proof.* i) In the algebraic case, the assertion follows from Theorem 3.2.12 iv). In the $C^*$-algebraic case, conditions i) and iii) of Definition 6.1.1 imply that $\Delta$ is coassociative, so that $(A, \Delta)$ is a $C^*$-bialgebra. Furthermore, condition i) implies that $u$ is a corepresentation matrix of $(A, \Delta)$. Now the assertion follows from conditions ii), iii) and Theorem 5.3.11 iii).

ii) We have $(f \otimes f) \circ \Delta_A = \Delta_B \circ f$ because $A$ is generated by the $u_{ij}$ and

$$(f \otimes f)(\Delta_A(u_{ij})) = \sum_k f(u_{ik}) \otimes f(u_{kj})$$
$$= \sum_k v_{ik} \otimes v_{kj} = \Delta_B(v_{ij}) = \Delta_B(f(u_{ij}))$$

for all $i, j$.

iii) Let $f\colon A \to B$ and $T$ be as in Definition 6.1.3. Then

$$\begin{aligned}(f \otimes f)(\Delta_A(u_{ij})) &= \sum_l f(u_{il}) \otimes f(u_{lj}) \\ &= \sum_{m,p,l,q,n} T^{-1}_{im} v_{mp} T_{pl} \otimes T^{-1}_{lq} v_{qn} T_{nj} \\ &= \sum_{m,k,n} T^{-1}_{im} v_{mk} \otimes v_{kn} T_{nj} \\ &= \sum_{m,n} T^{-1}_{im} \Delta_B(v_{mn}) T_{nj} = \Delta_B(f(u_{ij}))\end{aligned} \tag{6.1}$$

for all $i, j$, and hence $(f \otimes f) \circ \Delta_A = \Delta_B \circ f$. □

The next results characterize those compact quantum groups which can be regarded as compact matrix quantum groups:

**Lemma 6.1.5.** *Let $(A, \Delta)$ be a compact quantum group with a unitary corepresentation matrix $u \in M_n(A)$, where $n \in \mathbb{N}$. Then the following conditions are equivalent:*

i) *Every irreducible corepresentation of $(A, \Delta)$ is contained in some iterated tensor product of $u$, $\bar{u}$, and the trivial corepresentation.*
ii) *The matrix entries of $u$ generate $A$.*

*Proof.* This follows from the fact that every $*$-monomial in the matrix entries of $u$ appears as a matrix coefficient of some tensor product of the corepresentations $u$ and $\bar{u}$, and from Theorem 3.2.1 / 5.3.3 and Corollary 3.2.8 / 5.3.9. □

**Proposition 6.1.6.** *For every algebraic/$C^*$-algebraic compact quantum group $(A, \Delta)$, the following conditions are equivalent:*

i) *There exists a unitary corepresentation matrix $u \in M_n(A)$ for some $n \in \mathbb{N}$ such that every irreducible corepresentation of $(A, \Delta)$ is contained in some iterated tensor product of $u$, $\bar{u}$, and the trivial corepresentation.*
ii) *The following $*$-algebra is finitely generated: $A$/the subalgebra $A_0 \subseteq A$ of matrix coefficients of finite-dimensional corepresentations.*
iii) *There exists a unitary $u \in M_n(A)$ for some $n \in \mathbb{N}$ such that $(A, \Delta, u)$ is an algebraic/$C^*$-algebraic compact matrix quantum group.*

*Proof.* i) $\Rightarrow$ ii) This is evident.

ii) $\Rightarrow$ i) If $A$ / $A_0$ is finitely generated as a $*$-algebra, then there exist finite-dimensional unitary corepresentation matrices $u_1, \dots, u_k$ whose matrix entries generate $A$ / $A_0$ as a $*$-algebra (see Theorem 3.2.12 v)), and we can choose $u$ to be the block matrix with diagonal matrix blocks $u_1, \dots, u_k$.

iii) $\Leftrightarrow$ i) This follows directly from Lemma 6.1.5. □

**Remark 6.1.7.** Compact matrix groups can be reconstructed from their category of finite-dimensional corepresentations. This generalization of the Tannaka–Krein duality theory was proved by Woronowicz, see [195].

Algebraic and $C^*$-algebraic compact matrix quantum groups are related as follows. Recall from Section 5.4 that one can associate

- to every $C^*$-algebraic CQG $(A, \Delta)$ an algebraic CQG $(A_0, \Delta_0)$,
- to every algebraic CQG $(A, \Delta)$ a universal and a reduced $C^*$-algebraic CQG $(A_u, \Delta_u)$ and $(A_r, \Delta_r)$, respectively.

**Proposition 6.1.8.** i) *If $(A, \Delta, u)$ is a $C^*$-algebraic CMQG, then the coefficients of $u$ belong to $A_0$, and $(A_0, \Delta_0, u)$ is an algebraic CMQG.*

*If $(A, \Delta, u)$ is an algebraic CMQG, then $(A_u, \Delta_u, u)$ and $(A_r, \Delta_r, u)$ are $C^*$-algebraic CMQGs.*

ii) *If $(A, \Delta_A, u)$ and $(B, \Delta_B, v)$ are similar $C^*$-algebraic CMQGs, then the algebraic CMQGs $(A_0, \Delta_{A,0}, u)$ and $(B_0, \Delta_{B,0}, v)$ are similar.*

*If $(A, \Delta_A, u)$ and $(B, \Delta_B, v)$ are similar algebraic CMQGs, then the $C^*$-algebraic CMQGs $(A_u, \Delta_{A,u}, u)$ and $(B_u, \Delta_{B,u}, v)$ are similar.* □

Let us consider some examples of compact matrix quantum groups. We begin with examples related to classical groups:

**Example 6.1.9.** Let $G \subseteq \mathrm{U}_n(\mathbb{C})$ be a closed subgroup, where $n \in \mathbb{N}$. For $i, j = 1, \dots, n$, denote by $u_{ij} \colon C(G) \to \mathbb{C}$ the function $x \mapsto x_{ij}$. Then the $C^*$-bialgebra $C(G)$ together with the matrix $u = (u_{ij})_{i,j}$ is a $C^*$-algebraic compact matrix quantum group. Indeed, condition i) of Definition 6.1.1 holds because $\Delta(u_{ij})(x, y) = u_{ij}(xy) = \sum_k u_{ik}(x)u_{kj}(y)$ for all $x, y \in G$, condition ii) holds because $\bar{u}u^t = \overline{(uu^*)} = 1$ (here we use commutativity of $C(G)$), and condition iii) holds by the Stone–Weierstrass Theorem.

The Hopf $*$-algebra $\mathrm{Rep}(G)$ of all representative functions on $G$ contains all $u_{ij}$, and together with the matrix $u$, it is an algebraic compact matrix quantum group. The fact that condition ii) is satisfied in this case is well-known; this follows also from the fact that $\mathrm{Rep}(G)$ is the algebraic compact quantum group associated to $C(G)$ (see Proposition 6.1.8 and Example 5.4.2 i)).

Let $H \subseteq \mathrm{U}_n(\mathbb{C})$ be another closed subgroup. It is easy to see that the following conditions are equivalent:

i) $(C(G), \Delta_G, u^G) \simeq (C(H), \Delta_H, u^H)$;

ii) $(\mathrm{Rep}(G), \Delta_G, u^G) \simeq (\mathrm{Rep}(H), \Delta_H, u^H)$;

iii) $H = TGT^{-1}$ for some $T \in \mathrm{GL}_n(\mathbb{C})$.

Here, we indexed the comultiplications and the fundamental corepresentations of $C(G)$ and $C(H)$ by "$G$" and "$H$", respectively, to avoid ambiguities.

Note that every compact Lie group $G$ can be embedded as a closed subgroup into $\mathrm{U}_n(\mathbb{C})$ for some $n \in \mathbb{N}$ and thus gives rise to an example of a compact matrix quantum group.

**Example 6.1.10.** Let $G$ be a discrete group that is finitely generated by elements $x_1, \dots, x_n$. Then the group Hopf $*$-algebra $\mathbb{C}G$ together with the diagonal matrix $u = \mathrm{diag}(U_{x_1}, \dots, U_{x_n}) \in M_n(\mathbb{C}G)$ is an algebraic CMQG, and the universal group $C^*$-bialgebra $C^*(G)$ together with the matrix $u$ as well as the reduced group $C^*$-bialgebra $C^*_r(G)$ together with the matrix $\mathrm{diag}(L_{x_1}, \dots, L_{x_n})$ are $C^*$-algebraic CMQGs.

Every commutative compact matrix quantum group has the form described in Example 6.1.9:

**Proposition 6.1.11.** *Let $(A, \Delta, u)$ be an algebraic/$C^*$-algebraic compact matrix quantum group, where $u \in M_n(A)$, $n \in \mathbb{N}$. Assume that $A$ is commutative. Then there exist a closed compact subgroup $G \subseteq \mathrm{U}_n(\mathbb{C})$ and an isomorphism $\Phi\colon A \xrightarrow{\cong} \mathrm{Rep}(G)$ / $\Phi\colon A \xrightarrow{\cong} C(G)$ such that $(\Phi(u_{ij}))(x) = x_{ij}$ for all $x \in G$ and $i, j = 1, \dots, n$.*

*Proof.* By Proposition 6.1.8 and Example 5.4.2 i) and 5.4.6 ii), it suffices to consider the case that $(A, \Delta, u)$ is a $C^*$-algebraic CMQG. Denote by $\widehat{A}$ the spectrum of $A$, that is, the space of all $*$-homomorphisms $A \to \mathbb{C}$, and consider the map

$$\Psi\colon \widehat{A} \to M_n(\mathbb{C}), \quad \chi \mapsto (\chi(u_{ij}))_{i,j}.$$

This map is an embedding because the matrix entries of $u$ generate $A$, it is a homeomorphism onto a closed subset because $\widehat{A}$ is compact, and its image, which we denote by $G$, is contained in $\mathrm{U}_n(\mathbb{C})$ because $u$ is unitary.

Let $\chi, \chi' \in \widehat{A}$. Then $\chi'' := (\chi \otimes \chi') \circ \Delta \in \widehat{A}$, and $\Psi(\chi)\Psi(\chi') = \Psi(\chi'')$ because

$$(\Psi(\chi)\Psi(\chi'))_{ij} = \sum_k \chi(u_{ik})\chi'(u_{kj}) = (\chi \otimes \chi')(\Delta(u_{ij})) = \Psi(\chi'')_{ij} \quad \text{for all } i, j.$$

In particular, $GG \subseteq G$. Thus $G$ is a compact semigroup with cancellation and hence a group (see, for example, [63, Theorem II.9.16] or [105, Proposition 3.2]).

By the Gelfand Theorem (see Section 12.1), the map $\Phi\colon A \to C(G)$ given by $(\Phi(a))(x) := (\Psi^{-1}(x))(a)$ is an isomorphism, and $(\Phi(u_{ij}))(\Psi(\chi)) = \chi(u_{ij}) = (\Psi(\chi))_{ij}$ for all $\chi, i, j$. The claim follows. □

Finally, we give an example of a compact matrix quantum group that is neither commutative nor cocommutative. Its construction uses the following easy observation:

**Lemma 6.1.12.** *If $A$ is a $*$-algebra, $n \in \mathbb{N}$, and $u \in M_n(A)$ is unitary, then also the matrix $\Delta_n(u) := \left(\sum_k u_{ik} \otimes u_{kj}\right)_{i,j} \in M_n(A \otimes A)$ is unitary.* □

**Example 6.1.13.** For each $n \in \mathbb{N}$, one can define a *quantum permutation group* on $n$ letters as follows. Given a $*$-algebra $A$, call a unitary matrix $u \in M_n(A)$ *magic* if for each $i$ and $j$,

i) $u_{ij}$ is a projection,

ii) the projections $u_{i1}, \dots, u_{in}$ are orthogonal and $\sum_k u_{ik} = 1$, and

iii) the projections $u_{1j}, \dots, u_{nj}$ are orthogonal and $\sum_k u_{kj} = 1$.

Note that $\bar{u} = u$ for every magic unitary matrix $u$.

Denote by $A_s(n)$ the universal $C^*$-algebra with elements $u_{ij}$ $(1 \leq i, j \leq n)$ such that $u = (u_{ij})_{i,j}$ is a magic unitary matrix. A routine calculation shows that

$$\Delta_n(u) := \Big(\sum_k u_{ik} \otimes u_{kj}\Big)_{i,j} \in M_n(A_s(n) \otimes A_s(n))$$

is a magic unitary matrix again. By the universal property of $A_s(n)$, there exists a unique $*$-homomorphism $\Delta\colon A_s(n) \to A_s(n) \otimes A_s(n)$ such that $(A_s(n), \Delta, u)$ is a $C^*$-algebraic compact matrix quantum group. This is the *quantum permutation group* on $n$ letters. It was introduced by Wang [189] and thereafter studied by Banica and Collins [14], [15].

The quantum permutation group is related to the group $S(n)$ of permutations of $n$ letters as follows. Denote by $A_s^c(n)$ the universal commutative $C^*$-algebra with elements $u_{ij}^c$ $(1 \leq i, j \leq n)$ such that $u^c = (u_{ij}^c)_{i,j}$ is a magic unitary matrix. Then there exists a unique $*$-homomorphism $\Delta^c$ such that $(A_s^c(n), \Delta^c, u^c)$ is a $C^*$-algebraic CMQG, and a surjective morphism of $C^*$-algebraic CMQGs $A_s(n) \to A_s^c(n)$, given by $u_{ij} \mapsto u_{ij}^c$. By Proposition 6.1.11, $(A_s^c(n), \Delta^c, u^c)$ is isomorphic to the $C^*$-algebraic CMQG associated to the group

$$\begin{aligned}
&\{(\chi(u_{ij}^c))_{i,j} \in \mathrm{GL}_n(\mathbb{C}) \mid \chi \text{ is a character on } A_s^c(n)\} \\
&\quad = \{x \in \mathrm{GL}_n(\mathbb{C}) \mid x_{ij} \in \{0,1\},\ \textstyle\sum_k x_{ik} = 1,\ \sum_k x_{kj} = 1 \text{ for all } i, j\} \\
&\quad = \{x \in \mathrm{GL}_n(\mathbb{C}) \mid \text{ there is } \pi \in S(n) \text{ with } x_{ij} = \delta_{i,\pi(j)} \text{ for all } i, j\} \cong S(n).
\end{aligned}$$

## 6.2 The compact quantum group $SU_\mu(2)$

The quantum group $SU_\mu(2)$ is probably *the* fundamental example of a compact quantum group that is neither commutative nor cocommutative. It was introduced independently by Soibelman and Vaksman [169] as the dual of the $q$-deformed enveloping Lie algebra $U_q(\mathfrak{su}_2)$, and by Woronowicz [194] as a deformation of the compact Lie group $SU_2(\mathbb{C})$ or, more precisely, of the $C^*$-bialgebra $C(SU_2(\mathbb{C}))$. A comparison between these approaches can be found in [133].

We shall follow the approach of Woronowicz and present the construction of the compact quantum group $SU_\mu(2)$, describe the corepresentation theory, and deduce some results on the Haar state.

For further information, we refer to the following literature: A gentle introduction to the $q$-deformed enveloping Lie algebra $U_q(\mathfrak{su}_2)$, the Hopf algebra $SU_\mu(2)$, and their interrelation can be found in [79]; for a comprehensive treatment, see [24], [80]. An introduction to $q$-deformations, their relation to Poisson–Lie groups, and to functions on $q$-deformed quantum groups is given in [24], [84]. The harmonic analysis on $SU_\mu(2)$ and relations to $q$-special functions were studied in many papers, see, for example, [83], [108], [109], [169]; comprehensive treatments can be found in [80], [183].

Note that in the literature, the deformation parameter $\mu$ is usually denoted by $q$, and the quantum group $SU_\mu(2)$ is denoted by a variety of different symbols. Common notation for the underlying algebraic compact quantum group includes, for example, $SU_q(2)$, $\mathcal{O}_q(SU_2)$, $\mathbb{C}[SU_2]_q$.

### 6.2.1 Definition and first properties

Recall that the compact Lie group $SU_2(\mathbb{C})$ consists of all matrices of the form

$$g_{(\alpha,\gamma)} := \begin{pmatrix} \alpha & -\bar{\gamma} \\ \gamma & \bar{\alpha} \end{pmatrix}, \quad \text{where } \alpha, \gamma \in \mathbb{C} \text{ and } \bar{\alpha}\alpha + \bar{\gamma}\gamma = 1.$$

Define $a, c \in C(SU_2(\mathbb{C}))$ by $a(g_{(\alpha,\gamma)}) := \alpha$ and $c(g_{(\alpha,\gamma)}) := \gamma$. Then the $C^*$-bialgebra $C(SU_2(\mathbb{C}))$ together with the unitary matrix

$$u := \begin{pmatrix} a & -c^* \\ c & a^* \end{pmatrix} \in M_2(C(SU_2(\mathbb{C}))) \tag{6.2}$$

is a $C^*$-algebraic compact matrix quantum group (Example 6.1.9). The corresponding algebraic compact quantum group is the Hopf $*$-algebra of representative functions $\mathrm{Rep}(SU_2(\mathbb{C})) \subset C(SU_2(\mathbb{C}))$. As a $C^*$-algebra or $*$-algebra, respectively, $C(SU_2(\mathbb{C}))$ and $\mathrm{Rep}(SU_2(\mathbb{C}))$ are isomorphic to the universal unital commutative $C^*$-algebra or $*$-algebra generated by elements $a, c$ that satisfy

$a^*a + c^*c = 1$. A short calculation shows that the counit and antipode of $\mathrm{Rep}(\mathrm{SU}_2(\mathbb{C}))$ are given by

$$\epsilon(a) = 1, \quad \epsilon(c) = 0, \quad S(a) = a^*, \quad S(a^*) = a, \quad S(c) = -c, \quad S(c^*) = -c^*.$$

**Definition 6.2.1.** For $\mu \in [-1, 1]$, denote by $\mathrm{SU}_\mu(2)$ the universal unital $C^*$-algebra generated by elements $a, c$ subject to the condition that the following matrix is unitary:

$$u := \begin{pmatrix} a & -\mu c^* \\ c & a^* \end{pmatrix}.$$

If $u$ is as above, then

$$\begin{aligned} u^* &= \begin{pmatrix} a^* & c^* \\ -\mu c & a \end{pmatrix}, \\ u^*u &= \begin{pmatrix} a^*a + c^*c & -\mu a^*c^* + c^*a^* \\ -\mu ca + ac & \mu^2 cc^* + aa^* \end{pmatrix}, \\ uu^* &= \begin{pmatrix} aa^* + \mu^2 c^*c & ac^* - \mu c^*a \\ ca^* - \mu a^*c & cc^* + a^*a \end{pmatrix}. \end{aligned}$$

Hence $u$ is unitary if and only if $a$ and $c$ satisfy

$$\begin{gathered} a^*a + c^*c = 1, \qquad aa^* + \mu^2 c^*c = 1, \\ c^*c = cc^*, \qquad ac = \mu ca, \qquad ac^* = \mu c^*a. \end{gathered} \tag{6.3}$$

**Lemma 6.2.2.** *Let $A$ be a $*$-algebra and $v \in M_2(A)$. If $\mu \in [-1, 1] \setminus \{0\}$, then the following conditions are equivalent:*

i) $v = \begin{pmatrix} a' & -\mu c'^* \\ c' & a'^* \end{pmatrix}$ *for some $a', c' \in A$;*

ii) $v = F\bar{v}F^{-1}$, *where* $F = \begin{pmatrix} 0 & 1 \\ -\mu^{-1} & 0 \end{pmatrix}$.

*Proof.* A short calculation shows $F\bar{v}F^{-1} = \begin{pmatrix} v_{22}^* & -\mu v_{21}^* \\ -\mu^{-1} v_{12}^* & v_{11}^* \end{pmatrix}$. □

**Proposition 6.2.3.** *For each $\mu \in [-1, 1] \setminus \{0\}$, there exists a unique $*$-homomorphism $\Delta\colon \mathrm{SU}_\mu(2) \to \mathrm{SU}_\mu(2) \otimes \mathrm{SU}_\mu(2)$ such that $(\mathrm{SU}_\mu(2), \Delta, u)$ is a $C^*$-algebraic compact matrix quantum group, and $\Delta$ is determined by*

$$\Delta(a) = a \otimes a - \mu c^* \otimes c, \quad \Delta(c) = c \otimes a + a^* \otimes c. \tag{6.4}$$

*Proof.* The matrix $v \in M_2(\mathrm{SU}_\mu(2) \otimes \mathrm{SU}_\mu(2))$ given by $v_{ij} = \sum_k u_{ik} \otimes u_{kj}$ is unitary (Lemma 6.1.12) and has the form

$$v = \begin{pmatrix} a' & -\mu c'^* \\ c' & a'^* \end{pmatrix}, \quad \text{where} \quad \begin{aligned} a' &= a \otimes a - \mu c^* \otimes c, \\ c' &= c \otimes a + a^* \otimes c. \end{aligned}$$

The universal property of $\mathrm{SU}_\mu(2)$ yields a $*$-homomorphism

$$\Delta \colon \mathrm{SU}_\mu(2) \to \mathrm{SU}_\mu(2) \otimes \mathrm{SU}_\mu(2)$$

such that $\Delta(a) = a'$ and $\Delta(c) = c'$. Thus, condition i) of Definition 6.1.1 is satisfied. Condition ii) – invertibility of $\bar{u}$ – follows from Lemma 6.2.2, and condition iii) is trivially satisfied. □

Definition 6.2.1 and Proposition 6.2.3 yield an entire family $(\mathrm{SU}_\mu(2))_{\mu \in (0,1]}$ of compact matrix quantum groups which can be considered as a continuous deformation of $\mathrm{SU}_1(2) \cong C(\mathrm{SU}_2(\mathbb{C}))$ [17]. From now on, we fix $\mu$:

**Assumption.** The deformation parameter $\mu$ is some fixed element of $[-1, 1] \setminus \{0\}$.

**Remark 6.2.4.** Let $(A, \Delta_A)$ be a $C^*$-bialgebra with a unitary corepresentation matrix $v \in M_2(A)$. Denote by $(e_1, e_2)$ the standard basis of $\mathbb{C}^2$ and put $e_{kl} := e_k \otimes e_l \in \mathbb{C}^2 \otimes \mathbb{C}^2$ for $k, l = 1, 2$. Then the element $\det_\mu := e_{12} - \mu e_{21} \in \mathbb{C}^2 \otimes \mathbb{C}^2$ can be considered as a $\mu$-deformed determinant, and the following conditions are equivalent:

i) $v = \begin{pmatrix} a' & -\mu c'^* \\ c' & a'^* \end{pmatrix}$ for some $a', c' \in A$;

ii) the corepresentation $\delta \colon \mathbb{C}^2 \otimes \mathbb{C}^2 \to \mathbb{C}^2 \otimes \mathbb{C}^2 \otimes A$ corresponding to $v \boxtimes v$ preserves $\det_\mu$ in the sense that $\delta(\det_\mu) = \det_\mu \otimes 1$.

Indeed, writing $\delta$ and $\det_\mu$ in terms of the basis $e_{11}, e_{12}, e_{21}, e_{22}$, one can check:

$$\text{ii)} \Leftrightarrow \begin{pmatrix} v_{11} & & v_{12} & \\ & v_{11} & & v_{12} \\ v_{21} & & v_{22} & \\ & v_{21} & & v_{22} \end{pmatrix} \begin{pmatrix} v_{11} & v_{12} & & \\ v_{21} & v_{22} & & \\ & & v_{11} & v_{12} \\ & & v_{21} & v_{22} \end{pmatrix} \begin{pmatrix} 0 \\ 1 \\ -\mu \\ 0 \end{pmatrix} = \begin{pmatrix} 0 \\ 1 \\ -\mu \\ 0 \end{pmatrix}$$

$$\Leftrightarrow \begin{pmatrix} v_{11} & v_{12} & & \\ v_{21} & v_{22} & & \\ & & v_{11} & v_{12} \\ & & v_{21} & v_{22} \end{pmatrix} \begin{pmatrix} 0 \\ 1 \\ -\mu \\ 0 \end{pmatrix} = \begin{pmatrix} v_{11}^* & & v_{21}^* & \\ & v_{11}^* & & v_{21}^* \\ v_{12}^* & & v_{22}^* & \\ & v_{12}^* & & v_{22}^* \end{pmatrix} \begin{pmatrix} 0 \\ 1 \\ -\mu \\ 0 \end{pmatrix} \Leftrightarrow \text{i)}.$$

Denote by $\mathrm{SU}^0_\mu(2) \subset \mathrm{SU}_\mu(2)$ the $*$-subalgebra generated by $a$ and $c$, and by $\Delta_0$ the restriction of $\Delta$ to $\mathrm{SU}^0_\mu(2)$. Then $(\mathrm{SU}^0_\mu(2), \Delta_0, u)$ is the algebraic compact matrix quantum group associated to $(\mathrm{SU}_\mu(2), \Delta, u)$ (Proposition 6.1.8). By Remark 6.1.2 ii), the counit and the antipode of $(\mathrm{SU}^0_\mu(2), \Delta_0)$ are given by

$$\begin{gathered} \epsilon_0(a) = 1, \quad \epsilon_0(c) = 1, \\ S_0(a) = a^*, \quad S_0(a^*) = a, \quad S_0(c) = -\mu c, \quad S_0(c^*) = -\mu^{-1}c^*. \end{gathered} \tag{6.5}$$

For each $k \in \mathbb{Z}$ and $m, n \in \mathbb{N}$, put

$$a_{kmn} := \begin{cases} a^k c^{*m} c^n, & k \geq 0, \\ a^{*(-k)} c^{*m} c^n, & k < 0. \end{cases}$$

**Proposition 6.2.5.** *The family $(a_{kmn})_{k,m,n}$ is a basis of $\mathrm{SU}^0_\mu(2)$.*

*Proof.* Using equation (6.3), one can check that the family $(a_{kmn})_{k,m,n}$ spans $\mathrm{SU}^0_\mu(2)$. We show that this family is linearly independent. Let $H$ be a Hilbert space with orthonormal basis $(e_{r,s})_{r\in\mathbb{N},s\in\mathbb{Z}}$, and define $a', c' \in \mathcal{L}(H)$ by

$$a' e_{r,s} := \sqrt{1-\mu^{2r}}\, e_{r-1,s}, \qquad c' e_{r,s} := \mu^r e_{r,s+1} \qquad \text{for all } r \in \mathbb{N}, s \in \mathbb{Z}.$$

Straightforward calculations show that $a'$ and $c'$ satisfy the relations (6.3). Hence there exists a $*$-homomorphism $\pi \colon \mathrm{SU}_\mu(2) \to \mathcal{L}(H)$ mapping $a$ to $a'$ and $c$ to $c'$. A short calculation shows that for all $k \in \mathbb{Z}$, $m, n \in \mathbb{N}$, and $r \geq k$,

$$\pi(a_{kmn}) e_{r,0} = \Bigl( \prod_{l=0}^{|k|-1} \sqrt{1-\mu^{2r+\mathrm{sgn}(k)2l}} \Bigr) \mu^{r(m+n)} e_{r-k,n-m}.$$

Consider a non-trivial finite linear combination $x = \sum_{k,m,n} \lambda_{kmn} a_{kmn} \in \mathrm{SU}^0_\mu(2)$. We claim that $\pi(x) \neq 0$. Indeed, if $k, m, n$ are chosen such that $\lambda_{kmn} \neq 0$ and $\lambda_{k'm'n'} = 0$ whenever $m' + n' < m + n$, then

$$\lim_{r\to\infty} \frac{1}{\mu^{r(m+n)}} \langle e_{r-k,n-m} | \pi(x) e_{r,0} \rangle = \lambda_{kmn} \neq 0. \qquad \square$$

### 6.2.2 Corepresentations and their weights

The corepresentation theory of the compact quantum group $\mathrm{SU}_\mu(2)$ is very similar to the representation theory of the compact Lie group $\mathrm{SU}_2(\mathbb{C})$. It can be determined either via a differential calculus or via harmonic analysis and special $q$-functions. We shall follow the first approach, which is the original one of Woronowicz [194]; for the second approach, see, for example, [80, Section 4.3].

In this subsection, we formulate the main result of Woronowicz in terms of weight functions of corepresentations and draw some immediate conclusions concerning tensor products and characters of irreducible corepresentations. The proof of the main result, which involves the differential calculus mentioned above, is given in the next section.

Let us start with the definition of the weight function associated to a corepresentation of $\mathrm{SU}_\mu(2)$. This definition involves an analogue of the maximal torus of the compact Lie group $\mathrm{SU}_2(\mathbb{C})$:

**Proposition 6.2.6.** i) *Denote by $z \in C(\mathbb{T})$ the identity function. There exists a unique morphism of compact quantum groups $\pi_\mathbb{T} \colon \mathrm{SU}_\mu(2) \to C(T)$ such that $\pi_\mathbb{T}(a) = z$ and $\pi_\mathbb{T}(c) = 0$.*

ii) *For each $\zeta \in \mathbb{T}$, there exists a unique $*$-homomorphism $\theta_\zeta \colon \mathrm{SU}_\mu(2) \to \mathbb{C}$ such that $\theta_\zeta(a) = \zeta$ and $\theta_\zeta(c) = 0$. For all $\zeta, \zeta' \in \mathbb{T}$, one has $\theta_\zeta * \theta_{\zeta'} = \theta_{\zeta\zeta'}$.*

*Proof.* Immediate from the definition of $\mathrm{SU}_\mu(2)$ and equation (6.4). □

Let $v$ be a corepresentation operator of $\mathrm{SU}_\mu(2)$ on a finite-dimensional Hilbert space $H$. Then $(\mathrm{id} \otimes \pi_\mathbb{T})(v)$ is a corepresentation operator of $C(\mathbb{T})$, which corresponds to a representation

$$\rho_v \colon \mathbb{T} \to \mathcal{L}(H), \quad \zeta \mapsto (\mathrm{id} \otimes \theta_\zeta)(v),$$

(see Example 5.2.5). Since every representation of $\mathbb{T}$ is a direct sum of representations of the form $\zeta \mapsto \zeta^k$, where $k \in \mathbb{Z}$, the space $H$ decomposes into a direct sum

$$H \cong \bigoplus_{k \in \mathbb{Z}} H_k, \quad \text{where } H_k := \{\eta \in H \mid \rho_v(\zeta)\eta = \zeta^k \eta \text{ for all } \zeta \in \mathbb{T}\}.$$

For each $k \in \mathbb{Z}$, we call $H_k$ the $k$th *weight space* of $v$. If $H_k \neq 0$, we call $k$ a *weight* of $v$. The maximal $k \in \mathbb{Z}$ that is a weight for $v$ is called the *highest weight* of $v$. We shall see that up to unitary equivalence, $v$ is completely determined by the *weight function*

$$W_v \colon \mathbb{Z} \to \mathbb{N}, \quad k \mapsto \dim H_k.$$

**Lemma 6.2.7.** *Let $v$ and $w$ be finite-dimensional corepresentation operators of $(\mathrm{SU}_\mu(2), \Delta)$. Then for all $k \in \mathbb{Z}$,*

$$W_{v \boxplus w}(k) = W_v(k) + W_w(k), \quad W_{v \boxtimes w}(k) = \sum_{p+q=k} W_v(p) W_w(q).$$

*Proof.* For each $k \in \mathbb{Z}$, denote by $H_k$, $K_k$, $(H \oplus K)_k$, and $(H \otimes K)_k$ the $k$th weight space of $v$, $w$, $v \boxplus w$, and $v \boxtimes w$, respectively. The first equation above follows from the evident relation $(H \oplus K)_k = H_k \oplus K_k$. The second one will follow once we proved $(H \otimes K)_k = \bigoplus_{p+q=k} H_p \otimes K_q$. Writing $v = \sum_r f_r \otimes v_r \in \mathcal{L}(H) \otimes \mathrm{SU}_\mu(2)$ and $w = \sum_s g_s \otimes w_s \in \mathcal{L}(K) \otimes \mathrm{SU}_\mu(2)$, we find that for each $\zeta \in \mathbb{T}$,

$$\rho_{v \boxtimes w}(\zeta) = \sum_{r,s} (f_r \otimes g_s)\theta_\zeta(v_r w_s) = \sum_{r,s} f_r \theta_\zeta(v_r) \otimes g_s \theta_\zeta(w_s) = \rho_v(\zeta) \otimes \rho_w(\zeta).$$

The relation $\bigoplus_{p+q=k} H_p \otimes K_q \subseteq (H \otimes K)_k$ follows easily. Summing over $k$ and comparing dimensions, we find that this inclusion is an equality for each $k$. □

The key to the corepresentation theory of $\mathrm{SU}_\mu(2)$ is the following result:

**Theorem 6.2.8.** *Up to equivalence, there exists for each $n \in \mathbb{N}$ precisely one irreducible corepresentation matrix $u_{(n)} \in M_{n+1}(\mathrm{SU}_\mu(2))$, and the associated weight function $W_{(n)} := W_{u_{(n)}}$ is given by*

$$W_{(n)}(k) = \begin{cases} 1, & k \in \{-n, 2-n, \dots, n-2, n\}, \\ 0, & \textit{otherwise}. \end{cases} \tag{6.6}$$

The proof of this theorem is outlined in the next section.

**Corollary 6.2.9.** i) *Up to equivalence, every finite-dimensional corepresentation of $(\mathrm{SU}_\mu(2), \Delta)$ is completely determined by its weight function.*

ii) *For each $n \in \mathbb{N}$, the comultiplication $\Delta$ restricts to a corepresentation $\delta_{(n)}$ on $V_{(n)} := \operatorname{span}\{a^k c^{*l} \mid k + l = n\} \subset \mathrm{SU}^0_\mu(2)$ that is equivalent to $u_{(n)}$.*

iii) $u_{(1)} \simeq u$.

iv) $u_{(m)} \boxtimes u_{(n)} \simeq u_{(|m-n|)} \boxplus u_{(|m-n|+2)} \boxplus \cdots \boxplus u_{(m+n-2)} \boxplus u_{(m+n)}$ *for all $m, n \in \mathbb{N}$.*

*Proof.* i) Since every corepresentation is equivalent to a direct sum of irreducible ones (Theorem 3.2.1) and the weight function behaves additively (Lemma 6.2.7), it suffices to show that the coefficients $d_n$ of a finite linear combination $W = \sum_n d_n W_{(n)}$ can be reconstructed from $W$. But equation (6.6) implies that $d_n = W(n) - W(n+2)$ for all $n \in \mathbb{N}$.

ii) Equation (6.4) shows that $\Delta(V_{(n)}) \subseteq V_{(n)} \otimes \mathrm{SU}_\mu(2)$ and

$$(\mathrm{id} \otimes \pi_{\mathbb{T}})(\Delta(a^k c^{*l})) = a^k c^{*l} \otimes z^{k-l} \quad \text{for all } k, l \in \mathbb{N}.$$

Consequently, $\delta_{(n)}$ is a corepresentation with weight function $W_{(n)}$.

iii) This follows immediately from ii) and equation (6.4).

iv) It suffices to check that the weight functions of the left- and of the right-hand side are equal, and this can easily be verified using Lemma 6.2.7. □

The characters $\chi_{(n)} := \chi(u_{(n)})$ of the corepresentations $u_{(n)}$ ($n \in \mathbb{N}$) enjoy the following properties:

**Proposition 6.2.10.** i) $\chi_{(0)} = 1$, $\chi_{(1)} = a + a^*$, *and for all* $n, m \in \mathbb{N}$,

$$\chi_{(m)}\chi_{(n)} = \chi_{(|m-n|)} + \chi_{(|m-n|+2)} + \cdots + \chi_{(m+n-2)} + \chi_{(m+n)}. \tag{6.7}$$

ii) *For each* $n \in \mathbb{N}$,

$$\chi_{(n)} = \frac{\sin\left(\frac{n+1}{2}t\right)}{\sin\left(\frac{1}{2}t\right)}, \quad \textit{where } t := 2\arccos\left(\frac{a+a^*}{2}\right). \tag{6.8}$$

iii) *The unital homomorphism* $\mathbb{C}[X] \to \mathrm{SU}_\mu(2)$ *given by* $X \mapsto \chi_{(1)}$ *is injective.*

iv) *For each* $n \in \mathbb{N}$, *there exists a unique* $f_n \in \mathbb{C}[X]$ *of degree* $n$ *such that* $\chi_{(n)} = f_n(\chi_{(1)})$. *The family* $(f_n)_{n\in\mathbb{N}}$ *is a basis of* $\mathbb{C}[X]$.

*Proof.* i) $\chi_{(1)} = \chi(u) = a + a^*$ because $u_{(1)} \simeq u$. The decomposition formula for the product $\chi_{(m)}\chi_{(n)}$ follows from Corollary 6.2.9 iv) and Proposition 3.2.14.

ii) For each $n \in \mathbb{N}$, put $s_{(n)} := \sin((n+1)t/2)$. We have to show that $s_{(n)}/s_{(0)} = \chi_{(n)}$ for all $n \in \mathbb{N}$. Using the relations

$$\sin(2y) = 2\sin(y)\cos(y), \qquad \sin(x \pm y) = \sin(x)\cos(y) \pm \cos(x)\sin(y)$$

for $x = (n+1)t/2$, $y = t/2$, and inserting $2\cos(t/2) = a + a^* = \chi_{(1)}$, we find

$$s_{(1)} = 2s_{(0)}\cos(t/2) = s_{(0)}\chi_{(1)}, \quad s_{(n+1)} + s_{(n-1)} = 2s_{(n)}\cos(t/2) = s_{(n)}\chi_{(1)}.$$

On the other hand, equation (6.7) implies $\chi_{(n+1)} + \chi_{(n-1)} = \chi_{(n)}\chi_{(1)}$ for all $n \in \mathbb{N}$. Hence $\chi_{(n)} = s_{(n)}/s_{(0)}$ for all $n \in \mathbb{N}$.

iii) By Proposition 6.2.5, the powers $\chi_{(1)}^n = (a + a^*)^n$ ($n \in \mathbb{N}$) are linearly independent.

iv) Equation (6.7) shows that $\operatorname{span}\{\chi_{(n)} \mid n \in \mathbb{N}\} = \operatorname{span}\{\chi_{(1)}^n \mid n \in \mathbb{N}\}$. This relation implies the existence of polynomials $f_n \in \mathbb{C}[X]$ such that $f_n(\chi_{(1)}) = \chi_{(n)}$ for all $n \in \mathbb{N}$ and $\operatorname{span}\{f_n \mid n \in \mathbb{N}\} = \mathbb{C}[X]$. Uniqueness of $(f_n)_n$ follows from iii), and linear independence from the fact that the family $(\chi_{(n)})_n$ is linearly independent (see Proposition 3.2.15). □

**Proposition 6.2.11.** *For each* $f \in C_0(\mathbb{R})$,

$$h(f(\chi_{(1)})) = \frac{1}{2\pi}\int_{-2}^{2} f(x)\sqrt{4-x^2}\,dx.$$

*Proof.* By the Stone–Weierstrass Theorem, it suffices to prove the assertion for all the functions $f_n \in \mathbb{C}[X]$ $(n \in \mathbb{N})$ defined in Proposition 6.2.10 iv). By Proposition 3.2.15,

$$h(f_n(\chi_{(1)})) = h(\chi^*_{(0)}\chi_{(n)}) = \delta_{0,n}.$$

We substitute $t = 2\arccos(x/2)$, $x = 2\cos(t/2)$, use equation (6.8), and get

$$\begin{aligned}\frac{1}{2\pi}\int_{-2}^{2} f_n(x)\sqrt{4-x^2}dx &= \frac{1}{2\pi}\int_{-2\pi}^{0} \frac{\sin\left(\frac{n+1}{2}t\right)}{\sin\left(\frac{1}{2}t\right)} \cdot 2|\sin\left(\tfrac{1}{2}t\right)| \cdot \left(-\sin\left(\tfrac{1}{2}t\right)\right) dt \\ &= \frac{1}{\pi}\int_{-2\pi}^{0} \sin\left(\tfrac{n+1}{2}t\right)\sin\left(\tfrac{1}{2}t\right) dt = \delta_{0,n}. \qquad \square\end{aligned}$$

### 6.2.3 Corepresentations and differential calculi

In this subsection, we outline a proof of Theorem 6.2.8, following the approach of Woronowicz [194]. The basic idea behind his approach is to use a differential calculus for $\mathrm{SU}_\mu(2)$ in a similar way as one uses the Lie algebra $\mathfrak{su}_2$ to determine the irreducible representations of $\mathrm{SU}_2(\mathbb{C})$.

Let us briefly recall the classical situation. Every continuous representation $\pi$ of $\mathrm{SU}_2(\mathbb{C})$ is smooth and determined by the associated representation $D\pi$ of the Lie algebra $\mathfrak{su}_2$. The complexification of this Lie algebra is spanned by the matrices

$$H := \begin{pmatrix} 1 & 0 \\ 0 & -1 \end{pmatrix}, \quad X := \begin{pmatrix} 0 & 1 \\ 0 & 0 \end{pmatrix}, \quad Y := \begin{pmatrix} 0 & 0 \\ -1 & 0 \end{pmatrix},$$

which satisfy $[H,X] = 2X$, $[Y,H] = 2Y$, $[Y,X] = H$, see [55, Lecture 26], [61, Section 2.9]. Thus $\pi$ is determined by the infinitesimal generators $A_0 := D\pi(X)$, $A_1 := D\pi(H)$, $A_2 := D\pi(Y)$, and the preceding commutator relations and the unitarity of $\pi$ imply

$$\begin{gathered}[A_1, A_0] = 2A_0, \quad [A_2, A_1] = 2A_2, \quad [A_2, A_0] = A_1, \\ A_1^* = A_1, \quad -A_0^* = A_2.\end{gathered} \tag{6.9}$$

The analysis of these relations, which is straightforward and explained in almost every book on Lie groups or Lie algebras, leads directly to the classification of all irreducible representations of $\mathrm{SU}_2(\mathbb{C})$.

Let us now turn to the quantum group $\mathrm{SU}_\mu(2)$ and construct a replacement for the Lie algebra $\mathfrak{su}_2$. The basic ingredients are algebra homomorphisms $f_0, f_1, f_2 \colon \mathrm{SU}^0_\mu(2) \to \mathbb{C}$ and linear maps $\chi_0, \chi_1, \chi_2 \colon \mathrm{SU}^0_\mu(2) \to \mathbb{C}$, which are determined by

$$\begin{gathered}f_0(a) = f_2(a) = \mu^{-1}, \quad f_0(a^*) = f_2(a^*) = \mu, \quad f_1(a) = \mu^{-2}, \quad f_1(a^*) = \mu^2, \\ f_k(c) = f_k(c^*) = 0 \quad \text{for } k = 0, 1, 2,\end{gathered}$$

and

$$\chi_0(c^*) = -\mu^{-1}, \quad \chi_1(a^*) = -\mu^2, \quad \chi_1(a) = 1, \quad \chi_2(c) = -\mu,$$
$$\chi_k(x) = 0 \quad \text{for all other cases where } k \in \{0,1,2\},\ x \in \{a, a^*, c, c^*\},$$
$$\chi_k(xy) = \chi_k(x) f_k(y) + \epsilon_0(x)\chi_k(y) \quad \text{for all } x, y \in \mathrm{SU}^0_\mu(2).$$

For the precise construction of these maps, see [194].

**Theorem 6.2.12.** *Denote by $\Gamma$ the free left module over $\mathrm{SU}^0_\mu(2)$ with three generators $\omega_1$, $\omega_2$, $\omega_3$.*

i) *$\Gamma$ is a bimodule over $\mathrm{SU}^0_\mu(2)$ with respect to the right multiplication given by $\omega_k x := (f_k * x)\omega_k$ for all $k = 0, 1, 2$ and $x \in \mathrm{SU}^0_\mu(2)$.*

ii) *The map $d\colon \mathrm{SU}^0_\mu(2) \to \Gamma$ given by $x \mapsto \sum_k (\chi_k * x)\omega_k$ is a derivation, that is, $d(xy) = d(x)y + xd(y)$ for all $x, y \in \mathrm{SU}^0_\mu(2)$. Moreover, $d(x) = 0$ if and only if $x \in \mathbb{C}1$.*

iii) *There exists a unique involution on $\Gamma$ such that $(xd(y)z)^* = z^* d(y^*) x^*$ for all $x, y, z \in \mathrm{SU}^0_\mu(2)$. Moreover, $\omega_0^* = \mu\omega_2$, $\omega_1^* = -\omega_1$, $\omega_2^* = \mu^{-1}\omega_0$.*

*Proof.* See [194, Section 2]. □

The next step is the construction of an $n$-th order differential calculus for arbitrary $n \in \mathbb{N}$. Let $\Gamma^0 := \mathrm{SU}^0_\mu(2)$ and put $\Gamma^{\otimes n} := \Gamma \otimes_{\mathrm{SU}^0_\mu(2)} \cdots \otimes_{\mathrm{SU}^0_\mu(2)} \Gamma$ ($n$ factors $\Gamma$) for $n \geq 1$. Then $\Gamma^{\otimes} := \bigoplus_{n=0}^{\infty} \Gamma^{\otimes n}$ carries a natural structure of a $*$-algebra. Denote by $S \subset \Gamma^{\otimes}$ the $*$-ideal generated by the elements

$$\omega_0 \otimes \omega_0, \quad \omega_1 \otimes \omega_1, \quad \omega_2 \otimes \omega_2,$$
$$\omega_2 \otimes \omega_0 + \mu^2 \omega_0 \otimes \omega_2, \quad \omega_1 \otimes \omega_0 + \mu^4 \omega_0 \otimes \omega_1, \quad \omega_2 \otimes \omega_1 + \mu^4 \omega_1 \otimes \omega_2$$

(here, we omitted the subscripts at the tensor symbols), and put $\Gamma^\wedge := \Gamma^{\otimes}/S$. Since $S$ is spanned by homogeneous elements, $\Gamma^\wedge$ is a graded $*$-algebra again, that is, $\Gamma^\wedge$ is a direct sum of the subspaces $\Gamma^{\wedge n} = \Gamma^{\otimes n}/(\Gamma^{\otimes n} \cap S)$, where $n \in \mathbb{N}$, and $\Gamma^{\wedge n}\Gamma^{\wedge m} \subseteq \Gamma^{\wedge(n+m)}$, $(\Gamma^{\wedge n})^* = \Gamma^{\wedge n}$ for all $n, m \in \mathbb{N}$. Evidently, $\Gamma^{\wedge 0} = \mathrm{SU}^0_\mu(2)$ and $\Gamma^{\wedge 1} = \Gamma$. We denote the multiplication in $\Gamma^\wedge$ by the symbol "$\wedge$", and call elements $\omega \in \Gamma^{\wedge n}$, where $n \in \mathbb{N}$, *homogeneous* of degree $|\omega| = n$.

**Theorem 6.2.13.** i) *$\Gamma^\wedge$ is graded-commutative in the sense that $\omega \wedge \omega' = (-1)^{|\omega|\cdot|\omega'|}\omega' \wedge \omega$ for all homogeneous $\omega, \omega' \in \Gamma^\wedge$.*

ii) *The map $d\colon \mathrm{SU}^0_\mu(2) \to \Gamma^{\wedge 1}$ extends uniquely to a $*$-linear map $d\colon \Gamma^\wedge \to \Gamma^\wedge$ such that $d \circ d = 0$ and for all $n \in \mathbb{N}$ and all homogeneous $\omega, \omega' \in \Gamma^\wedge$,*

$$d(\Gamma^{\wedge n}) \subseteq \Gamma^{\wedge(n+1)}, \quad d(\omega \wedge \omega') = d(\omega) \wedge \omega' + (-1)^{|\omega|}\omega \wedge d(\omega').$$

iii) $d(\omega_0) = \omega_0 \wedge \omega_1/(\mu^{-2}+1)$, $d(\omega_1) = \mu\omega_0 \wedge \omega_2$, $d(\omega_2) = \omega_1 \wedge \omega_2/(\mu^{-2}+1)$.

Now we apply the differential calculus that we have just developed to the study of corepresentations. To every corepresentation matrix $v \in M_n(\mathrm{SU}^0_\mu(2))$, where $n \in \mathbb{N}$, we associate *infinitesimal generators*

$$A^v_k := (\chi_k(v_{ij}))_{i,j} \in M_n(\mathbb{C}), \quad \text{where } k = 0, 1, 2,$$

and an element $A^v := \sum_k A^v_k \omega_k \in M_n(\Gamma)$.

For each complex vector space $V$, let us identify $M_n(V)$ with $M_n(\mathbb{C}) \otimes V$.

**Proposition 6.2.14.** *Let $v \in M_n(\mathrm{SU}^0_\mu(2)) \cong M_n(\mathbb{C}) \otimes \mathrm{SU}^0_\mu(2)$ be a unitary corepresentation matrix.*

i) $(\mathrm{id} \otimes d)(v) = vA^v$, $(\mathrm{id} \otimes d)(v^*) = -A^v v^*$, *and* $A^v A^v = -(\mathrm{id} \otimes d)(A^v)$, $(A^v)^* = -A_v$.

ii) *Let $w \in M_m(\mathrm{SU}^0_\mu(2))$ be a corepresentation matrix and $T \in M_{m,n}(\mathbb{C})$. Then $Tv = wT \Leftrightarrow TA^v = A^w T$.*

*Proof.* i) These equations follow, in the order in which they are listed, from the relations

$$dv_{ij} = \sum_k (\chi_k * v_{ij})\omega_k = \sum_{k,l} v_{il}\chi_k(v_{lj})\omega_k = \sum_k (vA^v_k)_{ij}\omega_k \quad \text{for all } i, j,$$

$$0 = (\mathrm{id} \otimes d)(v^* v) = (\mathrm{id} \otimes d)(v^*)v + v^*(\mathrm{id} \otimes d)(v) = (\mathrm{id} \otimes d)(v^*)v + A^v,$$

$$0 = (\mathrm{id} \otimes d)^2(v) = (\mathrm{id} \otimes d)(v)A^v + v(\mathrm{id} \otimes d)(A^v) = v(A^v A^v + (\mathrm{id} \otimes d)(A^v)),$$

$$(A^v)^* v^* = (vA^v)^* = (\mathrm{id} \otimes d)(v)^* = (\mathrm{id} \otimes d)(v^*) = -A^v v^*.$$

ii) First, $wTv^* \in M_{m,n}(\mathbb{C})$ if and only if $TA^v = A^w T$ because both conditions are satisfied if and only if the following expression is 0:

$$(\mathrm{id} \otimes d)(wTv^*) = (\mathrm{id} \otimes d)(w)Tv^* + wT(\mathrm{id} \otimes d)(v^*) = w(A^w T - TA^v)v^*.$$

But if $wTv^* \in M_{m,n}(\mathbb{C})$, then $T = (\mathrm{id} \otimes \epsilon)(w)T(\mathrm{id} \otimes \epsilon)(v)^* = (\mathrm{id} \otimes \epsilon)(wTv^*) = wTv^*$ because $(\mathrm{id} \otimes \epsilon)(v) = 1$ and $(\mathrm{id} \otimes \epsilon)(w) = 1$. □

**Theorem 6.2.15.** *Let $v \in M_{n+1}(\mathrm{SU}^0_\mu(2))$ be a unitary corepresentation matrix. Then the infinitesimal generators $A^v_0, A^v_1, A^v_2$ satisfy*

$$\begin{aligned} \mu A^v_2 A^v_0 - \mu^{-1} A^v_0 A^v_2 &= A^v_1, \\ \mu^2 A^v_1 A^v_0 - \mu^{-2} A^v_0 A^v_1 &= (1+\mu^2) A^v_0, \\ \mu^2 A^v_2 A^v_1 - \mu^{-2} A^v_1 A^v_2 &= (1+\mu^2) A^v_2, \end{aligned} \qquad \begin{aligned} -\mu(A^v_0)^* &= A^v_2, \\ (A^v_1)^* &= A^v_1. \end{aligned}$$

*If $v$ is irreducible, there exists a basis $(\xi_{-n}, \xi_{-n+2}, \dots, \xi_{n-2}, \xi_n)$ of $\mathbb{C}^{n+1}$ such that*

$$A^v_0 \xi_k = -c_{k+1}\xi_{k+2}, \quad A^v_1 \xi_k = d_k \xi_k, \quad A^v_2 \xi_k = \mu c_k \xi_{k-2} \quad \textit{for all } k,$$

*where $d_k = (\mu^{-2k} - 1)/(\mu^{-2} - 1)$, $c_k = \sqrt{(\mu^{-k} - \mu^n)(\mu^{-n} - \mu^{2-k})/(\mu^{-2} - 1)}$ if $|\mu| < 1$, and $d_k = 2k$, $c_k = \mu((n+1)(n-k+1))^{1/2}$ if $|\mu| = 1$.*

*Proof.* The relations for the infinitesimal generators follow easily from Proposition 6.2.14. A careful analysis of these relations, which can be found in [194], leads to the explicit form given in the second part of the theorem. □

Note that for $\mu = 1$, the relations on the infinitesimal generators given above coincide with the relations given in equation (6.9).

The explicit form of an irreducible corepresentation given above allows us to compute the associated weight function and thereby prove the key result of the preceding section:

*Proof of Theorem* 6.2.8. Let $v \in M_{n+1}(\mathrm{SU}^0_\mu(2)) \cong M_{n+1}(\mathbb{C}) \otimes \mathrm{SU}^0_\mu(2)$ be an irreducible unitary corepresentation matrix. For each weight $k$ of $v$, denote by $p_k$ the orthogonal projection onto the $k$th weight space. Then $(\mathrm{id} \otimes \pi_{\mathbb{T}})(v) = \sum_k p_k \otimes z^k$.

Assume that $|\mu| < 1$. Define a linear map $\phi\colon \mathbb{C}[z, z^{-1}] \to \mathbb{C}$ by $z^k \mapsto (\mu^{-2k} - 1)/(\mu^{-2} - 1)$. We claim that $\chi_1 = \phi \circ \pi_{\mathbb{T}}$. This relation implies

$$A_1^v = (\mathrm{id} \otimes \chi_1)(v) = \sum_k p_k \phi(z^k) = \sum_k p_k (\mu^{-2k} - 1)/(\mu^{-2} - 1),$$

and comparing with the description of $A_1^v$ given in Theorem 6.2.15, we find

$$W_v(k) = \dim(\mathrm{Im}\, p_k) = \begin{cases} 1, & k \in \{-n, 2-n, \dots, n-2, n\}, \\ 0, & \text{otherwise.} \end{cases}$$

So it suffices to prove that $\chi_1 = \phi \circ \pi_{\mathbb{T}}$. We verify this equation for the basis elements $(a_{kpq})_{k,p,q}$ (see Proposition 6.2.5). Let $k \in \mathbb{Z}$ and $p, q \in \mathbb{N}$. By [194, Equation (2.5)], $\chi_1(a_{kpq}) = (f_1(a_{kpq}) - \epsilon_0(a_{kpq}))/(\mu^{-2} - 1)$. Using the multiplicativity of $f_1$ and $\epsilon_0$, we easily find $f_1(a_{kpq}) = \delta_{p,0}\delta_{q,0}\mu^{-2k}$ and $\epsilon_0(a_{kpq}) = \delta_{p,0}\delta_{q,0}$. Therefore,

$$\chi_1(a_{kpq}) = \delta_{p,0}\delta_{q,0}(\mu^{-2k} - 1)/(\mu^{-2} - 1) = \delta_{p,0}\delta_{q,0}\phi(z^k) = \phi(\pi_{\mathbb{T}}(a_{kpq})).$$

The case $|\mu| = 1$ is treated similarly. □

### 6.2.4 Modular properties of the Haar state

The results on the corepresentation theory obtained in the preceding sections allow us to describe the Haar state of $\mathrm{SU}_\mu(2)$ and the modular characters $(f_z)_{z \in \mathbb{C}}$ introduced in Theorem 3.2.19:

**Proposition 6.2.16.** i) *The automorphisms $(f_z)_{z \in \mathbb{C}}$ are determined by*

$$f_z(a) = |\mu|^{-z}, \quad f_z(a^*) = |\mu|^z, \quad f_z(c) = 0, \quad f_z(c^*) = 0 \quad \textit{for all } z \in \mathbb{C}.$$

ii) *For all* $z \in \mathbb{C}$,

$$f_z * a = |\mu|^{-z} a = a * f_z, \quad f_z * a^* = |\mu|^z a^* = a^* * f_z,$$
$$f_z * c = |\mu|^{-z} c, \quad c * f_z = |\mu|^z c, \quad f_z * c^* = |\mu|^z c^*, \quad c^* * f_z = |\mu|^{-z} c^*.$$

*Proof.* i) By equation (6.5), the matrix $S_0^2(u) := (S_0^2(u_{ij}))_{i,j}$ is equal to

$$\begin{pmatrix} a & \mu^{-2} c^* \\ \mu^2 c & a^* \end{pmatrix}.$$

A short calculation shows that the matrix $F := \operatorname{diag}(|\mu|^{-1}, |\mu|)$ satisfies $u = F^{-1} S_0^2(u) F$. Moreover, $\operatorname{Tr} F > 0$. By the proof of Theorem 3.2.19, $f_z(u_{ij}) = (F^z)_{ij}$ for $i, j = 1, 2$. The claim follows.

ii) Combine i) and equation (6.4). □

The next proposition describes the values of the Haar state on the basis elements $(a_{kmn})_{k,m,n}$, which were defined by

$$a_{kmn} := \begin{cases} a^k c^{*m} c^n, & k \geq 0, \\ a^{*(-k)} c^{*m} c^n, & k < 0. \end{cases}$$

**Theorem 6.2.17.** *The Haar state $h$ of* $\mathrm{SU}_\mu(2)$ *is given by*

$$h(a_{kmn}) = \delta_{k,0} \delta_{m,n} (1 - \mu^2)/(1 - \mu^{2m+2}). \tag{6.10}$$

*For every function* $f \in C(\sigma(c^* c))$,

$$h(f(c^* c)) = (1 - \mu^2) \sum_{k=0}^{\infty} \mu^{2k} f(\mu^{2k}). \tag{6.11}$$

*Proof.* Let $k \in \mathbb{Z}$, $m, n \in \mathbb{N}$, and $z \in \mathbb{C}$. Using Proposition 6.2.16 ii) and multiplicativity of $f_z$, we find

$$f_z * a_{kmn} = |\mu|^{(-k+m-n)z} a_{kmn}, \quad a_{kmn} * f_z = |\mu|^{(-k-m+n)z} a_{kmn}.$$

On the other hand,

$$h(f_z * a_{kmn}) = (h \otimes f_z)(\Delta(a_{kmn})) = f_z(1) h(a_{kmn}) = h(a_{kmn})$$

and similarly $h(a_{kmn} * f_z) = h(a_{kmn})$. Consequently, $h(a_{kmn}) = 0$ unless $k = 0$ and $m = n$.

It remains to compute the value of $h$ on the elements $b_m := c^{*m} c^m$ for all $m \in \mathbb{N}$. By Theorem 3.2.19 v),

$$\mu^{2m+2} h(a a^* b_m) = \mu^{2m+2} h(a^* b_m (f_1 * a * f_1)) = \mu^{2m} h(a^* b_m a),$$

and from equation (6.3), one easily deduces

$$\mu^{2m+2}aa^*b_m = \mu^{2m+2}b_m - \mu^{2m+4}b_{m+1}, \quad \mu^{2m}a^*b_m a = b_m - b_{m+1}.$$

Combining these relations, we find $h(b_m)(\mu^{2m+2} - 1) = h(b_{m+1})(\mu^{2m+4} - 1)$. Using the relation $h(b_0) = h(1) = 1$, we obtain equation (6.10). This equation implies that equation (6.11) holds whenever $f$ has the form $f(x) = x^m$ for some $m \in \mathbb{N}$. Since such functions are linearly dense in $C(\sigma(c^*c))$, equation (6.11) holds for all $f \in C(\sigma(c^*c))$. □

## 6.3 Products of compact quantum groups

The free product and the tensor product of compact quantum groups were introduced by Wang [185], [186], [187]. We briefly outline these general constructions, describe the irreducible corepresentations of a product in terms of the irreducible corepresentations of the two factors, and compare the algebraic and $C^*$-algebraic versions.

### Free products of compact quantum groups

Let $A$ and $B$ be unital algebras. The *free* (*unital*) *product* of $A$ and $B$ is a unital algebra $A * B$ with embeddings $\iota_A \colon A \hookrightarrow A * B$ and $\iota_B \colon B \hookrightarrow A * B$ such that for each unital algebra $C$ and each pair of unital homomorphisms $f \colon A \to C$ and $g \colon B \to C$, there exists a unique unital homomorphism $f * g \colon A * B \to C$ such that $(f * g) \circ \iota_A = f$ and $(f * g) \circ \iota_B = g$, that is, the following diagram commutes:

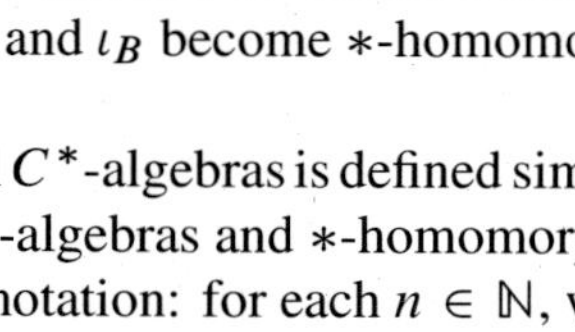


Routine arguments show that $A * B$ is uniquely determined up to isomorphism, that $A^{\mathrm{op}} * B^{\mathrm{op}} \cong (A * B)^{\mathrm{op}}$, and that $A * B$ carries a unique structure of a $*$-algebra such that the embeddings $\iota_A$ and $\iota_B$ become $*$-homomorphisms whenever $A$ and $B$ are $*$-algebras.

The free product of unital $C^*$-algebras is defined similarly – just replace algebras and homomorphisms by $C^*$-algebras and $*$-homomorphisms.

We adopt the following notation: for each $n \in \mathbb{N}$, we denote by $\tilde{\iota}_A \colon M_n(A) \to M_n(A * B)$ the map given by $(u_{ij})_{i,j} \mapsto (\iota_A(u_{ij}))_{i,j}$. Likewise, we define $\tilde{\iota}_B$.

Throughout this subsection, we treat the algebraic and the $C^*$-algebraic setting in parallel and use the symbol "⊗" to denote the algebraic tensor product of $*$-algebras and the minimal tensor product of $C^*$-algebras.

**Theorem 6.3.1.** *Let $(A, \Delta_A)$ and $(B, \Delta_B)$ be algebraic or $C^*$-algebraic compact quantum groups. Then there exists a unique comultiplication $\Delta$ on $A * B$ such that $(A, \Delta_A) * (B, \Delta_B) := (A * B, \Delta)$ is a compact quantum group and the maps $\iota_A$ and $\iota_B$ are morphisms of compact quantum groups.*

*Proof.* By the universal property of $A * B$, there exists a unique $*$-homomorphism $\Delta\colon A * B \to (A * B) \otimes (A * B)$ such that the following diagram commutes:

$$\begin{array}{ccccc} A & \xrightarrow{\iota_A} & A * B & \xleftarrow{\iota_B} & B \\ & \searrow^{(\iota_A \otimes \iota_A)\circ\Delta_A} & \downarrow{\scriptstyle \Delta} & \swarrow^{(\iota_B \otimes \iota_B)\circ\Delta_B} & \\ & & (A * B) \otimes (A * B). & & \end{array}$$

The coassociativity of $\Delta_A$ and $\Delta_B$ immediately implies that $\Delta$ is coassociative.

If $(A, \Delta_A)$ and $(B, \Delta_B)$ are algebraic compact quantum groups, then $(A * B, \Delta)$ is a Hopf $*$-algebra. Indeed, if $\epsilon_A$, $S_A$ and $\epsilon_B$, $S_B$ denote the counit and the antipode of $(A, \Delta_A)$ and $(B, \Delta_B)$, respectively, then the maps $\epsilon := \epsilon_A * \epsilon_B\colon A * B \to \mathbb{C}$ and $S := (\iota_{A^{\mathrm{op}}} \circ S_A) * (\iota_{B^{\mathrm{op}}} \circ S_B)\colon A * B \to A^{\mathrm{op}} * B^{\mathrm{op}} \cong (A * B)^{\mathrm{op}}$ are a counit and an antipode for $(A * B, \Delta)$, as one can easily check.

Let us show that $(A * B, \Delta)$ is a compact quantum group. The $C^*$-algebra/$*$-algebra $A * B$ is generated by $\iota_A(A)$ and $\iota_B(B)$, and by Theorem 3.2.12 or 5.3.11, respectively, $A$ and $B$ are generated by the entries of unitary corepresentation matrices. But if $u$ is a unitary corepresentation matrix of $(A, \Delta_A)$ or $(B, \Delta_B)$, then $\tilde{\iota}_A(u)$ or $\tilde{\iota}_B(u)$, respectively, is a unitary corepresentation matrix of $(A * B, \Delta)$. Applying Theorem 3.2.12 or Theorem 5.3.11 again, we find that $(A * B, \Delta)$ is a compact quantum group. □

**Corollary 6.3.2.** *If $(A, \Delta_A, u)$ and $(B, \Delta_B, v)$ are compact matrix quantum groups, then so is $(A * B, \Delta, \tilde{\iota}_A(u) \boxplus \tilde{\iota}_B(v))$.* □

The Haar state of $(A * B, \Delta)$ turns out to be the free product of the Haar states of $(A, \Delta_A)$ and $(B, \Delta_B)$. The free product of states was introduced by Voiculescu; it is a fundamental concept in free probability theory [184]. Let us briefly recall how it is defined. If $\phi$ and $\psi$ are states on $*$-algebras/$C^*$-algebras $A$ and $B$, respectively, then there exists a unique state $\phi * \psi$ on $A * B$, called the *free product* of $\phi$ and $\psi$, such that

i) $(\phi * \psi) \circ \iota_A = \phi$ and $(\phi * \psi) \circ \iota_B = \psi$;

ii) if $c_1, \dots, c_n$ are elements of $\ker \phi$ or $\ker \psi$ and no adjacent elements belong to the same algebra $A$ or $B$, then $c_1 \dots c_n \in \ker(\phi * \psi)$;

see [184, Chapter 1]. In the algebraic setting, the existence of $\phi * \psi$ is easy to show; in the setting of $C^*$-algebras, one can define a free product of the GNS-constructions for $\phi$ and $\psi$, which yields a GNS-construction for $\phi * \psi$. The image of $A * B$ under the GNS-representation for $\phi * \psi$ is called the *reduced free product* of $(A, \phi)$ and $(B, \psi)$; it is denoted by $(A, \phi) * (B, \psi)$.

**Proposition 6.3.3.** *Let $(A, \Delta_A)$ and $(B, \Delta_B)$ be compact quantum groups with Haar states $h_A$ and $h_B$, respectively. Then the Haar state of the compact quantum group $(A * B, \Delta)$ is equal to the free product $h_A * h_B$.*

*Proof.* We only need to show that $h := h_A * h_B$ is left-invariant. Let $(u^\alpha)_\alpha$ and $(v^\beta)_\beta$ be maximal families of non-trivial pairwise inequivalent irreducible corepresentation matrices of $(A, \Delta_A)$ and $(B, \Delta_B)$, respectively. Denote by $W \subset A * B$ the subset of all elements of the form

$$w = w^{\gamma_1}_{i_1 j_1} \cdots w^{\gamma_n}_{i_n j_n} \in A * B, \tag{6.12}$$

where $n \geq 1$, each $w^{\gamma_m}_{i_m j_m}$ is a matrix entry of some $\tilde{\iota}_A(u^\alpha)$ or $\tilde{\iota}_B(v^\beta)$, and no adjacent factors are taken from the same algebra $A$ or $B$.

By Theorem 3.2.12 or 5.3.11, the unit $1 \in A * B$ and the set $W$ span (a dense subspace of) $A * B$. Evidently, $h(1) = 1 = (\mathrm{id} \otimes h)(\Delta(1))$. We show that $h(w) = 0 = (\mathrm{id} \otimes h)(\Delta(w))$ for each $w \in W$, and then the claim follows.

By Corollary 3.2.7, $h_A(u^\alpha_{ij}) = 0 = h_B(v^\beta_{kl})$ for all $\alpha, i, j$ and $\beta, k, l$. Combining this relation with condition ii) in the definition of the free product of states, we get $h(w) = 0$ for each $w \in W$. Moreover, if $w$ is as in (6.12), then

$$(\mathrm{id} \otimes h)(\Delta(w)) = \sum_{k_1,\dots,k_n} (w^{\gamma_1}_{i_1 k_1} \cdots w^{\gamma_n}_{i_n k_n}) \cdot h(w^{\gamma_1}_{k_1 j_1} \cdots w^{\gamma_n}_{k_n j_n}) = 0. \qquad \square$$

The irreducible corepresentations of a free product are easily determined. Given a compact quantum group $(A, \Delta)$, let us call a maximal family of pairwise inequivalent non-trivial unitary irreducible corepresentation matrices a *representative family*.

**Proposition 6.3.4.** *Let $(A, \Delta_A)$ and $(B, \Delta_B)$ be compact quantum groups with representative families $(u^\alpha)_\alpha$ and $(v^\beta)_\beta$, respectively. Then a representative family for $(A * B, \Delta)$ is given by all corepresentation matrices of the form*

$$w^{\gamma_1} \boxtimes \cdots \boxtimes w^{\gamma_n}, \tag{6.13}$$

*where $n \geq 1$, each $w^{\gamma_m}$ belongs to $(\tilde{\iota}_A(u^\alpha))_\alpha$ or $(\tilde{\iota}_B(v^\beta))_\beta$, respectively, and no two adjacent factors are taken from the same family.*

*Proof.* Denote by $h_A$ and $h_B$ the Haar states of the compact quantum groups $(A, \Delta_A)$ and $(B, \Delta_B)$, and recall from Proposition 6.3.3 that the Haar state $h$ of $(A * B, \Delta)$ is equal to the free product $h_A * h_B$.

We show by induction on $n$ that for every corepresentation matrix $w$ as in (6.13), the associated character $\chi(w)$ satisfies $h(\chi(w)^* \chi(w)) = 1$; then $w$ is irreducible by Corollary 3.2.16.

For $n = 1$, the assertion is evident because $w = w^{\gamma_1}$ is irreducible. So, let $n > 1$ and assume that the assertion holds for $n-1$. By Proposition 3.2.14, $\chi(w) = \chi_1 \cdots \chi_n$, where $\chi_k := \chi(w^{\gamma_k})$ for $k = 1, \dots, n$, and by Corollary 3.2.7 and Proposition 3.2.15, $h(\chi_k) = 0 = h(\chi_k^*)$ for $k = 1, \dots, n-1$, and $h(\chi_1^* \chi_1 - 1) = 0$. Using condition ii) in the definition of the free product $h = h_A * h_B$, we find $h(\chi_n^* \cdots \chi_2^* (\chi_1^* \chi_1 - 1) \chi_2 \cdots \chi_n) = 0$, and inserting the induction hypothesis, we obtain

$$h(\chi(w)^* \chi(w)) = h(\chi_n^* \cdots \chi_2^* \chi_1^* \chi_1 \chi_2 \cdots \chi_n) = h(\chi_n^* \cdots \chi_2^* \chi_2 \cdots \chi_n) = 1.$$

Thus every corepresentation of the form given in (6.13) is irreducible. Evidently, the entries of the corepresentation matrices given in (6.13) include all elements of $A * B$ of the form given in (6.12). Therefore these matrix entries and the unit $1 \in A * B$ span (a dense subspace of) $A * B$. From Corollary 3.2.8 or 5.3.9, it follows that every non-trivial irreducible corepresentation matrix of $A * B$ has, up to equivalence, the form given in (6.13). □

Let us briefly clarify the relation between the algebraic and the $C^*$-algebraic free product. Recall from Section 5.4 that to every CQG $(A, \Delta)$, one can associate

- an algebraic CQG $(A_0, \Delta_0) = (A, \Delta)_0$,
- a universal and a reduced $C^*$-algebraic CQG $(A_u, \Delta_u) = (A, \Delta)_u$ and $(A_r, \Delta_r) = (A, \Delta)_r$, respectively.

**Proposition 6.3.5.** *Let $(A, \Delta_A)$ and $(B, \Delta_B)$ be compact quantum groups.*

i) $((A, \Delta_A) * (B, \Delta_B))_0 \cong (A, \Delta_A)_0 * (B, \Delta_B)_0$.

ii) $((A, \Delta_A) * (B, \Delta_B))_u \cong (A, \Delta_A)_u * (B, \Delta_B)_u$.

iii) *Denote by $h_A$ and $h_B$ the Haar states of $(A, \Delta_A)_r$ and $(B, \Delta_B)_r$, respectively. Then $(A * B)_r \cong (A_r, h_A) * (B_r, h_B)$.*

*Proof.* Assertion i) follows easily from Proposition 6.3.4, ii) from the definitions, and iii) from Proposition 6.3.3. □

**Corollary 6.3.6.** *Let $(A, \Delta_A)$ and $(B, \Delta_B)$ be CQGs, and denote by $h_A$, $h_B$ the Haar states of $A_r$, $B_r$, respectively. There exists a comultiplication $\Delta_r$ on $(A_r, h_A) * (B_r, h_B)$ such that $((A_r, h_A) * (B_r, h_B), \Delta_r)$ is a reduced $C^*$-algebraic CQG and the natural maps $A, B \to (A_r, h_A) * (B_r, h_B)$ are morphisms of CQGs.* □

**Tensor products of compact quantum groups.** To each pair of compact quantum groups, one can associate a tensor product which is a compact quantum group again. We briefly summarize this construction; the main results can be proved by similar techniques as used in the case of a free product.

We adopt the following notation. Let $A$ and $B$ be algebras/$C^*$-algebras. We denote by $A \underset{(\max)}{\otimes} B$ the algebraic tensor product/the maximal $C^*$-tensor product, and by $\iota_A, \iota_B \colon A \to A \underset{(\max)}{\otimes} B$ the natural maps given by $a \mapsto a \underset{(\max)}{\otimes} 1_B$ and $b \mapsto 1_A \underset{(\max)}{\otimes} b$, respectively. As before, the minimal tensor product of $C^*$-algebras $C$ and $D$ is denoted by $C \otimes D$. For each $n \in \mathbb{N}$, we denote by $\tilde{\iota}_A \colon M_n(A) \to M_n(A \underset{(\max)}{\otimes} B)$ the map given by $(u_{ij})_{i,j} \mapsto (\iota_A(u_{ij}))_{i,j}$. Similarly we define $\tilde{\iota}_B$. A *representative family* for a compact quantum group is a maximal family of pairwise inequivalent non-trivial unitary irreducible corepresentation matrices.

**Proposition 6.3.7.** *Let $(A, \Delta_A)$ and $(B, \Delta_B)$ be compact quantum groups.*

i) *There exists a unique comultiplication $\Delta$ on $A \underset{(\max)}{\otimes} B$ such that*

$$(A, \Delta_A) \underset{(\max)}{\otimes} (B, \Delta_B) := (A \underset{(\max)}{\otimes} B, \Delta)$$

*is a compact quantum group and $\iota_A$ and $\iota_B$ are morphisms of CQGs.*

ii) *Denote by $h_A$ and $h_B$ the Haar states of $(A, \Delta_A)$ and $(B, \Delta_B)$, respectively. Then the Haar state of $(A \underset{(\max)}{\otimes} B, \Delta)$ is $h_A \underset{(\max)}{\otimes} h_B$.*

iii) *If $(u^\alpha)_\alpha$ and $(v^\beta)_\beta$ are representative families for $(A, \Delta_A)$ and $(B, \Delta_B)$, respectively, then $(\tilde{\iota}_A(u^\alpha) \boxtimes \tilde{\iota}_B(v^\beta))_{\alpha,\beta}$ is a representative family for $(A \underset{(\max)}{\otimes} B, \Delta)$.*

iv) *If $(A, \Delta_A, u)$ and $(B, \Delta_B, v)$ are CMQGs, then $(A \underset{(\max)}{\otimes} B, \Delta, \tilde{\iota}_A(u) \boxplus \tilde{\iota}_B(v))$ is a CMQG.*

v) *In the notation introduced before Proposition* 6.3.5,

(a) $((A, \Delta_A) \underset{(\max)}{\otimes} (B, \Delta_B))_0 \cong (A, \Delta_A)_0 \underset{(\max)}{\otimes} (B, \Delta_B)_0$;

(b) $((A, \Delta_A) \underset{(\max)}{\otimes} (B, \Delta_B))_u \cong (A, \Delta_A)_u \underset{(\max)}{\otimes} (B, \Delta_B)_u$;

(c) $A_r \otimes B_r \cong (A \underset{(\max)}{\otimes} B)_r$. □

**Corollary 6.3.8.** *Let $(A, \Delta_A)$, $(B, \Delta_B)$ be CQGs. Then there exists a unique comultiplication $\Delta_r$ on $A_r \otimes B_r$ such that $(A_r \otimes B_r, \Delta_r)$ is a $C^*$-algebraic CQG and the natural maps $A, B \to A_r \otimes B_r$ are morphisms of CQGs.* □

## 6.4 The free unitary and the free orthogonal quantum groups

The free unitary quantum groups are universal compact matrix quantum groups which contain every other compact matrix quantum group as a quantum subgroup, very much like the unitary groups contain every other compact Lie group as a Lie subgroup. They were introduced by Wang [185], [186] and generalized by Wang and Van Daele [179]. Wang and Van Daele also introduced the free orthogonal quantum groups, whose definition was slightly modified by Banica [12]. In this section, we shall only give a brief overview of the definition and some fundamental properties.

**Definition and first properties.** The free unitary and the free orthogonal quantum groups are defined as follows:

**Definition 6.4.1.** i) For each $F \in \mathrm{GL}_n(\mathbb{C})$, we denote by $A_u(F)$ the universal unital $C^*$-algebra generated by elements $u_{ij}$ $(1 \leq i, j \leq n)$ such that the matrices $u = (u_{ij})_{i,j}$ and $F\bar{u}F^{-1}$ are unitary.

ii) We call a matrix $F \in \mathrm{GL}_n(\mathbb{C})$ *admissible* if $F\bar{F} \in \mathbb{R}1$. For each admissible $F \in \mathrm{GL}_n(\mathbb{C})$, we denote by $A_o(F)$ the universal unital $C^*$-algebra generated by elements $v_{ij}$ $(1 \leq i, j \leq n)$ such that the matrix $v = (v_{ij})_{i,j}$ is unitary and $v = F\bar{v}F^{-1}$.

**Remarks 6.4.2.** i) In Definition 6.4.1 i), the matrix $F\bar{u}F^{-1}$ is unitary if and only if

$$u^t(F^*F)\bar{u}(F^*F)^{-1} = 1 = (F^*F)\bar{u}(F^*F)^{-1}u^t.$$

In particular, $A_u(F) = A_u(F')$ whenever $F^*F = F'^*F'$. Therefore, some authors denote the $C^*$-algebra $A_u(F)$ defined above by $A_u(Q)$, where $Q = F^*F$.

ii) In Definition 6.4.1 ii), the unitary matrix $v$ satisfies $v = F\bar{v}F^{-1}$ if and only if $v^* = (F^*)^{-1}v^tF^*$, and this relations holds if and only if

$$v^tF^*v(F^*)^{-1} = 1 = F^*v(F^*)^{-1}v^t.$$

iii) Evidently, $A_o(F) = A_o(\lambda F)$ for each admissible $F$ and each $\lambda \in \mathbb{C}, \lambda \neq 0$.

The $C^*$-algebras $A_u(F)$ and $A_o(F)$ are compact matrix quantum groups:

**Proposition 6.4.3.** i) *For each $F \in \mathrm{GL}_n(\mathbb{C})$, the $C^*$-algebra $A_u(F)$ carries a unique comultiplication $\Delta_u$ that turns $(A_u(F), \Delta_u, u)$ into a $C^*$-algebraic compact matrix quantum group.*

ii) *For each admissible $F \in \mathrm{GL}_n(\mathbb{C})$, the $C^*$-algebra $A_o(F)$ carries a unique comultiplication $\Delta_o$ that turns $(A_o(F), \Delta_o, v)$ into a $C^*$-algebraic compact matrix quantum group.*

*Proof.* i) For each $w \in M_n(A_u(F))$, put $\Delta_n(w) := \left(\sum_k w_{ik} \otimes w_{kj}\right)_{i,j}$. By the universal property of $A_u(F)$, it suffices to show that $\Delta_n(u)$ and $F\overline{\Delta_n(u)}F^{-1}$ are unitary. But by Lemma 6.1.12, $\Delta_n(u)$ and $\Delta_n(F\bar{u}F^{-1})$ are unitary, and a similar calculation as in equation (6.1) shows that $\Delta_n(F\bar{u}F^{-1}) = F\overline{\Delta_n(u)}F^{-1}$.

ii) The proof is similar as for i). □

**Remark 6.4.4.** Let $(A, \Delta, w)$ be a compact matrix quantum group and assume that $w = F\bar{w}F^{-1}$ for some $F \in \mathrm{GL}_n(\mathbb{C})$. Then $w = F\overline{F\bar{w}F^{-1}}F^{-1} = (F\bar{F})w(F\bar{F})^{-1}$. If $F\bar{F} \notin \mathbb{C}1$, then by Schur's Lemma (Proposition 5.3.4), $w$ is reducible. Moreover, it is easy to see that $F\bar{F} \in \mathbb{C}1$ implies $F\bar{F} \in \mathbb{R}1$. These observations motivate the restriction to admissible $F \in \mathrm{GL}_n(\mathbb{C})$ in the definition of $A_o(F)$.

The quantum groups $A_u(F)$ are universal in the following sense:

**Proposition 6.4.5.** *Let $(A, \Delta, w)$ be a $C^*$-algebraic compact matrix quantum group, where $w \in M_n(A)$, $n \in \mathbb{N}$.*

i) *There exist an $F \in \mathrm{GL}_n(\mathbb{C})$ and a $*$-homomorphism $f : A_u(F) \to A$ such that $f(u_{ij}) = w_{ij}$ for all $i, j$. This $f$ is surjective and a morphism of compact matrix quantum groups.*

ii) *If $w$ is irreducible and $w \simeq \bar{w}$, then there exist an admissible $F \in \mathrm{GL}_n(\mathbb{C})$ and a $*$-homomorphism $f : A_o(F) \to A$ such that $f(v_{ij}) = w_{ij}$ for all $i, j$. This $f$ is surjective and a morphism of compact matrix quantum groups.*

*Proof.* i) By Theorem 5.3.3 i), $\bar{w}$ is equivalent to a unitary corepresentation matrix. Hence there exists an $F \in \mathrm{GL}_n(\mathbb{C})$ such that $F\bar{w}F^{-1}$ is unitary. Now the existence of $f$ follows from the universal property of $A_u(F)$.

ii) The proof is similar to the proof of i), see also Remark 6.4.4. □

**Definition 6.4.6.** For each $F \in \mathrm{GL}_n(\mathbb{C})$, we call $(A_u(F), \Delta_u, u)$ the *free unitary $C^*$-algebraic quantum group* with parameter $F$. For each admissible $F \in \mathrm{GL}_n(\mathbb{C})$, we call $(A_o(F), \Delta_o, v)$ the *free orthogonal $C^*$-algebraic quantum group* with parameter $F$.

The next result clarifies to which extent $(A_u(F), \Delta_u, u)$ and $(A_o(F), \Delta_o, v)$ depend on $F$. To avoid ambiguities, let us index the generators $u_{ij}$, $v_{ij}$ and the comultiplications $\Delta_u$, $\Delta_o$ by $F$.

**Proposition 6.4.7.** i) *Let $F \in \mathrm{GL}_n(\mathbb{C})$ and $F' = FU$ for some $U \in \mathrm{U}_n(\mathbb{C})$. Then $(A_u(F), \Delta_u^F, u^F) \simeq (A_u(F'), \Delta_u^{F'}, u^{F'})$.*

ii) *Up to similarity, $(A_u(F), \Delta_u^F, u^F)$ depends only on the list of eigenvalues (with multiplicity) of $F^*F$.*

iii) *Let $F \in \mathrm{GL}_n(\mathbb{C})$ be admissible and $U \in \mathrm{U}_n(\mathbb{C})$. Then $F' := UFU^t$ is admissible and $(A_o(F), \Delta_o^F, v^F) \simeq (A_o(F'), \Delta_o^{F'}, v^{F'})$.*

*Proof.* i) Since the matrices $\bar{U}, u^{F'}, F'\overline{u^{F'}}(F')^{-1}$ are unitary, so are $w := \bar{U}u^{F'}\bar{U}^*$ and $F\bar{w}F^{-1} = FU\overline{u^{F'}}U^*F^{-1}$. Consequently, we can define a $*$-homomorphism $f\colon A_u(F) \to A_u(F')$ such that $f(u^F_{ij}) = w_{ij}$ for all $i, j$. By similar arguments, we can define a $*$-homomorphism $A_u(F') \to A_u(F)$ that is inverse to $f$. Therefore, $f$ is an isomorphism.

ii) This follows directly from i).

iii) The matrix $F'$ is admissible because $F'\overline{F'} = UFU^t\bar{U}\bar{F}U^* = UF\bar{F}U^* \in \mathbb{R}1$. Moreover, the matrix $w := U^*v^{F'}U$ is unitary and satisfies $F\bar{w}F^{-1} = U^*F'\overline{v^{F'}}F'^{-1}U = U^*v^{F'}U = w$. Consequently, we can define a $*$-homomorphism $f\colon A_o(F) \to A_o(F')$ such that $f(v^F_{ij}) = w_{ij}$ for all $i, j$. By similar arguments, we can define a $*$-homomorphism $A_o(F') \to A_o(F)$ that is inverse to $f$. The claim follows. □

For each admissible matrix $F \in \mathrm{GL}_2(\mathbb{C})$, the compact matrix quantum group $(A_o(F), \Delta_o, v)$ coincides with the compact matrix quantum group $(\mathrm{SU}_\mu(2), \Delta, u)$ introduced in Section 6.2:

**Proposition 6.4.8.** *For each $\mu \in [-1,1] \setminus \{0\}$, the matrix $F := \left(\begin{smallmatrix} 0 & 1 \\ -\mu^{-1} & 0 \end{smallmatrix}\right)$ is admissible and $(\mathrm{SU}_\mu(2), \Delta, u) = (A_o(F), \Delta_o, v)$. Moreover,*

$$\begin{aligned}\{(A_o(F), \Delta_o, v) \mid F \in \mathrm{GL}_2(\mathbb{C}) \textit{ admissible}\}\\ = \{(\mathrm{SU}_\mu(2), \Delta, u) \mid \mu \in [-1,1] \setminus \{0\}\}.\end{aligned}$$

*Proof.* This follows from Lemma 6.2.2, Proposition 6.4.7 iii), and elementary linear algebra, see [13, Proposition 7]. □

**The corepresentation theory of the free orthogonal quantum group.** The corepresentation theory of the free orthogonal quantum group $(A_o(F), \Delta_o)$ is very similar to the corepresentation theory of $(\mathrm{SU}_\mu(2), \Delta)$. It was determined by Banica [12]. We shall outline the main results and the steps of the proofs.

We fix an admissible $F \in \mathrm{GL}_n(\mathbb{C})$, where $n \in \mathbb{N}$. To determine the corepresentation theory of $(A_o(F), \Delta_o, v)$, we shall use the results on the corepresentation theory of the $C^*$-algebraic compact matrix quantum group $(\mathrm{SU}_\mu(2), \Delta, u)$ that were obtained in Section 6.2. From now on, $u$ denotes the fundamental corepresentation matrix of $\mathrm{SU}_\mu(2)$.

**Proposition 6.4.9.** $\dim \mathrm{Hom}(v^{\boxtimes r}, v^{\boxtimes r}) \leq \dim \mathrm{Hom}(u^{\boxtimes r}, u^{\boxtimes r})$ *for all $r \in \mathbb{N}$.*

*Proof.* We outline the main steps of the proof; for details, see [12]. Let us fix some notation. We put $H := \mathbb{C}^n$ and define

- $I_p := \mathrm{id}_{H^{\otimes p}}$ for all $p \in \mathbb{N}$;
- $E\colon \mathbb{C} \to H \otimes H$ by $E(1) := \sum_{i,j} F_{ji} e_i \otimes e_j$;

- $V_{p,q} := I_p \otimes E \otimes I_q \colon H^{\otimes(p+q)} \to H^{\otimes(p+2+q)}$ for all $p, q$.

*Step* 1. Denote by $\mathbf{Z}$ the category with objects $H^{\otimes r}$ ($r \in \mathbb{N}$) and morphism sets $\mathbf{Z}(H^{\otimes r}, H^{\otimes s}) \subseteq \mathcal{L}(H^{\otimes r}, H^{\otimes s})$ consisting of all finite linear combinations of compositions of maps of the form $I_p$, $V_{p,q}$, $V_{p,q}^*$, where $p, q \in \mathbb{N}$. This category carries the structure of a concrete monoidal $W^*$-category [195], and as such corresponds to a compact matrix quantum group [195, Theorem 1.3] which is given by the same relations as $A_o(F)$. Hence, $\mathrm{Hom}(u^{\boxtimes r}, u^{\boxtimes s}) = \mathbf{Z}(H^{\otimes r}, H^{\otimes s})$ for all $r, s \in \mathbb{N}$.

*Step* 2. Using Step 1, one shows that $\mathrm{Hom}(u^{\boxtimes r}, u^{\boxtimes r})$ is generated (as a $*$-algebra) by the $I_r$ ($r \in \mathbb{N}$) and by the elements $f_s := I_{s-1} \otimes P \otimes I_{r-s-1}$, where $s = 1, \dots, r-1$ and $P := EE^*/\|E(1)\|^2$. These elements satisfy

i) $f_s^2 = f_s^* = f_s$ for all $1 \leq s \leq r-1$,

ii) $f_s f_t = f_t f_s$ for all $1 \leq s, t \leq r-1$ with $|s-t| \geq 2$,

iii) $\beta f_s f_t f_s = f_s$ for all $1 \leq s, t \leq r-1$ with $|s-t| = 1$,

$\beta = \|E(1)\|^4/\|F\bar{F}\|^2$. These are just the relations defining the Temperley–Lieb Algebra $A_{\beta,r}$, see [71]. Consequently, $\dim \mathrm{Hom}(u^{\boxtimes r}, u^{\boxtimes r}) \leq \dim A_{\beta,r}$. It is well known that the dimension of $A_{\beta,r}$ is the Catalan number $C_r = (2r)!/(r!(r+1)!)$ [71, Aside 4.1.4].

*Step* 3. By Corollary 3.2.16 and Proposition 6.2.11,

$$\dim \mathrm{Hom}(v^{\boxtimes r}, v^{\boxtimes r}) = h_{\mathrm{SU}}(\chi(v)^r \chi(v)^r) = \frac{1}{2\pi} \int_{-2}^{2} x^{2r} \sqrt{4 - x^2}\, dx,$$

where $h_{\mathrm{SU}}$ denotes the Haar state of $(\mathrm{SU}_\mu(2), \Delta)$. A routine calculation shows that this integral is equal to the Catalan number $C_r$; see, for example [12, Remarque after Corollaire 1]. Summarizing, we obtain

$$\dim \mathrm{Hom}(u^{\boxtimes r}, u^{\boxtimes r}) \leq \dim A_{\beta,r} = C_r = \dim \mathrm{Hom}(v^{\boxtimes r}, v^{\boxtimes r}). \qquad \square$$

Denote by $(u_{(r)})_{r \in \mathbb{N}}$ the irreducible corepresentation matrices of $(\mathrm{SU}_\mu(2), \Delta)$ given in Theorem 6.2.8, and by $(\chi_{(r)})_{r \in \mathbb{N}}$ the associated characters. By Proposition 6.2.10 iii), the subalgebra $\mathbb{C}[\chi(u)] \subset \mathrm{SU}_\mu(2)$ generated by $\chi(u) = \chi_{(1)}$ is just the polynomial algebra. Therefore we can define a unital homomorphism

$$\Phi \colon \mathbb{C}[\chi(u)] \to A_o(F), \quad \chi(u) \mapsto \chi(v).$$

**Theorem 6.4.10.** *For each $r \in \mathbb{N}$, there exists an irreducible corepresentation matrix $v_{(r)}$ of $(A_o(F), \Delta_o)$ such that $\chi(v_{(r)}) = \Phi(\chi_{(r)})$.*

*Proof.* For $r \leq 1$, the assertion is satisfied with $v_{(0)} = 1$ and $v_{(1)} = v$. We prove by induction on $r$ that there exists an irreducible corepresentation matrix $v_{(r)}$ such that $v_{(r-1)} \boxtimes v \simeq v_{(r-2)} \boxplus v_{(r)}$ and $\chi(v_{(r)}) = \Phi(\chi_{(r)})$ for all $r \geq 2$.

Assume that the assertion holds for all $s < r$. Using Frobenius reciprocity (Proposition 3.1.11), the relation $v \simeq \bar{v}$, and the induction hypothesis for $r-1$, we find

$$\operatorname{Hom}(v_{(r-2)}, v_{(r-1)} \boxtimes v) \cong \operatorname{Hom}(v_{(r-2)} \boxtimes v, v_{(r-1)}) \neq 0.$$

Since $v_{(r-2)}$ is irreducible, it follows from Theorem 5.3.3 that $v_{(r-1)} \boxtimes v \simeq v_{(r-2)} \boxplus v_{(r)}$ for some corepresentation matrix $v_{(r)}$, and by Proposition 3.2.14 and 6.2.10 i),

$$\chi(v_{(r)}) = \chi(v_{(r-1)})\chi(v) - \chi(v_{(r-2)}) = \Phi(\chi_{(r-1)}\chi_{(1)} - \chi_{(r-2)}) = \Phi(\chi_{(r)}).$$

We show that $\operatorname{Hom}(v_{(r)}, v_{(r)})$ has dimension one, and this implies that $v_{(r)}$ is irreducible. An inductive application of the formula $\chi_{(s)}\chi_{(1)} = \chi_{(s-1)} + \chi_{(s+1)}$ for $s < r$ (see Proposition 6.2.10 i)) shows that there exist numbers $a(r,s) \in \mathbb{N}$ such that

$$\chi(u^{\boxtimes r}) = \chi_{(1)}^r = \chi_{(r)} + \sum_{s<r} a(r,s)\chi_{(s)}.$$

Applying $\Phi$, we get

$$\chi(v^{\boxtimes r}) = \chi(v)^r = \chi(v_{(r)}) + \sum_{s<r} a(r,s)\chi(v_{(s)}). \tag{6.14}$$

By Corollary 3.2.16, $\dim \operatorname{Hom}(u^{\boxtimes r}, u^{\boxtimes r}) = 1 + \sum_{s<r} a(r,s)^2$, and using irreducibility of $v_{(s)}$ for $s < r$ and Proposition 6.4.9, we find

$$\begin{aligned}\dim \operatorname{Hom}(v_{(r)}, v_{(r)}) &\leq \dim \operatorname{Hom}(v^{\boxtimes r}, v^{\boxtimes r}) - \sum_{s<r} a(r,s)^2 \\ &\leq \dim \operatorname{Hom}(u^{\boxtimes r}, u^{\boxtimes r}) - \sum_{s<r} a(r,s)^2 = 1.\end{aligned}$$ □

**Remarks 6.4.11.** i) For $n = 2$, we have $v_{(r)} \in M_{r+1}(A_o(F))$, see Proposition 6.4.8 ii) and Theorem 6.2.8. For $n \geq 3$, one can show by induction that $v_{(r)} \in M_z(A_o(F))$ for $z := (x^{r+1} - y^{r+1})/(x-y)$, where $x$ and $y$ are the solutions of the equation $X^2 - nX + 1 = 0$.

ii) The proofs of Proposition 6.4.9 and Theorem 6.4.10 show that for each $r \in \mathbb{N}$, the algebra $\operatorname{Hom}(v^{\boxtimes r}, v^{\boxtimes r})$ is the Temperley–Lieb algebra $A_{\beta,r}$.

Knowing the characters of the corepresentation matrices $v_{(r)}$, we can draw the following conclusions:

**Corollary 6.4.12.** i) $v_{(0)} = 1$, $v_{(1)} = v$, *and for all* $r, s \in \mathbb{N}$,

$$v_{(r)} \boxtimes v_{(s)} \simeq v_{(|r-s|)} \boxplus v_{(|r-s|+2)} \boxplus \cdots \boxplus v_{(r+s-2)} \boxplus v_{(r+s)}.$$

ii) *Every irreducible corepresentation matrix of* $(A_o(F), \Delta_o)$ *is equivalent to* $v_{(r)}$ *for some* $r \in \mathbb{N}$.

iii) *The unital homomorphism* $\mathbb{C}[X] \to A_o(F)$ *given by* $X \mapsto \chi(v)$ *is injective.*

iv) $\chi(v)$ *is semi-circular with respect to the Haar state* $h$ *of* $(A_o(F), \Delta_o)$ *in the sense that*

$$h(f(\chi(v))) = \frac{1}{2\pi}\int_{-2}^{2} f(x)\sqrt{4-x^2}dx \quad \textit{for each } f \in C_0(\mathbb{R}).$$

v) $(A_o(F), \Delta_o)$ *is not amenable if* $n > 2$.

*Proof.* i) Clearly, $\chi(v_{(0)}) = 1$ and $\chi(v_{(1)}) = \Phi(\chi(u)) = \chi(v)$. Using the relation $\chi(v_{(t)}) = \Phi(\chi_{(t)})$ for $t = |r-s|, \dots, r+s$ and Proposition 6.2.10 i), we find

$$\chi(v_{(r)})\chi(v_{(s)}) = \chi(v_{(|r-s|)}) + \chi(v_{(|r-s|+2)}) + \dots + \chi(v_{(r+s-2)}) + \chi(v_{(r+s)}).$$

This relation, Proposition 3.2.14, and Corollary 3.2.16 imply the decomposition formula for $v_{(r)} \boxtimes v_{(s)}$.

ii) By Lemma 6.1.5, every irreducible corepresentation of $(A_o(F), \Delta_o)$ is contained in $v^{\boxtimes r}$ for some $r \in \mathbb{N}$, and formula (6.14) shows that $v^{\boxtimes r}$ is a direct sum of copies of the corepresentations $v_{(s)}$ for $s = 0, \dots, r$.

iii) By Proposition 3.2.15, the family $(\chi(v_{(r)}))_r$ is linearly independent, and equation (6.14) implies that so is the family $(\chi(v)^r)_r$.

iv) Using the Stone–Weierstrass Theorem, we see that it suffices to prove the assertion for $f \in C([-2,2])$ of the form $x \mapsto x^r$, where $r \in \mathbb{N}$. Denote by $h_{\mathrm{SU}}$ the Haar state of $\mathrm{SU}_\mu(2)$, and let $a(r,0)$ be as in formula (6.14). Then by Corollary 3.2.16 and Proposition 6.2.11,

$$\begin{aligned} h(\chi(v)^r) &= \dim \operatorname{Hom}(v_{(0)}, v^{\boxtimes r}) = a(r,0) \\ &= \dim \operatorname{Hom}(u_{(0)}, u^{\boxtimes r}) = h_{\mathrm{SU}}(\chi(u)^r) = \frac{1}{2\pi}\int_{-2}^{2} x^r\sqrt{4-x^2}dx. \end{aligned}$$

v) Assume that $n > 2$ and that $(A_o(F), \Delta_o)$ is amenable. Then the Haar state $h$ is faithful, and iv) implies that the spectrum of $\chi(v)$ is contained in the interval $[-2,2]$. Hence $n - \chi(v)$ is invertible. On the other hand, if $(A_o(F), \Delta_o)$ is amenable, the counit $\epsilon_0$ of the corresponding algebraic compact quantum group extends to a $*$-homomorphism $\epsilon\colon A_o(F) \to \mathbb{C}$, and $\epsilon(n - \chi(v)) = n - n = 0$ by Proposition 3.2.14, contradicting the invertibility of $n - \chi(v)$. □

**The corepresentation theory of the free unitary quantum group.** The corepresentation theory of the free unitary quantum group $(A_u(F), \Delta_u)$ was determined by Banica [13]. We shall only state the main results; for the proofs see [13].

Let us start with some preliminaries. We denote by $\mathbb{N} * \mathbb{N}$ the free coproduct of the monoid $\mathbb{N}$ with itself, by $\alpha, \beta \in \mathbb{N} * \mathbb{N}$ the two canonical generators, and by $e \in \mathbb{N} * \mathbb{N}$ the neutral element. Thus $\mathbb{N} * \mathbb{N}$ can be identified with the set of all words in $\alpha$ and $\beta$, where $e$ corresponds to the empty word and the multiplication is given by concatenation. Finally, we denote by $x \mapsto \bar{x}$ the unique antimultiplicative involution on $\mathbb{N} * \mathbb{N}$ defined by $\bar{e} := e, \bar{\alpha} := \beta, \bar{\beta} := \alpha$.

We fix an $F \in \mathrm{GL}_n(\mathbb{C})$, where $n \in \mathbb{N}$, and consider the free unitary quantum group $(A_u(F), \Delta_u, u)$.

**Theorem 6.4.13.** i) *There exists a family of pairwise inequivalent irreducible corepresentation matrices $(u_x)_{x \in \mathbb{N}*\mathbb{N}}$ of $(A_u(F), \Delta_u)$ such that*

$$u_e = 1, \quad u_\alpha = u, \quad u_\beta = \bar{u}, \quad \overline{u_x} = u_{\bar{x}}, \quad u_x \boxtimes u_y = \sum_{\substack{a,b,c \in \mathbb{N}*\mathbb{N} \\ x=ac,\, y=\bar{c}b}} u_{ab}$$

*for all $x, y \in \mathbb{N}$. Every irreducible corepresentation matrix of $(A_u(F), \Delta_u)$ is equivalent to $u_x$ for some $x \in \mathbb{N} * \mathbb{N}$.*

ii) *Denote by $\mathbb{C}\langle X, X^*\rangle$ the free unital $*$-algebra with generators $X$ and $X^*$. Then the unital $*$-homomorphism $\mathbb{C}\langle X, X^*\rangle \to A_u(F)$ given by $X \mapsto \chi(u)$ is injective.*

iii) *$\chi(u)$ is semi-circular with respect to the Haar state $h$ of $(A_u(F), \Delta_u)$ in the sense that*

$$h(f(\chi(u))) = \frac{1}{2\pi} \int_{-2}^{2} f(x)\sqrt{4 - x^2}\,dx \quad \textit{for each } f \in C_0(\mathbb{R}).$$

iv) *Assume that $F\bar{F} \in \mathbb{R}1$. Denote by $A_u(F)_r$ and $A_o(F)_r$ the reduced $C^*$-algebraic quantum groups associated to $A_u(F)$ and $A_o(F)$, respectively, by $h_{o,r}$ and $h_{\mathbb{T}}$ the Haar states of $A_o(F)_r$ and $C(\mathbb{T})$, respectively, and by $z \in C(\mathbb{T})$ the canonical generator. Then there exists an isomorphism of $C^*$-algebraic compact quantum groups $A_u(F)_r \to (C(\mathbb{T}), h_{\mathbb{T}}) * (A_o(F)_r, h_{o,r})$ such that $u_{ij} \mapsto z * v_{ij}$ for all $i, j$.*

# Chapter 7
# Multiplicative unitaries

Multiplicative unitaries are fundamental to the theory of quantum groups in the setting of $C^*$-algebras and von Neumann algebras, and to generalizations of Pontrjagin duality. Roughly, a multiplicative unitary is one single map that encodes all structure maps of a quantum group and of its generalized Pontrjagin dual simultaneously. In some more detail, the most important features of multiplicative unitaries are the following:

- To every "reasonable" quantum group, one can associate a multiplicative unitary (see Examples 7.1.4, 7.1.6, 7.1.7, and Theorems 8.3.1, 8.3.18).
- Out of every multiplicative unitary, one can construct a dual pair of von Neumann bialgebras and (if the unitary is well-behaved) a dual pair of "reduced" $C^*$-bialgebras (see Section 7.2). Moreover, one can define a dual pair of "universal" $C^*$-bialgebras, using the concept of a representation and a corepresentation of a multiplicative unitary [7, Appendice].

These two constructions facilitate the transition

- between the three guises of a quantum group (see Sections 4.3, 8.3.4 and Theorems 5.4.5, 7.2.14);
- from a quantum group (in form of a $C^*$-bialgebra/von Neumann bialgebra) to the dual quantum group. This is the first key step in generalized Pontrjagin duality (see Section 8.3.3). The second step in that duality – the identification of the bidual – can also be achieved via the associated multiplicative unitaries (see Theorem 8.3.15 and its proof).

Further important features of multiplicative unitaries are:

- Some examples of quantum groups are most easily constructed via a multiplicative unitary (see, e.g., Section 8.4).
- If a multiplicative unitary is modular, then one can construct from it antipodes on the associated "reduced" $C^*$-bialgebras (see Section 7.3.2). Under favorable circumstances, one can also construct Haar weights on these $C^*$-bialgebras (see [205] and [67, Section 1.4]).
- Multiplicative unitaries play a central rôle in the construction of reduced crossed products for coactions, in the construction of dual coactions, and in the proof of Baaj–Skandalis duality (see Chapter 9).

In the duality theory of locally compact groups and of crossed products, first examples of multiplicative unitaries were used for some time, before Baaj and Skandalis introduced the general definition and developed a beautiful and rich theory [7]. Building on their work, Woronowicz introduced the class of manageable multiplicative unitaries [201], which is particularly well adapted to operator-algebraic approaches to quantum groups.

## 7.1 The concept of a multiplicative unitary

Multiplicative unitaries are easily defined in terms of one short equation, but without a discussion of examples, the motivation for and the implications of this short equation are difficult to grasp. Therefore let us first consider an algebraic variant of a multiplicative unitary.

### 7.1.1 Motivation

Let $(A, \Delta)$ be a bialgebra. Consider the maps $T_1, T_2 \colon A \odot A \to A \odot A$ given by

$$T_1(a \odot b) = \Delta(a)(1 \odot b) \quad \text{and} \quad T_2(a \odot b) = (a \odot 1)\Delta(b),$$

respectively. By Theorem 1.3.18, $(A, \Delta)$ is a Hopf algebra if and only if $T_1$ and $T_2$ are isomorphisms, and in that case, the counit and the antipode of $(A, \Delta)$ can be reconstructed from $T_1$ or $T_2$. If $(A, \Delta)$ is a Hopf algebra, we can even reconstruct its multiplication and comultiplication from the map $T_1$ – thus all structure maps of an arbitrary Hopf algebra can be encoded by one single map. Let us explain in detail how this reconstruction works.

For arbitrary $a \in A,\ f \in A'$, consider the linear map $t_{f,a} \colon A \to A$ given by

$$t_{f,a}(b) := (f \odot \mathrm{id})(T_1(a \odot b)) \quad \text{for all } b \in A.$$

Inserting the definition of $T_1$, we find that $t_{f,a}(b) = (f \odot \mathrm{id})(\Delta(a))b$ for all $b \in A$. Denote by $\pi \colon A \to \mathrm{Hom}_{\Bbbk}(A)$ the left regular representation, that is, $\pi(c)b = cb$ for all $c, b \in A$. Then $t_{f,a} = \pi((f \odot \mathrm{id})(\Delta(a)))$, and

$$\mathrm{span}\{t_{f,a} \mid a \in A,\ f \in A'\} = \pi(A) \cong A \quad \text{as algebras},$$

because $A$ is spanned by elements of the form $(f \odot \mathrm{id})(\Delta(a))$, where $a \in A$ and $f \in A'$ (simply let $f = \epsilon$).

The comultiplication $\Delta$ can be recovered from $T_1$ as well: the formula

$$T_1((c \odot 1)(a \odot b)) = \Delta(ca)(1 \odot b) = \Delta(c)T_1(a \odot b)$$

shows that $(\pi \odot \pi)(\Delta(c)) = T_1(\pi(c) \odot \mathrm{id})T_1^{-1}$ for all $c \in A$.

Implicitly, we recovered simultaneously with the Hopf algebra $(A, \Delta)$ also the dual algebra $A'$. However, this algebra can also be reconstructed directly from the map $T_1$ in a similar way as $A$: For arbitrary $b \in A$, $f \in A'$, consider the linear map $t'_{f,b}\colon A \to A$ given by

$$t'_{f,b}(a) := (\mathrm{id} \odot f)(T_1(a \odot b)) \quad \text{for all } a \in A.$$

Inserting the definition of $T_1$ again, we find that $t'_{f,b}(a) = (\mathrm{id} \odot f(\cdot\, b))(\Delta(a)) = f(\cdot\, b) * a$ for all $a \in A$. For each $g \in A'$, denote by $\rho(g) \in \mathrm{Hom}_{\mathbb{k}}(A)$ the operator given by $a \mapsto g * a$. Then $t'_{f,b} = \rho(f(\cdot\, b))$. The map $\rho\colon A' \to \mathrm{Hom}_{\mathbb{k}}(A)$, $g \mapsto \rho(g)$, is an algebra homomorphism by Lemma 1.3.10 and easily seen to be injective. We have

$$\mathrm{span}\{t'_{f,b} \mid b \in A,\ f \in A'\} = \rho(A') \cong A' \quad \text{as algebras,}$$

because $A'$ is spanned by elements of the form $f(\cdot\, b)$, where $f \in A'$ and $b \in A$ (simply let $b = 1_A$).

The comultiplication on $A'$ can only be defined in form of the dual $\Delta_{A'}\colon A' \to (A \odot A)'$ of the multiplication map of $A$. However, we can extend the homomorphism $\rho \odot \rho\colon A' \odot A' \to \mathrm{Hom}_{\mathbb{k}}(A \odot A)$ to $(A \odot A)'$ by the formula

$$((\rho \odot \rho)(\omega))(a \odot b) := \sum a_{(1)} \odot b_{(1)}\omega(a_{(2)} \odot b_{(2)}),$$

and the calculation

$$\begin{aligned}
(\mathrm{id} \odot \rho(f))(T_1(a \odot b)) &= \sum a_{(1)} \odot a_{(2)}b_{(1)} f(a_{(3)}b_{(2)}) \\
&= \sum T_1(a_{(1)} \odot b_{(1)} f(a_{(2)}b_{(2)})) \\
&= \sum T_1(a_{(1)} \odot b_{(1)}\Delta_{A'}(f)(a_{(2)} \odot b_{(2)}))
\end{aligned}$$

shows that $(\rho \odot \rho)(\Delta_{A'}(f)) = T_1^{-1}(\mathrm{id} \odot \rho(f))T_1$ for all $f \in A'$.

Let us add that this reconstruction procedure extends to algebraic quantum groups after straightforward modifications.

### 7.1.2 Definition and examples

The definition of a multiplicative unitary involves the following leg notation:

**Notation 7.1.1.** Let $H$ be a Hilbert space and $T \in \mathcal{L}(H \otimes H)$. We define operators $T_{[12]}, T_{[23]}, T_{[13]} \in \mathcal{L}(H \otimes H \otimes H)$ by the formulas

$$\begin{gathered}
T_{[12]} := T \otimes \mathrm{id}_H, \quad T_{[23]} := \mathrm{id}_H \otimes T, \\
T_{[13]} := \Sigma_{[12]} T_{[23]} \Sigma_{[12]} = \Sigma_{[23]} T_{[12]} \Sigma_{[23]},
\end{gathered}$$

where $\Sigma \in \mathcal{L}(H \otimes H)$ denotes the flip $\xi \otimes \eta \mapsto \eta \otimes \xi$. Thus, the lower indices $i$, $j$ on $T_{[ij]}$ indicate the two factors of the tensor product $H \otimes H \otimes H$ on which $T$ acts.

More generally, given an operator $T \in \mathcal{L}(H^{\otimes k})$, an $n \geq k$, and indices $1 \leq i_1 < \cdots < i_k \leq n$, we define an operator $T_{[i_1 \dots i_k]} \in \mathcal{L}(H^{\otimes n})$ by letting $T$ act on the factors at the positions $i_1, \dots, i_k$. Clearly, this notation can also be extended to tuples of indices $(i_1, \dots, i_k)$ that are pairwise distinct but not necessarily ordered, for example, for $k = n = 2$, we put $T_{[21]} := \Sigma T \Sigma$.

Of course, the notation introduced above can also be applied to operators on algebraic tensor products of vector spaces and, more generally, to morphisms in any tensor category.

**Definition 7.1.2.** A *multiplicative unitary on a Hilbert space* $H$ is a unitary $V \in \mathcal{L}(H \otimes H)$ that satisfies

$$V_{[12]} V_{[13]} V_{[23]} = V_{[23]} V_{[12]}. \tag{7.1}$$

**Remarks 7.1.3.** i) Equation (7.1) holds if and only if the following diagram commutes:

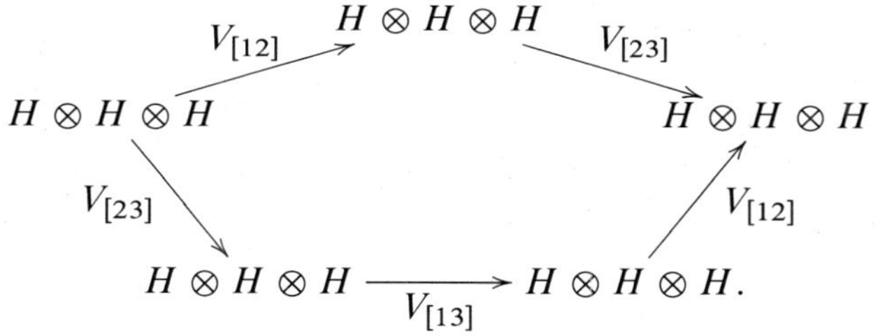


(7.2)

Therefore, (7.1) is called the *pentagon equation.*

ii) For every Hilbert space $H$, the identity $\mathrm{id}_{H \otimes H}$ is a multiplicative unitary on $H$.

iii) Given a multiplicative unitary $V$ on a Hilbert space $H$, we can construct the following new unitaries:

*Opposite unitary.* The operator $V^{\mathrm{op}} := \Sigma V^* \Sigma$ is a multiplicative unitary, called the *opposite* of $V$. It will play an important rôle later on.

*Unitary transformation.* If $K$ is a Hilbert space and $U \in \mathcal{L}(H, K)$ a unitary, then the operator $W := (U \otimes U) V (U^* \otimes U^*) \in \mathcal{L}(K \otimes K)$ is a multiplicative unitary on $K$. Two multiplicative unitaries $V$ and $W$ are called *equivalent* if they are related this way.

*Tensor product.* If $W$ is a multiplicative unitary on some Hilbert space $K$, then the operator $V \boxtimes W := V_{[13]} W_{[24]} \in \mathcal{L}(H \otimes K \otimes H \otimes K)$ is a multiplicative unitary on $H \otimes K$.

Let us consider some examples of multiplicative unitaries:

**Example 7.1.4.** Let $G$ be a locally compact group with left Haar measure $\lambda$ and right Haar measure $\lambda^{-1}$. For $\mu = \lambda, \lambda^{-1}$, we identify $L^2(G,\mu) \otimes L^2(G,\mu)$ with $L^2(G \times G, \mu \times \mu)$, where $\mu \times \mu$ denotes the product measure. Then the operators

$$V_G \in \mathcal{L}(L^2(G,\lambda^{-1}) \otimes L^2(G,\lambda^{-1})), \qquad W_G \in \mathcal{L}(L^2(G,\lambda) \otimes L^2(G,\lambda)),$$
$$(V_G\zeta)(x,y) := \zeta(xy,y), \qquad (W_G\zeta)(x,y) := \zeta(x,x^{-1}y),$$

respectively, are multiplicative unitaries. Indeed, $V_G$ and $W_G$ are isometric because $\lambda^{-1}$ and $\lambda$ are invariant with respect to right or left translations, and their images are dense because the maps $(x,y) \mapsto (xy,y)$ and $(x,y) \mapsto (x,x^{-1}y)$ are homeomorphisms of $G \times G$. The pentagon equations for $V_G$ and $W_G$ amount to associativity of the multiplication in $G$ and can be verified by straightforward calculations, for example,

$$(V_{G[23]}V_{G[12]}\zeta)(x,y,z) = (V_{G[12]}\zeta)(x,yz,z) = \zeta(x(yz),yz,z)$$

and

$$\begin{aligned}(V_{G[12]}V_{G[13]}V_{G[23]}\zeta)(x,y,z) &= (V_{G[13]}V_{G[23]}\zeta)(xy,y,z)\\ &= (V_{G[23]}\zeta)((xy)z,y,z) = \zeta((xy)z,yz,z)\end{aligned}$$

for all $\zeta \in L^2(G \times G \times G, \lambda^{-1}\times\lambda^{-1}\times\lambda^{-1}) \cong L^2(G,\lambda^{-1})^{\otimes 3}$ and all $x,y,z \in G$. Short calculations show that the multiplicative unitaries $V_G^{\mathrm{op}}$ and $W_G^{\mathrm{op}}$ act as follows:

$$(V_G^{\mathrm{op}}\zeta)(x,y) = \zeta(x,yx^{-1}) \quad \text{and} \quad (W_G^{\mathrm{op}}\zeta)(x,y) = \zeta(yx,y).$$

The multiplicative unitaries that arise from groups as in the previous example can be characterized as follows.

**Example 7.1.5.** A multiplicative unitary $V$ is called *commutative* if $V_{[13]}$ commutes with $V_{[23]}$, and *cocommutative* if $V_{[12]}$ commutes with $V_{[13]}$. One easily verifies that a multiplicative unitary $V$ is commutative/cocommutative if and only if $V^{\mathrm{op}}$ is cocommutative/commutative.

For every locally compact group $G$, the operator $V_G$ is commutative and the operator $W_G$ is cocommutative. Baaj and Skandalis showed that for every commutative multiplicative unitary $V$, there exist a locally compact group $G$ and a Hilbert space $K$ such that $V$ is equivalent to $V_G \boxtimes \mathrm{id}_{K\otimes K}$ [7, Theorem 2.2], [8].

To every algebraic quantum group, one can associate two multiplicative unitaries:

**Example 7.1.6.** Let $(A_0,\Delta_0)$ be an algebraic quantum group with positive right integral $\psi$. Recall that $\psi$ is faithful (Corollary 2.2.5). Denote by $H_\psi$ the

GNS-space for $\psi$, that is, the completion of $A_0$ with respect to the inner product $\langle a|b\rangle := \psi(a^*b)$. Then the map

$$T_1 \colon A_0 \odot A_0 \to A_0 \odot A_0, \quad a \odot b \mapsto \Delta(a)(1 \odot b),$$

extends to a multiplicative unitary $V_{A_0}$ on $H_\psi$:

- $T_1$ is isometric with respect to the inner product because $\psi$ is right invariant: for all $a, b, c, d \in A_0$

$$\begin{aligned}\langle T_1(a \odot b)|T_1(c \odot d)\rangle &= (\psi \odot \psi)((1 \odot b^*)\Delta_0(a^*c)(1 \odot d)) \\ &= \psi(a^*c)\psi(b^*d) = \langle a \odot b|c \odot d\rangle;\end{aligned}$$

- the image of $T_1$ is dense in $H_\psi \otimes H_\psi$ because it is equal to $A_0 \odot A_0$;
- $V_{A_0}$ satisfies the pentagon equation: for all $a, b, c \in A_0$,

$$\begin{aligned}&(V_{A_0})_{[12]}(V_{A_0})_{[13]}(V_{A_0})_{[23]}(a \odot b \odot c) \\ &= (\Delta_0^{(2)}(a))(1 \odot \Delta_0(b))(1 \odot 1 \odot c) \\ &= (V_{A_0})_{[23]}(V_{A_0})_{[12]}(a \odot b \odot c).\end{aligned}$$

If $\phi$ is a positive left integral on $(A_0, \Delta_0)$, then the inverse of the map

$$T_2^{\mathrm{op}} \colon A_0 \odot A_0 \to A_0 \odot A_0, \quad a \odot b \mapsto \Delta(b)(a \odot 1),$$

extends to a multiplicative unitary $W_{A_0}$ on the GNS-space $H_\phi$ for $\phi$. Indeed, $T_2^{\mathrm{op}}$ is isometric because $\phi$ is left-invariant, $T_2^{\mathrm{op}}$ has dense image because $(A_0, \Delta_0)$ is regular, and the pentagon equation for $W_{A_0}$ is a consequence of the equation $(T_2^{\mathrm{op}})_{[23]}(T_2^{\mathrm{op}})_{[13]}(T_2^{\mathrm{op}})_{[12]} = (T_2^{\mathrm{op}})_{[12]}(T_2^{\mathrm{op}})_{[23]}$ which is easily verified.

**Example 7.1.7.** Let $(A, \Delta)$ be a $C^*$-algebraic compact quantum group with Haar state $h$ and associated GNS-space $H_h$, and define $T_1, T_2^{\mathrm{op}} \colon A \otimes A \to A \otimes A$ by the same formulas as above. Then $T_1$ and $(T_2^{\mathrm{op}})^{-1}$ extend to multiplicative unitaries $V_A$ and $W_A$ on $H_h$; the unitary $V_A$ was used already in the proof of Theorem 5.4.5. This example is related to the preceding ones as follows:

- Consider the algebraic compact quantum group $(A_0, \Delta_0)$ associated to $(A, \Delta)$ (see Theorem 5.4.1). The restriction $h_0 := h|_{A_0}$ is a left- and a right-invariant integral on $(A_0, \Delta_0)$, and clearly $H_h = H_{h_0}$, $V_A = V_{A_0}$, $W_A = W_{A_0}$.
- If $G$ is a compact group and $A = C(G)$, then $H_h = L^2(G, \lambda)$, where $\lambda$ denotes the normalized Haar measure of $G$, and $V_A = V_G$, $W_A = W_G$, because

$$\begin{aligned}(T_1(g \otimes h))(x, y) &= (\Delta(g)(1 \otimes h))(x, y) \\ &= (\Delta(g))(x, y) \cdot (1 \otimes h)(x, y) = g(xy)h(y)\end{aligned}$$

and

$$(T_2^{\mathrm{op}}(g \otimes h))(x, y) = (\Delta(h)(g \otimes 1))(x, y) = g(x)h(xy)$$

for all $x, y \in G$ and $g, h \in C(G)$.

- If $G$ is a discrete group and $A = C_r^*(G)$, then $H_h = l^2(G)$ and $V_A = W_G$, $W_A = W_G^{\mathrm{op}}$. This can be deduced from the relations

$$\begin{aligned} T_1(L_x \otimes L_y) &= \Delta(L_x)(1 \otimes L_y) = L_x \otimes L_{xy}, \\ T_2^{\mathrm{op}}(L_x \otimes L_y) &= \Delta(L_y)(L_x \otimes 1) = L_{yx} \otimes L_y \quad \text{for all } x, y \in G. \end{aligned}$$

**Remark 7.1.8.** Those multiplicative unitaries that arise from $C^*$-algebraic compact quantum groups were characterized by Baaj and Skandalis, see [7, Paragraphe 4].

## 7.2 The legs of a multiplicative unitary

Out of every multiplicative unitary, we shall construct a dual pair of von Neumann bialgebras by a similar procedure like that presented in Section 7.1.1. If the multiplicative unitary satisfies an additional regularity or manageability condition, we can additionally construct a dual pair of bisimplifiable $C^*$-bialgebras that are weakly dense in the aforementioned von Neumann bialgebras. These von Neumann bialgebras and $C^*$-bialgebras are the *legs* of the multiplicative unitary.

Moreover, we construct a densely defined counit and antipode on these bialgebras, but it seems to be difficult to relate these maps to the usual axioms for the counit and the antipode of a Hopf algebra.

### 7.2.1 Definition and first properties

Let $V$ be a multiplicative unitary on a Hilbert space $H$. We define normal $*$-homomorphisms $\widehat{\Delta}_V, \Delta_V \colon \mathcal{L}(H) \to \mathcal{L}(H \otimes H)$ and subspaces $\widehat{A}_0(V), A_0(V) \subseteq \mathcal{L}(H)$ such that – under favorable conditions – suitable completions of $\widehat{A}_0(V)$ and $A_0(V)$, equipped with the comultiplications $\widehat{\Delta}_V$ or $\Delta_V$, turn out to be $C^*$-bialgebras or von Neumann bialgebras, respectively. These constructions are motivated by and should be compared with the reconstruction procedure discussed in Section 7.1.1. To simplify notation, we shall frequently denote the identity map $\mathrm{id}_H$ by 1, thinking of it as the unit in $\mathcal{L}(H)$.

**The comultiplications $\widehat{\Delta}_V$ and $\Delta_V$.** Consider the following normal $*$-homomorphisms:

$$\begin{aligned} \widehat{\Delta}_V \colon \mathcal{L}(H) &\to \mathcal{L}(H \otimes H), & \Delta_V \colon \mathcal{L}(H) &\to \mathcal{L}(H \otimes H), \\ T &\mapsto V^*(1 \otimes T)V, & T &\mapsto V(T \otimes 1)V^*. \end{aligned} \tag{7.3}$$

We shall usually omit the subscript "$V$" and write $\widehat{\Delta}$ and $\Delta$ for $\widehat{\Delta}_V$ and $\Delta_V$, respectively. Denote by "$\bar{\otimes}$" the von Neumann-algebraic tensor product. The normal $*$-homomorphisms

$$\widehat{\Delta} \mathbin{\bar{\otimes}} \mathrm{id},\ \Delta \mathbin{\bar{\otimes}} \mathrm{id} \colon \mathcal{L}(H \otimes H) \to \mathcal{L}(H \otimes H) \mathbin{\bar{\otimes}} \mathcal{L}(H) = \mathcal{L}(H \otimes H \otimes H)$$

and

$$\mathrm{id} \mathbin{\bar{\otimes}} \widehat{\Delta},\ \mathrm{id} \mathbin{\bar{\otimes}} \Delta \colon \mathcal{L}(H \otimes H) \to \mathcal{L}(H) \mathbin{\bar{\otimes}} \mathcal{L}(H \otimes H) = \mathcal{L}(H \otimes H \otimes H)$$

can easily be expressed in terms of $V$, for example,

$$(\widehat{\Delta} \mathbin{\bar{\otimes}} \mathrm{id})(R) = (V^* \otimes 1)(1 \otimes R)(V \otimes 1) \quad \text{for all } R \in \mathcal{L}(H \otimes H).$$

In succinct leg notation, we have for all $R \in \mathcal{L}(H \otimes H)$:

$$\begin{aligned} (\widehat{\Delta} \mathbin{\bar{\otimes}} \mathrm{id})(R) &= V_{[12]}^* R_{[23]} V_{[12]}, & (\Delta \mathbin{\bar{\otimes}} \mathrm{id})(R) &= V_{[12]} R_{[13]} V_{[12]}^*, \\ (\mathrm{id} \mathbin{\bar{\otimes}} \widehat{\Delta})(R) &= V_{[23]}^* R_{[13]} V_{[23]}, & (\mathrm{id} \mathbin{\bar{\otimes}} \Delta)(R) &= V_{[23]} R_{[12]} V_{[23]}^*. \end{aligned} \tag{7.4}$$

**Lemma 7.2.1.** $(\widehat{\Delta} \mathbin{\bar{\otimes}} \mathrm{id}) \circ \widehat{\Delta} = (\mathrm{id} \mathbin{\bar{\otimes}} \widehat{\Delta}) \circ \widehat{\Delta}$ *and* $(\Delta \mathbin{\bar{\otimes}} \mathrm{id}) \circ \Delta = (\mathrm{id} \mathbin{\bar{\otimes}} \Delta) \circ \Delta$.

*Proof.* Let $T \in \mathcal{L}(H)$. In leg notation, $(\widehat{\Delta} \mathbin{\bar{\otimes}} \mathrm{id})(\widehat{\Delta}(T))$ takes the form

$$(\widehat{\Delta} \mathbin{\bar{\otimes}} \mathrm{id})(\widehat{\Delta}(T)) = V_{[12]}^* \widehat{\Delta}(T)_{[23]} V_{[12]} = V_{[12]}^* V_{[23]}^* T_{[3]} V_{[23]} V_{[12]}.$$

By the pentagon equation, this is equal to $V_{[23]}^* V_{[13]}^* V_{[12]}^* T_{[3]} V_{[12]} V_{[13]} V_{[23]}$. But $V_{[12]}^* T_{[3]} V_{[12]} = T_{[3]} V_{[12]}^* V_{[12]} = T_{[3]}$, whence

$$\begin{aligned} (\widehat{\Delta} \mathbin{\bar{\otimes}} \mathrm{id})(\widehat{\Delta}(T)) &= V_{[23]}^* V_{[13]}^* T_{[3]} V_{[13]} V_{[23]} \\ &= V_{[23]}^* \widehat{\Delta}(T)_{[13]} V_{[23]} = (\mathrm{id} \mathbin{\bar{\otimes}} \widehat{\Delta})(\widehat{\Delta}(T)). \end{aligned}$$

The assertion concerning $\Delta$ follows from a similar calculation. □

**The algebras $\widehat{A}_0(V)$ and $A_0(V)$.** The subspaces $\widehat{A}_0(V)$ and $A_0(V)$ are obtained from $V$ by the application of slice maps, which are reviewed in Section 12.4. Denote by $\mathcal{L}(H)_*$ the set of all normal linear functionals on $\mathcal{L}(H)$. We shall be interested in the spaces

$$\widehat{A}_0(V) := \{\hat{a}_\omega \mid \omega \in \mathcal{L}(H)_*\}, \quad A_0(V) := \{a_\omega \mid \omega \in \mathcal{L}(H)_*\},$$

where for each $\omega \in \mathcal{L}(H)_*$,

$$\hat{a}_\omega := (\mathrm{id} \mathbin{\bar{\otimes}} \omega)(V) \in \mathcal{L}(H), \quad a_\omega := (\omega \mathbin{\bar{\otimes}} \mathrm{id})(V) \in \mathcal{L}(H).$$

In the following paragraphs, we shall frequently encounter calculations that involve slice maps. In particular, we use (analogues of) formulas (12.2)–(12.5) and relations like

$$\begin{aligned}(\mathrm{id} \mathbin{\bar{\otimes}} \omega)(X)(\mathrm{id} \mathbin{\bar{\otimes}} \upsilon)(Y) &= (\mathrm{id} \mathbin{\bar{\otimes}} \omega \mathbin{\bar{\otimes}} \upsilon)(X_{[12]}Y_{[13]}) \\ &= (\mathrm{id} \mathbin{\bar{\otimes}} \upsilon \mathbin{\bar{\otimes}} \omega)(X_{[13]}Y_{[12]}), \\ (\mathrm{id} \mathbin{\bar{\otimes}} \omega)(X)(\upsilon \mathbin{\bar{\otimes}} \mathrm{id})(Y) &= (\upsilon \mathbin{\bar{\otimes}} \mathrm{id} \mathbin{\bar{\otimes}} \omega)(X_{[23]}Y_{[12]}),\end{aligned}$$

which hold for all $X, Y \in \mathcal{L}(H \otimes H)$ and $\omega, \upsilon \in \mathcal{L}(H)_*$; see also Remark 12.4.5.

**Lemma 7.2.2.** *$\widehat{A}_0(V)$ and $A_0(V)$ are subalgebras of $\mathcal{L}(H)$, and each of the sets $\widehat{A}_0(V)H$, $\widehat{A}_0(V)^*H$, $A_0(V)H$ and $A_0(V)^*H$ is linearly dense in $H$.*

*Proof.* First, we show that $\widehat{A}_0(V)$ is an algebra. Let $\omega, \omega' \in \mathcal{L}(H)_*$. Then

$$\hat{a}_\omega \hat{a}_{\omega'} = (\mathrm{id} \mathbin{\bar{\otimes}} \omega)(V) \cdot (\mathrm{id} \mathbin{\bar{\otimes}} \omega')(V) = (\mathrm{id} \mathbin{\bar{\otimes}} \omega \mathbin{\bar{\otimes}} \omega')(V_{[12]}V_{[13]}).$$

We replace $V_{[12]}V_{[13]}$ by $V_{[23]}V_{[12]}V_{[23]}^*$ using the pentagon equation, and find

$$\hat{a}_\omega \hat{a}_{\omega'} = (\mathrm{id} \mathbin{\bar{\otimes}} \omega \mathbin{\bar{\otimes}} \omega')(V_{[23]}V_{[12]}V_{[23]}^*) = (\mathrm{id} \mathbin{\bar{\otimes}} \omega \mathbin{\bar{\otimes}} \omega')((\mathrm{id} \mathbin{\bar{\otimes}} \Delta)(V))$$

(see formula (7.4)). The linear functional $\omega'' := (\omega \mathbin{\bar{\otimes}} \omega') \circ \Delta$ is normal, and

$$\hat{a}_\omega \hat{a}_{\omega'} = (\mathrm{id} \mathbin{\bar{\otimes}} \omega'')(V) = \hat{a}_{\omega''} \in \widehat{A}_0(V).$$

Next, assume that $\zeta \in H$ is orthogonal to the set $\widehat{A}_0(V)H$. We show that $\zeta = 0$, and this implies that the set $\widehat{A}_0(V)H$ is linearly dense in $H$. For every $\xi, \eta \in H$, put $\omega_{\eta,\xi} := \langle \eta | \cdot \xi \rangle$. Then $\omega_{\eta,\xi} \in \mathcal{L}(H)_*$, and by assumption,

$$0 = \langle \zeta | \hat{a}_{\omega_{\eta,\xi}} \vartheta \rangle = \langle \zeta | (\mathrm{id} \mathbin{\bar{\otimes}} \omega_{\eta,\xi})(V)\vartheta \rangle = \langle \zeta \otimes \eta | V(\vartheta \otimes \xi) \rangle$$

for all $\eta, \xi, \vartheta \in H$. Since $V$ is surjective, this relation implies $\langle \zeta \otimes \eta | \vartheta' \otimes \xi' \rangle = 0$ for all $\eta, \vartheta', \xi' \in H$, and hence $\zeta = 0$.

The proofs of the remaining assertions are similar. □

The algebras $\widehat{A}_0(V)$ and $A_0(V)$ need not be closed with respect to the involution (see, for example, [9, Remark 4.5]). To describe the adjoints of elements $\hat{a}_\omega \in \widehat{A}_0(V)$ and $a_\omega \in A_0(V)$, we use the following simple result:

**Lemma 7.2.3.** i) *For every $\omega \in \mathcal{L}(H)_*$, the linear map $\omega^* \colon \mathcal{L}(H) \to \mathbb{C}$ given by $T \mapsto \overline{\omega(T^*)}$ is normal. The assignment $\omega \mapsto \omega^*$ defines an involution, that is, a conjugate-linear involutive map, on $\mathcal{L}(H)_*$.*

ii) $((\omega \mathbin{\bar{\otimes}} \mathrm{id})(R))^* = (\omega^* \mathbin{\bar{\otimes}} \mathrm{id})(R^*)$ *and* $((\mathrm{id} \mathbin{\bar{\otimes}} \omega)(R))^* = (\mathrm{id} \mathbin{\bar{\otimes}} \omega^*)(R^*)$ *for each $R \in \mathcal{L}(H \otimes H)$.*

*Proof.* This follows from standard arguments and calculations, see also formula (12.4). □

**Example 7.2.4.** For all $\eta, \xi \in H$, one has $(\omega_{\eta,\xi})^* = \omega_{\xi,\eta}$ because $(\omega_{\eta,\xi})^*(T) = \overline{\langle \eta | T^* \xi \rangle} = \langle \xi | T\eta \rangle$ for each $T \in \mathcal{L}(H)$.

The preceding lemma implies that for each $\omega \in \mathcal{L}(H)_*$,

$$(\hat{a}_\omega)^* = (\mathrm{id} \mathbin{\bar{\otimes}} \omega^*)(V^*) \quad \text{and} \quad (a_\omega)^* = (\omega^* \mathbin{\bar{\otimes}} \mathrm{id})(V^*). \tag{7.5}$$

In general, these elements need not belong to $\widehat{A}_0(V)$ or $A_0(V)$, respectively.

**The left-right symmetry.** The constructions introduced so far are symmetric in a sense that is made precise in the following lemma. As before, $\Sigma \in \mathcal{L}(H \otimes H)$ denotes the flip $\eta \otimes \xi \mapsto \xi \otimes \eta$. Note that $\mathrm{Ad}_\Sigma(S \otimes T) = T \otimes S$ for all $S, T \in \mathcal{L}(H)$ since $\Sigma(S \otimes T)\Sigma(\eta \otimes \xi) = \Sigma(S\xi \otimes T\eta) = T\eta \otimes S\xi$ for all $\eta, \xi \in H$.

**Lemma 7.2.5.** *We have*

$$\widehat{A}_0(V^{\mathrm{op}}) = A_0(V)^*, \qquad A_0(V^{\mathrm{op}}) = \widehat{A}_0(V)^*,$$
$$\widehat{\Delta}_{V^{\mathrm{op}}} = \mathrm{Ad}_\Sigma \circ \Delta_V, \qquad \Delta_{V^{\mathrm{op}}} = \mathrm{Ad}_\Sigma \circ \widehat{\Delta}_V.$$

*Proof.* The first and third equation follow from the fact that

$$\begin{aligned}(\mathrm{id} \mathbin{\bar{\otimes}} \omega)(V^{\mathrm{op}}) &= (\mathrm{id} \mathbin{\bar{\otimes}} \omega)(\Sigma V^* \Sigma) = (\omega \mathbin{\bar{\otimes}} \mathrm{id})(V^*) \\ &= ((\omega^* \mathbin{\bar{\otimes}} \mathrm{id})(V))^* = (a_{\omega^*})^*\end{aligned}$$

for every $\omega \in \mathcal{L}(H)_*$ (see equation (7.5)) and

$$\begin{aligned}\widehat{\Delta}_{V^{\mathrm{op}}}(T) &= V^{\mathrm{op}*}(1 \otimes T)V^{\mathrm{op}} = \Sigma V \Sigma(1 \otimes T)\Sigma V^* \Sigma \\ &= \Sigma V(T \otimes 1)V^* \Sigma = \mathrm{Ad}_\Sigma(\Delta_V(T))\end{aligned}$$

for every $T \in \mathcal{L}(H)$. The remaining assertions follow similarly. □

### 7.2.2 Well-behaved multiplicative unitaries

We shall primarily be interested in the following classes of multiplicative unitaries:

**Definition 7.2.6.** A multiplicative unitary $V$ on a Hilbert space $H$ is

i) *well-behaved* if the following conditions are satisfied:

(a) the following spaces are $C^*$-subalgebras of $\mathcal{L}(H)$:

$$\widehat{A}(V) := \|\cdot\|\text{-closure of } \widehat{A}_0(V), \quad A(V) := \|\cdot\|\text{-closure of } A_0(V);$$

(b) the restrictions of the $*$-homomorphisms $\widehat{\Delta}$ and $\Delta$ defined in equation (7.3) to $\widehat{A}(V)$ and $A(V)$, respectively, turn $\widehat{A}(V)$ and $A(V)$ into bisimplifiable $C^*$-bialgebras;

(c) $V \in M(\widehat{A}(V) \otimes A(V))$.

ii) *weakly well-behaved* if the following spaces are von Neumann algebras:

$$\widehat{A}_w(V) := \text{w-closure of } \widehat{A}_0(V), \quad A_w(V) := \text{w-closure of } A_0(V),$$

where "w-closure" denotes the closure with respect to the weak operator topology.

Informally, we call the spaces $\widehat{A}_0(V)$, $\widehat{A}(V)$, $\widehat{A}_w(V)$ and $A_0(V)$, $A(V)$, $A_\omega(V)$ together with the restrictions of the homomorphisms $\widehat{\Delta}$ and $\Delta$, respectively, the *left leg* and the *right leg* of $V$. In Section 7.3.1, we discuss two classes of (weakly) well-behaved multiplicative unitaries – the (weakly) regular and the manageable ones.

**Reformulation in terms of ket-bra-operators.** For later use in Section 7.3, we reformulate the definition of the spaces $\widehat{A}(V)$, $\widehat{A}_w(V)$ and $A(V)$, $A_w(V)$ in terms of ket-bra operators. More precisely, we combine the ket-bra notation with the leg notation as follows. For each $\eta, \xi \in H$, we define operators

$$|\eta\rangle_{[1]} : H \cong \mathbb{C} \otimes H \xrightarrow{|\eta\rangle \otimes \mathrm{id}} H \otimes H, \ \xi' \mapsto \eta \otimes \xi', \tag{7.6}$$

and

$$|\xi\rangle_{[2]} : H \cong H \otimes \mathbb{C} \xrightarrow{\mathrm{id} \otimes |\xi\rangle} H \otimes H, \ \eta' \mapsto \eta' \otimes \xi. \tag{7.7}$$

Evidently, the adjoints $\langle\eta|_{[1]} := |\eta\rangle_{[1]}{}^*$ and $\langle\xi|_{[2]} := |\xi\rangle_{[2]}{}^*$ are given by

$$\langle\eta|_{[1]}(\eta' \otimes \xi') = \langle\eta|\eta'\rangle\xi' \quad \text{and} \quad \langle\xi|_{[2]}(\eta' \otimes \xi') = \eta'\langle\xi|\xi'\rangle \tag{7.8}$$

for all $\eta', \xi' \in H$, respectively.

In terms of these maps, the operators $\hat{a}_\omega$ and $a_\omega$ that comprise the spaces $\widehat{A}_0(V)$ and $A_0(V)$ can be rewritten as follows. For a functional $\omega \in \mathcal{L}(H)_*$ of the form $\omega = \omega_{\xi',\xi} = \langle\xi'| \cdot \xi\rangle$, where $\xi, \xi' \in H$,

$$\hat{a}_\omega = (\mathrm{id} \mathbin{\bar{\otimes}} \omega_{\xi',\xi})(V) = \langle\xi'|_{[2]} V |\xi\rangle_{[2]},$$

and for a functional $\omega \in \mathcal{L}(H)_*$ of the form $\omega = \omega_{\eta',\eta} = \langle\eta'|\cdot\eta\rangle$, where $\eta, \eta' \in H$,

$$a_\omega = (\omega_{\eta',\eta} \mathbin{\bar{\otimes}} \mathrm{id})(V) = \langle\eta'|_{[1]} V |\eta\rangle_{[1]}.$$

**Lemma 7.2.7.** i) *The set* $\{\langle\xi'|_{[2]}V|\xi\rangle_{[2]} \mid \xi,\xi' \in H\}$ *is linearly dense in* $\widehat{A}(V)$ *with respect to the norm and in* $\widehat{A}_w(V)$ *with respect to the weak operator topology.*

ii) *The set* $\{\langle\eta'|_{[1]}V|\eta\rangle_{[1]} \mid \eta,\eta' \in H\}$ *is linearly dense in* $A(V)$ *with respect to the norm and in* $A_w(V)$ *with respect to the weak operator topology.*

*Proof.* i) Let $\omega \in \mathcal{L}(H)_*$. Then there exist sequences $(\xi_n)_n$ and $(\xi'_n)_n$ in $H$ that satisfy $\sum_n \|\xi_n\|^2, \sum_n \|\xi'_n\|^2 < \infty$, and $\omega(T) = \sum_n \langle\xi'_n|T\xi_n\rangle$ for each $T \in \mathcal{L}(H)$. Thus $\omega$ can be approximated in norm by finite linear combinations of maps of the form $\omega_{\xi',\xi}$, where $\xi,\xi' \in H$. Since $\|\operatorname{id} \mathbin{\overline{\otimes}} \upsilon\| \leq \|\upsilon\|$ for each $\upsilon \in \mathcal{L}(H)_*$ (Proposition 12.4.4), $(\operatorname{id} \mathbin{\overline{\otimes}} \omega)(V)$ can be approximated in norm by finite linear combinations of elements of the form $\langle\xi'|_{[2]}V|\xi\rangle_{[2]}$, where $\xi,\xi' \in H$.

ii) The proof is similar to the proof of i). □

**The setting of von Neumann algebras.** Evidently, every well-behaved multiplicative unitary is weakly well-behaved. But the definition of the term "weakly well-behaved" seems to be much weaker than the definition of the term "well-behaved": It only involves an analogue of condition i)(a) and not i)(b) or i)(c). We shall see that in the setting of von Neumann algebras, analogues of these conditions are automatically fulfilled.

Given a Hilbert space $K$ and a subset $C \subseteq \mathcal{L}(K)$, we denote by $C' := \{S \in \mathcal{L}(K) \mid ST = TS \text{ for all } T \in C\}$ the commutant of $C$.

**Proposition 7.2.8.** *Let $V$ be a multiplicative unitary on a Hilbert space $H$. Consider the following von Neumann algebras:*

$$\widehat{B}_w(V) := (\widehat{A}_0(V) + \widehat{A}_0(V)^*)'', \quad B_w(V) := (A_0(V) + A_0(V)^*)''.$$

i) *The maps $\widehat{\Delta}$ and $\Delta$ defined in equation (7.3) restrict to normal $*$-homomorphisms*

$$\widehat{\Delta}\colon \widehat{B}_w(V) \to \widehat{B}_w(V) \mathbin{\overline{\otimes}} \widehat{B}_w(V), \quad \Delta\colon B_w(V) \to B_w(V) \mathbin{\overline{\otimes}} B_w(V).$$

ii) $(\widehat{B}_w(V), \widehat{\Delta})$ *and* $(B_w(V), \Delta)$ *are von Neumann bialgebras.*

iii) $V \in \widehat{B}_w(V) \mathbin{\overline{\otimes}} B_w(V)$.

For the proof of this proposition, we use the following lemma.

**Lemma 7.2.9.** *For each $X \in \mathcal{L}(H \otimes H)$,*

$$X \in (\widehat{B}_w(V) \otimes 1)' \Leftrightarrow [X_{[12]}, V_{[13]}] = 0,$$

$$X \in (1 \otimes \widehat{B}_w(V))' \Leftrightarrow [X_{[12]}, V_{[23]}] = 0,$$

$$X \in (1 \otimes B_w(V))' \Leftrightarrow [X_{[23]}, V_{[13]}] = 0,$$

$$X \in (B_w(V) \otimes 1)' \Leftrightarrow [X_{[23]}, V_{[12]}] = 0.$$

*In particular,* $\widehat{B}_w(V)' = \{T \in \mathcal{L}(H) \mid [T \otimes 1, V] = 0\}$ *and* $B_w(V)' = \{T \in \mathcal{L}(H) \mid [1 \otimes T, V] = 0\}$.

*Proof.* We shall only prove the first equivalence; the others follow similarly. By definition, we have for all $\omega \in \mathcal{L}(H)_*$

$$X(\hat{a}_\omega \otimes 1) = (\mathrm{id} \mathbin{\bar{\otimes}} \mathrm{id} \mathbin{\bar{\otimes}} \omega)(X_{[12]} V_{[13]}),$$
$$(\hat{a}_\omega \otimes 1)X = (\mathrm{id} \mathbin{\bar{\otimes}} \mathrm{id} \mathbin{\bar{\otimes}} \omega)(V_{[13]} X_{[12]}).$$

Now the implication "$\Rightarrow$" follows immediately from the fact that maps of the form $\mathrm{id} \mathbin{\bar{\otimes}} \mathrm{id} \mathbin{\bar{\otimes}} \omega$, where $\omega \in \mathcal{L}(H)_*$, separate the elements of $\mathcal{L}(H \otimes H \otimes H)$. Conversely, if $X_{[12]} V_{[13]} = V_{[13]} X_{[12]}$, then $X_{[12]} V^*_{[13]} = V^*_{[13]} X_{[12]}$ and hence

$$\begin{aligned} X(\hat{a}^*_\omega \otimes 1) &= (\mathrm{id} \mathbin{\bar{\otimes}} \mathrm{id} \mathbin{\bar{\otimes}} \omega^*)(X_{[12]} V^*_{[13]}) \\ &= (\mathrm{id} \mathbin{\bar{\otimes}} \mathrm{id} \mathbin{\bar{\otimes}} \omega^*)(V^*_{[13]} X_{[12]}) = (\hat{a}^*_\omega \otimes 1)X \end{aligned}$$

for all $\omega \in \mathcal{L}(H)_*$ (compare equation (7.5)). □

*Proof of Proposition* 7.2.8. We prove the statements concerning $\widehat{B}_w(V)$; replacing $V$ by $V^{\mathrm{op}}$, we obtain the corresponding statements for $\widehat{B}_w(V^{\mathrm{op}}) = B_w(V)$.

i) We only need to show that for each $\omega \in \mathcal{L}(H)_*$ and $X \in (\widehat{B}_w(V) \mathbin{\bar{\otimes}} \widehat{B}_w(V))'$, the operator $\widehat{\Delta}(\hat{a}_\omega)$ commutes with $X$. By the pentagon equation,

$$\begin{aligned} \widehat{\Delta}(\hat{a}_\omega) = V^*(1 \otimes \hat{a}_\omega)V &= (\mathrm{id} \mathbin{\bar{\otimes}} \mathrm{id} \mathbin{\bar{\otimes}} \omega)(V^*_{[12]} V_{[23]} V_{[12]}) \\ &= (\mathrm{id} \mathbin{\bar{\otimes}} \mathrm{id} \mathbin{\bar{\otimes}} \omega)(V_{[13]} V_{[23]}). \end{aligned} \tag{7.9}$$

On the other hand, $X \in (1 \otimes \widehat{B}_w(V))' \cap (\widehat{B}_w(V) \otimes 1)'$, so that $X_{[12]}$ commutes with $V_{[13]} V_{[23]}$ by Lemma 7.2.9. Therefore,

$$\begin{aligned} X\widehat{\Delta}(\hat{a}_\omega) &= (\mathrm{id} \mathbin{\bar{\otimes}} \mathrm{id} \mathbin{\bar{\otimes}} \omega)(X_{[12]} V_{[13]} V_{[23]}) \\ &= (\mathrm{id} \mathbin{\bar{\otimes}} \mathrm{id} \mathbin{\bar{\otimes}} \omega)(V_{[13]} V_{[23]} X_{[12]}) = \widehat{\Delta}(\hat{a}_\omega)X. \end{aligned}$$

ii) This follows from i) and Lemma 7.2.1.

iii) We only need to show that $V$ commutes with $(\widehat{B}_w(V) \mathbin{\bar{\otimes}} B_w(V))'$. But if $X \in (\widehat{B}_w(V) \mathbin{\bar{\otimes}} B_w(V))'$, then $X \in (\widehat{B}_w(V) \otimes 1)' \cap (1 \otimes B_w(V))'$, so that $X_{[13]}$ commutes with $V_{[12]}$ and $V_{[23]}$ by Lemma 7.2.9, and then $X_{[13]}$ also commutes with $V_{[13]} = V^*_{[12]} V_{[23]} V_{[12]} V^*_{[23]}$. □

**Corollary 7.2.10.** *Let* $V$ *be a weakly well-behaved multiplicative unitary. Then the maps* $\widehat{\Delta}$ *and* $\Delta$ *restrict to normal* $*$*-homomorphisms*

$$\widehat{\Delta}\colon \widehat{A}_w(V) \to \widehat{A}_w(V) \mathbin{\bar{\otimes}} \widehat{A}_w(V) \quad \textit{and} \quad \Delta\colon A_w(V) \to A_w(V) \mathbin{\bar{\otimes}} A_w(V),$$

*and* $(\widehat{A}_w(V), \widehat{\Delta})$ *and* $(A_w(V), \Delta)$ *are von Neumann bialgebras.*

*Proof.* The assumption implies that $\widehat{A}_w(V)$ and $A_w(V)$ are equal to the von Neumann algebras $\widehat{B}_w(V)$ and $B_w(V)$ defined in the proposition above. Now the claim follows from that proposition. □

**The left/right-symmetry again.** The apparent symmetry in the definition of the left and of the right leg of a (weakly) well-behaved multiplicative unitary $V$ is related to the symmetry between $V$ and the opposite unitary $V^{\mathrm{op}}$ as follows:

**Proposition 7.2.11.** *If $V$ is a well-behaved multiplicative unitary, then the opposite multiplicative unitary $V^{\mathrm{op}}$ is well-behaved and*

$$(\widehat{A}(V^{\mathrm{op}}), \widehat{\Delta}_{V^{\mathrm{op}}}) = (A(V), \Delta_V)^{\mathrm{cop}}, \quad (A(V^{\mathrm{op}}), \Delta_{V^{\mathrm{op}}}) = (\widehat{A}(V), \widehat{\Delta}_V)^{\mathrm{cop}}.$$

*If $V$ is a weakly well-behaved multiplicative unitary, then the opposite multiplicative unitary $V^{\mathrm{op}}$ is weakly well-behaved and*

$$(\widehat{A}_w(V^{\mathrm{op}}), \widehat{\Delta}_{V^{\mathrm{op}}}) = (A_w(V), \Delta_V)^{\mathrm{cop}}, \quad (A_w(V^{\mathrm{op}}), \Delta_{V^{\mathrm{op}}}) = (\widehat{A}_w(V), \widehat{\Delta}_V)^{\mathrm{cop}}.$$

*Proof.* This follows immediately from Lemma 7.2.5. □

**Remark 7.2.12.** The proposition above shows that the symmetry between the two legs of a (weakly) well-behaved multiplicative unitary $V$ is not perfect. If we would replace either $\Delta$ by $\mathrm{Ad}_\Sigma \circ \Delta$ or $\widehat{\Delta}$ by $\mathrm{Ad}_\Sigma \circ \widehat{\Delta}$, then the passage between the left and the right leg would amount to the passage between $V$ and $V^{\mathrm{op}}$. This twist becomes relevant when we consider coactions of $C^*$-bialgebras and crossed products, see Sections 9.3 and 9.4.

### 7.2.3 Examples

Let us determine the legs of some of the multiplicative unitaries introduced in Section 7.1.2.

From the multiplicative unitary associated to a group, we recover the $C^*$-bialgebras and von Neumann bialgebras discussed in Section 4.2:

**Example 7.2.13.** Let $G$ be a locally compact group and consider the associated multiplicative unitary $W_G$ (see Example 7.1.4). Recall that $W_G$ acts on the Hilbert space $L^2(G, \lambda) \otimes L^2(G, \lambda) \cong L^2(G \times G, \lambda \times \lambda)$ via $(W_G \zeta)(x, y) = \zeta(x, x^{-1}y)$ for all $x, y \in G$ and $\zeta \in L^2(G \times G, \lambda \times \lambda)$, where $\lambda$ denotes the left Haar measure of $G$ as usual. We show that

$$\widehat{A}(W_G) \cong C_0(G) \quad \text{and} \quad A(W_G) = C_r^*(G) \quad \text{as } C^*\text{-bialgebras.} \tag{7.10}$$

This implies that $\widehat{A}_w(W_G) \cong L^\infty(G)$ and $A_w(W_G) = L(G)$ as von Neumann bialgebras. Moreover, it is easy to see that $W_G \in M(C_0(G) \otimes C_r^*(G))$, so that

$W_G$ is well-behaved (and weakly well-behaved); see also Example 7.3.4 iii) and Theorem 7.3.11. For the definition of the $C^*$-bialgebras $C_0(G)$, $C_r^*(G)$ and the von Neumann bialgebras $L^\infty(G)$, $L(G)$, see Example 4.2.1, 4.2.2, and 4.2.4.

Let us prove (7.10). We compute the operator $\hat{a}_\omega = (\mathrm{id} \mathbin{\bar{\otimes}} \omega)(W_G)$, where $\omega = \omega_{\xi',\xi} = \langle \xi' | \cdot \xi \rangle$ for some $\xi', \xi \in H$:

$$\begin{aligned}(\hat{a}_\omega \eta)(x) = (\langle \xi' |_{[2]} W_G | \xi \rangle_{[2]} \eta)(x) &= \int_G \overline{\xi'(y)}\, (W_G(\eta \otimes \xi))(x, y)\, d\lambda(y) \\ &= \int_G \overline{\xi'(y)} \eta(x) \xi(x^{-1} y)\, d\lambda(y) = f(x)\eta(x)\end{aligned}$$

for all $\eta \in L^2(G, \lambda)$ and $x \in G$, where the function $f : G \to \mathbb{C}$ is given by

$$f(x) = \int_G \overline{\xi'(y)} \xi(x^{-1} y)\, d\lambda(y) = \langle \xi' | L_x \xi \rangle \quad \text{for all } x \in G.$$

Here, $L_x \in \mathcal{L}(L^2(G, \lambda))$ denotes the left translation operator $\xi \mapsto \xi(x^{-1} \cdot)$ as usual. If we approximate $\xi', \xi$ in $L^2$-norm by elements of $C_c(G)$, the function $f$ gets approximated in sup-norm by elements of $C_c(G)$. Therefore $f$ belongs to $C_0(G)$. Denote by $\pi_M : C_0(G) \to \mathcal{L}(L^2(G, \lambda))$ the representation given by multiplication operators. The calculation above shows that $\hat{a}_\omega = \pi_M(f)$, and by Lemma 7.2.7,

$$\widehat{A}(W_G) \subseteq \pi_M(C_0(G)) \cong C_0(G).$$

If we replace $\xi$ by an approximate unit for the convolution algebra $C_c(G) \subseteq L^1(G, \lambda)$ (see Example 4.2.2) and choose $\xi' \in C_c(G)$, then $f$ converges in sup-norm to $\overline{\xi'}$. Therefore, the inclusion above is an equality.

Let us show that $\widehat{A}(W_G) \cong C_0(G)$ as $C^*$-bialgebras. For all $f \in C_0(G)$,

$$\widehat{\Delta}_{W_G}(\pi_M(f)) = (\pi_M \otimes \pi_M)(\Delta_{C_0(G)}(f)),$$

because

$$\begin{aligned}(W_G^*(1 \otimes \pi_M(f)) W_G \zeta)(x, y) &= ((1 \otimes \pi_M(f)) W_G \zeta)(x, xy) \\ &= f(xy)\, (W_G \zeta)(x, xy) = f(xy) \zeta(x, y)\end{aligned}$$

for all $x, y \in G$ and $\zeta \in L^2(G \times G, \lambda \times \lambda) \cong L^2(G, \lambda) \otimes L^2(G, \lambda)$.

To determine $A(W_G)$, we compute the operator $a_\omega = (\omega \mathbin{\bar{\otimes}} \mathrm{id})(W_G)$, where $\omega = \omega_{\eta',\eta} = \langle \eta' | \cdot \eta \rangle$ for some $\eta, \eta' \in H$:

$$\begin{aligned}(a_\omega \xi)(y) = (\langle \eta' |_{[1]} W_G | \eta \rangle_{[1]} \xi)(y) &= \int_G \overline{\eta'(x)} (W_G(\eta \otimes \xi))(x, y)\, d\lambda(x) \\ &= \int_G \overline{\eta'(x)} \eta(x)\, \xi(x^{-1} y)\, d\lambda(x) = (L(g)\xi)(y)\end{aligned}$$

for all $\xi \in L^2(G,\lambda)$ and $y \in G$, where $g = \overline{\eta}'\eta$ (pointwise product). Using Lemma 7.2.7 and the relation $\overline{L^2(G,\lambda)L^2(G,\lambda)} = L^1(G,\lambda)$, we conclude

$$A(W_G) = \|\cdot\|\text{-closure of } \{L(f) \mid f \in L^1(G,\lambda)\} = C_r^*(G)$$

as $C^*$-algebras.

Let us show that $A(W_G) = C_r^*(G)$ as $C^*$-bialgebras. As above, we denote by $L_z \in M(C_r^*(G))$, $z \in G$, the left translation operators. The operators

$$\Delta_{W_G}(L_z) = W_G(L_z \otimes 1)W_G^* \quad \text{and} \quad \Delta_{C_r^*(G)}(L_z) = L_z \otimes L_z$$

are equal for all $z \in G$, because

$$\begin{aligned}(W_G(L_z \otimes 1)W_G^*\zeta)(x,y) &= ((L_z \otimes 1)W_G^*\zeta)(x, x^{-1}y) \\ &= (W_G^*\zeta)(z^{-1}x, x^{-1}y) \\ &= \zeta(z^{-1}x, z^{-1}xx^{-1}y) = ((L_z \otimes L_z)\zeta)(x,y)\end{aligned}$$

for all $x, y \in G$ and $\zeta \in L^2(G \times G, \lambda \times \lambda) \cong L^2(G,\lambda) \otimes L^2(G,\lambda)$.

From the multiplicative unitary associated to an algebraic quantum group (see Example 7.1.6), we obtain a $C^*$-algebraic completion of the algebraic quantum group and a $C^*$-algebraic completion of its dual:

**Theorem 7.2.14.** *Let $(A_0, \Delta_0)$ be an algebraic quantum group with positive right integral $\psi$. Then the associated multiplicative unitary $V := V_{A_0}$ is well-behaved. In the notation of Example 7.1.6, we have:*

i) *For every $a \in A_0$, the map $A_0 \to A_0$ given by $d \mapsto ad$ extends to a bounded linear operator $\pi(a)$ on $H_\psi$. The map $\pi\colon A_0 \to \mathcal{L}(H_\psi)$, $a \mapsto \pi(a)$, is an injective $*$-homomorphism, and $A(V)$ is the norm-closure of $\pi(A_0)$.*

ii) *The space $\Delta_0(A_0) + A_0 \odot A_0 \subseteq M(A_0 \odot A_0)$ is a $*$-subalgebra, the $*$-homomorphism $\pi \odot \pi\colon A_0 \odot A_0 \to \mathcal{L}(H_\psi \otimes H_\psi)$ extends uniquely to a $*$-homomorphism on this subalgebra, and*

$$(\pi \odot \pi)(\Delta_0(a)) = V(\pi(a) \otimes 1)V^* = \Delta_V(\pi(a)) \quad \text{for all } a \in A_0.$$

*Denote by $(\widehat{A}_0, \widehat{\Delta}_0)$ the dual algebraic quantum group of $(A_0, \Delta_0)$.*

iii) *For every $\hat{a} \in \widehat{A}_0$, the map $A_0 \to A_0$ given by $d \mapsto (\mathrm{id} \odot \hat{a})(\Delta_0(d))$ extends to a bounded linear operator $\rho(\hat{a})$ on $H_\psi$. The map $\rho\colon \widehat{A}_0 \to \mathcal{L}(H_\psi)$, $\hat{a} \mapsto \rho(\hat{a})$, is an injective $*$-homomorphism, and $\widehat{A}(V)$ is equal to the norm-closure of $\rho(\widehat{A}_0)$.*

iv) *The space* $\widehat{\Delta}_0(\widehat{A}_0) + \widehat{A}_0 \odot \widehat{A}_0 \subseteq M(\widehat{A}_0 \odot \widehat{A}_0)$ *is a* $*$*-subalgebra, the* $*$*-homomorphism* $\rho \odot \rho \colon \widehat{A}_0 \odot \widehat{A}_0 \to \mathcal{L}(H_\psi \otimes H_\psi)$ *extends uniquely to a* $*$*-homomorphism on this* $*$*-subalgebra, and*

$$(\rho \odot \rho)(\widehat{\Delta}_0(\hat{a})) = V^*(1 \otimes \rho(\hat{a}))V = \widehat{\Delta}_V(\rho(\hat{a})) \quad \textit{for all } \hat{a} \in \widehat{A}_0.$$

*Proof.* We first prove the assertions i)–iv) and then show that $V$ is well-behaved.

i) Let us determine the operator $a_\omega = (\omega \mathbin{\overline{\otimes}} \mathrm{id})(V)$ for a functional $\omega$ of the form $\omega = \omega_{b,c} = \langle b| \cdot c\rangle$, where $b, c \in A_0$. For all $d \in A_0$,

$$\begin{aligned} a_\omega d &= \langle b|_{[1]} V |c\rangle_{[1]} d = \langle b|_{[1]} V(c \otimes d) \\ &= \sum \langle b|c_{(1)}\rangle c_{(2)} d = \sum \psi(b^* c_{(1)}) c_{(2)} d, \end{aligned}$$

that is, $a_\omega d = ad$, where $a := \sum \psi(b^* c_{(1)}) c_{(2)}$. Since the map $T_2$ (see Definition 2.1.9 on page 44) is surjective, the map $A_0 \odot A_0 \mapsto A_0$ given by $b \odot c \mapsto \sum \psi(b^* c_{(1)}) c_{(2)} = (\psi \odot \mathrm{id})((b^* \odot 1)\Delta_0(c))$ is surjective as well.

The existence of the map $\pi$ follows directly, and the density of $\pi(A_0)$ in $A(V)$ follows by Lemma 7.2.7. Evidently, $\pi$ is a $*$-homomorphism. It is injective, because $A_0$ is non-degenerate and $\psi$ is faithful (Corollary 2.2.5): For every non-zero $a \in A_0$, there exists a $d \in A_0$ such that $ad \neq 0$, and then $\|\pi(a)d\|^2 = \langle \pi(a)d|\pi(a)d\rangle = \psi(d^*a^*ad) \neq 0$.

ii) The first assertion follows easily from the fact that $A_0 \odot A_0 \subseteq M(A_0 \odot A_0)$ is an ideal, and the remaining assertions follow from the relations

$$V(b \otimes c) = \Delta_0(b)(1 \odot c) \quad \text{and} \quad V(\pi(a)b \otimes c) = \Delta_0(a)\Delta_0(b)(1 \odot c),$$

which hold for all $a, b, c \in A_0$.

iii) Let us compute the operator $\hat{a}_\omega = (\mathrm{id} \mathbin{\overline{\otimes}} \omega)(V)$, where $\omega = \omega_{b,c} = \langle b| \cdot c\rangle$ with $b, c \in A_0$. For all $d \in A_0$,

$$\begin{aligned} \hat{a}_\omega d &= \langle b|_{[2]} V |c\rangle_{[2]} d = \sum d_{(1)} \langle b|d_{(2)} c\rangle \\ &= \sum d_{(1)} \psi(b^* d_{(2)} c) = (\mathrm{id} \odot \hat{a})(\Delta_0(d)), \end{aligned}$$

where $\hat{a} = \psi(b^* \cdot c) \in (A_0)'$. Using the modular automorphism $\sigma'$ of $\psi$ (see Remarks 2.2.18), it is easy to see that the linear span of the subset $\{\psi(b^* \cdot c) \mid b, c \in A_0\} \subseteq (A_0)'$ coincides with $\widehat{A}_0$.

The existence of the map $\rho$ follows directly, and the density of $\rho(\widehat{A}_0)$ in $\widehat{A}(V)$ follows by Lemma 7.2.7. Next, we show that $\rho$ is a $*$-homomorphism. Let $\hat{a}, \hat{b} \in \widehat{A}_0$ and $c, d \in A_0 \subseteq H_\psi$. Since $\hat{a}\hat{b} = (\hat{a} \odot \hat{b}) \circ \Delta_0$ (see equation (2.10) on page 59),

$$\begin{aligned} \rho(\hat{a})\rho(\hat{b})d &= ((\mathrm{id} \odot \hat{a}) \circ \Delta_0 \circ (\mathrm{id} \odot \hat{b}) \circ \Delta_0)(d) \\ &= (\mathrm{id} \odot \hat{a} \odot \hat{b})(\Delta_0^{(2)}(d)) = \rho(\hat{a}\hat{b})d. \end{aligned}$$

Define $\check{a} \in \widehat{A}_0$ by $\check{a}(e) := \overline{\hat{a}(e^*)}$ for all $e \in A_0$. Then

$$\langle \rho(\hat{a})c|d\rangle = \sum \psi(c^*_{(1)}d)\overline{\hat{a}(c_{(2)})} = \sum \psi(c^*_{(1)}d)\check{a}(c^*_{(2)}).$$

Inserting Lemma 2.2.12 and the relation $\check{a} \circ S_0 = \hat{a}^*$ (Lemma 2.3.10), we find

$$\sum \psi(c^*_{(1)}d)\check{a}(c^*_{(2)}) = \sum \psi(c^* d_{(1)})\check{a}(S_0(d_{(2)})) = \langle c|\rho(\hat{a}^*)d\rangle.$$

Consequently, $\pi$ is a $*$-homomorphism.

Finally, since the set $(A_0 \odot 1)\Delta_0(A_0)$ spans $A_0 \odot A_0$, and since

$$\pi(e)\rho(\hat{a})d = (\mathrm{id} \odot \hat{a})((e \odot 1)\Delta_0(d)) \quad \text{for all } e \in A_0,$$

the $*$-homomorphism $\rho$ is injective.

iv) The first assertion follows easily from the fact that $\widehat{A}_0 \odot \widehat{A}_0 \subseteq M(\widehat{A}_0 \odot \widehat{A}_0)$ is an ideal, and the remaining assertions follow from the relations $V(b \otimes c) = \Delta_0(b)(1 \odot c)$ and

$$(\mathrm{id} \otimes \rho(\hat{a}))V(b \otimes c) = (\mathrm{id} \odot \mathrm{id} \odot \hat{a})(\Delta_0^{(2)}(b)(1 \odot \Delta_0(c))),$$

which hold for all $\hat{a} \in \widehat{A}_0$ and $b, c \in A_0$.

To complete the proof that $V$ is well-behaved, we need to show that $V$ belongs to $M(\widehat{A}(V) \otimes A(V))$. Let $a, b, c, d \in A_0$, and put $\hat{a} := \psi(\cdot\, a)$. Then

$$\begin{aligned} V(\rho(\hat{a}) \otimes \pi(b))(c \otimes d) &= \sum V(c_{(1)}\psi(c_{(2)}a) \otimes bd) \\ &= \sum c_{(1)} \odot c_{(2)}bd\psi(c_{(3)}a). \end{aligned}$$

Choose $b'_i, a'_i \in A_0$ such that $b \odot a = \sum_i \sum (a'_i)_{(1)}b'_i \odot (a'_i)_{(2)}$. Then

$$\begin{aligned} \sum c_{(1)} \odot c_{(2)}bd\psi(c_{(3)}a) &= \sum_i \sum c_{(1)} \odot c_{(2)}(a'_i)_{(1)}b'_i d\psi(c_{(3)}(a'_i)_{(2)}) \\ &= \sum_i \sum c_{(1)}\psi(c_{(2)}a'_i) \odot b'_i d \\ &= \sum_i (\rho(\hat{a}'_i) \otimes \pi(b'_i))(c \otimes d), \end{aligned}$$

where $\hat{a}'_i = \psi(\cdot\, a'_i)$. Consequently, $V(\rho(\widehat{A}_0) \odot \pi(A_0)) \subseteq \rho(\widehat{A}_0) \odot \pi(A_0)$, and a similar argument shows that $V^*(\rho(\widehat{A}_0) \odot \pi(A_0)) \subseteq \rho(\widehat{A}_0) \odot \pi(A_0)$. □

In Example 7.3.4 iv) and Theorem 7.3.11, we shall see again that the multiplicative unitary $V$ studied above is well-behaved. A detailed analysis of the preceding example can be found in [94].

### 7.2.4 The dual pairing, counit, and antipode of the legs

We now consider additional structural properties of the legs of a multiplicative unitary and construct a densely defined dual pairing, counit, and antipode. Up to minor modifications, the dual pairing satisfies all equations that characterize a dual pairing of Hopf $*$-algebras. Thus, this dual pairing gives a precise meaning to the vague idea that the two legs of a multiplicative unitary are dual to each other.

Before we introduce the pairing, let us recall some terminology. Let $X$ and $Y$ be complex vector spaces. We say that a bilinear map $(\cdot|\cdot)\colon X \times Y \to \mathbb{C}$ is *non-degenerate* if for every non-zero $x_0 \in X$ and every non-zero $y_0 \in Y$, there exist $y \in Y$ and $x \in X$ such that $(x_0|y) \neq 0$ and $(x|y_0) \neq 0$.

**Proposition 7.2.15.** *Let $V$ be a multiplicative unitary on a Hilbert space $H$. Then the bilinear map $(\cdot|\cdot)\colon \hat{A}_0(V) \times A_0(V) \to \mathbb{C}$ given by*

$$(\hat{a}_\omega|a_\upsilon) := (\upsilon \mathbin{\bar{\otimes}} \omega)(V) = \omega(a_\upsilon) = \upsilon(\hat{a}_\omega) \tag{7.11}$$

*is separately $\sigma$-weakly continuous, non-degenerate, and extends to bilinear maps*

$$\begin{aligned} (\cdot|\cdot]\colon \hat{A}_0(V) \times A_w(V) &\to \mathbb{C} \quad \text{and} \quad & [\cdot|\cdot)\colon \hat{A}_w(V) \times A_0(V) &\to \mathbb{C}, \\ (\hat{a}_\omega, a) &\mapsto \omega(a), & (\hat{a}, a_\upsilon) &\mapsto \upsilon(\hat{a}), \end{aligned} \tag{7.12}$$

*that are $\sigma$-weakly continuous in the second or first variable, respectively. For all $\omega, \omega' \in \mathcal{L}(H)_*$, $a \in A_w(V)$, and $\upsilon, \upsilon' \in \mathcal{L}(H)_*$, $\hat{a} \in \hat{A}_w(V)$,*

$$(\hat{a}_\omega \hat{a}_{\omega'}|a] = (\omega \mathbin{\bar{\otimes}} \omega')(\Delta(a)) \quad \text{and} \quad [\hat{a}|a_\upsilon a_{\upsilon'}) = (\upsilon \mathbin{\bar{\otimes}} \upsilon')(\hat{\Delta}(\hat{a})). \tag{7.13}$$

*Proof.* Evidently, formula (7.11) defines a bilinear map $(\cdot|\cdot)\colon \hat{A}_0(V) \times A_0(V) \to \mathbb{C}$ that is separately $\sigma$-weakly continuous. This bilinear map is non-degenerate because the natural pairing $\mathcal{L}(H) \times \mathcal{L}(H)_* \to \mathbb{C}$ is non-degenerate, and it extends uniquely to $\hat{A}_0(V) \times A_w(V)$ and $\hat{A}_w(V) \times A_0(V)$ because of the separate continuity. The first formula in (7.13) follows immediately from the proof of Lemma 7.2.2, and the second formula can be proved by similar calculations. □

Next, we construct a counit and an antipode on the legs of a weakly well-behaved multiplicative unitary $V$. In the formulation of the following proposition, we use the pairings defined in equation (7.12), the fact that the identity operator $\mathrm{id}_H$ belongs to $\hat{A}_w(V)$ and $A_w(V)$, and equation (7.5) from page 175.

**Proposition 7.2.16.** *Let $V$ be a weakly well-behaved multiplicative unitary on a Hilbert space $H$.*

i) *There exist algebra homomorphisms*

$$\begin{aligned} \hat{\epsilon}\colon \hat{A}_0(V) \to \mathbb{C}, &\quad \hat{a}_\omega \mapsto (\hat{a}_\omega|\,\mathrm{id}_H] = \omega(\mathrm{id}_H), \\ \epsilon\colon A_0(V) \to \mathbb{C}, &\quad a_\upsilon \mapsto [\mathrm{id}_H\,|a_\upsilon) = \upsilon(\mathrm{id}_H). \end{aligned}$$

ii) *There exist conjugate-linear algebra antihomomorphisms*

$$\widehat{S}\colon \widehat{A}_0(V) \to \widehat{A}_w(V),\ \hat{a}_\omega = (\mathrm{id}\mathbin{\overline{\otimes}}\omega)(V) \mapsto (\mathrm{id}\mathbin{\overline{\otimes}}\omega)(V^*) = (\hat{a}_{\omega^*})^*,$$
$$S\colon A_0(V) \to A_w(V),\ a_\upsilon = (\upsilon\mathbin{\overline{\otimes}}\mathrm{id})(V) \mapsto (\upsilon\mathbin{\overline{\otimes}}\mathrm{id})(V^*) = (a_{\upsilon^*})^*.$$

*For all* $\omega, \upsilon \in \mathcal{L}(H)_*$,

$$[\widehat{S}(\hat{a}_\omega)^*|a_\upsilon) = \overline{(\hat{a}_\omega|a_\upsilon{}^*]} \quad \textit{and} \quad (\hat{a}_\omega|S(a_\upsilon)^*] = \overline{[\hat{a}_\omega{}^*|a_\upsilon)}.$$

*If $V$ is well-behaved, then $\widehat{S}(\widehat{A}_0(V)) \subseteq \widehat{A}(V)$ and $S(A_0(V)) \subseteq A(V)$.*

*Proof.* We only prove the claims concerning $\hat{\epsilon}$ and $\widehat{S}$; for $\epsilon$ and $S$, the corresponding assertions follow similarly.

i) Equation (7.13) shows that for all $\omega, \omega' \in \mathcal{L}(H)_*$,

$$(\hat{a}_\omega\hat{a}_{\omega'}|\,\mathrm{id}_H] = (\omega\mathbin{\overline{\otimes}}\omega')(\Delta(\mathrm{id}_H)) = \omega(\mathrm{id}_H)\,\omega'(\mathrm{id}_H) = (\hat{a}_\omega|\,\mathrm{id}_H]\cdot(\hat{a}_{\omega'}|\,\mathrm{id}_H].$$

Therefore, $\hat{\epsilon}$ is a homomorphism.

ii) First, we find from equation (7.5) that

$$(\omega\mathbin{\overline{\otimes}}\mathrm{id})(V^*) = (a_{\omega^*})^* \in \widehat{A}_0(V)^* \subseteq \widehat{A}_w(V)^* = \widehat{A}_w(V)$$

for all $\omega \in \mathcal{L}(H)_*$. If $V$ is well-behaved, we can replace $\widehat{A}_w(V)$ by $\widehat{A}(V)$ in this equation.

Let us show that the assignment $\widehat{S}\colon \hat{a}_\omega \mapsto (\mathrm{id}\mathbin{\overline{\otimes}}\omega)(V^*)$ is well defined. By equation (7.5), we have for all $\omega, \upsilon \in \mathcal{L}(H)_*$

$$(\hat{a}_\omega|a_\upsilon{}^*] = \omega(a_\upsilon{}^*) = \omega((\upsilon^*\mathbin{\overline{\otimes}}\mathrm{id})(V^*)) = \upsilon^*((\mathrm{id}\mathbin{\overline{\otimes}}\omega)(V^*)).$$

Assume that $\hat{a}_\omega = 0$. Then the left-hand side of the equation above, and hence also the right-hand side, is zero. Since functionals of the form $\upsilon^*$, where $\upsilon \in \mathcal{L}(H)_*$, separate the elements of $\mathcal{L}(H)$, we must have $(\mathrm{id}\mathbin{\overline{\otimes}}\omega)(V^*) = 0$. Therefore, $\widehat{S}$ is well defined. Moreover, the calculation above shows that

$$\overline{(\hat{a}_\omega|a_\upsilon{}^*]} = \upsilon((\mathrm{id}\mathbin{\overline{\otimes}}\omega)(V^*)^*) = \upsilon(\widehat{S}(\hat{a}_\omega)^*) = [\widehat{S}(\hat{a}_\omega)^*|a_\upsilon).$$

Finally, let us prove that $\widehat{S}$ is an antihomomorphism. Given $\omega, \omega' \in \mathcal{L}(H)_*$, put $\omega'' := (\omega\mathbin{\overline{\otimes}}\omega')\circ\Delta$. Then $\hat{a}_\omega\hat{a}_{\omega'} = \hat{a}_{\omega''}$ by the proof of Lemma 7.2.2, and using the definition of $\widehat{S}$ and the pentagon equation, we find

$$\begin{aligned}
\widehat{S}(\hat{a}_{\omega'})\widehat{S}(\hat{a}_\omega) &= (\mathrm{id}\mathbin{\overline{\otimes}}\omega\otimes\omega')(V_{[13]}^*V_{[12]}^*) \\
&= (\mathrm{id}\mathbin{\overline{\otimes}}\omega\otimes\omega')(V_{[23]}V_{[12]}^*V_{[23]}^*) \\
&= (\mathrm{id}\mathbin{\overline{\otimes}}\omega\otimes\omega')((\mathrm{id}\mathbin{\overline{\otimes}}\Delta)(V^*)) \\
&= (\mathrm{id}\otimes\omega'')(V^*) = \widehat{S}(\hat{a}_{\omega''}) = \widehat{S}(\hat{a}_\omega\hat{a}_{\omega'}).
\end{aligned}$$

□

The antipodes $\widehat{S}$ and $S$ introduced above are unbounded and therefore difficult to handle. If $V$ is manageable or modular, then $\widehat{S}$ and $S$ extend to closed maps and can be described in terms of objects that are more tractable: a *unitary antipode*, which is an involutive $*$-antiautomorphism, and a *scaling group*, which is a one-parameter group of automorphisms, see Section 7.3.2.

It seems to be difficult to relate the maps $\hat{\epsilon}$, $\epsilon$ and $\widehat{S}$, $S$ constructed above to the usual axioms for the counit and the antipode of a Hopf algebra; several obstacles were discussed in the beginning of Chapter 4. In the following remark, we indicate some possible interpretations of these axioms.

**Remark 7.2.17.** Let $V$ be a weakly well-behaved multiplicative unitary on a Hilbert space $H$. Consider the maps $\hat{\epsilon}$ and $\widehat{S}$ constructed in Proposition 7.2.16.

In order to formulate an analogue of the counit identity $(\epsilon_0 \odot \mathrm{id}) \circ \Delta_0 = \mathrm{id} = (\mathrm{id} \odot \epsilon_0) \circ \Delta_0$ of a Hopf algebra $(A_0, \Delta_0)$ for the map $\hat{\epsilon}$, we need to extend the maps $\hat{\epsilon} \odot \mathrm{id}$ and $\mathrm{id} \odot \hat{\epsilon}$, which are initially defined on the algebraic tensor product $\widehat{A}_0(V) \odot \widehat{A}_0(V)$ only, to $\widehat{\Delta}(\widehat{A}_0(V))$. To do so, we approximate $\hat{\epsilon}$ pointwise by normal linear functionals as follows. Since $A_w(V)$ contains $\mathrm{id}_H$ and $A_0(V)$ is dense $A_w(V)$ with respect to the $\sigma$-weak topology, there exists a net $(\upsilon_\nu)_\nu$ in $\mathcal{L}(H)_*$ such that the elements $a_\nu := a_{\upsilon_\nu}$ converge $\sigma$-weakly to $\mathrm{id}_H$. Then

$$\hat{\epsilon}(\hat{a}_\omega) = (\hat{a}_\omega |\, \mathrm{id}_H] = \lim_\nu\, (\hat{a}_\omega | a_\nu) = \lim_\nu\, \upsilon_\nu(\hat{a}_\omega) \quad \text{for all } \omega \in \mathcal{L}(H)_*.$$

This equation suggests that we should extend the maps $\hat{\epsilon} \odot \mathrm{id}$ and $\mathrm{id} \odot \hat{\epsilon}$ to the image of $\widehat{\Delta}$ by approximating them pointwise by maps of the form $\upsilon_\nu \mathbin{\bar{\otimes}} \mathrm{id}$ and $\mathrm{id} \mathbin{\bar{\otimes}} \upsilon_\nu$, respectively. Thus we claim that for all $\omega \in \mathcal{L}(H)_*$,

$$(\upsilon_\nu \mathbin{\bar{\otimes}} \mathrm{id})(\widehat{\Delta}(\hat{a}_\omega)) \xrightarrow[\nu \to \infty]{\sigma\text{-weakly}} \hat{a}_\omega \quad \text{and} \quad (\mathrm{id} \mathbin{\bar{\otimes}} \upsilon_\nu)(\widehat{\Delta}(\hat{a}_\omega)) \xrightarrow[\nu \to \infty]{\sigma\text{-weakly}} \hat{a}_\omega.$$

Let us prove this claim. Inserting the relation $\widehat{\Delta}(\hat{a}_\omega) = (\mathrm{id} \mathbin{\bar{\otimes}} \mathrm{id} \mathbin{\bar{\otimes}} \omega)(V_{[13]} V_{[23]})$ (see equation (7.9) on page 178) into the left-hand sides, we find

$$\begin{aligned}
(\upsilon_\nu \mathbin{\bar{\otimes}} \mathrm{id})(\widehat{\Delta}(\hat{a}_\omega)) &= (\upsilon_\nu \mathbin{\bar{\otimes}} \mathrm{id} \mathbin{\bar{\otimes}} \omega)(V_{[13]} V_{[23]}) = (\mathrm{id} \mathbin{\bar{\otimes}} \omega)((1 \otimes a_\nu) V), \\
(\mathrm{id} \mathbin{\bar{\otimes}} \upsilon_\nu)(\widehat{\Delta}(\hat{a}_\omega)) &= (\mathrm{id} \mathbin{\bar{\otimes}} \upsilon_\nu \mathbin{\bar{\otimes}} \omega)(V_{[13]} V_{[23]}) = (\mathrm{id} \mathbin{\bar{\otimes}} \omega)(V (1 \otimes a_\nu)).
\end{aligned}$$

Now the claim follows from the fact that $(1 \otimes a_\nu)V$ and $V(1 \otimes a_\nu)$ converge $\sigma$-weakly to $V$ as $\nu$ tends to infinity and that $\mathrm{id} \mathbin{\bar{\otimes}} \omega$ is $\sigma$-weakly continuous.

Consider the antipode axiom $m_0 \circ (S_0 \odot \mathrm{id}) \circ \Delta_0 = \eta_0 \circ \epsilon_0 = m_0 \circ (\mathrm{id} \odot S_0) \circ \Delta_0$ of a Hopf algebra $(A_0, \Delta_0)$. The relations $\widehat{\Delta}(\hat{a}_\omega) = (\mathrm{id} \mathbin{\bar{\otimes}} \mathrm{id} \mathbin{\bar{\otimes}} \omega)(V_{[13]} V_{[23]})$ and $\widehat{S}((\mathrm{id} \mathbin{\bar{\otimes}} \omega')(V)) = (\mathrm{id} \mathbin{\bar{\otimes}} \omega')(V^*)$, $\omega' \in \mathcal{L}(H)_*$, suggest to think of the undefined expressions $(\widehat{S} \otimes \mathrm{id})(\widehat{\Delta}(\hat{a}_\omega))$ and $(\mathrm{id} \otimes \widehat{S})(\widehat{\Delta}(\hat{a}_\omega))$ as being equal to

$$(\mathrm{id} \mathbin{\bar{\otimes}} \mathrm{id} \mathbin{\bar{\otimes}} \omega)(V_{[13]}^* V_{[23]}) \quad \text{and} \quad (\mathrm{id} \mathbin{\bar{\otimes}} \mathrm{id} \mathbin{\bar{\otimes}} \omega)(V_{[13]} V_{[23]}^*),$$

respectively, for all $\omega \in \mathcal{L}(H)_*$. These identifications furthermore suggest to think of the expressions $m((\widehat{S} \otimes \mathrm{id})(\widehat{\Delta}(\hat{a}_\omega)))$ and $m((\mathrm{id} \otimes \widehat{S})(\widehat{\Delta}(\hat{a}_\omega)))$, which are not defined again, as being equal, respectively, to

$$(\mathrm{id} \mathbin{\overline{\otimes}} \omega)(V^*V) \quad \text{and} \quad (\mathrm{id} \mathbin{\overline{\otimes}} \omega)(VV^*).$$

Now, since $V$ is unitary, the last two operators are equal to $(\mathrm{id} \mathbin{\overline{\otimes}} \omega)(\mathrm{id}_{H \otimes H}) = \mathrm{id}_H\, \omega(\mathrm{id}_H) = \mathrm{id}_H\, \hat{\epsilon}(\hat{a}_\omega)$. Of course, these symbolic identifications and calculations are not precise and can only serve as a motivation.

We conclude this subsection with two examples:

**Example 7.2.18.** Consider the legs of the multiplicative unitary $W_G$ associated to a locally compact group $G$ (see Example 7.1.4 and 7.2.13). We calculate the dual pairing and the maps $\hat{\epsilon}$, $\widehat{S}$ and $\epsilon$, $S$ for operators of the form

$$\hat{a}_\omega = (\mathrm{id} \mathbin{\overline{\otimes}} \omega)(W_G), \text{ where } \omega = \omega_{\xi',\xi} \text{ for some } \xi, \xi' \in L^2(G,\lambda),$$

and

$$a_\upsilon = (\upsilon \mathbin{\overline{\otimes}} \mathrm{id})(W_G), \text{ where } \upsilon = \omega_{\eta',\eta} \text{ for some } \eta, \eta' \in L^2(G,\lambda).$$

The calculations in Example 7.2.13 showed that

$$\begin{aligned} \hat{a}_\omega &= \pi_M(f), \text{ where } f \in C_0(G) \text{ is given by } f(x) = \langle \xi' | L_x \xi \rangle \text{ for all } x \in G, \\ a_\upsilon &= L(g), \text{ where } g \in L^1(G,\lambda) \text{ is given by } g(x) = \overline{\eta'(x)}\eta(x) \text{ for all } x \in G. \end{aligned}$$

Here, $\pi_M$ denotes the representation by multiplication operators and $L$ denotes left convolution.

The dual pairing $(\cdot|\cdot)\colon \widehat{A}_0(W_G) \times A_0(W_G) \to \mathbb{C}$ is given by

$$(\hat{a}_\omega | a_\upsilon) = \upsilon(\hat{a}_\omega) = \langle \eta' | \pi_M(f)\eta \rangle = \int_G g(x) f(x)\, d\lambda(x).$$

The maps $\hat{\epsilon}$ and $\epsilon$ are given by

$$\hat{\epsilon}(\hat{a}_\omega) = \omega(\mathrm{id}) = \langle \xi' | \xi \rangle = f(e), \quad \epsilon(a_\upsilon) = \upsilon(\mathrm{id}) = \langle \eta' | \eta \rangle = \int_G g(x)\, d\lambda(x);$$

here, $e \in G$ denotes the unit. Finally, similar calculations as in Example 7.2.13 show that

$$\begin{aligned} \widehat{S}(\hat{a}_\omega) &= \pi_M(\tilde{f}), \text{ where } \tilde{f} \in C_0(G) \text{ is given by } \tilde{f}(x) = \langle \xi' | L_{x^{-1}} \xi \rangle = f(x^{-1}), \\ S(a_\upsilon) &= L(\tilde{g}), \text{ where } \tilde{g} \in L^1(G,\lambda) \text{ is given by } \tilde{g}(x) = g(x^{-1})\delta(x)^{-1}; \end{aligned}$$

here, $\delta$ denotes the modular function of $G$.

A comparison with Example 4.2.2 shows that the maps $\hat{\epsilon}$, $\widehat{S}$ and $\epsilon$, $S$ are given by the same formulas like the counit and the antipode of the $C^*$-bialgebras $C_0(G)$ and $C_r^*(G)$, respectively.

**Example 7.2.19.** Consider the legs of the multiplicative unitary $V := V_{A_0}$ associated to an algebraic quantum group $(A_0, \Delta_0)$ with positive right integral $\psi$ (see Example 7.1.6 and Theorem 7.2.14). We calculate the dual pairing and the maps $\hat{\epsilon}$, $\widehat{S}$ and $\epsilon$, $S$ for operators of the form

$$\hat{a}_\omega = (\mathrm{id} \mathbin{\overline{\otimes}} \omega)(V), \text{ where } \omega = \omega_{c,d} \text{ for some } c, d \in A_0,$$

and

$$a_\upsilon = (\upsilon \mathbin{\overline{\otimes}} \mathrm{id})(V), \text{ where } \upsilon = \omega_{e,f} \text{ for some } e, f \in A_0.$$

The calculations in the proof of Theorem 7.2.14 showed that

$$\hat{a}_\omega = \rho(\hat{a}), \text{ where } \hat{a} = \psi(c^* \cdot d) \in \widehat{A}_0,$$
$$a_\upsilon = \pi(a), \text{ where } a = \sum \psi(e^* f_{(1)}) f_{(2)} \in A_0;$$

for the definition of $\pi$ and $\rho$, see Theorem 7.2.14.

The dual pairing $(\cdot|\cdot)\colon \widehat{A}_0(V) \times A_0(V) \to \mathbb{C}$ is given by

$$(\hat{a}_\omega | a_\upsilon) = \omega(a_\upsilon) = \langle c|\pi(a)d\rangle = \psi(c^* a d) = \hat{a}(a);$$

thus it extends the natural pairing $\widehat{A}_0 \times A_0 \to \mathbb{C}$. The map $\hat{\epsilon}$ is given by

$$\hat{\epsilon}(\hat{a}_\omega) = \omega(\mathrm{id}) = \langle c|d\rangle = \psi(c^* d).$$

Using the modular automorphism of $\psi$ (Remark 2.2.18 i)) and the explicit formula for the counit $\hat{\epsilon}_0$ of $(\widehat{A}_0, \widehat{\Delta}_0)$ given in Proposition 2.3.3), it is easy to see that $\psi(c^* d) = \hat{\epsilon}_0(\hat{a})$. Thus the map $\hat{\epsilon}$ corresponds to the counit $\hat{\epsilon}_0$ of $\widehat{A}_0$. Likewise, the map $\epsilon$ defined in Proposition 7.2.16 corresponds to the counit $\epsilon_0$ of $(A_0, \Delta_0)$:

$$\epsilon(a_\upsilon) = \upsilon(\mathrm{id}) = \langle e|f\rangle = \psi(e^* f) = \epsilon_0(a).$$

Let us compute $\widehat{S}(\hat{a}_\omega) = (\hat{a}_{\omega^*})^*$ and $S(a_\upsilon) = (a_{\upsilon^*})^*$. Using the relations $\omega^* = \omega_{d,c}$, $\upsilon^* = \omega_{f,e}$ and the calculations in the proof of Theorem 7.2.14, we find

$$\widehat{S}(\hat{a}_\omega) = \rho(\hat{b})^*, \text{ where } \hat{b} = \psi(d^* \cdot c),$$
$$S(a_\upsilon) = \pi(b)^*, \text{ where } b = \sum \psi(f^* e_{(1)}) e_{(2)}.$$

Denote by $\widehat{S}_0$ and $S_0$ the antipode of $(\widehat{A}_0, \widehat{\Delta}_0)$ and $(A_0, \Delta_0)$, respectively. By definition of the involution on $\widehat{A}_0$,

$$\hat{b}^*(x) = \overline{\psi(d^* S_0(x)^* c)} = \psi(c^* S_0(x) d) = \hat{a}(S_0(x)) \quad \text{for all } x \in A_0,$$

and hence $\widehat{S}(\rho(\hat{a})) = \rho(\widehat{S}_0(\hat{a}))$. Using positivity of $\psi$ and Lemma 2.2.12, we find

$$b^* = (\psi \otimes \mathrm{id})(\Delta(e^*)(f \otimes 1)) = (\psi \otimes S_0)((e^* \otimes 1)\Delta(f)) = S_0(a),$$

and therefore $S(\pi(a)) = \pi(S_0(a))$.

## 7.3 Classes of well-behaved multiplicative unitaries

Evidently, it is desirable to have simple general criteria that tell whether a given multiplicative unitary is (weakly) well-behaved or not. The first such criterion – regularity – was introduced by Baaj and Skandalis [7]; variants were studied in [4], [5], [44]. Some examples of multiplicative unitaries are regular, but other important examples are not (see Example 7.3.4 v)) – regularity is too restrictive to cover all multiplicative unitaries that arise from $C^*$-algebraic quantum groups.

In [201], Woronowicz introduced a *manageability* condition that is particularly well adapted to $C^*$-algebraic quantum groups. In particular, the multiplicative unitary of every locally compact quantum group is manageable, see Section 8.3 or [91], [110]. For some examples of quantum groups, it is easier to construct a multiplicative unitary which is not manageable but satisfies a more general *modularity* condition that was introduced by Soltan and Woronowicz in [142].

In this section, we discuss all these criteria and show that each of them is sufficient for the well-behavior of the multiplicative unitary under consideration.

### 7.3.1 Regular multiplicative unitaries

Let $V$ be a multiplicative unitary on a Hilbert space $H$. We consider the space

$$\mathcal{C}_0(V) := \{(\omega \mathbin{\bar{\otimes}} \mathrm{id})(V\Sigma) \mid \omega \in \mathcal{L}(H)_*\} \subseteq \mathcal{L}(H),$$

where $\Sigma \in \mathcal{L}(H \otimes H)$ denotes the flip $\eta \otimes \xi \mapsto \xi \otimes \eta$ as usual. Put

$$\mathcal{C}(V) := \|\cdot\|\text{-closure of } \mathcal{C}_0(V), \qquad \mathcal{C}_w(V) := w\text{-closure of } \mathcal{C}_0(V),$$

where $w$-closure denotes the closure with respect to the weak operator topology. Note that $(\omega \mathbin{\bar{\otimes}} \mathrm{id})(V\Sigma) = ((\mathrm{id} \mathbin{\bar{\otimes}} \omega) \circ \mathrm{Ad}_\Sigma)(V\Sigma) = (\mathrm{id} \mathbin{\bar{\otimes}} \omega)(\Sigma V)$ for all $\omega \in \mathcal{L}(H)_*$.

**Definition 7.3.1.** A multiplicative unitary $V$ on a Hilbert space $H$ is called

- *regular* if $\mathcal{K}(H) = \mathcal{C}(V)$,
- *semi-regular* if $\mathcal{K}(H) \subseteq \mathcal{C}(V)$,
- *weakly regular* if $\mathcal{L}(H) = \mathcal{C}_w(V)$.

**Remark 7.3.2.** If $V$ is regular, then it is also semi-regular. Moreover, if $V$ is semi-regular, then it is also weakly regular, because in that case, $\mathcal{L}(H) = w\text{-closure of } \mathcal{K}(H) \subseteq w\text{-closure of } \mathcal{C}(V) = \mathcal{C}_w(V)$.

Before we explain the relevance of the conditions introduced above, let us reformulate the definition of the spaces $\mathcal{C}(V)$ and $\mathcal{C}_w(V)$ in terms of ket-bra operators and consider some examples. Recall the maps

$$|\zeta\rangle_{[1]}, |\zeta\rangle_{[2]} \colon H \to H \otimes H \quad \text{and} \quad \langle\zeta|_{[1]}, \langle\zeta|_{[2]} \colon H \otimes H \to H, \quad \text{where } \zeta \in H,$$

that were introduced in the formulas (7.6)–(7.8) on page 176. In terms of these maps, we can rewrite an operator $(\omega \mathbin{\bar{\otimes}} \mathrm{id})(V\Sigma) \in \mathcal{C}_0(V)$, where $\omega = \omega_{\eta',\xi} = \langle \eta' | \cdot \xi\rangle$ with $\eta', \xi \in H$, as follows:

$$\begin{aligned}(\omega \mathbin{\bar{\otimes}} \mathrm{id})(V\Sigma) &= \langle \eta'|_{[1]} V\Sigma |\xi\rangle_{[1]} \\ &= \langle \eta'|_{[1]} V |\xi\rangle_{[2]} \\ &= \langle \eta'|_{[2]} \Sigma V |\xi\rangle_{[2]} = (\mathrm{id} \mathbin{\bar{\otimes}} \omega)(\Sigma V).\end{aligned}$$

**Lemma 7.3.3.** *Let $V$ be a multiplicative unitary on a Hilbert space $H$. Then the set $\{\langle \eta'|_{[1]} V|\xi\rangle_{[2]} \mid \eta', \xi \in H\} \subseteq \mathcal{L}(H)$ is linearly dense in $\mathcal{C}(V)$ with respect to the norm and in $\mathcal{C}_w(V)$ with respect to the weak operator topology.*

*Proof.* The proof is virtually the same as the proof of Lemma 7.2.7. □

**Examples 7.3.4.** i) For every Hilbert space $H$, the operator $\mathrm{id}_{H\otimes H}$ is regular. This follows from the relation

$$(\langle \eta'|_{[1]} \mathrm{id}_{H\otimes H} |\xi\rangle_{[2]})\eta = \langle \eta'|_{[1]}(\eta \otimes \xi) = \langle \eta' | \eta\rangle \xi \quad \text{for all } \eta, \xi, \eta' \in H.$$

ii) If $V$ is a (weakly) regular multiplicative unitary on a Hilbert space $H$, then its opposite $V^{\mathrm{op}} = \Sigma V^* \Sigma$ is (weakly) regular again. Indeed, by Lemma 7.2.3, we have for every $\omega \in \mathcal{L}(H)_*$

$$(\omega \mathbin{\bar{\otimes}} \mathrm{id})(V^{\mathrm{op}}\Sigma) = (\omega \mathbin{\bar{\otimes}} \mathrm{id})(\Sigma V^*) = ((\omega^* \mathbin{\bar{\otimes}} \mathrm{id})(V\Sigma))^*,$$

and hence $\mathcal{C}_0(V^{\mathrm{op}}) = \mathcal{C}_0(V)^*$.

iii) The multiplicative unitaries $V_G$ and $W_G$ associated to a locally compact group $G$ (see Example 7.1.4) are regular. We prove this for $W_G$; for $V_G$, the proof is similar. Let $\eta', \xi \in L^2(G, \lambda)$. The operator $\langle \eta'|_{[1]} W_G |\xi\rangle_{[2]}$ is determined by

$$\begin{aligned}(\langle \eta'|_{[1]} W_G |\xi\rangle_{[2]} \eta)(y) &= \int_G \overline{\eta'(x)} (W_G(\eta \otimes \xi))(x, y)\, d\lambda(x) \\ &= \int_G \overline{\eta'(x)} \eta(x) \xi(x^{-1}y)\, d\lambda(x) \quad \text{for all } \eta \in L^2(G, \lambda).\end{aligned}$$

Thus, $\langle \eta'|_{[1]} W_G |\xi\rangle_{[2]}$ is an integral operator with kernel $K_{\eta',\xi} \colon G \times G \to \mathbb{C}$ given by $K_{\eta',\xi}(y, x) = \overline{\eta'(x)} \xi(x^{-1}y)$. It is easy to see that $K_{\eta',\xi}$ belongs to $L^2(G \times G, \lambda \times \lambda)$, and the space of all integral operators with kernel in $L^2(G \times G, \lambda \times \lambda)$ is a dense subspace of $\mathcal{K}(L^2(G, \lambda))$ [28, II, Proposition 4.7]. Combining these observations with Lemma 7.3.3, we find that $\mathcal{C}(W_G)$ is contained in $\mathcal{K}(L^2(G, \lambda))$. A standard argument shows that the linear span of functions of the form $K_{\eta',\xi}$, where $\eta', \xi \in L^2(G, \lambda)$, is dense in $L^2(G \times G, \lambda \times \lambda)$, and therefore, $\mathcal{C}(W_G) = \mathcal{K}(L^2(G, \lambda))$.

iv) The multiplicative unitaries $V_{A_0}$ and $W_{A_0}$ associated to an algebraic quantum group $(A_0, \Delta_0)$ (see Example 7.1.6) are regular. We prove this for $V := V_{A_0}$; the proof for $W_{A_0}$ is similar. Let us use the notation introduced in Example 7.1.6 and Theorem 7.2.14. An operator of the form $\langle c|_{[1]}V|b\rangle_{[2]}$, where $b, c \in A_0$, acts on an element $a \in A_0$ as follows:

$$\langle c|_{[1]}V|b\rangle_{[2]}\, a = \langle c|_{[1]}V(a \otimes b) = \sum \langle c|a_{(1)}\rangle a_{(2)}b = \sum \psi(c^* a_{(1)})a_{(2)}b.$$

Using the modular automorphism $\sigma'$ of $\psi$ (see Remark 2.2.18 i)), we can rewrite the right-hand side in the form

$$\sum \psi(a_{(1)}\sigma'(c^*))a_{(2)}b = (\psi \odot \mathrm{id})(\Delta_0(a)(\sigma'(c^*) \odot b)).$$

Since the map $A_0 \odot A_0 \to A_0 \odot A_0$, $c' \odot b' \mapsto \Delta(c')(1 \odot b')$, is surjective (see Definition 2.1.9), we can write $\sigma'(c^*) \odot b = \sum_i \Delta_0(c'_i)(1 \odot b'_i)$ with $c'_i, b'_i \in A_0$. We insert this equation into the expression above, use right-invariance of $\psi$, and find

$$\langle c|_{[1]}V|b\rangle_{[2]}\, a = \sum_i (\psi \odot \mathrm{id})(\Delta_0(ac'_i)(1 \odot b'_i)) = \sum_i \psi(ac'_i)b'_i.$$

Define $c''_i \in A_0$ by $c''^*_i := \sigma'^{-1}(c'_i)$. Then $\psi(ac'_i) = \psi(c''^*_i a) = \langle c''_i|a\rangle$, and

$$\langle c|_{[1]}V|b\rangle_{[2]}\, a = \sum_i \langle c''_i|a\rangle b'_i.$$

This equation shows that the operator $\langle c|_{[1]}V|b\rangle_{[2]}$ is compact. By Lemma 7.3.3, $\mathcal{C}(V) \subseteq \mathcal{K}(H_\psi)$. Reversing the preceding transformations, we find that this inclusion is an equality.

v) The bicrossed product construction [9] yields examples of multiplicative unitaries that are semi-regular but not regular, and examples that are not even semi-regular, see [9, Theorem 3.11 and Section 4].

vi) The multiplicative unitary of the quantum group $E_\mu(2)$ is semi-regular but not regular, see [4], [5] or Section 8.4.2.

Semi-regularity, regularity, and weak regularity have interesting implications on the legs of a multiplicative unitary. Most importantly, every (weakly) regular multiplicative unitary is (weakly) well-behaved. The proof of this result, which is given at the end of this subsection, and the proofs of the other results discussed in the remainder of this subsection all use similar techniques:

Given a multiplicative unitary $V$ on a Hilbert space $H$, we perform purely algebraic manipulations on operators between threefold tensor products of the Hilbert spaces $H$ and $\mathbb{C}$ like, for example, $H \otimes H \otimes H$, $\mathbb{C} \otimes H \otimes H$, $H \otimes \mathbb{C} \otimes \mathbb{C}$. Of course, we can always neglect the factor $\mathbb{C}$ wherever it occurs, but the algebraic manipulations will be easier to follow if we do not make such identifications during the calculations. To streamline the presentation, we adopt the following notation:

- We extend the leg notation to operators between different spaces in a similar way as in formulas (7.6)–(7.8) on page 176; for example, we denote the operator

$$\mathrm{id}_{\mathbb{C}} \otimes |\xi\rangle \otimes \mathrm{id}_H : \mathbb{C} \otimes \mathbb{C} \otimes H \to \mathbb{C} \otimes H \otimes H$$

by $|\xi\rangle_{[2]}$ and its adjoint by $\langle\xi|_{[2]}$. The symbols $|\xi\rangle_{[2]}$ and $\langle\xi|_{[2]}$ will also denote the operators $\mathrm{id}_H \otimes |\xi\rangle \otimes \mathrm{id}_H$, $\mathrm{id}_H \otimes |\xi\rangle \otimes \mathrm{id}_{\mathbb{C}}$, $\mathrm{id}_{\mathbb{C}} \otimes |\xi\rangle \otimes \mathrm{id}_{\mathbb{C}}$ and their respective adjoints; from the context, it will be clear which of these operators we refer to.

- We put

$$|H\rangle := \{|\xi\rangle \mid \xi \in H\} = \mathcal{K}(\mathbb{C}, H)$$

and

$$\langle H| := \{\langle\eta| \mid \eta \in H\} = \mathcal{K}(H, \mathbb{C}),$$

and apply the leg notation to the spaces $\langle H|$ and $|H\rangle$ in the obvious way.

- We indicate identifications of tensor products that involve $\mathbb{C}$ as a factor and identifications of operators on such tensor products by the symbol "$\equiv$"; for example, $H \otimes \mathbb{C} \equiv H \equiv \mathbb{C} \otimes H$ and $\mathcal{L}(H \otimes \mathbb{C}, \mathbb{C} \otimes H) \equiv \mathcal{L}(H)$.

Let us illustrate this notation by some examples that will be used later on.

- $\mathcal{K}(H) \equiv \mathcal{K}(H \otimes \mathbb{C}, \mathbb{C} \otimes H) = \big[|H\rangle_{[2]}\langle H|_{[1]}\big]$.

- By Lemma 7.3.3,

$$\mathcal{C}(V) \equiv \big[\langle H|_{[1]} V |H\rangle_{[2]}\big] \equiv \big[\langle H|_{[1]} V \Sigma |H\rangle_{[1]}\big] \equiv \big[\langle H|_{[2]} \Sigma V |H\rangle_{[2]}\big].$$

In particular, $V$ is regular if and only if

$$\big[\langle H|_{[1]} V |H\rangle_{[2]}\big] = \big[|H\rangle_{[2]}\langle H|_{[1]}\big]. \tag{7.14}$$

- By Lemma 7.2.7, $\widehat{A}(V) \equiv \big[\langle H|_{[2]} V |H\rangle_{[2]}\big] \subseteq \mathcal{L}(H \otimes \mathbb{C}) \equiv \mathcal{L}(H)$.

The following results explain the relevance of Definition 7.3.1. In the proofs, we permanently use Lemma 12.5.3. First, we study the space $\mathcal{C}(V)$.

**Lemma 7.3.5.** *Let $V$ be a multiplicative unitary on a Hilbert space $H$. Then the set $\mathcal{C}(V)\mathcal{C}(V)$ is linearly dense in $\mathcal{C}(V)$; in particular, $\mathcal{C}(V)$ and $\mathcal{C}_w(V)$ are algebras. Furthermore, $\mathcal{C}(V)H$ and $\mathcal{C}(V)^*H$ are linearly dense in $H$.*

*Proof.* With respect to the identification $\mathcal{L}(H) \equiv \mathcal{L}(H \otimes \mathbb{C} \otimes \mathbb{C}, \mathbb{C} \otimes \mathbb{C} \otimes H)$,

$$\begin{aligned}[\mathcal{C}(V)\mathcal{C}(V)] &\equiv \big[\langle H|_{[2]}V_{[23]}|H\rangle_{[3]}\langle H|_{[1]}V_{[12]}|H\rangle_{[2]}\big] \\ &= \big[\langle H \otimes H|_{[12]}V_{[23]}V_{[12]}|H \otimes H\rangle_{[23]}\big] \\ &= \big[\langle H \otimes H|_{[12]}V_{[12]}V_{[13]}V_{[23]}|H \otimes H\rangle_{[23]}\big] \\ &= \big[\langle H \otimes H|_{[12]}V_{[13]}|H \otimes H\rangle_{[23]}\big] = \big[\langle H|_{[1]}V_{[13]}|H\rangle_{[3]}\big] \equiv \mathcal{C}(V).\end{aligned}$$

Moreover,

$$\begin{aligned}[\mathcal{C}(V)H] &= [\langle H|_{[1]}V(H \otimes H)] = [\langle H|_{[1]}(H \otimes H)] = H, \\ [\mathcal{C}(V)^*H] &= [\langle H|_{[2]}V^*(H \otimes H)] = [\langle H|_{[2]}(H \otimes H)] = H.\end{aligned}$$ □

**Proposition 7.3.6.** *If $V$ is a semi-regular multiplicative unitary, then $\mathcal{C}(V)$ is a $C^*$-algebra.*

For the proof of this proposition, we use the following result:

**Lemma 7.3.7.** *Let $H$ be a Hilbert space and $C \subseteq \mathcal{L}(H)$ a norm-closed subalgebra such that $C^*C \subseteq C$ and $CC^* \subseteq C$. Then $C$ is a $C^*$-algebra.*

*Proof.* We only need to prove $C^* \subseteq C$. For every $c \in C$, there exists a sequence of polynomials $(p_n)_n$ such that $c = \lim_n cp_n(c^*c)$, and then $c^* = \lim_n p_n(c^*c)c^* \in [C^*CC^*] \subseteq C$. □

*Proof of Proposition* 7.3.6. Let $H$ denote the underlying Hilbert space of $V$. By definition,

$$\begin{aligned}[\mathcal{C}(V)\mathcal{C}(V)^*] &\equiv \big[\langle H|_{[1]}V_{[13]}|H\rangle_{[3]}\langle H|_{[2]}V^*_{[12]}|H\rangle_{[1]}\big] \\ &\subseteq \mathcal{L}(\mathbb{C} \otimes H \otimes \mathbb{C}, \mathbb{C} \otimes \mathbb{C} \otimes H).\end{aligned}$$

Since $V$ is semi-regular, $\big[|H\rangle_{[3]}\langle H|_{[2]}\big] \subseteq \big[\langle H|_{[2]}V_{[23]}|H\rangle_{[3]}\big]$. Thus

$$\begin{aligned}[\mathcal{C}(V)\mathcal{C}(V)^*] &\subseteq \big[\langle H|_{[1]}V_{[13]}\langle H|_{[2]}V_{[23]}|H\rangle_{[3]}V^*_{[12]}|H\rangle_{[1]}\big] \\ &= \big[\langle H \otimes H|_{[12]}V_{[13]}V_{[23]}V^*_{[12]}|H \otimes H\rangle_{[13]}\big].\end{aligned}$$

By the pentagon equation, $V_{[13]}V_{[23]}V^*_{[12]} = V^*_{[12]}V_{[23]}$. Therefore

$$\begin{aligned}[\mathcal{C}(V)\mathcal{C}(V)^*] &\subseteq \big[\langle H \otimes H|_{[12]}V^*_{[12]}V_{[23]}|H \otimes H\rangle_{[13]}\big] \\ &= \big[\langle H \otimes H|_{[12]}V_{[23]}|H \otimes H\rangle_{[13]}\big] \\ &= \big[\langle H|_{[2]}V_{[23]}|H\rangle_{[3]}\big] \equiv \mathcal{C}(V).\end{aligned}$$

A similar calculation shows that $[\mathcal{C}(V)^*\mathcal{C}(V)]$ is contained in $\mathcal{C}(V)$. Now the assertion follows from Lemma 7.3.5 and Lemma 7.3.7. □

For a general multiplicative unitary $V$, the space $\mathcal{C}(V)$ need not be closed under involution. If it is, then also $\widehat{A}(V)$ and $A(V)$ are closed under involution, as we shall see in the next lemma, but there exist multiplicative unitaries for which this is not the case [9, Remark 4.5].

**Lemma 7.3.8.** *Let $V$ be a multiplicative unitary that satisfies $\mathcal{C}(V)^* = \mathcal{C}(V)$. Then $\widehat{A}(V)$ and $A(V)$ are $C^*$-algebras.*

*Proof.* Let $H$ denote the underlying Hilbert space of $V$. By Lemma 7.2.2, $\widehat{A}_0(V)$ and $A_0(V)$ are algebras; thus we only need to show that $\widehat{A}(V)^* = \widehat{A}(V)$ and $A(V)^* = A(V)$. Let us prove the first equation. By assumption, the space

$$\widehat{B} := \big[\langle H|_{[2]} V(1 \otimes \mathcal{C}(V))V^*|H\rangle_{[2]}\big] \subseteq \mathcal{L}(H \otimes \mathbb{C})$$

is self-adjoint. Inserting the definition of $\mathcal{C}(V)$ and using the pentagon equation, we find

$$\begin{aligned}\widehat{B} &\equiv \big[\langle H|_{[3]} V_{[13]} \langle H|_{[2]} V_{[23]} |H\rangle_{[3]} V^*_{[12]} |H\rangle_{[2]}\big] \\ &= \big[\langle H \otimes H|_{[23]} V_{[13]} V_{[23]} V^*_{[12]} |H \otimes H\rangle_{[23]}\big] \\ &= \big[\langle H \otimes H|_{[23]} V^*_{[12]} V_{[23]} |H \otimes H\rangle_{[23]}\big] \\ &= \big[\langle H \otimes H|_{[23]} V^*_{[12]} |H \otimes H\rangle_{[23]}\big] \equiv \big[\langle H|_{[2]} V^*_{[12]} |H\rangle_{[2]}\big] = \widehat{A}(V)^*.\end{aligned}$$

Since $\widehat{B}$ was self-adjoint, so is $\widehat{A}(V)$. The assertion concerning $A(V)$ can be proved similarly. Alternatively, apply the preceding proof to $V^{\mathrm{op}}$, and use Example 7.3.4 ii) and Lemma 7.2.5. □

**Corollary 7.3.9.** *Let $V$ be a semi-regular multiplicative unitary. Then $\widehat{A}(V)$ and $A(V)$ are $C^*$-algebras.*

*Proof.* Combine Proposition 7.3.6 with Lemma 7.3.8. □

The following result is due to Enock [44, Proposition 3.12], who proved it in a much more general setting, see Theorem 10.3.18 in Section 10.3.

**Theorem 7.3.10.** *If $V$ is a multiplicative unitary that satisfies $\mathcal{C}_w(V)^* = \mathcal{C}_w(V)$, then $V$ is weakly well-behaved. In particular, every weakly regular multiplicative unitary is weakly well-behaved.*

*Proof.* Similar calculations as in the proof of Lemma 7.3.8 show that $\widehat{A}_w(V)$ is closed under involution – simply let $[\cdot]$ denote the weakly closed linear span instead of the norm closed linear span, and use Lemma 12.5.3. Since $\widehat{A}_w(V)$ acts non-degenerately on $H$ (Lemma 7.2.2), it is a von Neumann algebra. Replacing $V$ by $V^{\mathrm{op}}$, which is regular by Example 7.3.4 ii), we find that also $A_w(V)$ is a von Neumann algebra; here, we use Lemma 7.2.5. □

The main result of this subsection is the following theorem of Baaj and Skandalis [7, Théorème 3.8]:

**Theorem 7.3.11.** *Every regular multiplicative unitary is well-behaved.*

We divide the proof into several steps. From now on, let $V$ be a regular multiplicative unitary on a Hilbert space $H$. To simplify notation, we put $\widehat{A} := \widehat{A}(V)$.

**Lemma 7.3.12.** *$\widehat{A}$ is a $C^*$-algebra.*

*Proof.* This follows immediately from Corollary 7.3.9, but let us give an alternative proof that is more direct. By definition,

$$[\widehat{A}\widehat{A}^*] \equiv \big[\langle H|_{[3]} V_{[13]} |H\rangle_{[3]} \langle H|_{[2]} V^*_{[12]} |H\rangle_{[2]}\big] \subseteq \mathcal{L}(H \otimes \mathbb{C} \otimes \mathbb{C}).$$

Since $V$ is regular, we can replace $|H\rangle_{[3]}\langle H|_{[2]}$ by $\langle H|_{[2]} V_{[23]} |H\rangle_{[3]}$ in the middle:

$$\begin{aligned}[\widehat{A}\widehat{A}^*] &\equiv \big[\langle H|_{[3]} V_{[13]} \langle H|_{[2]} V_{[23]} |H\rangle_{[3]} V^*_{[12]} |H\rangle_{[2]}\big] \\ &= \big[\langle H \otimes H|_{[23]} V_{[13]} V_{[23]} V^*_{[12]} |H \otimes H\rangle_{[23]}\big].\end{aligned}$$

We use the pentagon equation to replace $V_{[13]} V_{[23]} V^*_{[12]}$ by $V^*_{[12]} V_{[23]}$, and obtain

$$\begin{aligned}[\widehat{A}\widehat{A}^*] &\equiv \big[\langle H \otimes H|_{[23]} V^*_{[12]} V_{[23]} |H \otimes H\rangle_{[23]}\big] \\ &= \big[\langle H \otimes H|_{[23]} V^*_{[12]} |H \otimes H\rangle_{[23]}\big] \equiv \widehat{A}^*.\end{aligned}$$

This equation implies $\widehat{A}^* = \widehat{A}$ and $\widehat{A}\widehat{A} \subseteq \widehat{A}$. □

**Lemma 7.3.13.** i) $[(\widehat{A} \otimes \langle H|)V] = [\widehat{A} \otimes \langle H|] \subseteq \mathcal{L}(H \otimes H, H \otimes \mathbb{C})$.

ii) $[V(\widehat{A} \otimes |H\rangle)] = [\widehat{A} \otimes |H\rangle] \subseteq \mathcal{L}(H \otimes \mathbb{C}, H \otimes H)$.

*Proof.* i) By definition,

$$\begin{aligned}[(\widehat{A} \otimes \langle H|)V] &\equiv \big[\langle H|_{[2]} V_{[12]} |H\rangle_{[2]} \langle H|_{[3]} V_{[13]}\big] \\ &\subseteq \mathcal{L}(H \otimes \mathbb{C} \otimes \mathbb{C}, H \otimes \mathbb{C} \otimes H),\end{aligned}$$

and by the pentagon equation,

$$\begin{aligned}[(\widehat{A} \otimes \langle H|)V] &\equiv \big[\langle H \otimes H|_{[23]} V_{[12]} V_{[13]} |H\rangle_{[2]}\big] \\ &= \big[\langle H \otimes H|_{[23]} V_{[23]} V_{[12]} V^*_{[23]} |H\rangle_{[2]}\big] \\ &= \big[\langle H \otimes H|_{[23]} V_{[12]} V^*_{[23]} |H\rangle_{[2]}\big].\end{aligned}$$

We move $\langle H|_{[3]}$ to the right of $V_{[12]}$, use regularity of $V$, and find

$$\begin{aligned}[(\widehat{A} \otimes \langle H|)V] &\equiv \big[\langle H|_{[2]} V_{[12]} \langle H|_{[3]} V^*_{[23]} |H\rangle_{[2]}\big] \\ &= \big[\langle H|_{[2]} V_{[12]} |H\rangle_{[2]} \langle H|_{[3]}\big] \equiv [\widehat{A} \otimes \langle H|].\end{aligned}$$

ii) Multiply the equation in i) by $V^*$ on the right, take adjoints, and identify $\widehat{A}^*$ with $\widehat{A}$ using Lemma 7.3.12. □

**Lemma 7.3.14.** $[(\widehat{A} \otimes 1)\widehat{\Delta}(\widehat{A})] = [\widehat{A} \otimes \widehat{A}] = [\widehat{\Delta}(\widehat{A})(1 \otimes \widehat{A})]$.

*Proof.* The definition of $\widehat{\Delta}$ and $\widehat{A}$ together with the pentagon equation imply

$$\begin{aligned}\widehat{\Delta}(\widehat{A}) = V^*(1 \otimes \widehat{A})V &\equiv \big[V^*_{[12]}\langle H|_{[3]}V_{[23]}|H\rangle_{[3]}V_{[12]}\big]\\ &= \big[\langle H|_{[3]}V^*_{[12]}V_{[23]}V_{[12]}|H\rangle_{[3]}\big]\\ &= \big[\langle H|_{[3]}V_{[13]}V_{[23]}|H\rangle_{[3]}\big] \subseteq \mathcal{L}(H \otimes H \otimes \mathbb{C}).\end{aligned}$$

We insert this relation into the left- and right-hand side of the equation that we want to prove, apply the previous lemma, and obtain

$$\begin{aligned}[(\widehat{A} \otimes 1)\widehat{\Delta}(\widehat{A})] &\equiv \big[\widehat{A}_{[1]}\langle H|_{[3]}V_{[13]}V_{[23]}|H\rangle_{[3]}\big]\\ &= \big[\widehat{A}_{[1]}\langle H|_{[3]}V_{[23]}|H\rangle_{[3]}\big] \equiv [\widehat{A} \otimes \widehat{A}]\end{aligned}$$

and

$$\begin{aligned}[\widehat{\Delta}(\widehat{A})(1 \otimes \widehat{A})] &\equiv \big[\langle H|_{[3]}V_{[13]}V_{[23]}|H\rangle_{[3]}\widehat{A}_{[2]}\big]\\ &= \big[\langle H|_{[3]}V_{[13]}|H\rangle_{[3]}\widehat{A}_{[2]}\big] \equiv [\widehat{A} \otimes \widehat{A}].\end{aligned}$$ □

*Proof of Theorem* 7.3.11. If $V$ is a regular multiplicative unitary, then $\widehat{A}(V)$ is a $C^*$-algebra by Lemma 7.3.12, and $(\widehat{A}(V), \widehat{\Delta})$ is a bisimplifiable $C^*$-bialgebra by Lemma 7.2.1 and Lemma 7.3.14. Replacing $V$ by $V^{\mathrm{op}}$, we obtain the respective statements for $\widehat{A}(V^{\mathrm{op}}) = A(V)$, see also Lemma 7.2.5.

It remains to show that $V \in M(\widehat{A}(V) \otimes A(V))$. Lemma 7.3.13 implies that $V$ is a multiplier of $\widehat{A}(V) \otimes \mathcal{K}(H)$, and a similar argument shows that $V$ is a multiplier of $\mathcal{K}(H) \otimes A(V)$. Therefore, $V_{[13]} = V^*_{[12]}V_{[23]}V_{[12]}V^*_{[23]}$ belongs to $M(\widehat{A}(V) \otimes \mathcal{K}(H) \otimes A(V))$. The claim follows. □

For completeness, we cite the following result of Baaj [5, Théorème 3.12]. The proof involves similar techniques as presented above. Recall that a symmetry in a $C^*$-algebra is a self-adjoint unitary.

**Theorem 7.3.15.** *Let $V$ be a multiplicative unitary on a Hilbert space $H$ and assume that*

i) *$V$ is semi-biregular in the sense that $\mathcal{K}(H)$ is contained in*

$$\overline{\mathrm{span}}\{(\omega \mathbin{\overline{\otimes}} \mathrm{id})(V\Sigma) \mid \omega \in \mathcal{L}(H)_*\}$$

*and*

$$\overline{\mathrm{span}}\{(\mathrm{id} \mathbin{\overline{\otimes}} \omega)(V\Sigma) \mid \omega \in \mathcal{L}(H)_*\};$$

ii) *$V$ is balanced in the sense that there exists a symmetry $U \in \mathcal{L}(H)$ such that $\Sigma(U \otimes 1)V(U \otimes 1)\Sigma$ is a multiplicative unitary.*

*Then $V$ is well-behaved.*

### 7.3.2 Manageable and modular multiplicative unitaries

The manageability condition and the more general modularity condition were introduced by Woronowicz [201] and by Soltan and Woronowicz [142], respectively. They are particularly well adapted to unitaries associated to quantum groups and can be considered as modifications of the regularity condition discussed in the previous section.

We formulate the manageability and the modularity condition and show that every modular multiplicative unitary is well-behaved, building on the proofs given in the preceding section. Moreover, we discuss the polar decomposition of the antipode for a modular multiplicative unitary, but omit the proof. All results are taken from [201] and [142].

The manageability and the modularity condition involve the conjugate of a Hilbert space and unbounded self-adjoint operators; a standard references for the latter are, for example, [28], [75], [136]. We denote the conjugate of a Hilbert space $H$ by $\overline{H}$ and the canonical conjugate-linear isomorphism $H \to \overline{H}$ by $\xi \mapsto \bar{\xi}$; thus, $\lambda\bar{\eta} = \overline{\bar{\lambda}\eta}$ and $\langle\bar{\eta}|\bar{\xi}\rangle = \langle\xi|\eta\rangle$ for all $\eta, \xi \in H$ and $\lambda \in \mathbb{C}$. For each $T \in \mathcal{L}(H)$, the map $\bar{\xi} \mapsto \overline{T\xi}$ defines an operator $\overline{T} \in \mathcal{L}(\overline{H})$, and the map $T \mapsto \overline{T}$ is a conjugate-linear $*$-isomorphism $\mathcal{L}(H) \to \mathcal{L}(\overline{H})$.

The precise formulation of the manageability and the modularity condition may appear a bit mysterious at first glance; therefore we begin with a motivation. Let $V$ be a multiplicative unitary on a Hilbert space $H$. Then the regularity condition can be expressed in terms of the map

$$\Phi\colon \mathcal{B}(H)_* \to \mathcal{B}(H), \quad \omega \mapsto (\omega \mathbin{\bar{\otimes}} \mathrm{id})(V\Sigma).$$

Indeed, $V$ is regular if and only if $\Phi(\mathcal{B}(H)_*)$ is a dense subset of $\mathcal{K}(H)$. Assume that $H$ has finite dimension. Then $\overline{H} \odot H$ can be identified with $\mathcal{B}(H)_*$ via $\bar{\eta} \odot \xi \equiv \langle\eta| \cdot \xi\rangle$, and with $\mathcal{B}(H)$ via $\bar{\eta} \odot \xi \equiv |\xi\rangle\langle\eta|$. With respect to these identifications, $\Phi$ corresponds to the linear map $\widetilde{V}\colon \overline{H} \odot H \to \overline{H} \odot H$ given by

$$\widetilde{V}(\overline{\eta'} \odot \xi) = \sum_i \overline{\eta''_i} \odot \xi''_i \;\Leftrightarrow\; \langle\eta'|_{[1]}V|\xi\rangle_{[2]} = \sum_i |\xi''_i\rangle\langle\eta''_i|.$$

If $\eta', \xi, \xi''_i, \eta''_i \in H$ are as above and $\eta, \xi' \in H$, then

$$\begin{aligned}\langle\eta' \odot \xi'|V(\eta \odot \xi)\rangle &= \langle\xi'|\langle\eta'|_{[1]}V|\xi\rangle_{[2]}\eta\rangle = \sum_i \langle\xi'|\xi''_i\rangle\langle\eta''_i|\eta\rangle \\ &= \sum_i \langle\bar{\eta} \odot \xi'|\overline{\eta''_i} \odot \xi''_i\rangle = \langle\bar{\eta} \odot \xi'|\widetilde{V}(\overline{\eta'} \odot \xi)\rangle.\end{aligned}$$

The existence of such a relation, modified by an additional scaling operator, is the key idea of modularity and manageability:

**Definition 7.3.16.** A multiplicative unitary $V$ on a Hilbert space $H$ is *modular* if there exist positive self-adjoint (not necessarily bounded) operators $\widehat{Q}$ and $Q$ on $H$ and a unitary operator $\widetilde{V}$ on $\overline{H} \otimes H$ such that

i) $\ker \widehat{Q} = \ker Q = \{0\}$;

ii) $V^*(\widehat{Q} \otimes Q)V = \widehat{Q} \otimes Q$;

iii) for all $\eta, \eta' \in H$ and $\xi \in \mathrm{Dom}(Q^{-1})$, $\xi' \in \mathrm{Dom}(Q)$,

$$\langle \eta' \otimes \xi' | V(\eta \otimes \xi) \rangle = \langle \bar{\eta} \otimes Q\xi' | \widetilde{V}(\bar{\eta'} \otimes Q^{-1}\xi) \rangle. \tag{7.15}$$

We call $V$ *manageable* if we can choose $\widehat{Q} = Q$.

The opposite of a modular or manageable multiplicative unitary is modular or manageable again:

**Proposition 7.3.17.** *Let $V$ be a modular/manageable multiplicative unitary on a Hilbert space $H$ and $Q, \widehat{Q}, \widetilde{V}$ as in the definition above.*

i) $\widetilde{V}^*(\overline{\widehat{Q}} \otimes Q^{-1})\widetilde{V} = \overline{\widehat{Q}} \otimes Q^{-1}$.

ii) *For all $\xi, \xi' \in H$ and $\eta' \in \mathrm{Dom}(\widehat{Q}^{-1})$, $\eta \in \mathrm{Dom}(\widehat{Q})$,*

$$\langle \eta' \otimes \xi' | V(\eta \otimes \xi) \rangle = \big\langle \overline{\widehat{Q}\eta} \otimes \xi' \big| \widetilde{V}(\overline{\widehat{Q}^{-1}\eta'} \otimes \xi) \big\rangle.$$

iii) *The multiplicative unitary $V^{\mathrm{op}} = \Sigma V^* \Sigma$ is modular/manageable.*

*Proof.* i) Let $t \in \mathbb{R}$. Since $V^*(\widehat{Q} \otimes Q)V = \widehat{Q} \otimes Q$, the unitary $\widehat{Q}^{it} \otimes Q^{it}$ commutes with $V$. Therefore the left-hand side of equation (7.15) does not change when we replace $\eta', \xi', \eta, \xi$ by $\widehat{Q}^{it}\eta'$, $Q^{it}\xi'$, $\widehat{Q}^{it}\eta$, $Q^{it}\xi$, and

$$\big\langle \overline{\widehat{Q}^{it}\eta} \otimes Q^{it}Q\xi' \big| \widetilde{V}(\overline{\widehat{Q}^{it}\eta'} \otimes Q^{it}Q^{-1}\xi) \big\rangle = \langle \bar{\eta} \otimes Q\xi' | \widetilde{V}(\bar{\eta'} \otimes Q^{-1}\xi) \rangle$$

for all $\eta, \eta' \in H$ and $\xi \in \mathrm{Dom}(Q^{-1})$, $\xi' \in \mathrm{Dom}(Q)$. Consequently, $\widetilde{V}$ commutes with $\overline{\widehat{Q}^{it}} \otimes Q^{it} = (\overline{\widehat{Q}})^{-it} \otimes Q^{it}$ and, since $t \in \mathbb{R}$ was arbitrary, also with $\overline{\widehat{Q}} \otimes Q^{-1}$.

ii) For $\xi \in \mathrm{Dom}(Q^{-1})$ and $\xi' \in \mathrm{Dom}(Q)$, the formula follows directly from i) and equation (7.15), and for arbitrary $\xi, \xi' \in H$ by a simple continuity argument.

iii) The relation $V^*(\widehat{Q} \otimes Q)V = \widehat{Q} \otimes Q$ implies $(V^{\mathrm{op}})^*(Q \otimes \widehat{Q})V^{\mathrm{op}} = Q \otimes \widehat{Q}$, and by ii), we have for all $\xi, \xi' \in H$ and $\eta' \in \mathrm{Dom}(\widehat{Q}^{-1})$, $\eta \in \mathrm{Dom}(\widehat{Q})$

$$\begin{aligned}
\langle V^{\mathrm{op}}(\xi' \otimes \eta') | \xi \otimes \eta \rangle &= \langle \eta' \otimes \xi' | V(\eta \otimes \xi) \rangle \\
&= \big\langle \overline{\widehat{Q}\eta} \otimes \xi' \big| \widetilde{V}(\overline{\widehat{Q}^{-1}\eta'} \otimes \xi) \big\rangle \\
&= \big\langle \Sigma \widetilde{V} \Sigma(\bar{\xi} \otimes \widehat{Q}^{-1}\eta') \big| \bar{\xi'} \otimes \widehat{Q}\eta \big\rangle.
\end{aligned}$$

We put $\widetilde{V^{\mathrm{op}}} := \overline{\Sigma \widetilde{V} \Sigma}$, conjugate the equation above, and find

$$\langle \xi \otimes \eta | V^{\mathrm{op}}(\xi' \otimes \eta') \rangle = \big\langle \overline{\xi'} \otimes \hat{Q}\eta \big| \widetilde{V^{\mathrm{op}}}(\bar{\xi} \otimes \hat{Q}^{-1}\eta') \big\rangle. \qquad \square$$

The first main result concerning modular unitaries is the following theorem of Woronowicz and Soltan [201, Theorem 1.5], [142, Theorem 2.3]:

**Theorem 7.3.18.** *Every modular multiplicative unitary is well-behaved.*

Before we give the proof, let us state the second main result concerning modular unitaries.

For each well-behaved multiplicative unitary $V$, we defined an antipode $S\colon A_0(V) \to A(V)$, where $A_0(V) = \{(\omega \mathbin{\bar{\otimes}} \mathrm{id})(V) \mid \omega \in \mathcal{L}(H)_*\} \subseteq A(V)$, by

$$S((\omega \mathbin{\bar{\otimes}} \mathrm{id})(V)) = (\omega \mathbin{\bar{\otimes}} \mathrm{id})(V^*) \quad \text{for each } \omega \in \mathcal{L}(H)_*,$$

see Section 7.2.4. If $V$ is modular, much more information about this antipode is available. Most importantly, it can de described in terms of a $*$-antiautomorphism $R\colon A(V) \to A(V)$ called the *unitary antipode*, and of a one-parameter group of automorphisms $(\tau_t)_t$ of $A(V)$ called the *scaling group*. We only state the result and refer to [201, Theorem 1.5] and [142, Theorem 2.3] for the proof. For background on one-parameter groups and their analytic generators, see Section 8.1.3.

**Theorem 7.3.19.** *Let $V$ be a modular multiplicative unitary on a Hilbert space $H$ and $Q$, $\hat{Q}$, $\widetilde{V}$ as in Definition* 7.3.16.

i) *The map $S\colon A_0(V) \to A(V)$ extends to a closed linear map $S\colon \mathrm{Dom}(S) \subseteq A(V) \to A(V)$.*

ii) $\mathrm{Dom}(S) \subseteq A(V)$ *is a subalgebra and $S(ab) = S(b)S(a)$ for all $a, b \in \mathrm{Dom}(S)$.*

iii) $S(\mathrm{Dom}(S)) = \mathrm{Dom}(S)^*$ *and $S(S(a)^*)^* = a$ for all $a \in \mathrm{Dom}(S)$.*

iv) *The map $S$ admits the following polar decomposition: $S = R \circ \tau_{i/2}$, where $R$ is a $*$-antiautomorphism of $A(V)$ and $\tau_{i/2}$ is the analytic generator of a one-parameter group $\tau = (\tau_t)_{t \in \mathbb{R}}$ of $*$-automorphisms of the $C^*$-algebra $A(V)$.*

v) *$R$ commutes with $\tau_t$ for all $t \in \mathbb{R}$; in particular,* $\mathrm{Dom}(S) = \mathrm{Dom}(\tau_{i/2})$.

vi) *$R$ and $\tau$ are uniquely determined by $S$.*

vii) *$\Delta \circ \tau_t = (\tau_t \otimes \tau_t) \circ \Delta$ for all $t \in \mathbb{R}$, and $\Delta \circ R = \Sigma \circ (R \otimes R) \circ \Delta$, where $\Sigma\colon A \otimes A \to A \otimes A$ denotes the flip map.*

viii) *Let* $\widetilde{V}$ *and* $Q$ *be the operators related to* $V$ *as in Definition* 7.3.16. *Then*

(a) $\tau_t(a) = Q^{2it} a Q^{-2it}$ *for all* $t \in \mathbb{R}$, $a \in A(V)$.

(b) *Denote by* $V^{\top \otimes R}$ *the image of* $V$ *under the extension of the* $*$*-antihomomorphism* $\widehat{A}(V) \otimes A(V) \to \mathcal{L}(\bar{H}) \otimes A(V)$, $X \otimes Y \mapsto \bar{X}^* \otimes R(Y)$, *to the multiplier algebra. Then* $\widetilde{V}^* = V^{\top \otimes R}$.

**Remark 7.3.20.** Replacing the unitary $V$ by its opposite $V^{\mathrm{op}}$, we obtain a closed extension of the map $\widehat{S}\colon \widehat{A}_0(V) \to \widehat{A}(V)$ introduced in Proposition 7.2.16 with a similar polar decomposition as above.

In the remainder of this subsection, we prove Theorem 7.3.18. We fix a modular multiplicative unitary $V$ on a Hilbert space $H$ and choose $\widehat{Q}$, $Q$, $\widetilde{V}$ as in Definition 7.3.16.

The strategy of the proof is to establish a "$Q$-weighted regularity" for $V$ (Lemma 7.3.23) and to follow the proof of Theorem 7.3.11, replacing usual regularity by the modified version. The formulation of this $Q$-weighted regularity involves Hilbert–Schmidt operators; two standard references are [75, Section 2.7] and [113, Section 2.4]. Put

$$\mathcal{HS}_Q(H) := \{\rho \in \mathcal{L}(H) \mid \rho H \subseteq \mathrm{Dom}(Q^{-1}) \text{ and } Q^{-1}\rho \text{ is of Hilbert–Schmidt class}\}.$$

This is a Banach space with respect to the $Q$-weighted Hilbert–Schmidt-norm

$$\|\rho\|_Q := \|Q^{-1}\rho\|_2 = \big(\mathrm{Tr}(\rho^* Q^{-2}\rho)\big)^{1/2};$$

here, $\|\cdot\|_2$ denotes the usual Hilbert–Schmidt norm, and Tr denotes the usual trace. The space $\mathcal{HS}_Q(H)$ can be described as follows:

**Lemma 7.3.21.** *An operator* $\rho \in \mathcal{L}(H)$ *belongs to* $\mathcal{HS}_Q(H)$ *if and only if there exists a vector* $\Psi(\rho) \in \bar{H} \otimes H$ *such that*

$$\langle \xi' | \rho \eta \rangle = \langle \bar{\eta} \otimes Q\xi' | \Psi(\rho) \rangle \quad \text{for all } \xi' \in \mathrm{Dom}(Q),\ \eta \in H.$$

*The correspondence* $\rho \mapsto \Psi(\rho)$ *is a linear isometric bijection* $\mathcal{HS}_Q(H) \to \bar{H} \otimes H$.

*Proof.* This follows easily from the well-known bijective correspondence between $\bar{H} \otimes H$ and the space of all Hilbert–Schmidt-class operators on $H$, see [75, Proposition 2.6.9]. □

**Example 7.3.22.** For each $\eta' \in H$ and $\xi \in \mathrm{Dom}(Q^{-1})$, the operator $\rho := |\xi\rangle\langle\eta'|$ belongs to $\mathcal{HS}_Q(H)$: its image is evidently contained in $\mathrm{Dom}(Q^{-1})$, and $Q^{-1}|\xi\rangle\langle\eta'| = |Q^{-1}\xi\rangle\langle\eta'|$ is a Hilbert–Schmidt operator. Since

$$\langle \xi' | \rho\eta \rangle = \langle \xi' | \xi \rangle \langle \eta' | \eta \rangle = \langle \bar{\eta} \otimes Q\xi' | \bar{\eta'} \otimes Q^{-1}\xi \rangle \quad \text{for all } \xi' \in \mathrm{Dom}(Q),\ \eta \in H,$$

the condition of the previous lemma is satisfied with $\Psi(\rho) = \bar{\eta}' \otimes Q^{-1}\xi$.

The linear span of all operators of the form $|\xi\rangle\langle\eta'|$, where $\eta' \in H$ and $\xi \in \mathrm{Dom}(Q^{-1})$, is dense in $\mathcal{HS}_Q(H)$ because $\bar{H} \odot \mathrm{Dom}(Q)$ is dense in $\bar{H} \otimes H$.

The next lemma shows that $V$ satisfies a $Q$-weighted regularity condition:

**Lemma 7.3.23.** *For each $\eta' \in H$ and $\xi \in \mathrm{Dom}(Q^{-1})$, the operator $\rho := \langle\eta'|_{[1]} V|\xi\rangle_{[2]}$ belongs to $\mathcal{HS}_Q(H)$, and the linear span of all operators of this form is dense in $\mathcal{HS}_Q(H)$.*

*Proof.* Let $\eta', \xi, \rho$ be as above. Then for all $\xi' \in \mathrm{Dom}(Q)$ and $\eta \in H$,

$$\langle\xi'|\rho\eta\rangle = \langle\eta' \otimes \xi'|V(\eta \otimes \xi)\rangle = \langle\bar{\eta} \otimes Q\xi'|\widetilde{V}(\bar{\eta}' \otimes Q^{-1}\xi)\rangle.$$

This relation shows that the condition of Lemma 7.3.21 is satisfied with $\Psi(\rho) = \widetilde{V}(\bar{\eta}' \otimes Q^{-1}\xi) \in \bar{H} \otimes H$. The linear span of all operators of this form is dense in $\mathcal{HS}_Q(H)$ because $\widetilde{V}$ is unitary and $\bar{H} \odot \mathrm{Dom}(Q)$ is dense in $\bar{H} \otimes H$. □

Given a subset $X \subseteq \mathcal{HS}_Q(H) \subseteq \mathcal{L}(H)$, we denote by $[X]_Q \subseteq \mathcal{HS}_Q(H)$ and $[X] \subseteq \mathcal{L}(H)$ the closure of the linear span of $X$, taken with respect to the $Q$-weighted Hilbert–Schmidt norm or the usual operator norm. Example 7.3.22 and Lemma 7.3.23 imply the following analogue of equation (7.14) (page 192):

$$[|\,\mathrm{Dom}(Q^{-1})\rangle\langle H|]_Q = \mathcal{HS}_Q(H) = \big[\langle H|_{[1]}\, V\,|\,\mathrm{Dom}(Q^{-1})\rangle_{[2]}\big]_Q. \tag{7.16}$$

The following proposition allows us to adopt the proof of Theorem 7.3.11 to Theorem 7.3.18:

**Proposition 7.3.24.** i) *Let $\xi' \in \mathrm{Dom}(Q)$ and $\rho \in \mathcal{HS}_Q(H)$. Then the operator $T := \langle\xi'|_{[2]} V(1 \otimes \rho) \in \mathcal{L}(H \otimes H, H \otimes \mathbb{C})$ satisfies $\|T\| \le \|Q\xi'\| \cdot \|\rho\|_Q$.*

ii) $[\langle H|_{[2]}\, V(1 \otimes X)] = [\langle H|_{[2]}\, V(1 \otimes [X]_Q)]$ *for each subset $X \subseteq \mathcal{HS}_Q(H)$.*

*Proof.* i) Let $\eta, \eta' \in H$ and $\xi \in \mathrm{Dom}(Q^{-1})$. Then $\rho\xi \in \mathrm{Dom}(Q^{-1})$ and

$$\langle\eta'|T(\eta \otimes \xi)\rangle = \langle\eta' \otimes \xi'|V(\eta \otimes \rho\xi)\rangle = \langle\bar{\eta} \otimes Q\xi'|\widetilde{V}(\bar{\eta}' \otimes Q^{-1}\rho\xi)\rangle.$$

Since $\|\widetilde{V}\| = 1$ and $\|Q^{-1}\rho\| \le \|Q^{-1}\rho\|_2 = \|\rho\|_Q$, the equation above implies

$$|\langle\eta'|T(\eta \otimes \xi)\rangle| \le \|\eta\| \cdot \|Q\xi'\| \cdot \|\eta'\| \cdot \|\rho\|_Q \cdot \|\xi\|.$$

Since $\mathrm{Dom}(Q^{-1})$ is dense in $H$, the inequality extends to all $\xi \in H$. Using an orthonormal basis $(e_i)_i$ of $H$ and the decomposition $H \otimes H = \bigoplus_i \mathbb{C}e_i \otimes H$, it is easy to deduce that $|\langle\eta'|T\zeta\rangle| \le \|Q\xi'\| \cdot \|\eta'\| \cdot \|\rho\|_Q \cdot \|\zeta\|$ for all $\eta' \in H$ and $\zeta \in H \otimes H$.

ii) By statement i), $[\langle \xi'|_{[2]} V(1 \otimes X)] = [\langle \xi'|_{[2]} V(1 \otimes [X]_Q)]$ for every $\xi' \in \mathrm{Dom}(Q)$, and since $\mathrm{Dom}(Q)$ is dense in $H$,

$$\begin{aligned}[\langle H|_{[2]} V(1 \otimes X)] &= [\langle \mathrm{Dom}(Q)|_{[2]} V(1 \otimes X)] \\ &= [\langle \mathrm{Dom}(Q)|_{[2]} V(1 \otimes [X]_Q)] \\ &= [\langle H|_{[2]} V(1 \otimes [X]_Q)]. \end{aligned}$$

□

*Proof of Theorem* 7.3.18. The proof proceeds along the same line as the proof of Theorem 7.3.11; let us briefly indicate the necessary modifications. The first step is to show that $\widehat{A} = \widehat{A}(V)$ is a $C^*$-algebra. Since $\mathrm{Dom}(Q^{-1})$ is dense in $H$,

$$\begin{aligned}[\widehat{A}\widehat{A}^*] &= [\langle H|_{[3]} V_{[13]} |H\rangle_{[3]} \langle H|_{[2]} V_{[12]}^* |H\rangle_{[2]}] \\ &= \big[\langle H|_{[3]} V_{[13]} |\,\mathrm{Dom}(Q^{-1})\rangle_{[3]} \langle H|_{[2]} V_{[12]}^* |H\rangle_{[2]}\big]. \end{aligned}$$

By Proposition 7.3.24 ii) and equation (7.16), we can replace the middle term $|\,\mathrm{Dom}(Q^{-1})\rangle_{[3]} \langle H|_{[2]}$ by $\langle H|_{[2]} V_{[23]} |\,\mathrm{Dom}(Q^{-1})\rangle_{[3]}$:

$$[\widehat{A}\widehat{A}^*] = \big[\langle H|_{[3]} V_{[13]} \langle H|_{[2]} V_{[23]} |\,\mathrm{Dom}(Q^{-1})\rangle_{[3]} V_{[12]}^* |H\rangle_{[2]}\big].$$

Now we replace $\mathrm{Dom}(Q^{-1})$ by $H$ again and proceed as in Lemma 7.3.12.

The rest of the proof of Theorem 7.3.11 carries over similarly. □

# Chapter 8
# Locally compact quantum groups

The theory of locally compact quantum groups developed by Kustermans and Vaes [87], [88], [91], [158] provides a comprehensive framework for the study of quantum groups in the setting of $C^*$-algebras and von Neumann algebras. It includes a far-reaching generalization of the classical Pontrjagin duality of locally compact abelian groups that covers all locally compact groups.

The theory developed by Kustermans and Vaes builds on work of Kac, Vainerman, Enock, Schwartz, Baaj, Skandalis, Van Daele, Woronowicz, and many others. It is technically demanding, and the proofs of the main results are long and involved. Therefore we only present a survey and refer the reader to the original articles [87], [91], [93], [158] for details. Other surveys are [88], [89], [92]. In our presentation, we shall profit from the material developed in the preceding chapters of this book, in particular in Chapters 2 and 7, which motivates many constructions in the theory of locally compact quantum groups. Some background on $C^*$-algebras and von Neumann algebras used in this chapter is summarized in the appendix.

Every locally compact quantum group appears in several guises:

- as a locally compact quantum group in the setting of von Neumann algebras,
- as a reduced $C^*$-algebraic quantum group, and
- as a universal $C^*$-algebraic quantum group,

see also Section 4.3. The first and second of these variants have a similar flavor; they will form the topic of this chapter. For a discussion of the third variant, see [87].

## 8.1 The concept of a locally compact quantum group

A locally compact quantum group is a von Neumann bialgebra or $C^*$-bialgebra equipped with a left and a right Haar weight. These Haar weights are analogues of the left and the right Haar measure on a locally compact group and of the left and the right integral of an algebraic quantum group. As indicated in Chapter 4, these Haar weights are absolutely fundamental for the development of the theory. Unfortunately, up to now, the existence of Haar weights on a given $C^*$-bialgebra or von Neumann bialgebra can not be deduced from a reasonable set of assumptions but has to be postulated as an axiom. The situation improves if one starts from a multiplicative unitary, see [205] and [67, Section 1.4]. A fundamental tool for the treatment of the Haar weights is the celebrated Tomita–Takesaki theory.

We begin this section with a brief introduction to weights on $C^*$-algebras and von Neumann algebras. Then we define locally compact quantum groups in the setting of von Neumann algebras and summarize the main results of the Tomita–Takesaki theory that are fundamental to the theory of locally compact quantum groups. In the setting of $C^*$-algebras, similar results do not hold in general and have to be incorporated into the definition of the Haar weights in form of additional assumptions. The section closes with the definition of reduced $C^*$-algebraic quantum groups.

### 8.1.1 Weights

The prototypical example of a Haar weight on a von Neumann bialgebra or $C^*$-bialgebra is integration of functions on a locally compact group $G$ with respect to the Haar measure $\lambda$. Unless $G$ is compact, not every element of the von Neumann algebra $L^\infty(G,\lambda)$ or the $C^*$-algebra $C_0(G)$ is integrable, and the integral is well defined only on a dense subspace or on the cone of positive functions. Likewise, a weight on a $C^*$-algebra or von Neumann algebra is first defined on the cone of positive elements and then extended to a linear map on the subset of integrable elements. The precise definition subsumes von Neumann algebras as special $C^*$-algebras:

**Definition 8.1.1.** Let $A$ be a $C^*$-algebra. Denote by $A^+$ the set of all positive elements of $A$. A *weight on $A$* is a function $\phi\colon A^+ \to [0,\infty]$ satisfying

i) $\phi(a+b) = \phi(a) + \phi(b)$ for all $a, b \in A^+$;

ii) $\phi(ra) = r\phi(a)$ for all $r \in [0,\infty)$ and $a \in A^+$.

The weight $\phi$ is called *faithful* if $\phi(a) \neq 0$ for each non-zero $a \in A^+$.

Let $\phi$ be a weight on a $C^*$-algebra $A$. We use the following standard notation:

- $\mathcal{M}_\phi^+ := \{a \in A^+ \mid \phi(a) < \infty\}$ is the set of all positive $\phi$-integrable elements;
- $\mathcal{N}_\phi := \{a \in A \mid \phi(a^*a) < \infty\}$ is the set of all $\phi$-square-integrable elements;
- $\mathcal{M}_\phi := \operatorname{span} \mathcal{M}_\phi^+ = \operatorname{span} \mathcal{N}_\phi^* \mathcal{N}_\phi$ is the set of all $\phi$-integrable elements.

It is easy to check that $\mathcal{N}_\phi$ is a left ideal and that $\mathcal{M}_\phi$ is a $*$-subalgebra of $A$ [121, Lemma 5.1.2]. Moreover, $\phi$ extends uniquely to a linear functional on $\mathcal{M}_\phi$ [121, Lemma 5.1.2], which we denote by $\phi$ again.

The theory of weights subsumes classical integration theory and is generally considered as non-commutative integration theory.

**Example 8.1.2.** Let $X$ be a locally compact space with a Borel measure $\mu$. Then integration with respect to $\mu$ defines a weight on the von Neumann algebra $L^\infty(X,\mu)$, which we denote by $\mu$ again:

$$\mu(f) := \int_X f d\mu \quad \text{for all } f \in L^\infty(X,\mu)^+.$$

Evidently, $\mathcal{M}_\mu = L^1(X,\mu) \cap L^\infty(X,\mu)$ and $\mathcal{N}_\mu = L^2(X,\mu) \cap L^\infty(X,\mu)$.

Unbounded weights are difficult to handle. To retain some control, one usually imposes some of the following conditions:

**Definition 8.1.3.** A weight $\phi$ on a $C^*$-algebra $A$ is

- *lower semi-continuous* if the subset $\{a \in A^+ \mid \phi(a) \leq \lambda\} \subset A$ is closed for every $\lambda \in \mathbb{R}^+$;
- *densely defined* if $\mathcal{M}_\phi^+$ is dense in $A^+$ (or, equivalently, if $\mathcal{N}_\phi$ or $\mathcal{M}_\phi$ is dense in $A$);
- *proper* if $\phi$ is non-zero, densely defined, and lower semi-continuous.

A weight $\phi$ on a von Neumann algebra $M$ is

- *normal* if the subset $\{a \in M^+ \mid \phi(a) \leq \lambda\} \subset M$ is $\sigma$-weakly closed for every $\lambda \in \mathbb{R}^+$;
- *semi-finite* if $\mathcal{M}_\phi^+$ is dense in $M^+$ (or, equivalently, if $\mathcal{N}_\phi$ or $\mathcal{M}_\phi$ is dense in $M$) with respect to the $\sigma$-weak topology;
- *n.s.f.* if $\phi$ is normal, semi-finite, and faithful.

### 8.1.2 Locally compact quantum groups in the setting of von Neumann algebras

A locally compact quantum group in the setting of von Neumann algebras is simply a von Neumann bialgebra equipped with a left invariant and a right invariant n.s.f. weight. The definition involves the cone of positive normal linear functionals on a von Neumann algebra $M$, which we denote by $M_*^+$, and slice maps (see Section 12.4).

**Definition 8.1.4.** Let $(M,\Delta)$ be a von Neumann bialgebra. A weight $\phi$ on $M$ is

- *left-invariant* if $\phi((\omega \mathbin{\bar{\otimes}} \mathrm{id})(\Delta(a))) = \omega(1)\phi(a)$ for all $a \in \mathcal{M}_\phi^+$, $\omega \in M_*^+$;
- *right-invariant* if $\phi((\mathrm{id} \mathbin{\bar{\otimes}} \omega)(\Delta(a))) = \omega(1)\phi(a)$ for all $a \in \mathcal{M}_\phi^+$, $\omega \in M_*^+$.

A *locally compact quantum group* (*in the setting of von Neumann algebras*) is a von Neumann bialgebra $(M, \Delta)$ that has a left-invariant n.s.f. weight and a right-invariant n.s.f. weight.

For a motivation of the invariance conditions, see Section 2.2.1. As in the case of algebraic quantum groups, uniqueness of the Haar weights can be shown once the existence is settled; however, the proof is much more involved and depends on a fair amount of the theory of locally compact quantum groups.

**Theorem 8.1.5.** *For every locally compact quantum group in the setting of von Neumann algebras, the left-invariant n.s.f. weight and the right-invariant n.s.f. weight are unique up to positive constants.*

*Proof.* This is stated in [158, Section 1.14]; a proof is given in the setting of $C^*$-algebras, and that proof carries over to the setting of von Neumann algebras. □

**Definition 8.1.6.** The invariant n.s.f. weights of a locally compact quantum group in the setting of von Neumann algebras (which are uniquely determined up to some constant) are called its *left* and its *right Haar weight*, respectively.

Classical examples of locally compact quantum groups are the von Neumann bialgebras associated to a locally compact group:

**Example 8.1.7.** For every locally compact group $G$, the von Neumann bialgebras $L^\infty(G)$ and $L(G)$ introduced in Example 4.2.4 are locally compact quantum groups. Let us describe their Haar weights.

The left Haar weight $\phi$ and the right Haar weight $\psi$ of $L^\infty(G)$ are given by

$$\phi(f) = \int_G f d\lambda, \quad \psi(f) = \int_G f d\lambda^{-1} \quad \text{for all } f \in L^\infty(G)^+,$$

where $\lambda$ and $\lambda^{-1}$ denote the left and the right Haar measure of $G$, respectively. Invariance of $\phi$ and of $\psi$ follows immediately from the translation invariance of $\lambda$ and $\lambda^{-1}$, see also Section 2.2.1.

The von Neumann bialgebra $L(G)$ has a Haar weight $\hat{\phi}$ that is both left- and right-invariant; it is given by

$$\hat{\phi}(T^*T) = \begin{cases} (\|f\|_2)^2, & \text{there exists } f \in L^2(G,\lambda) \text{ such that} \\ & (T\xi)(y) = \int_G f(x)\xi(x^{-1}y)d\lambda(x) \\ & \text{for all } \xi \in C_c(G),\ y \in G, \\ \infty, & \text{otherwise,} \end{cases}$$

for every $T \in L(G)^+$, see [121, Section 7.2.7] or [148, Section VII.3]. A short calculation shows

$$\hat{\phi}(L(g)) = g(e) \quad \text{for all } g \in C_c(G) \text{ such that } L(g) \in L(G)^+.$$

### 8.1.3 The modular automorphism group of a weight

Every n.s.f. weight on a von Neumann algebra has a modular automorphism group that measures the deviation of the weight from being a trace. This automorphism group is reviewed in the following paragraphs. Standard references are, for example, [21], [76], [121], [144], [147], [150].

The Haar weights $\phi$ and $\psi$ of a locally compact quantum group $(M, \Delta)$ need not be traces, that is, we can neither expect $\phi(a^*b) = \phi(ba^*)$ for all $a, b \in \mathcal{N}_\phi$ nor $\psi(a^*b) = \psi(ba^*)$ for all $a, b \in \mathcal{N}_\psi$. Recall that for every left integral $\phi_0$ on an algebraic quantum group $(A_0, \Delta_0)$, there exists a modular automorphism $\sigma_0$ such that $\phi_0(a^*b) = \phi(b\sigma_0(a^*))$ for all $a, b \in A_0$ (Theorem 2.2.17). A deep and fundamental result of the celebrated Tomita–Takesaki theory says that every n.s.f. weight on a von Neumann algebra admits a similar modular automorphism group. This modular automorphism group is essential for the development of a satisfying theory of locally compact quantum groups.

Let us proceed to the precise definitions and statements. The modular automorphism group of an n.s.f. weight is a particular instance of the following concept:

**Definition 8.1.8.** A *one-parameter group of $*$-automorphisms* on a $C^*$-algebra $A$ is a family $\alpha = (\alpha_t)_{t \in \mathbb{R}}$ of $*$-automorphisms of $A$ that satisfies $\alpha_s \circ \alpha_t = \alpha_{s+t}$ for all $s, t \in \mathbb{R}$. The one-parameter group $\alpha$ is called

- *norm-continuous* if for every $a \in A$, the map $\mathbb{R} \to A$ given by $t \mapsto \alpha_t(a)$ is continuous;
- *strongly continuous* if $A$ is a von Neumann algebra and for every $a \in A$, the map considered above is continuous with respect to the strong operator topology on $A$.

**Remark 8.1.9.** The term "strongly continuous one-parameter group" is frequently also used for one-parameter groups of automorphisms of general Banach spaces that are norm-continuous in a similar sense as above.

We shall be interested in analytic extensions of one-parameter groups. These extensions involve analytic functions that are defined on horizontal strips on the complex plane of the form

$$I(z) := \{y \in \mathbb{C} : |\Im y| \le |\Im z|\} \subset \mathbb{C}, \quad \text{where } z \in \mathbb{C}.$$

Given a $C^*$-algebra/von Neumann algebra $A$, let us call a function $f : I(z) \to A$ *norm-regular/strongly regular* if it is

i) analytic on the interior of $I(z)$, that is, for every $y_0$ in the interior of $I(z)$, the limit $\lim_{y \to y_0}(f(y) - f(y_0))/(y - y_0)$ exists w.r.t. the norm topology;

ii) norm-bounded on $I(z)$;

iii) norm-continuous/$\sigma$-strongly-$*$ continuous on $I(z)$.

For the definition of the $\sigma$-strong-$*$ topology, see Section 12.3. From the Identity Theorem of complex analysis, it follows that two functions $f, g\colon I(z) \to A$ that are norm-regular/strongly regular and coincide on $\mathbb{R} \subseteq I(z)$ must coincide on $I(z)$.

**Definition 8.1.10.** Let $\alpha$ be a norm-continuous/strongly continuous one-parameter group of automorphisms on a $C^*$-algebra/von Neumann algebra $A$. For each $z \in \mathbb{C} \setminus \mathbb{R}$, put

$$\begin{aligned}\mathrm{Dom}(\alpha_z) := \{x \in A \mid {}&\text{there exists } f\colon I(z) \to A \text{ norm-regular/strongly}\\ &\text{regular such that } f(t) = \alpha_t(x) \text{ for all } t \in \mathbb{R}\},\end{aligned}$$

and define $\alpha_z\colon \mathrm{Dom}(\alpha_z) \to A$ by $\alpha_z(x) := f(z)$, where $f$ is as above. Note that the function $f$ is uniquely determined and $\alpha_z$ is well defined.

The family $(\alpha_z)_{z\in\mathbb{C}}$ is called the *analytic extension of* $\alpha$, and the elements of $\bigcap_{z\in\mathbb{C}} \mathrm{Dom}(\alpha_z)$ are called *analytic*.

**Proposition 8.1.11.** *Let $\alpha$ be a norm-continuous/strongly continuous one-parameter group of automorphisms on a $C^*$-algebra/von Neumann algebra $A$, and let $y, z \in \mathbb{C}$.*

i) *The analytic elements are dense in $A$ with respect to the norm/$\sigma$-strong-$*$ topology.*

ii) $\mathrm{Dom}(\alpha_z) \subseteq A$ *is a subalgebra and* $\alpha_z\colon \mathrm{Dom}(\alpha_z) \to A$ *is an algebra homomorphism.*

iii) $\mathrm{Dom}(\alpha_z)^* = \mathrm{Dom}(\alpha_{\bar{z}})$ *and* $\alpha_{\bar{z}}(a^*) = \alpha_z(a)^*$ *for all* $a \in \mathrm{Dom}(\alpha_z)$.

iv) $\alpha_z \circ \alpha_t = \alpha_t \circ \alpha_z = \alpha_{z+t}$ *for all* $t \in \mathbb{R}$.

v) $\mathrm{Dom}(\alpha_y \circ \alpha_z) = \mathrm{Dom}(\alpha_{y+z}) \cap \mathrm{Dom}(\alpha_z)$ *and* $\alpha_y(\alpha_z(a)) = \alpha_{y+z}(a)$ *for all* $a \in \mathrm{Dom}(\alpha_y \circ \alpha_z)$. *If $y$ and $z$ lie on the same side of the real axis, then* $\alpha_y \circ \alpha_z = \alpha_{y+z}$.

vi) *If* $y \in I(z)$, *then* $\mathrm{Dom}(\alpha_z) \subseteq \mathrm{Dom}(\alpha_y)$.

vii) $\alpha_z$ *is injective,* $\operatorname{Im}\alpha_z = \mathrm{Dom}(\alpha_{-z})$, *and* ${\alpha_z}^{-1} = \alpha_{-z}$.

viii) $\alpha_z$ *is closed with respect to the norm topology/$\sigma$-strong-$*$ topology.*

For the definition of closed maps, see Section 8.2.

*Proof.* Statements ii)–viii) follow without much work from the definition, see the Preprint [86]. For the proof of statement i), see also [121, Section 8.12]. □

**Remark 8.1.12.** Every norm-continuous/strongly continuous one-parameter group $\alpha$ on a $C^*$-algebra/von Neumann algebra is uniquely determined by the map $\alpha_i$, which is called the (*analytic*) *generator* of $\alpha$.

Given the preceding definitions, we can formulate the main result of Tomita–Takesaki theory:

**Theorem 8.1.13.** *Let $\phi$ be an n.s.f. weight on a von Neumann algebra $M$. There exists a unique strongly continuous one-parameter group of automorphisms $\sigma$ on $M$ such that*

i) *$\sigma$ leaves $\phi$ invariant, that is, $\phi \circ \sigma_t = \phi$ for all $t \in \mathbb{R}$;*

ii) *$\phi(x^*x) = \phi\left(\sigma_{i/2}(x)\sigma_{i/2}(x)^*\right)$ for all $x \in \operatorname{Dom}(\sigma_{i/2})$.*

*Furthermore, $\phi(ax) = \phi(x\sigma_{-i}(a))$ in the following two situations:*

a) *$a \in \operatorname{Dom}(\sigma_{-i})$ and $x \in \mathcal{M}_\phi$; in that case, $ax$ and $x\sigma_{-i}(a)$ belong to $\mathcal{M}_\phi$;*

b) *$x \in \mathcal{N}_\phi \cap \mathcal{N}_\phi^*$, $a \in \mathcal{N}_\phi^* \cap \operatorname{Dom}(\sigma_{-i})$, and $\sigma_{-i}(a) \in \mathcal{N}_\phi$.*

*The one-parameter group $\sigma$ is called the* modular automorphism group *of $\phi$.*

For an illustration, consider the following simple example:

**Example 8.1.14.** Let $\phi$ be an n.s.f. weight on the von Neumann algebra $M_n(\mathbb{C})$, where $n \in \mathbb{N}$. Elementary linear algebra shows that there exists a positive definite matrix $\delta \in M_n(\mathbb{C})$ such that

$$\phi(x) = \operatorname{Tr}(x\delta) = \operatorname{Tr}(\delta^{1/2}x\delta^{1/2}) = \operatorname{Tr}(\delta x) \quad \text{for all } x \in M_n(\mathbb{C})^+,$$

where $\operatorname{Tr}\colon M_n(\mathbb{C}) \to \mathbb{C}$ denotes the usual trace. Note that $\phi$ extends to a positive linear map on $M_n(\mathbb{C})$ by the same formula as above. We claim that the modular automorphism group $\sigma$ of $\phi$ is given by

$$\sigma_t(x) = \delta^{it}x\delta^{-it} \quad \text{for all } x \in M_n(\mathbb{C}),\ t \in \mathbb{R}.$$

Note that $\delta^{it}$ is unitary for each $t \in \mathbb{R}$, so that $\sigma_t$ is a $*$-automorphism.

Let us prove the claim. The weight $\phi$ is invariant with respect to $\sigma$ because

$$\phi(\sigma_t(x)) = \phi(\delta^{it}x\delta^{-it}) = \operatorname{Tr}(\delta^{it}x\delta^{-it}\delta) = \operatorname{Tr}(x\delta^{-it}\delta\delta^{it}) = \operatorname{Tr}(x\delta) = \phi(x)$$

for all $x \in M_n(\mathbb{C})$. Evidently, the analytic extension of $\sigma$ is given by $\sigma_z(x) = \delta^{iz}x\delta^{-iz}$ for all $x \in M_n(\mathbb{C})$ and $z \in \mathbb{C}$. Therefore, $\sigma_{i/2}(x) = \delta^{-1/2}x\delta^{1/2}$ and

$$\begin{aligned}\phi(\sigma_{i/2}(x)\sigma_{i/2}(x)^*) &= \phi(\delta^{-1/2}x\delta^{1/2}\cdot\delta^{1/2}x^*\delta^{-1/2})\\ &= \operatorname{Tr}(x\delta x^*) = \operatorname{Tr}(x^*x\delta) = \phi(x^*x)\end{aligned}$$

for all $x \in M_n(\mathbb{C})$. These equations show that $\sigma$ satisfies conditions i) and ii) of Theorem 8.1.13. Furthermore, $\phi(ax) = \phi(x\sigma_{-i}(a))$ for all $a, x \in M_n(\mathbb{C})$:

$$\phi(ax) = \operatorname{Tr}(ax\delta) = \operatorname{Tr}(x\delta a) = \operatorname{Tr}(x\delta a\delta^{-1}\delta) = \operatorname{Tr}(x\sigma_{-i}(a)\delta) = \phi(x\sigma_{-i}(a)).$$

### 8.1.4 Reduced $C^*$-algebraic quantum groups

For the Haar weights of a reduced $C^*$-algebraic quantum group, the existence of a modular automorphism group has to be assumed.

**Definition 8.1.15.** A proper weight $\phi$ on a $C^*$-algebra $A$ is a *KMS-weight* if there exists a norm-continuous one-parameter group $\sigma$ on $A$ such that conditions i) and ii) of Theorem 8.1.13 hold. As in the setting of von Neumann algebras, the one-parameter group $\sigma$ is called a *modular automorphism group* for $\phi$.

References for KMS-weights on $C^*$-algebras are, for example, [90], [91], [121], [150], [158].

Every KMS-weight on a $C^*$-algebra satisfies the analogue of the second part of Theorem 8.1.13:

**Theorem 8.1.16.** *Let $\phi$ be a KMS-weight on a $C^*$-algebra $A$. If $\phi$ is faithful, then the modular automorphism group of $\phi$ is uniquely determined. Moreover, $\phi(ax) = \phi(x\sigma_{-i}(a))$ in the following two situations:*

a) *$a \in \mathrm{Dom}(\sigma_{-i})$ and $x \in \mathcal{M}_\phi$; in that case, $ax$ and $x\sigma_{-i}(a)$ belong to $\mathcal{M}_\phi$;*

b) *$x \in \mathcal{N}_\phi \cap \mathcal{N}_\phi^*$, $a \in \mathcal{N}_\phi^* \cap \mathrm{Dom}(\sigma_{-i})$, and $\sigma_{-i}(a) \in \mathcal{N}_\phi$.*

Given a $C^*$-algebra $A$, we denote by $A_+^*$ the cone of all positive linear functionals on $A$.

**Definition 8.1.17.** A *reduced $C^*$-algebraic quantum group* is a $C^*$-bialgebra $(A, \Delta)$ that satisfies the following conditions:

i) the following two sets are linearly dense subsets of $A$:

$$\{(\omega \otimes \mathrm{id})(\Delta(a)) \mid \omega \in A_+^*,\ a \in A\}, \quad \{(\mathrm{id} \otimes \omega)(\Delta(a)) \mid \omega \in A_+^*,\ a \in A\};$$

ii) there exists a faithful KMS-weight $\phi$ on $A$ which is *left-invariant* in the sense that $\phi((\omega \otimes \mathrm{id})(\Delta(a))) = \omega(1)\phi(a)$ for all $\omega \in A_+^*$ and $a \in \mathcal{M}_\phi^+$;

iii) there exists a KMS-weight $\psi$ on $A$ which is *right-invariant* in the sense that $\psi((\mathrm{id} \otimes \omega)(\Delta(a))) = \omega(1)\psi(a)$ for all $\omega \in A_+^*$ and $a \in \mathcal{M}_\psi^+$.

**Remarks 8.1.18.** i) In conditions ii) and iii), $\omega \otimes \mathrm{id}$ and $\mathrm{id} \otimes \omega$ denote slice maps as defined in Proposition 12.4.1. To define $\omega(1)$, one extends $\omega$ to the unitization (or multiplier algebra) of $A$, see Corollary 12.1.2, and finds (as for any positive functional) $\omega(1) = \|w\|$. Finally, note that $(\omega \otimes \mathrm{id})(\Delta(a))$ and $(\mathrm{id} \otimes \omega)(\Delta(a))$ belong to $A$ by condition i), so that we can apply $\phi$ or $\psi$, respectively.

ii) The KMS-condition on the weights in ii) and iii) can be replaced by a condition that first seems to be weaker, but turns out to be equivalent: the weights $\phi$ and $\psi$ need only be approximate KMS-weights [91].

iii) If $(A, \Delta)$ is a bisimplifiable $C^*$-bialgebra, then condition i) above holds. This follows easily from Proposition 12.4.3. Conversely, every reduced $C^*$-algebraic quantum group is a bisimplifiable $C^*$-bialgebra, as we shall see in Proposition 8.3.2.

iv) For locally compact quantum groups in the setting of von Neumann algebras, an analogue of condition i) is automatically satisfied, see [93, Proposition 1.4] or [158, Proposition 1.14.5].

v) The KMS-weight $\psi$ in ii) turns out to be faithful, see Remark 8.3.9.

As in the setting of von Neumann algebras, one can show:

**Proposition 8.1.19.** *For every reduced $C^*$-algebraic quantum group, the faithful left-invariant KMS-weight and the right-invariant KMS-weight are unique up to positive constants.*

*Proof.* The proof involves a large amount of the theory of locally compact quantum groups, see [91, Theorems 7.14, 7.15] or [158, Theorems 1.10.1, 1.10.2]. □

**Definition 8.1.20.** The invariant KMS-weights of a reduced $C^*$-algebraic quantum group (which are uniquely determined up to some constant) are called its *left* and its *right Haar weight*, respectively.

Classical examples of reduced $C^*$-algebraic quantum groups are the $C^*$-bialgebras associated to a locally compact group:

**Example 8.1.21.** For every locally compact group $G$, the $C^*$-bialgebras $C_0(G)$ and $C_r^*(G)$ introduced in Example 4.2.2 are reduced $C^*$-algebraic quantum groups. The Haar weights on $C_0(G)$ coincide with the restrictions of the Haar weights on $L^\infty(G)$; trivially, they are traces. The Haar weight on $C_r^*(G)$ coincides with the restriction of the Haar weight on $L(G)$. Its modular automorphism group can be described as follows [150, VII, Proposition 3.1]: Denote by $\delta$ the modular function of $G$ (see Section 2.2.3), and consider the function $\delta^{it}$ for each $t \in \mathbb{R}$ as a unitary multiplication operator on $L^2(G, \lambda)$. Then $\sigma_t(x) = \delta^{it} x \delta^{-it}$ for all $x \in C_r^*(G)$ and $t \in \mathbb{R}$.

**Example 8.1.22.** Every reduced $C^*$-algebraic compact quantum group $(A, \Delta)$ is a reduced $C^*$-algebraic quantum group. To prove this, we only need to show that the Haar state $h$ of $(A, \Delta)$ is a KMS-state. Denote by

- $(H, \Lambda, \pi)$ the GNS-construction for $h$;
- $(u^\alpha)_\alpha$ a maximal family of pairwise inequivalent irreducible unitary corepresentation matrices of $(A, \Delta)$;
- $(A_0, \Delta_0)$ the Hopf $*$-algebra of matrix elements of finite-dimensional corepresentations, that is, $A_0 = \operatorname{span}\{u^\alpha_{kl} \mid \alpha, k, l\}$;
- $(f_z)_{z \in \mathbb{C}}$ the family of characters on $A_0$ defined in Theorem 3.2.19;

- $(\sigma_z^0)_{z\in\mathbb{C}}$ the family of automorphisms (not necessarily $*$-automorphisms) of $A_0$ given by $\sigma_z^0(a) = \sum f_{iz}(a_{(1)})a_{(2)}f_{iz}(a_{(3)})$; in the notation of Corollary 3.2.20, $\sigma_z^0 = \rho_{iz,iz}$.

We show that the family $(\sigma_t^0)_{t\in\mathbb{R}}$ extends to a norm-continuous one-parameter group $\sigma$ on $A$ and that $\sigma$ is a modular automorphism group for $h$. The extension is implemented by a one-parameter group of unitaries on $H$ which is constructed as follows. Consider the map

$$\nabla_0 \colon \Lambda(A_0) \to \Lambda(A_0), \quad \Lambda(a) \mapsto \Lambda(\sigma_{-i}^0(a)).$$

Let $\alpha$ be arbitrary. Then $H_\alpha := \operatorname{span}\{\Lambda(u_{kl}^\alpha) \mid k,l\}$ has finite dimension and $\nabla_0(H_\alpha) \subseteq H_\alpha$ because

$$\sigma_{-i}^0(u_{kl}^\alpha) = \sum_{m,n} f_1(u_{km}^\alpha)u_{mn}^\alpha f_1(u_{nl}^\alpha) \quad \text{for all } k,l$$

by equation (5.6). By Corollary 3.2.20, we have for all $a,b \in A_0$

$$\begin{aligned}\langle \Lambda(b)|\nabla_0\Lambda(a)\rangle &= h(b^*\sigma_{-i}^0(a)) = (h \circ \sigma_{-i}^0)(\sigma_i^0(b^*)a) \\ &= h(\sigma_{-i}^0(b)^*a) = \langle \nabla_0\Lambda(b)|\Lambda(a)\rangle.\end{aligned}$$

Thus $\nabla_0|_{H_\alpha}$ is self-adjoint.

Since $H = \bigoplus_\alpha H_\alpha$ (Proposition 3.2.6, 3.2.9), there exists a unique self-adjoint operator $\nabla$ on $H$ such that $\nabla|_{H_\alpha} = \nabla_0|_{H_\alpha}$ for all $\alpha$. A short calculation shows that $\nabla^{iz}\pi(a)\nabla^{-iz} = \pi(\sigma_z^0(a))$ for all $a \in A_0$, $z \in \mathbb{C}$. Now we can conclude:

- Since $\nabla^{it}$ is unitary for each $t \in \mathbb{R}$, we can extend for each $t \in \mathbb{R}$ the $*$-automorphism $\sigma_t^0$ of $A_0$ to a $*$-automorphism $\sigma_t$ of $A$ by the formula $\pi(\sigma_t(a)) = \nabla^{it}\pi(a)\nabla^{-it}$.
- Since $A_0$ is dense in $A$ and $\sigma_s^0 \circ \sigma_t^0 = \sigma_{s+t}^0$ for all $s,t \in \mathbb{R}$, we have $\sigma_s \circ \sigma_t = \sigma_{s+t}$ for all $s,t \in \mathbb{R}$.
- Since for each $b \in A_0$, the map $\mathbb{C} \to \mathbb{C}, z \mapsto f_z(b)$, is holomorphic, the map $\mathbb{R} \to A_0 \hookrightarrow A$ given by $t \mapsto \sigma_t^0(a) = \sigma_t(a)$ is norm-continuous for each $a \in A_0$. Since each $\sigma_t$ has norm 1, it follows that for each $a \in A$, the map $\mathbb{R} \to A$ given by $t \mapsto \sigma_t(a)$ is norm-continuous.

Thus we obtain a norm-continuous one-parameter group $\sigma = (\sigma_t)_{t\in\mathbb{R}}$ on $A$.

Let us show that $\sigma$ satisfies conditions i) and ii) of Theorem 8.1.13. By Corollary 3.2.20, we have $h(\sigma_t(a)) = h(a)$ for all $a \in A_0$ and hence for all $a \in A$. Let $a \in \operatorname{Dom}(\sigma_{i/2})$. By Proposition 8.1.11,

$$h(\sigma_{i/2}(a)\sigma_{i/2}(a)^*) = h(\sigma_{-i/2}(a^*)^*\sigma_{-i/2}(a^*)) = \langle \nabla^{1/2}\Lambda(a^*)|\nabla^{1/2}\Lambda(a^*)\rangle.$$

We need to show that this is equal to $h(a^*a) = \langle\Lambda(a)|\Lambda(a)\rangle$. Since the map $\nabla^{1/2}$ and the map $\Lambda(A_0) \to \Lambda(A_0)$, $\Lambda(a) \mapsto \Lambda(a^*)$, are compatible with the decomposition $H = \bigoplus_\alpha H_\alpha$, we may assume that $a \in A_0$. But for $a \in A_0$, Corollary 3.2.20 implies

$$\begin{aligned} h(\sigma_{i/2}(a)\sigma_{i/2}(a)^*) &= h(\sigma^0_{i/2}(a)\sigma^0_{i/2}(a)^*) = h(\sigma^0_{i/2}(a)\sigma^0_{-i/2}(a^*)) \\ &= h(\sigma^0_{-i/2}(a^*)\sigma^0_{-i/2}(a)) = h(a^*a). \end{aligned}$$

More generally, every algebraic quantum group gives rise to a reduced $C^*$-algebraic quantum group:

**Example 8.1.23.** Let $(A_0, \Delta_0)$ be an algebraic quantum group with dual $(\widehat{A}_0, \widehat{\Delta}_0)$. In Example 7.1.6 and Theorem 7.2.14, we associated to $(A_0, \Delta_0)$ a well-behaved multiplicative unitary $V$ whose right and left leg $(A(V), \Delta)$ and $(\widehat{A}(V), \widehat{\Delta})$ were completions of the algebraic quantum group $(A_0, \Delta_0)$ and its dual $(\widehat{A}_0, \widehat{\Delta}_0)$, respectively. Kustermans and Van Daele showed that these $C^*$-bialgebras are reduced $C^*$-algebraic quantum groups [94].

Further examples of locally compact quantum groups are discussed in Section 8.4.

## 8.2 Additional prerequisites

Before we can proceed in the theory of locally compact quantum groups, we need to review several preliminary concepts and tools related to unbounded linear maps and weights.

**Closed and densely defined operators.** We will often encounter linear maps that are not continuous and not everywhere defined, as, for example, the analytic extensions of one-parameter groups. Let us fix some related terminology.

Given topological vector spaces $E$ and $F$, a densely defined linear map from $E$ to $F$ is a linear map $T\colon \operatorname{Dom}(T) \to F$ whose domain of definition $\operatorname{Dom}(T)$ is a dense subspace of $E$. We write $T\colon \operatorname{Dom}(T) \subseteq E \to F$ in order to indicate the domain and range of $T$. Such a map $T$ is called *closed* if it satisfies the following equivalent conditions:

i) the graph $G(T) := \{(e, T(e)) \mid e \in \operatorname{Dom}(T)\}$ is a closed subspace of $E \oplus F$;

ii) if $(e_\nu)_\nu$ is a net in $\operatorname{Dom}(T)$ that converges to some element $e \in E$ and the net $(T(e_\nu))_\nu$ converges to some element $f \in F$, then $e \in \operatorname{Dom}(T)$ and $T(e) = f$.

A subspace $E_0 \subseteq \mathrm{Dom}(T)$ is called a *core* for $T$ if the subset $\{(e, T(e)) \mid e \in E_0\} \subseteq G(T)$ is dense. Evidently, every densely defined closed map is completely determined by the restriction to some core.

The composition of densely defined linear maps is defined like the usual composition of partially defined maps; it need not be densely defined. Furthermore, the composition of closed maps need not be closed.

**Constructions with weights ([90], [91], [150], [158]).** We need to define several constructions with weights such as formation of tensor products or slice maps and extension to multiplier algebras. The general idea is that all constructions that behave nicely for positive linear functionals on $C^*$-algebras or normal positive linear functionals on von Neumann algebras can be carried over to lower semi-continuous or normal weights, respectively, by means of the following approximation result:

**Lemma 8.2.1.** *A weight $\phi$ on a $C^*$-algebra $A$ is lower semi-continuous if and only if*

$$\phi(x) = \sup_{\omega \in \mathcal{F}_\phi} \omega(x) \quad \textit{for all } x \in A^+, \textit{ where } \mathcal{F}_\phi = \{\omega \in A^*_+ \mid \omega \leq \phi \textit{ on } A^+\}.$$

*A semi-finite weight $\phi$ on a von Neumann algebra $A$ is normal if and only if the condition above holds with $A^*_+$ replaced by $A^+_*$ in the definition of $\mathcal{F}_\phi$.*

Let $\phi$ and $\psi$ be proper weights on $C^*$-algebras $A$ and $B$, respectively, and put $\mathcal{G}_\phi := (0,1) \cdot \mathcal{F}_\phi \subseteq \mathcal{F}_\phi$. The set $\mathcal{G}_\phi$ is naturally ordered; therefore it can be used as the index set of a net.

- The *extension* of $\phi$ to the $C^*$-algebra $M(A)$ is the weight $\bar{\phi}$ defined by $\bar{\phi}(x) := \sup_{\omega \in \mathcal{F}_\phi} \bar{\omega}(x)$ for all $x \in M(A)^+$. Here, $\bar{\omega}$ denotes the unique strictly continuous extension of a functional $\omega \in A^*_+$ to $M(A)$, see Corollary 12.1.2.

- The *tensor product* of $\phi$ and $\psi$ is the weight $\phi \otimes \psi$ on $A \otimes B$ defined by

  $$(\phi \otimes \psi)(x) = \sup\{(\omega \otimes \theta)(x) \mid \omega \in \mathcal{F}_\phi,\ \theta \in \mathcal{F}_\psi\}$$

  for all $x \in (A \otimes B)^+$. This weight is proper.

- The *slice map* $\phi \otimes \mathrm{id}_B$ on $M(A \otimes B)$ is defined as follows. Put

  $$\bar{\mathcal{M}}^+_{\phi \otimes \mathrm{id}} := \{x \in M(A \otimes B)^+ \mid \operatorname*{s-lim}_{\omega \in \mathcal{G}_\phi} (\omega \otimes \mathrm{id})(x) \text{ exists in } M(B)\}$$

  and

  $$(\phi \otimes \mathrm{id}_B)(x) := \operatorname*{s-lim}_{\omega \in \mathcal{G}_\phi} (\omega \otimes \mathrm{id})(x) \quad \text{for all } x \in \bar{\mathcal{M}}^+_{\phi \otimes \mathrm{id}},$$

where "s-lim" denotes the strict limit. Furthermore, let

$$\bar{\mathcal{M}}_{\phi\otimes\mathrm{id}} := \operatorname{span} \bar{\mathcal{M}}^{+}_{\phi\otimes\mathrm{id}}, \quad \bar{\mathcal{N}}_{\phi\otimes\mathrm{id}} := \{x \in M(A \otimes B) \mid x^*x \in \bar{\mathcal{M}}^{+}_{\phi\otimes\mathrm{id}}\}.$$

Then $\phi \otimes \mathrm{id}_B$ extends to a linear map $\phi \otimes \mathrm{id}_B : \bar{\mathcal{M}}_{\phi\otimes\mathrm{id}} \to M(B)$.

Likewise, one can define a slice map $\mathrm{id}_A \otimes \psi$.

Similarly, one defines tensor products and slice maps of n.s.f. weights on von Neumann algebras.

**GNS-construction for weights ([90], [91], [150], [158]).** For a locally compact quantum group, the Hilbert space associated to the left Haar weight plays a fundamental rôle. This space is a particular example of the following construction:

**Definition 8.2.2.** Let $\phi$ be a weight on a $C^*$-algebra $A$. A *GNS-construction* for $\phi$ is a triple $(H_\phi, \Lambda_\phi, \pi_\phi)$ consisting of a Hilbert space $H_\phi$, a linear map $\Lambda_\phi \colon \mathcal{N}_\phi \to H_\phi$ with dense image, and a representation $\pi_\phi \colon A \to \mathcal{L}(H_\phi)$ such that for all $a, b \in \mathcal{N}_\phi$ and $c \in A$,

$$\langle \Lambda_\phi(b) | \Lambda_\phi(a) \rangle = \phi(b^*a) \quad \text{and} \quad \pi_\phi(c)\Lambda_\phi(b) = \Lambda_\phi(cb).$$

It is easy to see that for each weight, there exists a GNS-construction, and that this construction is unique up to a unitary transformation. In general, the GNS-map $\Lambda_\phi$ of a weight $\phi$ is unbounded, but one still has some control:

**Proposition 8.2.3.** i) *Let $\phi$ be a lower semi-continuous weight a $C^*$-algebra with GNS-construction $(H_\phi, \Lambda_\phi, \pi_\phi)$. Then the map $\Lambda_\phi$ is closed with respect to the norm, and the representation $\pi_\phi$ is non-degenerate.*

ii) *Let $\phi$ be a normal and semi-finite weight on a von Neumann algebra $M$ with GNS-construction $(H_\phi, \Lambda_\phi, \pi_\phi)$. Then the map $\Lambda_\phi$ is closed with respect to the $\sigma$-weak topology on $M$ and the weak topology on $H_\phi$, and the representation $\pi_\phi$ is normal and unital.*

The Tomita–Takesaki theory [147] provides a detailed description of the GNS-construction of an n.s.f. weight:

**Theorem 8.2.4.** *Let $\phi$ be an n.s.f. weight on a von Neumann algebra $M$ with GNS-construction $(H_\phi, \Lambda_\phi, \pi_\phi)$ and modular automorphism group $\sigma$.*

i) *There exists a unique closed conjugate-linear operator $T$ on $H_\phi$ such that $\Lambda_\phi(\mathcal{N}_\phi \cap \mathcal{N}_\phi^*)$ is a core for $T$ and $T\Lambda_\phi(x) = \Lambda_\phi(x^*)$ for all $x \in \mathcal{N}_\phi \cap \mathcal{N}_\phi^*$.*

ii) *The operator $\nabla := T^*T$ is strictly positive. There exists a unique antiunitary $J$ on $H_\phi$ such that $T = J\nabla^{1/2}$. One has $J = J^*$, $J^2 = 1$, and $J\nabla^{it}J = \nabla^{it}$, $J\nabla^t J = \nabla^{-t}$ for all $t \in \mathbb{R}$.*

iii) $J\pi_\phi(M)J = \pi_\phi(M)'$.

iv) $\Lambda_\phi(\sigma_t(x)) = \nabla^{it}\Lambda_\phi(x)$ *for all* $x \in \mathcal{N}_\phi$ *and* $t \in \mathbb{R}$.

v) $J\Lambda_\phi(x) = \Lambda_\phi(\sigma_{i/2}(x)^*)$ *for all* $x \in \mathcal{N}_\phi \cap \mathrm{Dom}(\sigma_{i/2})$.

vi) *If* $x \in \mathcal{N}_\phi$ *and* $y \in \mathrm{Dom}(\sigma_{i/2})$, *then* $xy \in \mathcal{N}_\phi$ *and*

$$\Lambda_\phi(xy) = J\pi_\phi(\sigma_{i/2}(y))^* J\Lambda_\phi(x).$$

For KMS-weights on $C^*$-algebras, the following partial analogue holds:

**Theorem 8.2.5.** *Let $\phi$ be a KMS-weight on a $C^*$-algebra $A$ with GNS-construction $(H_\phi, \Lambda_\phi, \pi_\phi)$ and modular automorphism group $\sigma$. Then conditions* i)–ii) *and* iv)–vi) *of Theorem* 8.2.4 *hold, and* $J\pi_\phi(A)J \subseteq \pi_\phi(A)'$.

The operators $J$ and $\nabla$ above are called the *modular conjugation* and the *modular operator* of $\phi$, respectively. Usually, the modular operator is denoted by the symbol $\Delta$, which we reserve for the comultiplication.

## 8.3 Main properties

For every locally compact quantum group, one can construct:

- a *counit* in the form of a densely defined unbounded operator; since it only plays a minor rôle in the theory, we will not discuss it;
- an *antipode* in the form of a densely defined unbounded operator and a polar decomposition of this antipode into a *scaling group* and a *unitary antipode*;
- a *modular element* that relates the right to the left Haar weight;
- a *dual locally compact quantum group* – this generalizes Pontrjagin duality to the class of all locally compact quantum groups;
- a *manageable multiplicative unitary* which plays a fundamental rôle for the development of the theory.

Furthermore, one can associate to every locally compact quantum group in the setting of von Neumann algebras a reduced $C^*$-algebraic quantum group and vice versa. These transitions establish a bijective correspondence between the two classes of bialgebras, see also Section 4.3. We focus on reduced $C^*$-algebraic quantum groups; the setting of von Neumann algebras is very similar.

All constructions discussed in this section are technical and involved; therefore we only indicate some main ideas. The interested reader who is willing to spend some time and energy is referred to the original article [91] and to the thesis [158] for proofs and details.

Throughout this section, let $(A, \Delta)$ be a reduced $C^*$-algebraic quantum group with left Haar weight $\phi$ and right Haar weight $\psi$. For motivation, we shall frequently consider an algebraic quantum group $(A_0, \Delta_0)$ with left integral $\phi_0$ and right integral $\psi_0$. Let us stress that the algebraic quantum group and the reduced $C^*$-algebraic quantum group are not assumed to be related in any way. As before, we denote the algebraic tensor product by "$\odot$".

### 8.3.1 The multiplicative unitary

The multiplicative unitary of a locally compact quantum group plays a central rôle in the theory. Most importantly, it facilitates the transition between the setting of $C^*$-algebras and the setting of von Neumann algebras, and the construction of the dual of a locally compact quantum group.

To motivate the construction of the multiplicative unitary, we recall the definition of the multiplicative unitary $W_{A_0}$ of an algebraic quantum group $(A_0, \Delta_0)$ (Example 7.1.6). The underlying Hilbert space of $W_{A_0}$ is the GNS-space for the left integral on $(A_0, \Delta_0)$, and the adjoint $(W_{A_0})^*$ is given by

$$a \odot b \mapsto \Delta_0(b)(a \odot 1), \quad \text{where } a, b \in A_0. \tag{8.1}$$

How can this construction be adapted to the reduced $C^*$-algebraic quantum group $(A, \Delta)$? The underlying Hilbert space should now be the GNS-space $H_\phi$ of the left Haar weight $\phi$ of $(A, \Delta)$, and formula (8.1) should be rewritten as

$$\Lambda_\phi(a) \otimes \Lambda_\phi(b) \mapsto (\Lambda_\phi \otimes \Lambda_\phi)(\Delta(b)(a \otimes 1)), \quad \text{where } a, b \in \mathcal{N}_\phi.$$

To make sense of the right-hand side, we need to extend the map

$$\Lambda_\phi \odot \Lambda_\phi \colon \mathcal{N}_\phi \odot \mathcal{N}_\phi \to H_\phi \odot H_\phi \subseteq H_\phi \otimes H_\phi$$

to the subspace $\Delta(\mathcal{N}_\phi)(\mathcal{N}_\phi \otimes 1)$ of $A \otimes A$. This can be done as follows. Consider the weight $\phi \otimes \phi$ on $A \otimes A$. Evidently, $\mathcal{N}_\phi \odot \mathcal{N}_\phi \subseteq \mathcal{N}_{\phi\otimes\phi}$, and one can show that

- the map $\Lambda_\phi \odot \Lambda_\phi$ extends uniquely to a closed linear map $\Lambda_\phi \otimes \Lambda_\phi \colon \mathcal{N}_{\phi\otimes\phi} \to H_\phi \otimes H_\phi$ for which $\mathcal{N}_\phi \odot \mathcal{N}_\phi$ is a core;
- $\Delta(\mathcal{N}_\phi)(\mathcal{N}_\phi \otimes 1) \subseteq \mathcal{N}_{\phi\otimes\phi}$; this follows from the left-invariance of $\phi$.

**Theorem 8.3.1.** *Let $(A, \Delta)$ be a reduced $C^*$-algebraic quantum group with left Haar weight $\phi$ and associated GNS-construction $(H_\phi, \Lambda_\phi, \pi_\phi)$. Then there exists a multiplicative unitary $W$ on $H_\phi$ such that*

$$W^*(\Lambda_\phi(a) \otimes \Lambda_\phi(b)) = (\Lambda_\phi \otimes \Lambda_\phi)(\Delta(b)(a \otimes 1)) \quad \textit{for all } a, b \in \mathcal{N}_\phi.$$

*Proof.* The detailed proof is given in [91, Theorem 3.16, Proposition 3.18] and [158, Theorem 1.3.1]; be prepared for a long reading. Using the left-invariance of the left Haar weight $\phi$, it is not difficult to show that the formula above defines an isometry $W^*$. The hard part of the proof is to show that $W^*$ has dense image; this step involves the right Haar weight. The pentagon equation for $W$ follows easily from the coassociativity of the comultiplication $\Delta$. □

Recall that every multiplicative unitary gives rise to a left leg $(\widehat{A}(W), \widehat{\Delta}_W)$ and a right leg $(A(W), \Delta_W)$ (see Section 7.2).

**Proposition 8.3.2.** *The representation $\pi_\phi$ defines an isomorphism of the $C^*$-bialgebras $(A, \Delta)$ and $(\widehat{A}(W), \widehat{\Delta}_W)$:*

$$\pi_\phi(A) = \overline{\operatorname{span}}\{(\mathrm{id} \mathbin{\bar{\otimes}} \omega)(W) \mid \omega \in \mathcal{L}(H_\phi)_*\} = \widehat{A}(W),$$
$$(\pi_\phi \otimes \pi_\phi)(\Delta(a)) = W^*(1 \otimes \pi_\phi(a))W = \widehat{\Delta}_W(\pi_\phi(a)) \quad \textit{for all } a \in A.$$

*Proof.* See [91, Equation (4.2), Proposition 3.17] or [158, Proposition 1.3.4]. □

Later, we shall see that $W$ is manageable (Proposition 8.3.10) and that its right leg $(A(W), \Delta_W)$ is a reduced $C^*$-algebraic quantum group again, namely, the coopposite of the generalized Pontrjagin dual of $(A, \Delta)$ (Section 8.3.3).

### 8.3.2 The antipode and modular properties

At first sight, it may be difficult to guess which construction known from the theory of algebraic quantum groups can be used to define the antipode of a reduced $C^*$-algebraic quantum group. However, the antipode $S_0$ and the left integral $\phi_0$ of an algebraic quantum group $(A_0, \Delta_0)$ are related by the equation

$$S_0((\mathrm{id} \odot \phi_0)(\Delta_0(b^*)(1 \odot a))) = (\mathrm{id} \odot \phi_0)((1 \odot b^*)\Delta_0(a))$$

for all $a, b \in A_0$ (see Lemma 2.2.12).

To adapt this formula to the reduced $C^*$-algebraic quantum group $(A, \Delta)$, we use the slice map $\mathrm{id}_A \bar{\otimes} \phi$ defined on page 214. To apply this slice map to products of the form $\Delta(b^*)(1 \otimes a)$ and $(1 \otimes b^*)\Delta(a)$, we need to impose restrictions on $b$ and $a$. The following relations imply that it suffices to assume $a, b \in \mathcal{N}_\phi$:

$$\Delta(\mathcal{N}_\phi) \subseteq \bar{\mathcal{N}}_{\mathrm{id} \bar{\otimes} \phi}, \quad M(A) \odot \mathcal{N}_\phi \subseteq \bar{\mathcal{N}}_{\mathrm{id} \bar{\otimes} \phi}, \quad (\bar{\mathcal{N}}_{\mathrm{id} \bar{\otimes} \phi})^* \bar{\mathcal{N}}_{\mathrm{id} \bar{\otimes} \phi} \subseteq \bar{\mathcal{M}}_{\mathrm{id} \bar{\otimes} \phi}.$$

The first of these inclusions can be deduced from the left-invariance of $\phi$, and the second and third inclusion are obvious.

Alternatively, we can identify $(A, \Delta)$ with the left leg of the multiplicative unitary $W$ (Theorem 8.3.1) and define an antipode as in Proposition 7.2.16. Both approaches turn out to be equivalent:

**Theorem 8.3.3.** *Let $(A, \Delta)$ be a reduced $C^*$-algebraic quantum group with left Haar weight $\phi$, right Haar weight $\psi$, and multiplicative unitary $W$. There exists a unique closed densely defined linear map $S\colon \mathrm{Dom}(S) \subseteq A \to A$ that satisfies the following conditions:*

i) $\mathrm{span}\{(\mathrm{id} \otimes \phi)(\Delta(b^*)(1 \otimes a)) \mid a, b \in \mathcal{N}_\phi\} \subseteq A$ *is a core for $S$ and*
$$S((\mathrm{id} \otimes \phi)(\Delta(b^*)(1 \otimes a))) = (\mathrm{id} \otimes \phi)((1 \otimes b^*)\Delta(a)) \text{ for all } a, b \in \mathcal{N}_\phi.$$

ii) $\mathrm{span}\{(\psi \otimes \mathrm{id})((b^* \otimes 1)\Delta(a)) \mid a, b \in \mathcal{N}_\psi\} \subseteq A$ *is a core for $S$ and*
$$S((\psi \otimes \mathrm{id})((b^* \otimes 1)\Delta(a))) = (\psi \otimes \mathrm{id})(\Delta(b^*)(a \otimes 1)) \text{ for all } a, b \in \mathcal{N}_\psi.$$

iii) $\{(\mathrm{id} \,\bar{\otimes}\, \omega)(W) \mid \omega \in \mathcal{L}(H_\phi)_*\} \subseteq \pi_\phi(A) \cong A$ *is a core for $S$ and*
$$S((\mathrm{id} \,\bar{\otimes}\, \omega)(W)) = (\mathrm{id} \,\bar{\otimes}\, \omega)(W^*) \text{ for all } \omega \in \mathcal{L}(H_\phi)_*.$$

*Proof.* See [91, Proposition 5.24, Corollary 5.35, Proposition 8.3] or [158, Propositions 1.6.4, 1.6.17, 1.8.6]. □

The preceding theorem characterizes the antipode, but the actual construction and the polar decomposition of the antipode are given in the next theorem. The starting point of this construction is an operator $G$ on $H_\phi$ which, roughly, satisfies $G\Lambda_\phi(c) = \Lambda_\phi(S(c^*))$ for suitable $c \in \mathcal{N}_\phi$. To define $G$ without reference to the antipode $S$, we insert the equation of condition 8.3.3 ii) into the desired equation $G\Lambda_\phi(c) = \Lambda_\phi(S(c^*))$: for suitable $a, b \in \mathcal{N}_\psi$, the operator $G$ acts by
$$\Lambda_\phi((\psi \otimes \mathrm{id})(\Delta(b^*)(a \otimes 1))) \mapsto \Lambda_\phi((\psi \otimes \mathrm{id})(\Delta(a^*)(b \otimes 1))).$$
To be able to apply the GNS-map $\Lambda_\phi$ to the image of the slice map $\psi \otimes \mathrm{id}$ on both sides above, we need to impose restrictions on the elements $a$ and $b$. If $a, b \in \mathcal{N}_\phi^* \mathcal{N}_\psi$, then also $a, b \in \mathcal{N}_\psi$ because $\mathcal{N}_\psi$ is a left ideal, and hence the slice map $\psi \otimes \mathrm{id}$ can be applied in both sides of the equation above (see the discussion before Theorem 8.3.3). Using a Fubini-type theorem and left-invariance of $\phi$, one can furthermore show that in this case, the images $(\psi \otimes \mathrm{id})(\Delta(a^*)(b \otimes 1))$ and $(\psi \otimes \mathrm{id})(\Delta(b^*)(a \otimes 1))$ belong to $\mathcal{N}_\phi$.

**Theorem 8.3.4.** *Let $(A, \Delta)$ be a reduced $C^*$-algebraic quantum group with right Haar weight $\psi$ and left Haar weight $\phi$, and let $(H_\phi, \Lambda_\phi, \pi_\phi)$ be a GNS-construction for $\phi$.*

i) *There exists a unique closed densely defined conjugate-linear operator $G$ on $H_\phi$ such that* $\mathrm{span}\{\Lambda_\phi((\psi \otimes \mathrm{id})(\Delta(b^*)(a \otimes 1))) \mid a, b \in \mathcal{N}_\phi^* \mathcal{N}_\psi\} \subseteq H_\phi$ *is a core for $G$ and*
$$G\Lambda_\phi((\psi \otimes \mathrm{id})(\Delta(b^*)(a \otimes 1))) = \Lambda_\phi((\psi \otimes \mathrm{id})(\Delta(a^*)(b \otimes 1)))$$
*for all $a, b \in \mathcal{N}_\phi^* \mathcal{N}_\psi$. This operator satisfies $G^2 = \mathrm{id}$.*

ii) *The operator* $N := G^*G$ *is strictly positive. There exists a unique anti-unitary* $I$ *on* $H_\phi$ *such that* $G = IN^{1/2}$. *Moreover,* $I = I^*$, $I^2 = 1$, *and* $INI = N^{-1}$.

iii) *There exists a unique norm-continuous one-parameter group* $\tau$ *of* $*$*-automorphisms of* $A$ *such that* $\pi_\phi(\tau_t(a)) = N^{-it}\pi_\phi(a)N^{it}$ *for all* $t \in \mathbb{R}$, $a \in A$.

iv) *There exists a unique* $*$*-antiautomorphism* $R$ *of* $A$ *such that* $\pi_\phi(R(a)) = I\pi_\phi(a)^*I$ *for all* $a \in A$.

v) *The operator* $S := R\tau_{-i/2}$ *satisfies the conditions of Theorem* 8.3.3.

*Proof.* See [91, Propositions 3.22, 5.11, 5.20] or [158, Propositions 1.4.1, 1.4.3, 1.4.14]. □

Let us point out the parallel between the operators $N$, $I$ and the one-parameter group $\tau$ obtained above with the modular operator, the modular conjugation, and the modular automorphism group of a weight $\omega$ in Tomita–Takesaki theory: The former are constructed from the map $G$ in a similar way like the latter from the map $\Lambda_\omega(a) \mapsto \Lambda_\omega(a^*)$, where $a \in \mathcal{N}_\omega \cap \mathcal{N}_\omega^*$, and its polar decomposition.

**Definition 8.3.5.** The one-parameter group $\tau$ and the maps $R$ and $S$ are called the *scaling group*, the *unitary antipode*, and the *antipode* of $(A, \Delta)$, respectively.

**Remark 8.3.6.** We shall see in Proposition 8.3.10 that the multiplicative unitary $W$ of $(A, \Delta)$ is manageable. Therefore, an antipode for $(A, \Delta)$ and a polar decomposition can also be obtained from Theorem 7.3.19. Comparing Theorem 7.3.19 with Theorem 8.3.3 and 8.3.4, we see that the antipode and its polar decomposition constructed above differ from the antipode and the polar decomposition constructed in Theorem 7.3.19 only in the sign for the parameter of the scaling group. However, the constructions in this paragraph can not be avoided because they are used in the proof of manageability of $W$.

**Example 8.3.7.** Let us compute the scaling group and unitary antipode of a reduced $C^*$-algebraic compact quantum group $(A, \Delta)$. We shall use the same notation as in Example 8.1.22. In particular, $(A_0, \Delta_0)$ denotes the associated algebraic compact quantum group and $S_0$ its antipode.

First, we determine $G$. Evidently, $\Lambda(A_0) \subseteq \mathrm{Dom}(G)$. By Lemma 2.2.12,

$$(h \odot \mathrm{id})(\Delta(b^*)(a \odot 1)) = S_0((h \odot \mathrm{id})((b^* \odot 1)\Delta(a))) \quad \text{for all } a, b \in A_0,$$

and consequently $G\Lambda(a) = \Lambda(S_0(a^*))$ for all $a \in A_0$ as expected. Moreover, Remark 3.1.10 iii) implies that $GH_\alpha \subseteq H_\alpha$.

Let us determine $G^*$. Using the relation $h \circ S_0 = h$ (Proposition 2.2.6 ii)), we find

$$\begin{aligned}\langle \Lambda(a)|G\Lambda(b)\rangle = h(a^*S_0(b^*)) &= h(S_0(b^*S_0^{-1}(a^*))) \\ &= h(b^*S_0^{-1}(a^*)) = \langle \Lambda(b)|\Lambda(S_0^{-1}(a^*))\rangle\end{aligned}$$

for all $a, b \in A_0$. Since $G$ is conjugate-linear, the equation above implies $G^*\Lambda(a) = \Lambda(S_0^{-1}(a^*))$ for all $a \in A_0$.

Using the relation $* \circ S_0 \circ * = S_0^{-1}$ (Proposition 1.3.28), we find

$$N\Lambda(a) = G^*G\Lambda(a) = \Lambda(S_0^{-1}(S_0(a^*)^*)) = \Lambda(S_0^{-2}(a)) \text{ for all } a \in A_0.$$

For each $z \in \mathbb{C}$, define $\tau_z^0 \colon A_0 \to A_0$ by $a \mapsto f_{-iz} * a * f_{iz}$. Then $S_0^2 = \tau_{-i}^0$ (Theorem 3.2.19 iv)) and hence $N\Lambda(a) = \Lambda(\tau_i^0(a))$ for all $a \in A$. Since $\tau_z^0\tau_{z'}^0 = \tau_{z+z'}^0$ for all $z, z' \in \mathbb{C}$ (Corollary 3.2.20),

$$N^{-it}\Lambda(a) = \Lambda(\tau_t^0(a)) \quad \text{for all } a \in A_0,\ t \in \mathbb{R},$$

and since $\tau_z^0$ is an algebra automorphism for all $z \in \mathbb{C}$,

$$\pi(\tau_t(a)) = N^{-it}\pi(a)N^{it} = \pi(\tau_t^0(a)) = \pi(f_{-it} * a * f_{it})$$

for all $a \in A_0$, $t \in \mathbb{R}$. Moreover, the unitary antipode of $(A, \Delta)$ is given by

$$R(a) = S(\tau_{i/2}(a)) = S(f_{1/2} * a * f_{-1/2}) \quad \text{for all } a \in A_0.$$

By now we have associated quite a number of structure maps to the reduced $C^*$-algebraic quantum group $(A, \Delta)$. Evidently, it is useful to collect as many relations between these maps as possible. Let us denote by $\sigma^\phi$ and $\sigma^\psi$ the modular automorphism groups of $\phi$ and $\psi$, respectively.

**Proposition 8.3.8.** i) *$\tau$, $R$ and $S$ commute in the sense that $S \circ R = R \circ S$ and $\tau_t \circ R = R \circ \tau_t$, $\tau_t \circ S = S \circ \tau_t$ for all $t \in \mathbb{R}$;*

ii) *$\tau$, $\sigma^\phi$ and $\sigma^\psi$ commute in a similar sense as above;*

iii) *$R^2 = \mathrm{id}_A$ and $S^2 = \tau_{-i}$;*

iv) *$S$ is injective and $S^{-1} = \tau_{i/2} \circ R$;*

v) *for all $x, y \in \mathrm{Dom}(S)$, we have $xy \in \mathrm{Dom}(S)$ and $S(xy) = S(y)S(x)$;*

vi) *for all $x \in \mathrm{Dom}(S)$, we have $S(x)^* \in \mathrm{Dom}(S)$ and $S(S(x)^*)^* = x$;*

vii) *$\Delta \circ R = \Sigma \circ (R \otimes R) \circ \Delta$, where $\Sigma$ denotes the extension of the flip $a \otimes b \mapsto b \otimes a$;*

viii) *for all $t \in \mathbb{R}$,*

$$\Delta \circ \sigma_t^\phi = (\tau_t \otimes \sigma_t^\phi) \circ \Delta, \quad \Delta \circ \sigma_t^\psi = (\sigma_t^\psi \otimes \tau_{-t}) \circ \Delta,$$

$$(\tau_t \otimes \tau_t) \circ \Delta = \Delta \circ \tau_t = (\sigma_t^\phi \otimes \sigma_{-t}^\psi) \circ \Delta;$$

ix) *$\phi \circ R$ is a right Haar weight and $\psi \circ R$ a left Haar weight;*

x) *there exists a* scaling constant $\nu > 0$ *such that*

$$\phi \circ \tau_t = \nu^{-t}\phi, \quad \psi \circ \tau_t = \nu^{-t}\psi, \quad \phi \circ \sigma_t^\psi = \nu^t\phi, \quad \psi \circ \sigma_t^\phi = \nu^{-t}\psi.$$

*Proof.* See [91, Propositions 5.22, 5.23, 5.26, 6.8] or [158, Corollary 1.4.18, Propositions 1.4.20, 1.4.21, Theorem 1.8.1]. □

**Remark 8.3.9.** By statement ix) and Proposition 8.1.19, the right Haar weight of a reduced $C^*$-algebraic quantum group is faithful.

Like every locally compact group and every algebraic quantum group, the reduced $C^*$-algebraic quantum group $(A, \Delta)$ has a *modular element* $\delta$ which relates the left and the right Haar weight [91, Section 7], [158, Section 1.9]. This modular element is an unbounded multiplier of the $C^*$-algebra $A$, more precisely, an affiliated element (see Section 8.4.1). Roughly, $\delta$ is the Radon–Nikodym derivative of $\psi$ with respect to $\phi$, and $\psi = \phi(\delta^{1/2} \cdot \delta^{1/2})$. The following relations hold:

$$\Delta(\delta) = \delta \otimes \delta, \quad R(\delta) = \delta^{-1},$$

and for all $t \in \mathbb{R}$, $a \in A$,

$$\tau_t(\delta) = \delta, \quad \sigma_t^{\phi}(\delta) = \nu^t \delta = \sigma_t^{\psi}(\delta), \quad \sigma_t^{\psi}(a) = \delta^{it} \sigma_t^{\phi}(a) \delta^{-it};$$

see [91, Proposition 7.12] or [158, Proposition 1.9.11]. For the precise definition of the expressions above, see Section 8.4.1.

### 8.3.3 The duality of locally compact quantum groups

To every locally compact quantum group, one can associate a dual locally compact quantum group. Furthermore, one can show that the bidual – the dual of the dual – is naturally isomorphic to the initial locally compact quantum group. This is a far-reaching generalization of Pontrjagin duality which covers all locally compact groups. The most difficult step in this generalized Pontrjagin duality is the construction of the Haar weights on the dual of a locally compact quantum group. The bidual can then quite easily be identified via the associated multiplicative unitary. Let us turn to the details.

As a $C^*$-bialgebra, the dual of the reduced $C^*$-algebraic quantum group $(A, \Delta)$ is constructed out of the multiplicative unitary $W$ defined in Theorem 8.3.1: It is the pair $(\widehat{A}, \widehat{\Delta}) := (A(W), \Delta_W)^{\mathrm{cop}}$, in detail,

$$\widehat{A} := \overline{\operatorname{span}}\{(\omega \mathbin{\overline{\otimes}} \mathrm{id})(W) \mid \omega \in \mathcal{L}(H_\phi)_*\} = A(W) \tag{8.2}$$

and

$$\widehat{\Delta}(\hat{a}) := \Sigma W(\hat{a} \otimes 1) W^* \Sigma = (\mathrm{Ad}_\Sigma \circ \Delta_W)(\hat{a}) \quad \text{for all } \hat{a} \in \widehat{A}, \tag{8.3}$$

where $\Sigma \in \mathcal{L}(H \otimes H)$ denotes the flip $\eta \otimes \xi \mapsto \xi \otimes \eta$.

**Proposition 8.3.10.** *The multiplicative unitary $W$ is manageable.*

*Proof.* See [91, Proposition 6.10] or [158, Proposition 1.8.3]. The operators $Q$ and $\widehat{Q}$ figuring in Definition 7.3.16 are related to the scaling group $\tau$ and the scaling constant $\nu$ of $(A, \Delta)$: $Q = \widehat{Q} = P^{1/2}$, where $P$ is the strictly positive operator on $H$ defined by $P^{it}\Lambda_\phi(a) = \nu^{1/2}\Lambda_\phi(\tau_t(a))$ for all $a \in \mathcal{N}_\phi$ and $t \in \mathbb{R}$. □

**Corollary 8.3.11.** *$(\widehat{A}, \widehat{\Delta})$ is a bisimplifiable $C^*$-algebra.*

*Proof.* This follows from Proposition 8.3.10 and Theorem 7.3.18. □

To prove that the $C^*$-bialgebra $(\widehat{A}, \widehat{\Delta})$ is a reduced $C^*$-algebraic quantum group, we need to construct a left and a right Haar weight on it. It turns out to be easier to identify the associated GNS-constructions instead of the Haar weights themselves. These GNS-constructions can be obtained from the densely defined dual pairing of the $C^*$-algebras $A = \widehat{A}(W)$ and $\widehat{A} = A(W)$ introduced in Proposition 7.2.15 by an approach that is motivated by the Plancherel Theorem for algebraic quantum groups (Theorem 2.3.11).

Let us first illustrate the approach for an algebraic quantum group $(A_0, \Delta_0)$ with left integral $\phi_0$. Because of differing conventions, the left integral $\phi_0$ gives rise to a right integral $\hat{\psi}_0$ on the dual multiplier Hopf $*$-algebra $(\widehat{A}_0, \widehat{\Delta}_0)$, whereas the left Haar weight $\phi$ of $(A, \Delta)$ will give rise to a left Haar weight $\hat{\phi}$ on $(\widehat{A}, \widehat{\Delta})$.

The Plancherel Theorem 2.3.11 states that

$$\hat{\psi}_0(\phi_0(\cdot\, a_1)^*\phi_0(\cdot\, a_2)) = \phi_0(a_1^* a_2) \quad \text{for all } a_1, a_2 \in A_0.$$

We can rewrite this equation in terms of

- the GNS-map $\Lambda_{\phi_0} : A_0 \to H_{\phi_0}$ of $\phi_0$,
- the GNS-map $\Lambda_{\hat{\psi}_0} : \widehat{A}_0 \to H_{\hat{\psi}_0}$ of $\hat{\psi}_0$, and
- the natural pairing $(\cdot|\cdot) : \widehat{A}_0 \times A_0 \to \mathbb{C}$ given by $(\hat{a}, b) \mapsto \hat{a}(b)$

as follows. For $i = 1, 2$, put $\hat{a}_i := \phi_0(\cdot\, a_i)$ and $\xi_i := \Lambda_{\phi_0}(a_i)$. Then

$$\langle \Lambda_{\hat{\psi}_0}(\hat{a}_1)|\Lambda_{\hat{\psi}_0}(\hat{a}_2)\rangle = \hat{\psi}_0(\hat{a}_1^*\hat{a}_2) = \phi_0(a_1^* a_2) = \langle \Lambda_{\phi_0}(a_1)|\Lambda_{\phi_0}(a_2)\rangle = \langle \xi_1|\xi_2\rangle,$$

and for $i = 1, 2$, the vector $\xi_i$ can be characterized in terms of $\hat{a}_i$ by the relation

$$\langle \Lambda_{\phi_0}(b)|\xi_i\rangle = \langle \Lambda_{\phi_0}(b)|\Lambda_{\phi_0}(a_i)\rangle = \phi_0(b^* a_i) = (\hat{a}_i|b^*) \quad \text{for all } b \in A_0.$$

Thus, the GNS-map $\Lambda_{\hat{\psi}_0}$ can be constructed out of the dual pairing $\widehat{A}_0 \times A_0 \to \mathbb{C}$ and the GNS-map $\Lambda_{\phi_0}$ without knowledge of $\hat{\psi}_0$: put $H_{\hat{\psi}_0} := H_{\phi_0}$ and

$$\Lambda_{\hat{\psi}_0}(\hat{a}) := \xi \;\Leftrightarrow\; (\hat{a}|b^*) = \langle \Lambda_{\phi_0}(b)|\xi\rangle \quad \text{for all } b \in A_0.$$

Let us adapt this approach to the reduced $C^*$-algebraic quantum group $(A, \Delta)$. We define a dual pairing on a dense subspace of $\widehat{A} \times A$. Forget that $\widehat{A}_0$ denoted an algebraic quantum group before, and put $\widehat{A}_0 := \{(\omega \mathbin{\bar{\otimes}} \mathrm{id})(W) \mid \omega \in \mathcal{L}(H_\phi)_*\} \subseteq \widehat{A}$. Consider the pairing

$$(\cdot|\cdot]\colon \widehat{A}_0 \times A \to \mathbb{C}, \; ((\omega \mathbin{\bar{\otimes}} \mathrm{id})(W)|b] := \omega(\pi_\phi(b)).$$

This is just the dual pairing of $A_0(W) = \widehat{A}_0$ and $\widehat{A}(W) = \pi_\phi(A) \cong A$ defined in Proposition 7.2.15; in particular, it is well defined. Put

$$D_0(\widehat{\Lambda}) := \{\hat{a} \in \widehat{A}_0 \mid \text{there is } \xi \in H_\phi \text{ with } (\hat{a}|b^*] = \langle \Lambda_\phi(b)|\xi\rangle \text{ for all } b \in \mathcal{N}_\phi\},$$

and consider the map $\widehat{\Lambda}\colon D_0(\widehat{\Lambda}) \to H_\phi$ given by

$$\widehat{\Lambda}(\hat{a}) := \xi \;\Leftrightarrow\; (\hat{a}|b^*] = \langle \Lambda_\phi(b)|\xi\rangle \text{ for all } b \in \mathcal{N}_\phi.$$

**Theorem 8.3.12.** i) *The map $\widehat{\Lambda}$ extends uniquely to a closed densely defined linear map $\widehat{\Lambda}\colon \mathrm{Dom}(\widehat{\Lambda}) \subseteq \widehat{A} \to H_\phi$ for which $D_0(\widehat{\Lambda})$ is a core.*

ii) *There exists a unique KMS-weight $\hat{\phi}$ on $\widehat{A}$ such that $\mathcal{N}_{\hat{\phi}} = \mathrm{Dom}(\widehat{\Lambda})$ and $\hat{\phi}(a^*b) = \langle \widehat{\Lambda}(a)|\widehat{\Lambda}(b)\rangle_{H_\phi}$ for all $a, b \in \mathcal{N}_{\hat{\phi}}$. Moreover, there exists a unique representation $\pi_{\hat{\phi}}\colon \widehat{A} \to \mathcal{L}(H_\phi)$ such that $(H_\phi, \widehat{\Lambda}, \pi_{\hat{\phi}})$ is a GNS-construction for $\hat{\phi}$.*

iii) *The weight $\hat{\phi}$ is faithful and left-invariant with respect to $\widehat{\Delta}$.*

*Proof.* See [91, Propositions 8.13, 8.14, 8.15] or [158, Proposition 1.11.12, Theorems 1.11.13, 1.11.14]. □

A right Haar weight on $(\widehat{A}, \widehat{\Delta})$ can be constructed by a similar procedure or by means of the unitary antipode of $(A, \Delta)$. Thus, we arrive at the following important theorem.

**Theorem 8.3.13** ([91, Theorem 8.20], [158, Theorem 1.11.19]). *Let $(A, \Delta)$ be a reduced $C^*$-algebraic quantum group. Then the $C^*$-bialgebra $(\widehat{A}, \widehat{\Delta})$ defined by* (8.2) *and* (8.3) *is a reduced $C^*$-algebraic quantum group.*

**Definition 8.3.14.** The reduced $C^*$-algebraic quantum group $(\widehat{A}, \widehat{\Delta})$ is called the *reduced dual* of $(A, \Delta)$.

The structure maps of $(A, \Delta)$ and of $(\widehat{A}, \widehat{\Delta})$ are related by many equations, see [91, Section 8] or [158, Sections 1.11, 1.13].

The next theorem identifies the reduced dual $(\widehat{\widehat{A}}, \widehat{\widehat{\Delta}})$ of $(\widehat{A}, \widehat{\Delta})$. This result is a far-reaching generalization of the classical Pontrjagin duality of locally compact abelian groups and may be considered as the most important result in the theory of locally compact quantum groups.

**Theorem 8.3.15.** *Let $(A, \Delta)$ be a reduced $C^*$-algebraic quantum group. Then the $C^*$-bialgebras $(\hat{\hat{A}}, \hat{\hat{\Delta}})$ and $(A, \Delta)$ are isomorphic.*

In contrast to Theorem 8.3.13, Theorem 8.3.15 follows quite easily: One need not construct an explicit isomorphism between the $C^*$-bialgebras $(\hat{\hat{A}}, \hat{\hat{\Delta}})$ and $(A, \Delta)$, but only compares the associated multiplicative unitaries.

**Proposition 8.3.16.** *The multiplicative unitary $\widehat{W}$ associated to the reduced dual $(\hat{A}, \hat{\Delta})$ coincides with $\Sigma W^* \Sigma$.*

*Proof.* This result follows without much work from Theorem 8.3.12, see [91, Proposition 8.16] or [158, Proposition 1.11.15]. □

*Proof of Theorem* 8.3.15. By the previous proposition, the multiplicative unitary associated to $(\hat{\hat{A}}, \hat{\hat{\Delta}})$ is equal to $\hat{\hat{W}} = \Sigma \widehat{W}^* \Sigma = W$. By Proposition 8.3.2, the $C^*$-bialgebras $(\hat{\hat{A}}, \hat{\hat{\Delta}})$ and $(A, \Delta)$ can be identified with the left legs of $\hat{\hat{W}}$ and $W$, respectively, and therefore they are isomorphic. □

### 8.3.4 Passage between the different levels

The theory of locally compact quantum groups in the setting of von Neumann algebras is very similar to the theory of reduced $C^*$-algebraic quantum groups; both theories describe the same objects from different points of view. The equivalence of both theories can be considered as a generalization of Weil's theorem [190, Appendice] which establishes a bijective correspondence between measurable groups equipped with translation-invariant measures and locally compact groups. In the following section, we sketch how this correspondence extends to locally compact quantum groups and how one can pass back and forth between the setting of $C^*$-algebras and the setting of von Neumann algebras.

Let $(A, \Delta)$ be a reduced $C^*$-algebraic quantum group. We shall associate to $(A, \Delta)$ a locally compact quantum group $(\tilde{A}, \tilde{\Delta})$ in the setting of von Neumann algebras. As before, we denote by $(H_\phi, \Lambda_\phi, \pi_\phi)$ the GNS-construction for the left Haar weight $\phi$, and by $W$ the multiplicative unitary associated to $(A, \Delta)$ (Theorem 8.3.1). Since $W$ is manageable (Proposition 8.3.10), it is well-behaved (Theorem 7.3.18), and hence the pair $(\tilde{A}, \tilde{\Delta})$ given by

$$\tilde{A} := \pi_\phi(A)'' = \hat{A}_w(W), \qquad \tilde{\Delta} := \hat{\Delta}_W : x \mapsto W^*(1 \otimes x)W,$$

is a von Neumann bialgebra (Corollary 7.2.10). The left Haar weight $\phi$ of $(A, \Delta)$ can be extended to a left Haar weight $\tilde{\phi}$ on $(\tilde{A}, \tilde{\Delta})$ as follows. Put $\mathcal{F}_\phi = \{\omega \in A^*_+ \mid \omega(a) \leq \phi(a) \text{ for all } a \in A_+\}$. One shows:

- for every $\omega \in \mathcal{F}_\phi$, there exists a unique $\tilde{\omega} \in \mathcal{L}(H_\phi)_*$ such that $\tilde{\omega}(\pi_\phi(a)) = \omega(a)$ for all $a \in A$;

- the formula $\tilde{\phi}(x) = \sup_{\omega \in \mathcal{F}_\phi} \tilde{\omega}(x)$, $x \in \tilde{A}^+$, defines an n.s.f. weight on $\tilde{A}$ which is left-invariant with respect to $\tilde{\Delta}$.

A similar procedure, applied to the right Haar weight of $(A, \Delta)$, yields a right Haar weight on $(\tilde{A}, \tilde{\Delta})$. Thus, we arrive at the following result:

**Theorem 8.3.17** ([158, Section 1.14]). *The von Neumann bialgebra $(\tilde{A}, \tilde{\Delta})$ is a locally compact quantum group in the setting of von Neumann algebras.*

Conversely, let $(M, \Delta)$ be a locally compact quantum group in the setting of von Neumann algebras. We shall associate to $(M, \Delta)$ a reduced $C^*$-algebraic quantum group $(A, \Delta)$. This can be done by a similar procedure as above:

- identify $(M, \Delta)$ with the left leg of a multiplicative unitary $W$ in a similar way as in Theorem 8.3.1 and Proposition 8.3.2,
- take the $C^*$-bialgebra $(A, \Delta)$ corresponding to the left leg of the multiplicative unitary $W$, and
- restrict the Haar weights of $(M, \Delta)$ to $(A, \Delta)$.

Let us state the main steps precisely. Denote by $(H_\phi, \Lambda_\phi, \pi_\phi)$ the GNS-construction for the left Haar weight $\phi$ of $(M, \Delta)$.

**Theorem 8.3.18.** *There exists a multiplicative unitary $W \in \mathcal{L}(H_\phi \otimes H_\phi)$ such that $W^*(\Lambda_\phi(a) \otimes \Lambda_\phi(b)) = (\Lambda_\phi \otimes \Lambda_\phi)(\Delta(b)(a \otimes 1))$ for all $a, b \in \mathcal{N}_\phi$. This multiplicative unitary is manageable.*

*Proof.* See [93, Theorem 1.2] or [158, Theorem 1.14.2]; manageability is proved similarly as in the setting of $C^*$-algebras. □

Here, the right-hand side of the equation characterizing $W^*$ has to be defined carefully; this can be done as in the setting of $C^*$-algebras, see the discussion before Theorem 8.3.1. By Theorem 7.3.18, the unitary $W$ is well-behaved, whence the pair $(A, \Delta)$ given by

$$A := \overline{\operatorname{span}}\{(\omega \mathbin{\overline{\otimes}} \mathrm{id})(W) \mid \omega \in \mathcal{L}(H_\phi)_*\} = \widehat{A}(W),$$
$$\Delta := \widehat{\Delta}_W : a \mapsto W^*(1 \otimes a)W,$$

is a bisimplifiable $C^*$-bialgebra. Since $A$ is a $C^*$-subalgebra of $\pi_\phi(M) \cong M$, we can restrict the Haar weights of $(M, \Delta)$ from $\pi_\phi(M)^+$ to $A^+$.

**Theorem 8.3.19** ([93, Proposition 1.6], [158, Theorem 1.14.7] ). *The restriction of the left/right Haar weight of $(M, \Delta)$ from $\pi_\phi(M)^+$ to $A^+$ is a faithful KMS-weight which is left-invariant/right-invariant with respect to $\Delta$. In particular, the $C^*$-bialgebra $(A, \Delta)$ is a reduced $C^*$-algebraic quantum group.*

The transition from a locally compact quantum group in the setting of von Neumann algebras to the associated reduced $C^*$-algebraic quantum group and the reverse transition preserve the associated multiplicative unitaries. Thus we find:

**Theorem 8.3.20.** *Theorem* 8.3.17 *and* 8.3.19 *set up a bijective correspondence* (*up to isomorphism*) *between all locally compact quantum groups in the setting of von Neumann algebras and all reduced $C^*$-algebraic quantum groups.*

*Proof.* See [93, Section 1.2] or [158, Section 1.14]. □

## 8.4 Examples of locally compact quantum groups

Like the general theory of locally compact quantum groups, the construction of examples has a long history, is highly non-trivial, and often involves special techniques. The development of the general theory and the construction of examples happened in parallel, often influencing and stimulating each other.

Presently, the following examples are known. First, there exist two general constructions that produce new quantum groups out of given ones:

- the bicrossed product construction [7], [157], [9], [162], [10], [166], and
- the quantum double and the double crossed product construction [7], [110], [11].

In a purely algebraic framework, these constructions were introduced by Kac [74] and Drinfeld [37], respectively, and studied by many authors; see, for example, [107, Chapters 6, 7]. We shall not discuss these constructions here.

Second, there exist several quantum versions of classical groups, like

- the quantum group $E_\mu(2)$ [197], [198], [196], [180], [4], [5], [117], [118], [67],
- the quantum $az + b$ group [204], [127], [141], [178], [205], [124],
- the quantum $ax + b$ group [207], [126], [178], [205], [203], [135],
- the quantum group $\widetilde{\mathrm{SU}}(1,1)$ [81], [82],
- the quantum group $\mathrm{GL}_2(\mathbb{C})$ [125], [123].

The construction of these examples requires a lot of hard work. Therefore we shall only give an overview over the first two examples and refer to the original literature for proofs and further details.

Roughly, the construction of the examples listed above can be described by the following recipe:

1. Start with a classical group $G$ of matrices, look at the Hopf $*$-algebra of polynomial functions on $G$, and describe this $*$-algebra in terms of generators and relations.

2. Deform the relations by some complex number $\mu$, consider the $*$-algebra $A_0$ generated by the generators and the deformed relations, and try to define a comultiplication $\Delta_0\colon A_0 \to A_0 \odot A_0$ such that $(A_0, \Delta_0)$ becomes a Hopf $*$-algebra.
3. Represent the generators of $A_0$ by (possibly) unbounded closed operators on a Hilbert space $H$, or, equivalently, construct a (universal) $C^*$-algebra $A$ with generators (suitably interpreted) and relations as $A_0$.
4. Construct a comultiplication $\Delta$ on $A$ that agrees with $\Delta_0$ on the generators.
5. Find a left and a right Haar weight on $(A, \Delta)$.

Sometimes, one can proceed after Steps 1–3 as follows:

4′. Construct a Hopf $*$-algebra $(\widehat{A}_0, \widehat{\Delta}_0)$ that is dual to $(A_0, \Delta_0)$ by applying Steps 1–3 to a dual of the Hopf $*$-algebra of polynomial functions on $G$ (with the same parameter $\mu$ and Hilbert space $H$).

5′. Using a magic formula that involves special functions of the operators introduced in Steps 3 and 4′, construct a (manageable/modular) multiplicative unitary $W$ whose right (or left) leg is the desired $C^*$-bialgebra $(A, \Delta)$.

6′. Construct the Haar weights on $(A, \Delta)$ from the unitary $W$, see [205] and [67, Section 1.4].

The first approach was developed for $E_\mu(2)$ and $\widetilde{\mathrm{SU}}_\mu(1,1)$ in [197], [4] and [81], respectively. The second approach is particularly elegant; it was developed for the quantum $az+b$- and $ax+b$-group in [204], [127], [141] and [207], [126], [205], respectively, and chosen in [67] for $E_\mu(2)$.

Let us add several comments on the individual steps:

- The deformation of the relations in Step 2 can sometimes be motivated by the analysis of a natural coaction of $(A_0, \Delta_0)$, see, for example, the case $E_\mu(2)$ (Section 8.4.2).
- The unbounded operators constructed in Step 3 do (of course) not belong to the $C^*$-algebra $A$ and are only *affiliated* with it, see Section 8.4.1.
- The construction of the comultiplication $\Delta$ in Step 4 is often the most difficult step and requires a detailed analysis of tensor products of unbounded operators.
- Usually, the relations on the generators of $A_0$ obtained in Step 2 have to be complemented by additional *spectral conditions* depending on $\mu$.
- Steps 4, 5, and 5′ often involve advanced usage of *special functions*, which itself constitutes a nice but non-trivial piece of mathematics.

### 8.4.1 $C^*$-algebras generated by unbounded elements

In the following sections, we want to describe $C^*$-algebras of "functions on locally compact quantum groups that vanish at infinity" in terms of generators, thought of as "quantum coordinate functions", and relations. In general, these generators will be unbounded because the quantum groups are non-compact; in particular, they can not belong to the $C^*$-algebra itself. Therefore we have to discuss unbounded elements that are affiliated with a $C^*$-algebra and clarify in which sense such elements can generate a $C^*$-algebra.

**Affiliated elements of $C^*$-algebras.** The affiliation relation was introduced by Baaj [3] and Woronowicz [197]. We shall follow the treatment in [197] but omit most of the proofs.

Let us start with an informal motivation. Briefly, an element $T$ should be affiliated with a $C^*$-algebra $A$ if bounded continuous functions of $T$ belong to $M(A)$. In particular, we expect $z(T) \in M(A)$ for the function

$$z \colon \mathbb{C} \to \mathbb{C}, \quad \lambda \mapsto \lambda(1 + |\lambda|^2)^{-1/2}.$$

Since $1 - |z(\lambda)|^2 = (1 + |\lambda|^2)^{-1}$ for all $\lambda \in \mathbb{C}$, we should have

$$z(T) = T(1 - |z(T)|^2)^{1/2}.$$

**Definition 8.4.1.** Let $A$ be a $C^*$-algebra and $T\colon \operatorname{Dom}(T) \subseteq A \to A$ a densely defined linear map. We say that $T$ is *affiliated* with $A$ and write $T \;\eta\; A$ if there exists a $z_T \in M(A)$ such that $\|z_T\| \leq 1$ and for all $x, y \in A$,

$$x \in \operatorname{Dom}(T) \text{ and } y = Tx$$

$$\Leftrightarrow$$

$$\text{there exists } a \in A \text{ with } x = (1 - z_T^* z_T)^{1/2} a \text{ and } y = z_T a.$$

In this case, the element $z_T$, which is uniquely determined by $T$, is called the *$z$-transform* of $T$. The set of all affiliated elements of $A$ is denoted by $A^\eta$.

**Remarks 8.4.2.** i) Beware that the sum and the product of two affiliated elements of a $C^*$-algebra need not be densely defined.

ii) There exists the notion of an affiliated element of a von Neumann algebra which differs from the notion introduced above. If $M$ is a von Neumann algebra and $T$ is an affiliated element of $M$ in the $C^*$-algebraic sense, then $T \in M$ (see Proposition 8.4.3 iii)).

**Proposition 8.4.3.** *Let $A$ be a $C^*$-algebra and $T \;\eta\; A$.*

i) *$T$ is a closed linear map,* $\operatorname{Dom}(T) \subseteq A$ *is a right ideal, and* $T(ab) = T(a)b$ *for all* $a \in \operatorname{Dom}(T)$ *and* $b \in A$.

ii) *If* $\|T\| < \infty$*, then* $T \in M(A)$.

iii) *If* $A$ *is unital, then* $T \in A$.

*Proof.* Statement i) follows easily from the definition; for ii) and iii), see [197, Proposition 1.3]. □

**Example 8.4.4.** Let $X$ be a locally compact space. It is easy to see that for each $g \in C(X)$, pointwise multiplication of functions by $g$ defines an affiliated element of the $C^*$-algebra $C_0(X)$, and every affiliated element of the $C^*$-algebra $C_0(X)$ is obtained this way.

Non-degenerate $*$-homomorphisms can be extended to affiliated elements. The precise formulation of this result involves the concept of a core (see Section 8.2).

**Proposition 8.4.5** ([197, Theorem 1.2]). *Let* $A, B$ *be* $C^*$*-algebras,* $\phi\colon A \to M(B)$ *a non-degenerate* $*$*-homomorphism, and* $T \;\eta\; A$.

i) *There exists a* $\phi(T) \;\eta\; B$ *such that* $\phi(\mathrm{Dom}(T))B$ *is a core of* $\phi(T)$ *and* $\phi(T)(\phi(a)b) = \phi(Ta)b$ *for all* $a \in \mathrm{Dom}(T)$ *and* $b \in B$.

ii) *The* $z$*-transforms of* $T$ *and* $\phi(T)$ *satisfy* $z_{\phi(T)} = \phi(z_T)$.

iii) *If* $C$ *is a* $C^*$*-algebra and* $\psi\colon B \to M(C)$ *is a non-degenerate* $*$*-homomorphism, then* $(\psi \circ \phi)(T) = \psi(\phi(T))$.

Each affiliated element has an adjoint:

**Proposition 8.4.6** ([197, Theorem 1.4]). *Let* $A$ *be a* $C^*$*-algebra and* $T \;\eta\; A$.

i) *There exists a unique* $T^* \;\eta\; A$ *such that for all* $a, b \in A$,

$$a \in \mathrm{Dom}(T^*) \text{ and } b = T^*a \;\Leftrightarrow\; a^*(Tx) = b^*x \text{ for all } x \in \mathrm{Dom}(T).$$

ii) *The* $z$*-transforms of* $T$ *and* $T^*$ *satisfy* $z_{(T^*)} = (z_T)^*$.

iii) *If* $B$ *is a* $C^*$*-algebra and* $\phi\colon A \to M(B)$ *is a non-degenerate* $*$*-homomorphism, then* $\phi(T^*) = \phi(T)^*$.

We shall be interested primarily in affiliated elements that are normal in the following sense:

**Lemma 8.4.7** ([197, pages 406, 407]). *Let* $A$ *be a* $C^*$*-algebra and* $T \;\eta\; A$*. Then* $z_T$ *is normal if and only if* $\mathrm{Dom}(T) = \mathrm{Dom}(T^*)$ *and* $(Ta)^*(Ta) = (T^*a)^*(T^*a)$ *for all* $a \in \mathrm{Dom}(T)$.

**Definition 8.4.8.** Let $A$ be a $C^*$-algebra and $T \;\eta\; A$. We call $T$ *normal* if $z_T$ is normal.

Multiplication by the identify function $\mathrm{id}_{\mathbb{C}} \in C(\mathbb{C})$ defines a normal affiliated element $m$ of $C_0(\mathbb{C})$ (see Example 8.4.4) which is universal in the following sense:

**Proposition 8.4.9** ([197, Theorem 1.5]). *Let $A$ be a $C^*$-algebra and $T \,\eta\, A$. Then there exists a unique non-degenerate $*$-homomorphism $\phi_T \colon C_0(\mathbb{C}) \to M(A)$ such that $\phi_T(m) = T$.*

**Definition 8.4.10.** Let $A$ be a $C^*$-algebra, $T \,\eta\, A$ normal, and $\phi_T \colon C_0(\mathbb{C}) \to M(A)$ as above. Then the *spectrum* of $T$ is the set

$$\sigma(T) := \{\lambda \in \mathbb{C} \mid f(\lambda) = 0 \text{ for all } f \in \ker \phi_T\} \subseteq \mathbb{C}.$$

**Proposition 8.4.11** ([197, Theorem 1.6]). *Let $A$ be a $C^*$-algebra, $T \,\eta\, A$ normal, and define $m_T \,\eta\, C_0(\sigma(T))$ by $(m_T f)(\lambda) = \lambda f(\lambda)$. Then there exists a unique injective non-degenerate $*$-homomorphism $\psi_T \colon C_0(\sigma(T)) \to M(A)$ such that $\psi_T(m_T) = T$.*

Using the $*$-homomorphism $\psi_T$, we can define the functional calculus for $T$:

**Notation 8.4.12.** Let $A$ be a $C^*$-algebra, $T \,\eta\, A$ normal, and $\psi_T$ as above. For each $f \in C_b(\sigma(T))$, we define $f(T) \in M(A)$ by $f(T) := \psi_T(f)$.

To describe the action of the comultiplication on the generators of our quantum groups, we need to form tensor products of affiliated elements:

**Proposition 8.4.13** ([206, Theorem 6.1]). *Let $A, B$ be $C^*$-algebras and $S \,\eta\, A$, $T \,\eta\, B$. Then there exists a unique $S \otimes T \,\eta\, A \otimes B$ such that the algebraic tensor product* $\mathrm{Dom}(S) \odot \mathrm{Dom}(T) \subseteq A \otimes B$ *is a core for $S \otimes T$ and $(S \otimes T)(a \otimes b) = Sa \otimes Tb$ for all $a \in \mathrm{Dom}(S)$, $b \in \mathrm{Dom}(T)$. Moreover, $(S \otimes T)^* = S^* \otimes T^*$.*

For later use, we note the following simple result:

**Proposition 8.4.14** ([197, Proposition 0.2]). *Let $A$ be a $C^*$-algebra and $a \in M(A)$. If $aA$ and $a^*A$ are dense in $A$, then there exists a unique unitary $u \in M(A)$ such that $a = u|a|$.*

**$C^*$-algebras generated by affiliated elements.** Given a $C^*$-algebra $A$ and elements $T_1, \dots, T_n \in A$, it is clear what we mean when we say that $A$ is generated by $T_1, \dots, T_n$. When $T_1, \dots, T_n$ are no longer elements of the $C^*$-algebra $A$ but only affiliated with it, the interpretation of this statement is no longer obvious. Woronowicz introduced the following definition [200, Definition 3.1]:

**Definition 8.4.15.** Let $A$ be a $C^*$-algebra and $T_1, \dots, T_n \,\eta\, A$. We say that $A$ *is generated by* $T_1, \dots, T_n$ if for each Hilbert space $H$, each non-degenerate $C^*$-algebra $B \subseteq \mathcal{L}(H)$, and each non-degenerate $*$-homomorphism $\pi \colon A \to \mathcal{L}(H)$, the relation $\pi(T_1), \dots, \pi(T_n) \,\eta\, B$ implies $\overline{\pi(A)B} = B$, that is, $\pi$ restricts to a non-degenerate $*$-homomorphism $A \to M(B)$.

**Remark 8.4.16.** Given a $C^*$-algebra $A$ and elements $T_1, \dots, T_n \eta A$, it is not clear whether there exists a $C^*$-subalgebra $B \subseteq M(A)$ such that $T_1, \dots, T_n \eta B$ and such that $B$ is generated by $T_1, \dots, T_n$.

The following result illustrates the concept introduced above:

**Proposition 8.4.17.** *Let $A$, $B$ be $C^*$-algebras, $\phi\colon B \hookrightarrow M(A)$ a non-degenerate embedding, and $S_1, \dots, S_n \eta B$. If $\phi(S_1), \dots, \phi(S_n)$ generate $A$, then $\phi(B) = A$.*

*Proof.* Let $\pi$ be a faithful non-degenerate representation of $A$ on a Hilbert space $H$. Then $\pi(\phi(S_1)), \dots, \pi(\phi(S_n)) \eta \pi(\phi(B))$ and hence $\overline{\pi(A)\pi(\phi(B))} = \pi(\phi(B))$. On the other hand, $\overline{A\phi(B)} = A$. Since $\pi$ is faithful, the claim follows. □

If one wants to find out whether a $C^*$-algebra is generated by certain affiliated elements, the following criterion is useful:

**Proposition 8.4.18** ([200, Theorem 3.3]). *Let $A$ be a $C^*$-algebra, $T_1, \dots, T_n \eta A$, and assume that*

i) *$T_1, \dots, T_n$ separate the representations of $A$: if $\pi_1$ and $\pi_2$ are non-degenerate representations of $A$ on a Hilbert space $H$, then $\pi_1 = \pi_2$ if and only if $\pi_1(T_j) = \pi_2(T_j)$ for $j = 1, \dots, n$;*

ii) *there exist $r_1, \dots, r_k \in \{(1 + T_j^* T_j)^{-1}, (1 + T_j T_j^*)^{-1} \mid j = 1, \dots, n\} \subseteq M(A)$ such that $r_1 \dots r_k \in A$.*

*Then $A$ is generated by $T_1, \dots, T_n$.*

The following examples illustrate this criterion in simple cases:

**Examples 8.4.19.** i) Let $A$ be a unital $C^*$-algebra and $T_1, \dots, T_n \in A$. It is easy to see that if the space of all $*$-polynomials in 1 and $T_1, \dots, T_n$ is dense in $A$, then the conditions of Proposition 8.4.18 hold and $A$ is generated by $T_1, \dots, T_n$ in the sense of Definition 8.4.15.

Conversely, if $A$ is a $C^*$-algebra generated by elements $T_1, \dots, T_n \eta A$ and $\|T_1\|, \dots, \|T_n\| < \infty$, then $A$ is unital, $T_1, \dots, T_n \in A$, and the space of all $*$-polynomials in 1 and $T_1, \dots, T_n$ is dense in $A$. Indeed, in that case, $T_1, \dots, T_n \in M(A)$ (Proposition 8.4.3 ii)), and if $B \subseteq M(A)$ denotes the closure of the space of all $*$-polynomials in 1 and $T_1, \dots, T_n$, then $B = A$ by Proposition 8.4.17.

ii) Let $X$ be a locally compact space and $T_1, \dots, T_n \in C(X)$. Then the $C^*$-algebra $C_0(X)$ is generated by $T_1, \dots, T_n$ if and only if $T_1, \dots, T_n$ separate the points of $X$ and $\lim_{x \to \infty} \sum_{i=1}^n |T_i(x)|^2 = \infty$. The "if" part follows easily from Proposition 8.4.18; for the "only if" part, see [200, Example 2 after Theorem 3.3].

Further results and applications related to the presentation of $C^*$-algebras in terms of generators and relations can be found in [200].

## 8.4.2 The quantum groups $E_\mu(2)$ and $\widehat{E}_\mu(2)$

The quantum group $E_\mu(2)$ was constructed by Woronowicz via the first approach outlined in the introduction to this section: He starts from a double cover of the group of motions of the Euclidean plane, deforms the associated Hopf $*$-algebra of polynomials functions, and constructs a $C^*$-bialgebra $E_\mu(2)$ with similar generators and relations like that Hopf $*$-algebra. These relations are complemented by additional spectral conditions on the generators which ensure that the comultiplication is well defined [197].

Woronowicz also studied the corepresentation theory of the $C^*$-bialgebra $E_\mu(2)$, constructed the Pontrjagin dual $\widehat{E}_\mu(2)$ as a $C^*$-bialgebra [196], and studied its corepresentation theory in a joint work with Van Daele [180].

Another fundamental contribution was made by Baaj, who found the Haar weights on the $C^*$-bialgebras $E_\mu(2)$ and $\widehat{E}_\mu(2)$ [4], [5]; see also [118]. In particular, his results imply that $E_\mu(2)$ and $\widehat{E}_\mu(2)$ are locally compact quantum groups. Moreover, Baaj studied the multiplicative unitary associated to $E_\mu(2)$ and showed that this unitary is not regular but only semi-regular.

A comprehensive description of the locally compact quantum groups $E_\mu(2)$ and $\widehat{E}_\mu(2)$ was finally given by Jacobs in his PhD thesis [67].[1]

Let us add that the quantum groups $E_\mu(2)$ and $\mathrm{SU}_\mu(2)$ are related via a contraction procedure studied in [117], [199].

Throughout this subsection, the parameter $\mu$ is an element of $(0, 1)$.

**The classical group, its Hopf $*$-algebra, and the deformed Hopf $*$-algebra.** Consider the group $E(2) \subset \mathrm{GL}_2(\mathbb{C})$ consisting of all matrices of the form

$$g_{(v,n)} = \begin{pmatrix} v & n \\ 0 & \bar{v} \end{pmatrix}, \qquad \text{where } v \in \mathbb{T},\ n \in \mathbb{C}.$$

$E(2)$ acts on $\mathbb{C}$ via $g_{(v,n)}z := v^2 z + vn$, and $\{g_{(1,0)}, g_{(-1,0)}\}$ is the kernel of this action. Thus, $E(2)$ is the unique connected double cover of the group of rotations and dilations of the Euclidean plane.

The $*$-subalgebra $A_0 \subseteq C(E(2))$ generated by the coordinate functions $g_{(v,n)} \mapsto v$ and $g_{(v,n)} \mapsto n$ is isomorphic to the universal unital commutative $*$-algebra generated by a unitary $v$ and an element $n$. It carries the structure of a Hopf $*$-algebra; the structure maps can be read off from equation (8.5) below for $\mu = 1$.

Denote by $A_{0,\mu}$ the universal unital $*$-algebra generated by elements $v, n$ such that

$$v \text{ is unitary}, \quad n \text{ is normal}, \quad vn = \mu nv. \tag{8.4}$$

[1] In [67], the generators $v, n$ of $E_\mu(2)$ are denoted by $c, d$.

Then $A_{0,\mu}$ can be equipped with the structure of a Hopf $*$-algebra, where

$$\left.\begin{aligned} \Delta_0(v) = v \odot v, \quad \Delta_0(n) = v \odot n + n \odot v^*, \\ \epsilon_0(v) = 1, \quad \epsilon_0(n) = 0, \\ S_0(v) = v^*, \quad S_0(v^*) = v, \quad S_0(n) = -n/\mu, \quad S_0(n^*) = -n^*/\mu. \end{aligned}\right\} \tag{8.5}$$

The deformation of $A_0$ into $A_{0,\mu}$ can be related to a natural deformation of the $*$-algebra $\mathbb{C}[z,z^*]$ into a "non-commutative quantum plane" $\mathbb{C}_\mu[z,z^*]$ as follows. For a Hopf $*$-algebra $(B_0, \Delta_{B_0})$, the analogue of a group action on a $*$-algebra $C$ is a left coaction on $C$, which is a $*$-homomorphism $\delta\colon C \to B_0 \odot C$ that satisfies $(\mathrm{id} \odot \delta) \circ \delta = (\Delta_{B_0} \odot \mathrm{id}) \circ \delta$ (see also Section 9.2). The action of $E(2)$ on $\mathbb{C}$ induces a (unital) right coaction of $A_0$ on $\mathbb{C}[z,z^*]$ via

$$z \mapsto v^2 \odot z + vn \odot 1. \tag{8.6}$$

Denote by $\mathbb{C}_\mu[z,z^*]$ the universal unital $*$-algebra generated by an element $z$ subject to the relation $z^*z = \mu^2 zz^*$. One can easily check that if $B_0$ is a $*$-algebra with a unitary $v$ and a normal element $n$, then formula (8.6) defines a $*$-homomorphism $B_0 \to B_0 \odot \mathbb{C}_\mu[z,z^*]$ if and only if $n^*nv = \mu^2 vn^*n$ and $nv^2 = \mu^2 v^2 n$. In particular, formula (8.6) defines a right coaction of $(A_{0,\mu}, \Delta_0)$ on $\mathbb{C}_\mu[z,z^*]$.

**Commutation relations, spectral conditions, and the $C^*$-bialgebra $E_\mu(2)$.** The following discussion involves actions of groups on $C^*$-algebras and associated crossed products, which are reviewed in Section 9.1.

Assume that $H$ is a Hilbert space and that $v, n \,\eta\, \mathcal{L}(H)$ satisfy the relations in (8.4). Then

$$vf(n)v^* = f(vnv^*) = f(\mu n) \quad \text{for all } f \in C_0(\mathbb{C}), \tag{8.7}$$

whence $\mu\sigma(n) = \sigma(n)$. Given a subset $Y \subseteq \mathbb{C}$ satisfying $\mu Y = Y$, denote by $\alpha\colon \mathbb{Z} \to \mathrm{Aut}(C_0(Y))$ the action given by $\alpha_k(f) = f(\mu^k \cdot)$. Equation (8.7) and the universal property of the crossed product imply that there exists a $*$-homomorphism

$$\rho\colon C_0(\sigma(n)) \rtimes_\alpha \mathbb{Z} \to \mathcal{L}(H), \quad f \rtimes U_k \mapsto f(n)v^k.$$

These considerations show that a universal $C^*$-algebra $A$ with elements $v, n \,\eta\, A$ satisfying (8.4) can be constructed as follows. Denote by $m \,\eta\, C_0(\mathbb{C})$ the multiplication operator given by $(mf)(z) = zf(z)$, and put

$$A := C_0(\mathbb{C}) \rtimes_\alpha \mathbb{Z}, \quad v := 1 \rtimes U_1 \in M(A), \quad n := m \rtimes 1 \,\eta\, A;$$

here, $(\cdot) \rtimes 1\colon C_0(\mathbb{C}) \hookrightarrow A$, $f \mapsto f \rtimes 1$, denotes the natural inclusion.

Unfortunately, $A$ can *not* be equipped with a comultiplication $\Delta$ such that

$$\Delta(v) = v \otimes v \quad \text{and} \quad \Delta(n) = v \otimes n + n \otimes v^*.$$

This fundamental fact was discovered by Woronowicz [197]. The precise statement of his result, which is given below, involves irreducible representations of $A$. Standard arguments show that each irreducible representation $\pi$ of $A$ on a Hilbert space $H$ has one of the following two forms:

i) $\sigma(\pi(n)) = 0$, $\dim H = 1$, and $\pi(v) = z$ for some $z \in \mathbb{T}$; or

ii) $\sigma(\pi(n)) = t\mu^{\mathbb{Z}}$ for a non-zero $t \in \mathbb{C}$, and there exists an orthonormal basis $(e_k)_k$ such that $\pi(n)e_k = t\mu^k e_k$ and $\pi(v)e_k = e_{k+1}$ for all $k \in \mathbb{N}$.

**Theorem 8.4.20** ([197, Theorem 3.1]). *Let $\pi_1$ and $\pi_2$ be infinite-dimensional irreducible representations of $A$. Put $N := \pi_1(v) \otimes \pi_2(n) + \pi_1(n) \otimes \pi_2(v)^*$ and $\Sigma_i := \sigma(\pi_i(n))\mathbb{T}$, where $i = 1, 2$.*

i) *If $\Sigma_1 \neq \Sigma_2$, then $N$ is closed but not normal and has no normal extension.*

ii) *If $\Sigma_1 = \Sigma_2$, then $N$ is closeable, its closure $\widetilde{N}$ is normal, and $\sigma(\widetilde{N}) = \Sigma_1$.*

The theorem above implies that the desired comultiplication can only be defined if we supplement the defining relations (8.4) by a restriction on the spectrum $\sigma(n)$, which amounts to replacing $A$ by the quotient $\sigma(n) \rtimes_\alpha \mathbb{Z}$. The usual choice is

$$\sigma(n) \subseteq \overline{\Gamma}_\mu, \quad \text{where } \Gamma_\mu := \{\mu^k z \mid k \in \mathbb{Z}, z \in \mathbb{T}\} \subseteq \mathbb{C}. \tag{8.8}$$

Denote by $m \,\eta\, C_0(\overline{\Gamma}_\mu)$ the multiplication operator $(mf)(z) = zf(z)$. Then similar arguments as above show that

$$E_\mu(2) := C_0(\overline{\Gamma}_\mu) \rtimes_\alpha \mathbb{Z}, \quad v := 1 \rtimes U_1 \in M(E_\mu(2)), \quad n := m \rtimes 1 \,\eta\, E_\mu(2)$$

is the universal example of a $C^*$-algebra with affiliated elements $v, n$ satisfying the algebraic relations (8.4) and the spectral condition (8.8). Moreover, $E_\mu(2)$ is generated by $v, n$ in the sense of Definition 8.4.15, as can easily be seen from Proposition 8.4.18. Note that $f \rtimes U_k = f(n)v^k$ for all $f \in C_0(\overline{\Gamma}_\mu)$ and $k \in \mathbb{Z}$.

**Theorem 8.4.21** ([197, Theorems 3.3, 3.4]). *There exists a unique non-degenerate $*$-homomorphism $\Delta\colon E_\mu(2) \to M(E_\mu(2) \otimes E_\mu(2))$ such that $\Delta(v) = v \otimes v$ and $\Delta(n) = v \otimes n + n \otimes v^*$. Moreover, $(E_\mu(2), \Delta)$ is a $C^*$-bialgebra.*

**The Haar weight and the antipode of $E_\mu(2)$.** A left- and right-invariant Haar weight on the $C^*$-bialgebra $(E_\mu(2), \Delta)$ was found by Baaj [4], [5]; we briefly outline his construction.

There exists a positive contraction

$$\rho\colon E_\mu(2) = C_0(\overline{\Gamma}_\mu) \rtimes_\alpha \mathbb{Z} \to C_0(\overline{\Gamma}_\mu), \quad f \rtimes U_k \mapsto f\delta_{k,0}.$$

Indeed, if we put $C := C_0(\overline{\Gamma}_\mu)$ and denote by $(e_k)_k$ the standard basis of $l^2(\mathbb{Z})$ and by $j_0 \in \mathcal{L}_C(C, l^2(\mathbb{Z}) \otimes C)$ the map $c \mapsto e_0 \otimes c$, then $\rho$ coincides with the map $C \rtimes_\alpha \mathbb{Z} = C \rtimes_{\alpha,r} \mathbb{Z} \subseteq \mathcal{L}_C(l^2(\mathbb{Z}) \otimes C) \to \mathcal{L}_C(C), T \mapsto j_0^* T j_0$.

Define a weight $\nu$ on $C_0(\overline{\Gamma}_\mu)$ by

$$\nu(f) := \sum_{n\in\mathbb{Z}} \mu^{2n} \int_0^1 f(\mu^n e^{2\pi i t})dt \quad \text{for } f \in C_0(\overline{\Gamma}_\mu),\ f \geq 0.$$

**Theorem 8.4.22.** *The composition $\phi := \nu \circ \rho$ is a left and a right Haar weight on $(E_\mu(2), \Delta)$, and $(E_\mu(2), \Delta)$ is a unimodular locally compact quantum group. The modular automorphism group $\sigma$ of $\phi$ is uniquely determined by $\sigma_t(v) = \mu^{-2it} v$ and $\sigma_t(n) = n$ for all $t \in \mathbb{R}$.*

*Proof.* The assertions concerning $\phi$ and $(\sigma_t)_t$ are proved in [5, Proposition 4.1, Théorème 4.2], see also [67, Section 2.7]. The fact that $(E_\mu(2), \Delta)$ is a locally compact quantum group is proved in [67, Sections 2.4 and 2.5]. □

A short calculation shows that for each $x = \sum_k f_k(n)v^k \in E_\mu(2)$,

$$\phi(x^*x) = \sum_{k,n} \mu^{2n} \int_0^1 |f_k(\mu^{n-k} e^{2\pi i t})|^2 dt.$$

Next, we describe the unitary antipode, the scaling group, and the antipode of $(E_\mu(2), \Delta)$. For all $k, n \in \mathbb{Z}$, put $S_k := \{z \in \overline{\Gamma}_\mu \mid |z| = \mu^k\}$ and define $h_{k,n}\colon S_k \to \mathbb{C}$ by $z \mapsto z^n$. Put

$$\begin{aligned} D_1 &:= \{f \in C_0(\overline{\Gamma}_\mu) \mid f \text{ is constant on every } S_k\}, \\ D_2 &:= \{f \in C_0(\overline{\Gamma}_\mu) \mid \text{there exists } n \in \mathbb{Z} \text{ such that } f|_{S_k} = h_{k,n} \\ &\qquad\qquad \text{or } f|_{S_k} = 0 \text{ for all } k \in \mathbb{Z}\}. \end{aligned}$$

**Theorem 8.4.23.** i) *The unitary antipode $R$ of $(E_\mu(2), \Delta)$ is uniquely determined by $R(v) = v^*$ and $R(n) = -n$.*

ii) *The scaling group $\tau$ of $(E_\mu(2), \Delta)$ is uniquely determined by $\tau_t(v) = v$ and $\tau_t(n) = \mu^{-2it} n$ for all $t \in \mathbb{R}$.*

iii) *The $*$-subalgebra $D \subset E_\mu(2)$ generated by the set $D_1 \cup D_2$ is a core for $S$.*

*Proof.* See [67, Section 2.6]. □

Short calculations show that for each $x = \sum_k f_k(n)v^k \in E_\mu(2)$ and $t \in \mathbb{R}$,

$$R(x) = \sum_k f_k(-\mu^{-k}n)v^{-k}, \quad \tau_t(x) = \sum_k f_k(\mu^{-2it}n)v^k,$$

and if $x \in D$, then

$$S(x) = \sum_k f_k(-\mu^{-k-1}n)v^{-k}.$$

In particular, the antipode $S$ is unbounded, $S^2 \neq \mathrm{id}$, and $S \circ * \neq * \circ S$ [67, Proposition 2.6.32].

**The classical dual group, its Hopf $*$-algebra, and the deformed Hopf $*$-algebra.** The dual of the locally compact quantum group $(E_\mu(2), \Delta)$ turns out to be a deformation of the group $\widehat{E}(2) \subset \mathrm{GL}_2(\mathbb{C})$ that consists of all matrices of the form

$$\hat{g}_{(a,b)} = \begin{pmatrix} a^{-1} & 0 \\ b & a \end{pmatrix}, \quad \text{where } a \in (0,\infty),\ b \in \mathbb{C}.$$

The action of $\widehat{E}(2)$ on $\mathbb{C}$ given by $\hat{g}_{(a,b)}z := a^2z + ba$ identifies $\widehat{E}(2)$ with the group generated by all translations and dilations of the Euclidian plane.

Denote by $\widehat{E}(2)^{\pm} \subset \mathrm{GL}_2(\mathbb{C})$ the group consisting of all matrices $\hat{g}_{(a,b)}$ as above, but with $a \in \mathbb{R}\setminus\{0\}$ arbitrary. The $*$-subalgebra $\widehat{A}_0 \subset C(\widehat{E}(2)^{\pm})$ generated by the coordinate functions $\hat{g}_{(a,b)} \mapsto a$ and $\hat{g}_{(a,b)} \mapsto b$ is isomorphic to the universal unital commutative $*$-algebra generated by a self-adjoint invertible element $a$ and an element $b$, and carries the structure of a Hopf $*$-algebra, where

$$\left.\begin{aligned} \widehat{\Delta}_0(a) = a \odot a, \quad \widehat{\Delta}_0(b) = b \odot a^{-1} + a \odot b, \\ \hat{\epsilon}_0(a) = 1, \quad \hat{\epsilon}_0(b) = 0, \\ \widehat{S}_0(a) = a^{-1}, \quad \widehat{S}_0(a^*) = (a^*)^{-1}, \quad \widehat{S}_0(b) = -b, \quad \widehat{S}_0(b^*) = -b^*. \end{aligned}\right\} \tag{8.9}$$

Denote by $\widehat{A}_{0,\mu}$ the universal unital $*$-algebra generated by elements $a$ and $b$ satisfying

$$a \text{ is self-adjoint and invertible}, \quad b \text{ is normal}, \quad ab = \mu ba. \tag{8.10}$$

This is a Hopf $*$-algebra with respect to the comultiplication and counit given in (8.9); the antipode is given by

$$\widehat{S}_0(a) = a^{-1}, \quad \widehat{S}_0(a^*) = (a^*)^{-1}, \quad \widehat{S}_0(b) = -b/\mu, \quad \widehat{S}_0(b^*) = -b^*/\mu.$$

There exists a dual pairing $(\cdot|\cdot)$ between the Hopf $*$-algebras $(A_{0,\mu}, \Delta_0)$ and $(\widehat{A}_{0,\mu}, \widehat{\Delta}_0)$, given by

$$(v|a) = \mu^{1/2}, \quad (n|a) = 0, \quad (v|b) = 0, \quad (n|b) = 1.$$

**Commutation relations, spectral conditions, and the $C^*$-bialgebra $\widehat{E}_\mu(2)$.** When we shift from Hopf $*$-algebras to $C^*$-algebras, it is appropriate to replace (8.10) by

$$\begin{aligned} &a > 0 \text{ is self-adjoint}, && b \text{ is normal}, \\ &a, |b| \text{ strongly commute}, && ab = \mu ba; \end{aligned} \tag{8.11}$$

(recall that two self-adjoint operators are said to commute strongly if all of their spectral projections commute). Moreover, one imposes a spectral condition on the joint spectrum $\sigma(a, |b|)$:

$$\sigma(a, |b|) \subseteq \bar{\Omega}_\mu, \quad \text{where } \Omega_\mu := \{(\mu^s, \mu^r) \mid s \in \tfrac{1}{2}\mathbb{Z},\ r \in \tfrac{1}{2}\mathbb{Z},\ r + s \in \mathbb{Z}\}. \tag{8.12}$$

Let us note that in [180] and [196], the generator $a$ is replaced by the element $N = -2\log_\mu a$, so that $a = \mu^{-\frac{1}{2}N}$.

Assume that $H$ is a Hilbert space and that $a, b \;\eta\; \mathcal{L}(H)$ satisfy (8.11), (8.12). Let $b = u|b|$ be the polar decomposition, and put $H_0 := uH = (\ker b)^\perp$. Then

$$ug(a, |b|)u^* = g(uau^*, |b|) = g(\mu^{-1}a, |b|) \quad \text{in } \mathcal{L}(H_0) \text{ for each } g \in C_0(\Omega_\mu).$$

Denote by $\beta$ the action of $\mathbb{Z}$ on $C_0(\bar{\Omega}_\mu)$ given by $\beta_k(g)(x, y) = g(\mu^{-k}x, y)$, and observe that $\beta$ preserves $C_0(\Omega_\mu) \subseteq C_0(\bar{\Omega}_\mu)$. Put

$$\widehat{E}_\mu(2) := C_0(\Omega_\mu) \rtimes_\beta \mathbb{Z} + C_0(\bar{\Omega}_\mu) \rtimes_\beta 1 \subset C_0(\bar{\Omega}_\mu) \rtimes_\beta \mathbb{Z}.$$

The equation above and the universal property of the crossed product imply that there exists a $*$-homomorphism

$$C_0(\Omega_\mu) \rtimes_\beta \mathbb{Z} \to \mathcal{L}(H_0), \quad g \rtimes U_k \mapsto g(a, |b|)u^k,$$

which naturally extends to a $*$-homomorphism $\widehat{E}_2(\mu) \to \mathcal{L}(H)$.

Denote by $m_x, m_y \;\eta\; C_0(\bar{\Omega}_\mu)$ the multiplication operators given by

$$(m_x g)(x, y) := xg(x, y), \quad (m_y g)(x, y) := yg(x, y).$$

The preceding arguments show that

$$\widehat{E}_\mu(2) \text{ as above}, \quad a := m_x \rtimes 1 \;\eta\; \widehat{E}_\mu(2), \quad b := (1 \rtimes U_1)(m_y \rtimes 1) \;\eta\; \widehat{E}_\mu(2)$$

is the universal example of a $C^*$-algebra with affiliated elements $a, b$ satisfying the algebraic relations (8.11) and the spectral condition (8.12). Proposition 8.4.18 can be used to show that $\widehat{E}_\mu(2)$ is generated by $a, b$ in the sense of Definition 8.4.15.

**Theorem 8.4.24.** *There exists a unique non-degenerate $*$-homomorphism $\widehat{\Delta}\colon \widehat{E}_\mu(2) \to M(\widehat{E}_\mu(2) \otimes \widehat{E}_\mu(2))$ such that $\widehat{\Delta}(a) = a \otimes a$ and $\widehat{\Delta}(b) = b \otimes a^{-1} + a \otimes b$. $(\widehat{E}_\mu(2), \widehat{\Delta})$ is a $C^*$-bialgebra and as such isomorphic to the coopposite of the dual of the locally compact quantum group $(E_\mu(2), \Delta)$. In particular, $(\widehat{E}_\mu(2), \widehat{\Delta})$ is a locally compact quantum group.*

*Proof.* The existence of the comultiplication follows from results in [198]. The fact that $(\widehat{E}_\mu(2), \widehat{\Delta})^{\mathrm{cop}}$ is the Pontrjagin dual of $(E_\mu(2), \Delta)$ was first proved (not in the setting of locally compact quantum groups) in [196], see also [180]. For the proof of the last assertions, see [67, Sections 2.4, 2.5, 2.8]. □

**The Haar weights and the antipode of $\widehat{E}_\mu(2)$.** The Haar weights of $(\widehat{E}_\mu(2), \widehat{\Delta})$ were first constructed by Baaj [4], [5]. They can be described as follows. Similarly as for $E_\mu(2)$, there exists a positive contraction

$$\hat{\rho}\colon \widehat{E}_\mu(2) = C_0(\overline{\Omega}_\mu) \rtimes_\beta \mathbb{Z} \to C_0(\overline{\Omega}_\mu), \quad g \rtimes U_k \mapsto g\delta_{k,0}.$$

Define weights $\nu_{\hat{\phi}}$ and $\nu_{\hat{\psi}}$ on $C_0(\overline{\Omega}_\mu)$ by

$$\nu_{\hat{\phi}}(g) := \sum_{(\mu^s,\mu^r)\in\Omega_\mu} \mu^{r+s} g(\mu^s, \mu^r), \quad \nu_{\hat{\psi}}(g) := \sum_{(\mu^s,\mu^r)\in\Omega_\mu} \mu^{r-s} g(\mu^s, \mu^r).$$

**Theorem 8.4.25.** i) *The left Haar weight $\hat{\phi}$ and the right Haar weight $\hat{\psi}$ on $(\widehat{E}_\mu(2), \widehat{\Delta})$ are (up to a positive factor) given by $\hat{\phi} = \nu_{\hat{\phi}} \circ \hat{\rho}$ and $\hat{\psi} = \nu_{\hat{\psi}} \circ \hat{\rho}$.*

ii) *The modular automorphism groups $\sigma^{\hat{\phi}}$ and $\sigma^{\hat{\psi}}$ of $\hat{\phi}$ and $\hat{\psi}$, respectively, are uniquely determined by $\sigma_t^{\hat{\phi}}(a) = a$, $\sigma_t^{\hat{\phi}}(b) = \mu^{-2it} b$ and $\sigma_t^{\hat{\psi}}(a) = a$, $\sigma_t^{\hat{\psi}}(b) = \mu^{2it} b$ for all $t \in \mathbb{R}$.*

iii) *The modular element of $(\widehat{E}_\mu(2), \widehat{\Delta})$ is given by $\hat{\delta} = a^4$.*

*Proof.* See [5, Proposition 4.20, Théorème 4.21] or [67, Section 2.7]. □

A short calculation shows that for each $x = \sum_k g_k(a, |b|)u^k$,

$$\hat{\phi}(x^*x) = \sum_{k\in\mathbb{Z}} \sum_{(\mu^s,\mu^r)\in\Omega_\mu} \mu^{2(r-s+k)} |g_k(\mu^s, \mu^r)|^2,$$

$$\hat{\psi}(x^*x) = \sum_{k\in\mathbb{Z}} \sum_{(\mu^s,\mu^r)\in\Omega_\mu} \mu^{2(r+s-k)} |g_k(\mu^s, \mu^r)|^2.$$

The antipode of $(\widehat{E}_\mu(2), \widehat{\Delta})$ can be described as follows:

**Theorem 8.4.26.** i) *The unitary antipode* $\widehat{R}$ *of* $(\widehat{E}_\mu(2), \widehat{\Delta})$ *is uniquely determined by* $R(a) = a^{-1}$ *and* $R(b) = -b$.

ii) *The scaling group* $\hat{\tau}$ *of* $(\widehat{E}_\mu(2), \widehat{\Delta})$ *is uniquely determined by* $\hat{\tau}_t(a) = a$ *and* $\hat{\tau}_t(b) = \mu^{-2it} b$ *for all* $t \in \mathbb{R}$.

iii) *The* $*$*-algebra* $\widehat{D} \subset \widehat{E}_\mu(2)$ *generated by the set* $\{g(a, |b|)u^k,\ h(a, |b|) \mid g \in C_0(\Omega_\mu), k \in \mathbb{N} \setminus \{0\},\ h \in C_0(\overline{\Omega}_\mu)\}$ *is a core for* $\widehat{S}$.

*Proof.* See [67, Section 2.6]. □

Short calculations show that for each $x = \sum_k g_k(a, |b|)u^k \in \widehat{E}_\mu(2), t \in \mathbb{R}$,

$$\widehat{R}(x) = \sum_k g_k(\mu^k a^{-1}, |b|)(-u)^k, \quad \hat{\tau}_t(x) = \sum_k \mu^{-2itk} g_k(a, |b|)u^k,$$

and if $x \in \widehat{D}$, then

$$\widehat{S}(x) = \sum_k (-\mu)^{-k} g_k(\mu^k a^{-1}, |b|)u^k.$$

In particular, the antipode $\widehat{S}$ is unbounded, $\widehat{S}^2 \neq \mathrm{id}$, and $\widehat{S} \circ * \neq * \circ \widehat{S}$ [67, Proposition 2.6.58].

**The multiplicative unitary of $E_\mu(2)$ and $\widehat{E}_\mu(2)$.** We close this subsection with some brief remarks on the multiplicative unitary of $(E_\mu(2), \Delta)$. Denote by $H_\phi$ the GNS-space for the Haar weight $\phi$ on $(E_\mu(2), \Delta)$, and by $W \in \mathcal{L}(H_\phi \otimes H_\phi)$ the multiplicative unitary associated to $(E_\mu(2), \Delta)$ (see Theorem 8.3.1).

**Theorem 8.4.27.** *W is semi-regular but not regular.*

*Proof.* See [5, Proposition 4.7] or [67, Proposition 2.8.23]. □

Recall that we can identify $E_\mu(2) = \widehat{A}(W)$ and $\widehat{E}_\mu(2) = A(W)$ with $C^*$-subalgebras of $\mathcal{L}(H_\phi)$ and consider $W$ as an element of $M(E_\mu(2) \otimes \widehat{E}_\mu(2)) \subset \mathcal{L}(H_\phi \otimes H_\phi)$ (see Section 8.3 and Theorem 7.3.18). As such, $W$ can elegantly be described in terms of the generators $v, n \;\eta\; E_\mu(2)$ and $a, b \;\eta\; \widehat{E}_\mu(2)$. This description involves the *quantum exponential function* $F_\mu \colon \overline{\Gamma}_\mu \to \mathbb{C}$, defined by

$$F_\mu(z) = \begin{cases} \displaystyle\prod_{k=0}^{\infty} \frac{1 + \mu^{2k}\bar{z}}{1 + \mu^{2k}z}, & z \in \overline{\Gamma}_\mu \setminus \{-\mu^{-2k} \mid k \in \mathbb{N}\}, \\ -1, & z \in \{-\mu^{-2k} \mid k \in \mathbb{N}\}, \end{cases} \tag{8.13}$$

and the bicharacter $\chi_\mu \colon \mu^{\mathbb{Z}/2} \times \mathbb{T} \to \mathbb{T}$, defined by

$$\chi_\mu(\mu^{k/2}, z) := z^k \quad \text{for all } k \in \mathbb{Z},\ z \in \mathbb{T}. \tag{8.14}$$

Note that the product in (8.13) converges since $0 < \mu < 1$. These functions were introduced by Woronowicz; they play a central rôle in the study of the quantum groups $E_\mu(2)$ and $\widehat{E}_\mu(2)$ and of the associated multiplicative unitary. It is easy to see that

- $|F_\mu(z)| = 1$ and $F_\mu(\bar{z}) = \overline{F_\mu(z)}$ for all $z \in \Gamma_\mu$, and
- $F_\mu$ is continuous.

To explain the terminology "quantum exponentional function", we cite the following result; for further details, see [198].

**Theorem 8.4.28** ([198]). *Let $H$ be a Hilbert space and let $R$ and $S$ be normal operators on $H$ with polar decompositions $R = U|R|$, $S = V|S|$. Assume that*

i) $\sigma(R), \sigma(S) \subseteq \overline{\Gamma}_\mu$,

ii) *$V$ and $U$ commute,*

iii) *$|R|$ and $|S|$ strongly commute, that is, all of their spectral projections commute,*

iv) $U^*|S|U = \mu|S|$ *and* $V|R|V^* = \mu|R|$.

*Then $R + S$ admits a normal extension $R \dotplus S$, and $F_\mu(R \dotplus S) = F_\mu(R)F_\mu(S)$.*

The multiplicative unitary $W \in M(E_\mu(2) \otimes \widehat{E}_\mu(2))$ can be described in terms of the functions $F_\mu$, $\chi_\mu$ and the operators $vn \otimes ab, v \otimes 1, 1 \otimes a$ η $E_\mu(2) \otimes \widehat{E}_\mu(2)$ as follows:

**Lemma 8.4.29.** *$vn \otimes ab$ is normal and $\sigma(vn \otimes ab) \subseteq \overline{\Gamma}_\mu$.*

**Theorem 8.4.30.** *$W = F_\mu(vn \otimes ab)\chi_\mu(v \otimes 1, 1 \otimes a)$ in $M(E_\mu(2) \otimes \widehat{E}_\mu(2))$.*

*Proof.* See [67, Section 2.8.2, in particular Proposition 2.8.11, Lemma 2.8.18, Lemma 2.8.20]. □

A variation of the formula for $W$ given above first appeared in [196]. An instructive motivation for this formula can be found in [67, Section 2.1].

### 8.4.3 The quantum $az + b$ group

The quantum $az+b$ group was constructed by Woronowicz in [204] via the second approach outlined in the introduction to this section: He starts from the classical $az+b$ group, deforms the associated Hopf $*$-algebra of polynomial functions, and formulates an appropriate set of relations and spectral conditions on the generators of the quantum $az+b$ group. Here, the deformation parameter $\mu$ either is a root of unity or an element of $(0, 1)$. From a representation of the generators as operators on a Hilbert space, he directly constructs a manageable multiplicative unitary whose right leg is the quantum $az+b$ group (on the level of $C^*$-bialgebras). This construction involves an analogue of the quantum exponential function encountered in the study of $E_\mu(2)$ [198]. The Pontrjagin dual of the quantum $az+b$ group turns out to be isomorphic to the quantum $az+b$ group again.

The Haar weights on the quantum $az+b$ group were found by Van Daele [178], see also [127], [205]. He observed that in the root of unity case, the scaling constant is not equal to 1 – the quantum $az+b$ group was the first example for this phenomenon. In his PhD thesis [141], Soltan constructed the quantum $az+b$ group for new values of the deformation parameter $\mu$, extending the results of Woronowicz and Van Daele.

**The classical group, its Hopf $*$-algebra, and the deformed Hopf $*$-algebra.** The $az+b$ group is the group of affine transformations of the plane $\mathbb{C}$. It can identified with the subgroup $G \subset \mathrm{GL}_2(\mathbb{C})$ consisting of all matrices of the form

$$g_{(a,b)} := \begin{pmatrix} a & b \\ 0 & 1 \end{pmatrix}, \qquad \text{where } a, b \in \mathbb{C},\ a \neq 0;$$

the action on the plane $\mathbb{C}$ being given by $g_{(a,b)}z = az + b$.

The $*$-subalgebra $A_0 \subset C(G)$ generated by the coordinate functions $g_{(a,b)} \mapsto a$ and $g_{(a,b)} \mapsto b$ is isomorphic to the universal unital commutative $*$-algebra generated by an invertible element $a$ and an element $b$, and carries the structure of a Hopf $*$-algebra, where

$$\left.\begin{aligned} \Delta_0(a) = a \odot a, \quad \Delta_0(b) &= a \odot b + b \odot 1, \\ \epsilon_0(a) = 1, \quad \epsilon_0(b) &= 0, \\ S_0(a) = a^{-1}, \qquad S_0(b) &= -a^{-1}b, \\ S_0(a^*) = (a^*)^{-1}, \quad S_0(b^*) &= -(a^*)^{-1}b^*. \end{aligned}\right\} \tag{8.15}$$

Fix a non-zero $\mu \in \mathbb{C}$, and denote by $A_{0,\mu}$ the universal unital $*$-algebra generated by elements $a, b$ satisfying

$$a \text{ is normal and invertible}, \quad b \text{ is normal}, \quad ab = \mu^2 ba, \quad ab^* = b^*a. \tag{8.16}$$

The $*$-algebra $A_{0,\mu}$ can be equipped with the structure of a Hopf $*$-algebra, where the structure maps are given by equation (8.15) again. Moreover, there exists a dual pairing $(\cdot|\cdot)$ of the Hopf $*$-algebra $A_{0,\mu}$ with itself such that

$$(a|a) = \mu^2, \quad (b|a) = 0, \quad (a|b) = 0, \quad (b|b) = t,$$

where $t \in \mathbb{C}$ can be chosen arbitrarily.

**The commutation relations in the setting of $C^*$-algebras.** When we shift from Hopf $*$-algebras to $C^*$-algebras, it is appropriate to replace (8.16) by

$$\left.\begin{aligned} a, b \text{ are normal}, \quad \ker a &= \{0\}, \\ \mathrm{Phase}(a) b \mathrm{Phase}(a)^* &= \mu b, \\ |a|^{it} b |a|^{-it} &= \mu^{it} b \text{ for all } t \in \mathbb{R}. \end{aligned}\right\} \tag{8.17}$$

Here, $\mathrm{Phase}(a)$ denotes the unitary in the polar decomposition of $a$, see Proposition 8.4.14.

Assume that $H$ is a Hilbert space and that $a, b \,\eta\, \mathcal{L}(H)$ satisfy (8.17). For each pair $(k,t) \in \mathbb{Z} \times \mathbb{R}$, define a character $\theta_{(k,t)}$ on $\mathbb{C}_\times := \mathbb{C} \setminus \{0\}$ by

$$\theta_{(k,t)}(z) := \mathrm{Phase}(z)^k |z|^{it} \quad \text{for all } z \in \mathbb{C}_\times.$$

Note that the map $\mathbb{Z} \times \mathbb{R} \to \widehat{\mathbb{C}}_\times$, $(k,t) \mapsto \theta_{(k,t)}$, is a group isomorphism. Now, the last two relations in (8.17) imply

$$\theta_{(k,t)}(a) b \theta_{(k,t)}(a)^* = \mu^{k+it} b \quad \text{for all } (k,t) \in \mathbb{Z} \times \mathbb{R}. \tag{8.18}$$

As in the analysis of $E_\mu(2)$, we deduce the existence of an action $\gamma\colon \mathbb{Z} \times \mathbb{R} \to \mathrm{Aut}(C_0(\sigma(b)))$, $\gamma_{(k,t)}(f) = f(\mu^{k+it}\,\cdot)$, and of a $*$-homomorphism

$$M(C_0(\sigma(b)) \rtimes_\gamma (\mathbb{Z} \times \mathbb{R})) \to \mathcal{L}(H), \quad f \rtimes U_{(k,t)} \mapsto f(b)\theta_{(k,t)}(a).$$

Denote by $m_\times \,\eta\, C^*(\mathbb{Z} \times \mathbb{R})$ the element corresponding to the identity map on $\mathbb{C}_\times$ with respect to the isomorphism $C^*(\mathbb{Z} \times \mathbb{R}) \cong C_0(\mathbb{C}_\times)$ of Proposition 4.2.3, so that $\theta_{(k,t)}(m_\times) = U_{(k,t)}$ for all $(k,t) \in \mathbb{Z} \times \mathbb{R}$. Define $m \,\eta\, C_0(\mathbb{C})$ by $(mf)(z) = zf(z)$. Then the considerations above show that a universal $C^*$-algebra with affiliated elements $a, b$ satisfying (8.17) is given by

$$C_0(\mathbb{C}) \rtimes_\gamma (\mathbb{Z} \times \mathbb{R}), \quad a := 1 \rtimes m_\times, \quad b := m \rtimes 1. \tag{8.19}$$

**Spectral conditions, the group $\Gamma_\mu$, and the $C^*$-algebra $A$.** Similarly as in the case of the quantum group $E_\mu(2)$, the $C^*$-algebra in (8.19) can *not* be equipped with a comultiplication $\Delta$ such that

$$\Delta(a) = a \otimes a, \quad \Delta(b) = a \otimes b + b \otimes 1.$$

To remedy this defect, one imposes additional spectral conditions on the generators $a, b$ of the form

$$\sigma(a), \sigma(b) \subseteq \overline{\Gamma}_\mu, \quad \text{where } \Gamma_\mu := \{\mu^{k+it} \mid (k,t) \in \mathbb{Z} \times \mathbb{R}\} \subseteq \mathbb{C}. \tag{8.20}$$

The shape of $\Gamma_\mu$ and the analysis of the relations (8.17) and (8.20) depends very much on the value of $\mu$. The following cases have been studied:

*Case* (1). $\mu \in (0,1)$ ([204], [127], [178], [205]); here, $\Gamma_\mu$ is a union of circles whose radii form a geometric progression: $\Gamma_\mu = \bigcup_{k \in \mathbb{Z}} \mu^k \mathbb{T}$;

*Case* (2). $\mu = \exp(\frac{2\pi i}{N})$, where $N \in 2\mathbb{N}$ ([204], [178], [205]); here, $\Gamma_\mu$ is a union of straight lines dividing $\mathbb{C}$ into $N$ sectors: $\Gamma_\mu = \bigcup_{k=0}^{N-1} \exp(2\pi i \frac{k}{N})\mathbb{R}_+$;

*Case* (3). $\mu = \exp(\rho^{-1})$, where $\Re\rho < 0$ and $\Im\rho \in 2\mathbb{Z} \setminus \{0\}$ ([141], [205]); here, $\Gamma_\mu$ is a union of $N$ logarithmic spirals dividing $\mathbb{C}$ into $N$ sectors: $\Gamma_\mu = \bigcup_{k=0}^{N-1} \mu^k \{\mu^{it} \mid t \in \mathbb{R}\}$.

From now on, we assume that $\mu$ has one of the forms listed above. In all three cases, $\Gamma_\mu$ is a multiplicative subgroup of $\mathbb{C}_\times$ and self-dual. Explicitly, the self-duality can be expressed in terms of a so-called *Fresnel function*

$$\alpha\colon \Gamma_\mu \to \mathbb{T}, \quad \alpha(\mu^{k+it}) := \begin{cases} \mu^{itk}, & \text{case (1)}, \\ \mu^{(k^2-t^2)/2}, & \text{case (2)}, \\ e^{i\Im\frac{(k+it)^2}{2\rho}}, & \text{case (3)}, \end{cases}$$

and the function

$$\chi_\mu\colon \Gamma_\mu \times \Gamma_\mu \to \mathbb{T}, \quad (x,y) \mapsto \frac{\alpha(xy)}{\alpha(x)\alpha(y)}.$$

Straightforward calculations show that $\chi_\mu$ is symmetric a bicharacter,

$$\chi_\mu(x,y) = \chi_\mu(y,x), \quad \chi_\mu(x,yz) = \chi_\mu(x,y)\chi_\mu(x,z) \quad \text{for all } x,y,z \in \Gamma_\mu,$$

that the map $\Gamma_\mu \mapsto \widehat{\Gamma}_\mu$ given by $x \mapsto \chi_\mu(x,\cdot)$ is a group isomorphism, and that

$$\chi_\mu(x, \mu^{k+it}) = \operatorname{Phase}(x)^k |x|^{it} = \theta_{(k,t)}(x) \quad \text{for all } (k,t) \in \mathbb{Z} \times \mathbb{R},\ x \in \Gamma_\mu. \tag{8.21}$$

The universal $C^*$-algebra with affiliated elements $a, b$ satisfying (8.17) and (8.20) can be constructed as follows. Define

$$\gamma\colon \Gamma_\mu \to \operatorname{Aut}(C_0(\overline{\Gamma}_\mu)), \quad m \mathrel{\eta} C_0(\overline{\Gamma}_\mu), \quad m_\times \mathrel{\eta} C^*(\mathbb{Z} \times \mathbb{R})$$

(similarly) as above, and denote by $\phi_*\colon C^*(\mathbb{Z}\times\mathbb{R})\to M(C^*(\Gamma_\mu))$ the $*$-homomorphism induced by the homomorphism $\phi\colon \mathbb{Z}\times\mathbb{R}\to\Gamma_\mu$, $(k,t)\mapsto\mu^{k+it}$. Then similar considerations as in the preceding paragraph show that the triple

$$A := C_0(\overline{\Gamma}_\mu) \rtimes_\gamma \Gamma_\mu, \quad a := 1 \rtimes \phi_*(m_\times)\ \eta\ A, \quad b := m \rtimes 1\ \eta\ A$$

has the desired universality property. Moreover, $A$ is generated by $a, a^{-1}, b$ in the sense of Definition 8.4.15 [204, Proposition 4.1], [141, Proposition 6.1]. Note that equation (8.21) implies

$$U_{\mu^{k+it}} = \phi_*(U_{(k,t)}) = \phi_*(\theta_{(k,t)}(m_\times)) = \theta_{(k,t)}(\phi_*(m_\times)) = \chi_\mu(\phi_*(m_\times), \mu^{k+it})$$

for all $(k,t)\in\mathbb{Z}\times\mathbb{R}$. Therefore, $a\ \eta\ A$ can also be characterized by the relation

$$1 \rtimes U_x = \chi_\mu(a,x) \quad \text{for all } x\in\Gamma_\mu.$$

**Quantum exponential function, a multiplicative unitary, and the $C^*$-bialgebra structure on $A$.** The comultiplication on $A$ is defined via a multiplicative unitary $W$ as in the second approach outlined in the introduction.

Similarly as in the case of $E_\mu(2)$, the definition of $W$ involves a (continuous) *quantum exponential function* $F_\mu\colon\overline{\Gamma}_\mu\to\mathbb{T}$, which is characterized by

$$\text{case (1), (3):}\qquad F_\mu(z) = \prod_{k=0}^{\infty}\frac{1+\overline{\mu^{2k}z}}{1+\mu^{2k}z} \quad \text{if } z\in\overline{\Gamma}_\mu\setminus\{-\mu^{-2k}\mid k\in\mathbb{N}\},$$

$$\text{case (2):}\qquad F_\mu(\mu^l r) = \begin{cases} \dfrac{f_0(\mu r)}{1+r}\displaystyle\prod_{k=1}^{l/2}\frac{1+\mu^{2k}r}{1+\overline{\mu^{2k}r}}, & 2\mid l,\ r>0,\\[2ex] f_0(r)\displaystyle\prod_{k=0}^{(l-1)/2}\frac{1+\mu^{2k+1}r}{1+\overline{\mu^{2k+1}r}}, & 2\nmid l,\ r>0,\end{cases}$$

$$\text{where } f_0(z) = \exp\left(\frac{1}{\pi i}\int_0^\infty\frac{\log(1+t^{-N/2})}{t+z^{-1}}\,dt\right),$$

see [204, Section 1] and [141, Proposition 2.3]. In all cases, $F_\mu$ satisfies an analogue of Theorem 8.4.28 which explains the terminology [204, Theorem 2.6], [141, Theorem 3.5].

Let us call a non-degenerate representation $\pi$ of $A$ on a Hilbert space $H$ *proper* if it satisfies the following equivalent conditions:

i) $\pi(b)\ \eta\ \mathcal{L}(H)$ is invertible,

ii) $\ker\pi(b)=0$,

iii) the restriction of $\pi$ to $C_0(\Gamma_\mu) \rtimes_\gamma \Gamma_\mu \subset A$ is non-degenerate.

**Theorem 8.4.31.** *Let $\pi$ be a proper non-degenerate representation of $A$ on a Hilbert space $H$. Put $a_0 := \pi(a)$ and $b_0 := \pi(b)$.*

i) *There exists a proper non-degenerate representation $\hat{\pi}\colon A \to \mathcal{L}(H)$ such that*

$$\hat{\pi}(a) = b_0^{-1}, \quad \hat{\pi}(b) = \begin{cases} b_0^{-1}a_0, & \textit{case (1), (3)}, \\ a_0 b_0^{-1}, & \textit{case (2)}. \end{cases}$$

*Put $\hat{a}_0 := \hat{\pi}(a)$ and $\hat{b}_0 := \hat{\pi}(b)$.*

ii) $W := F_\mu(\hat{b}_0 \otimes b_0)\chi_\mu(\hat{a}_0 \otimes 1, 1 \otimes a_0)$ *is a modular multiplicative unitary.*

iii) *We have*

$$A(W) = \pi(A), \quad \Delta_W(a_0) = a_0 \otimes a_0, \quad \Delta_W(b_0) = a_0 \otimes b_0 + b_0 \otimes 1,$$
$$\widehat{A}(W) = \hat{\pi}(A), \quad \widehat{\Delta}_W(\hat{a}_0) = \hat{a}_0 \otimes \hat{a}_0, \quad \widehat{\Delta}_W(\hat{b}_0) = \hat{b}_0 \otimes \hat{a}_0 + 1 \otimes \hat{b}_0.$$

*Proof.* The proofs are very similar in all cases and scattered over several papers.

Case (1): i) is contained in [204, Theorem A.1]; for ii), iii) see [127, Section 3].

Case (2): i) follows from [204, Propositions 2.1, 2.2], see also [178, Lemma 3.9]; ii) is stated in [142, Section 5], see also [204, Proposition 3.2]; iii) follows similarly as in case (3), see also [204, Sections 6, 7] and [204, Proposition 3.2].

Case (3): i) follows from [141, Theorem 3.2, Corollary 3.3], ii) [141, Proposition 5.2, Corollary 5.7], iii) [141, Propositions 5.2, 6.2, 6.3]. □

Now we can choose a non-degenerate faithful representation $\pi$ of $C_0(\Gamma_\mu) \rtimes_\gamma \Gamma_\mu$ on some Hilbert space $H$, extend it to a faithful representation $\pi$ of $A \subset M(C_0(\Gamma_\mu) \rtimes_\gamma \Gamma_\mu)$, apply Theorem 8.4.31, and find:

**Corollary 8.4.32.** *There exists a (unique) non-degenerate $*$-homomorphism $\Delta\colon A \to M(A \otimes A)$ such that $\Delta(a) = a \otimes a$ and $\Delta(b) = a \otimes b + b \otimes 1$. Moreover, $(A, \Delta)$ is a bisimplifiable $C^*$-bialgebra.* □

**The antipode and the Haar weights of $A$.** From the modular multiplicative unitary $W$ constructed above, we obtain a unitary antipode $R$ and a scaling group $(\tau_t)_t$ (see Theorem 7.3.19):

**Theorem 8.4.33.** i) *The unitary antipode $R$ of $(A, \Delta)$ is the unique $*$-antiautomorphism $R\colon A \to A$ satisfying $R(a) = a^{-1}$, $R(b) = -\mu a^{-1} b$.*

ii) *The scaling group*[2] $(\tau_t)_t$ *of* $(A, \Delta)$ *is given by* $\tau_t(x) = |a|^{2it} x |a|^{-2it}$; *in particular,* $\tau_t(a) = a$ *and* $\tau_t(b) = \mu^{2it} b$ *for each* $t \in \mathbb{R}$.

*Proof.* Case (1): [127, Section 3].

Case (2): [204, Proposition 3.3, Section 6]; Woronowicz proves the assertions starting from a modified manageable multiplicative unitary instead of $W$, but the argument can be adapted to $W$.

Case (3): [141, Section 6.2]. □

A right Haar weight on $(A, \Delta)$ was first found by Van Daele [178]:

**Theorem 8.4.34.** i) *There exists a right Haar weight* $\psi$ *on* $(A, \Delta)$ *such that for all* $c \in A$ *of the form* $c = g(a) f(b)$, *where* $g \in C_0(\Gamma_\mu)$, $f \in C_0(\bar{\Gamma}_\mu)$,

$$\psi(c^* c) = \int_{\Gamma_\mu} |g(x)|^2 d\lambda(x) \int_{\bar{\Gamma}_\mu} |f(x)|^2 |x|^2 d\lambda(x);$$

*here,* $\lambda$ *denotes the Haar measure of* $\Gamma_\mu$.

ii) *The modular automorphism group* $(\sigma_t^\psi)_t$ *of* $\psi$ *is given by*

$$\sigma_t^\psi(x) = |b|^{2it} x |b|^{-2it};$$

*in particular,* $\sigma_t^\psi(a) = \mu^{-2it} a$ *and* $\sigma_t^\psi(b) = b$ *for each* $t \in \mathbb{R}$.

*Proof.* i) See [205, Theorem 3.1, Equation (3.8)]; the original references are [178, Theorems 4.5, 5.11].

ii) This follows from [205, Equation (3.7)]; for cases (1), (2), see also [178, Theorem 4.4, Proposition 5.6]. □

A left Haar weight on $(A, \Delta)$ is given by $\phi := \psi \circ R$. Consequently, we find:

**Theorem 8.4.35.** $(A, \Delta)$ *is a locally compact quantum group.*

The quantum $az + b$ group was the first example of a locally compact quantum group with scaling constant not equal to 1 (in cases (2), (3)):

**Proposition 8.4.36.** $\psi \circ \tau_t = |\mu^{-4it}| \psi$; *in particular, the scaling constant* $\nu$ *of* $(A, \Delta)$ *is not* 1.

*Proof.* Case (1): [178, Proposition 4.12]; case (2): [178, Section 5]; case (3) (and also cases (1), (2)): [205, Section 3]. □

[2]Here, we adopt the sign convention for $\tau$ chosen by Woronowicz, which differs from the convention chosen by Kustermans and Vaes.

# Part III

# Selected topics

# Chapter 9

# Coactions on $C^*$-algebras, reduced crossed products, and duality

The topic of this chapter are coactions of $C^*$-bialgebras on $C^*$-algebras. Such coactions generalize actions of groups on one side and Fell bundles on groups on the other side; in particular, they provide a unified approach to these two concepts. We review group actions in Section 9.1 and introduce the concept of a coaction in Section 9.2.

Like actions of a group, every coaction of a $C^*$-bialgebra can be encoded by a crossed product. The construction of this crossed product involves not only the $C^*$-bialgebra of the coaction but also a second $C^*$-bialgebra that is dual to the initial one in a suitable sense. In the classical setting of a group action, such a pair of dual $C^*$-bialgebras appears as follows. An action of a locally compact group $G$ on a $C^*$-algebra $C$ corresponds to a coaction of the $C^*$-bialgebra $C_0(G)$ on $C$, and the associated (reduced) crossed product $C \rtimes_{(r)} G$ can be considered as a twisted tensor product of $C$ and the $C^*$-bialgebra $C^*_{(r)}(G)$. For the convenience of the reader, we review the reduced crossed product construction in Section 9.1.

As a framework for the construction of reduced crossed products for coactions of $C^*$-bialgebras, Baaj and Skandalis introduced the concept of a Kac system [6]. A Kac system consists of a regular multiplicative unitary $V$ and an additional symmetry, subject to a number of axioms. The multiplicative unitary $V$ gives rise to two $C^*$-bialgebras $(\widehat{A}(V), \widehat{\Delta})$ and $(A(V), \Delta)$ (see Chapter 7), which are dual to each other in a suitable sense. Now the reduced crossed product for a coaction of the $C^*$-bialgebra $(\widehat{A}(V), \widehat{\Delta})$ or $(A(V), \Delta)$ on a $C^*$-algebra $C$ can be considered as a twisted tensor product of $C$ and the $C^*$-algebra $A(V)$ or $\widehat{A}(V)$, respectively. These concepts and constructions are explained in Sections 9.3 and 9.4.

The main result of this chapter is the Baaj–Skandalis duality theorem which generalizes Takesaki–Takai duality. Let us briefly recall the latter. For every action $\alpha$ of a locally compact abelian group $G$ on a $C^*$-algebra $C$, the associated crossed product $C \rtimes_{\alpha,r} C$ carries a dual action $\hat{\alpha}$ of the dual group $\widehat{G}$, and the Takesaki–Takai duality theorem identifies the bidual action $\hat{\hat{\alpha}}$ of $\widehat{\widehat{G}} \cong G$ on $C \rtimes_{\alpha,r} G \rtimes_{\hat{\alpha},r} \widehat{G}$ with a stabilization of the initial action $\alpha$ on $C$. In particular, the iterated reduced crossed product is just a stabilization of $C$. Thus, Takesaki–Takai duality extends the Pontrjagin duality of locally compact abelian groups to actions of such groups on $C^*$-algebras. A short summary is given in Section 9.1.

The duality theorem of Baaj and Skandalis extends the result of Takesaki and Takai as follows: In the framework of Kac systems, the reduced crossed product of a coaction of a $C^*$-bialgebra carries a dual coaction of the dual $C^*$-bialgebra, and the theorem of Baaj and Skandalis identifies the dual of that coaction (that is, the bidual coaction) with a stabilization of the initial coaction. On the level of the underlying $C^*$-algebras, this means that the iterated reduced crossed product is a stabilization of the initial $C^*$-algebra. These constructions and results are presented in Sections 9.4 and Section 9.5.

The definitions and results presented in Sections 9.3–9.5 are taken from the fundamental article of Baaj and Skandalis [7]. The presentation, however, has been adapted to the purpose of this book.

Let us briefly comment on the setting of the theory presented in this chapter. The most comprehensive framework for the generalization of group actions, the crossed product construction, and Takesaki–Takai duality to quantum groups in the setting of $C^*$-algebras or von Neumann algebras is the theory of locally compact quantum groups introduced in Chapter 8. In that framework, one can work on the level of

- locally compact quantum groups in the setting of von Neumann algebras,
- reduced $C^*$-algebraic quantum groups, or
- universal $C^*$-algebraic quantum groups.

Similarly, one can consider

- coactions of von Neumann bialgebras on von Neumann algebras and associated crossed products,
- coactions of $C^*$-bialgebras on $C^*$-algebras and associated reduced crossed products, and
- coactions of $C^*$-bialgebras on $C^*$-algebras and associated full/universal crossed products.

In the first setting, a rich theory with many important results, applications, and examples was developed by Vaes [159], [160], [161] and Vaes and Vainerman [163], building on work of Enock and Schwartz [41], [46] and Enock and Nest [45]. The theory presented in this chapter fits into the second setting. A full/universal crossed product construction was introduced by Vergnioux [182].

## 9.1 Actions of groups and Takesaki–Takai duality

This section gives a brief review on group actions on $C^*$-algebras and associated reduced crossed products. The definitions and results presented here are classical, therefore we omit the proofs. Reference are, for example, [121], [191].

Throughout this section, $G$ denotes a locally compact group with left Haar measure $\lambda$ and modular function $\delta$.

**Definition 9.1.1.** A (*strongly continuous*) *action* of $G$ on a $C^*$-algebra $C$ is a group homomorphism $\alpha\colon G \to \operatorname{Aut}(C)$, $x \mapsto \alpha_x$, such that for every $c \in C$, the map $G \to C$ given by $x \mapsto \alpha_x(c)$ is norm-continuous. We call $C$ a $G$-$C^*$-algebra if $\alpha$ is understood.

Let $\alpha$ and $\beta$ be actions of $G$ on $C^*$-algebras $C$ and $D$, respectively. A non-degenerate $*$-homomorphism $\rho\colon C \to M(D)$ is called *covariant with respect to* $\alpha$ *and* $\beta$ if $\rho(\alpha_x(c)) = \beta_x(\rho(c))$ for all $c \in C$ and $x \in G$. Here, $\beta_x$ is extended to a $*$-automorphism of $M(C)$ for each $x \in G$.

Actions on $C^*$-algebras are usually studied in terms of covariant representations:

**Definition 9.1.2.** Let $\alpha$ be an action of $G$ on a $C^*$-algebra $C$. A (*non-degenerate*) *covariant representation* of $(C, \alpha)$ is a triple $(H, \pi, u)$ consisting of a Hilbert space $H$, a (non-degenerate) representation $\pi$ of $C$ on $H$, and a unitary representation $u$ of $G$ on $H$ such that

$$\pi(\alpha_x(c)) = u(x)\pi(c)u(x)^* \quad \text{for all } c \in C,\ x \in G. \tag{9.1}$$

**Remark 9.1.3.** In the situation above, the representation $u$ induces a strongly continuous action $\mathrm{Ad}_u$ of $G$ on the $C^*$-algebra $\mathcal{K}(H)$ via

$$\mathrm{Ad}_u(x)\colon T \mapsto u(x)Tu(x)^* \quad \text{for all } x \in G,\ T \in \mathcal{K}(H),$$

and condition (9.1) is equivalent to the condition that the $*$-homomorphism $\pi : C \to \mathcal{L}(H) \cong M(\mathcal{K}(H))$ is covariant with respect to $\alpha$ and $\mathrm{Ad}_u$.

Let $\alpha$ be an action of $G$ on a $C^*$-algebra $C$. Then the space $C_c(G; C)$ of $C$-valued continuous functions on $G$ with compact support is a $*$-algebra with respect to the operations

$$(f * g)(x) := \int_G f(y)\alpha_y(g(y^{-1}x))d\lambda(y), \quad f^*(x) := \alpha_x(f(x^{-1}))^* \cdot \delta(x)^{-1}.$$

We shall define the reduced crossed product of the action $\alpha$ in terms of a left regular representation of $C_c(G; C)$, using the language of $C^*$-modules (for a short summary, see Section 12.2). We consider the Hilbert space $L^2(G, \lambda)$ and the $C^*$-algebra $C$ as $C^*$-modules over $\mathbb{C}$ and $C$, respectively, and form the tensor product $L^2(G, \lambda) \otimes C$, which is a $C^*$-module over $C$ again. This $C^*$-module can also be described as follows. It is easy to see that the space $C_c(G; C)$ is a pre-$C^*$-module over $C$ with respect to the operations

$$\langle f|g\rangle := \int_G f(x)^* g(x)d\lambda(x) \text{ and } (fc)(x) := f(x)c, \text{ where } f, g \in C, c \in C.$$

Denote the completion of this pre-$C^*$-module by $L^2(G;C)$. Then the natural embedding of the algebraic tensor product $C_c(G) \odot C$ into $C_c(G;C)$ extends to an isomorphism of $C^*$-modules $L^2(G,\lambda) \otimes C \xrightarrow{\cong} L^2(G;C)$, as one can easily check.

**Proposition 9.1.4.** *Let $\alpha$ be an action of $G$ on a $C^*$-algebra $C$. Then there exist non-degenerate injective $*$-homomorphisms*

$$\begin{aligned} C &\to \mathcal{L}_C(L^2(G;C)), & c &\mapsto c \rtimes_r 1, \\ C_r^*(G) &\to \mathcal{L}_C(L^2(G;C)), & L(h) &\mapsto 1 \rtimes_r L(h), \\ C_c(G;C) &\to \mathcal{L}_C(L^2(G;C)), & f &\mapsto L_\alpha(f), \end{aligned}$$

*such that for all $x \in G$ and $g \in C_c(G;C)$,*

$$\begin{aligned} ((c \rtimes_r 1)g)(x) &= {\alpha_x}^{-1}(c)g(x), \\ ((1 \rtimes_r L(h))g)(x) &= \int_G h(y)g(y^{-1}x)d\lambda(y), \\ (L_\alpha(f)g)(x) &= \int_G {\alpha_x}^{-1}(f(y))g(y^{-1}x)d\lambda(y). \end{aligned}$$

*Proof.* This can be checked by straightforward calculations; see, for example, [121], [191]. □

**Definition 9.1.5.** Let $\alpha$ be an action of $G$ on a $C^*$-algebra $C$. The associated *reduced crossed product* is the $C^*$-subalgebra $C \rtimes_{\alpha,r} G \subseteq \mathcal{L}_C(L^2(G;C))$ given by

$$C \rtimes_{\alpha,r} G := \overline{\operatorname{span}}\{L_\alpha(f) \mid f \in C_c(G;C)\}.$$

The reduced crossed product can also be constructed without reference to $C^*$-modules (see, for example, [121]), but the definition given above is more direct and more convenient.

**Proposition 9.1.6.** *Let $\alpha$ be an action of $G$ on a $C^*$-algebra $C$. Then*

$$C \rtimes_r 1 \subset M(C \rtimes_{\alpha,r} G), \qquad 1 \rtimes_r C_r^*(G) \subset M(C \rtimes_{\alpha,r} G),$$
$$\overline{\operatorname{span}}((C \rtimes_r 1)(1 \rtimes_r C^*(G))) = C \rtimes_{\alpha,r} G = \overline{\operatorname{span}}((1 \rtimes_r C^*(G))(C \rtimes_r 1)).$$

*Proof.* Again, this can be checked by straightforward calculations and arguments. □

The reduced crossed product has the following simple functorial property:

**Proposition 9.1.7.** *Let $\alpha$ and $\beta$ be actions of $G$ on $C^*$-algebras $C$ and $D$, respectively, and let $\rho\colon C \to M(D)$ be a non-degenerate covariant $*$-homomorphism. Then there exists a unique non-degenerate $*$-homomorphism*

$$\rho \rtimes_r \mathrm{id}\colon C \rtimes_{\alpha,r} G \to M(D \rtimes_{\beta,r} G)$$

*such that* $(\rho \rtimes_r \mathrm{id})(L_\alpha(f)) = L_\beta(\rho \circ f)$ *for all* $f \in C_c(G; C)$.

*Proof.* Uniqueness is evident, let us prove existence. The $C^*$-module $L^2(G; D) \cong L^2(G,\lambda) \otimes D$ can be identified with the internal tensor product $L^2(G,\lambda) \otimes C \otimes_\rho D \cong L^2(G; C) \otimes_\rho D$ because $C \otimes_\rho D \cong D$, and the map

$$\begin{aligned} \mathcal{L}_C(L^2(G;C)) \to \mathcal{L}_D(L^2(G;C) \otimes_\rho D) \cong \mathcal{L}_D(L^2(G;D)),\\ T \mapsto T \otimes_\rho \mathrm{id}_D, \end{aligned}$$

is a $*$-homomorphism (see Section 12.2). An easy calculation shows that $L_\alpha(f) \otimes_\rho \mathrm{id}_D = L_\beta(\rho \circ f)$ for all $f \in C_c(G; C)$. □

In the remaining part of this section, we focus on the case that $G$ is abelian. Then the reduced crossed product $C \rtimes_{\alpha,r} G$ carries a natural dual action $\hat{\alpha}$ of the dual group $\widehat{G}$, and the assignment $C \mapsto C \rtimes_{\alpha,r} G$ extends to a functor from $G$-$C^*$-algebras to $\widehat{G}$-$C^*$-algebras.

**Proposition 9.1.8.** *Let $\alpha$ be an action of $G$ on a $C^*$-algebra $C$, where $G$ is abelian. For every continuous character $\chi$ on $G$, there exists an automorphism $\hat{\alpha}_\chi$ of $C_c(G; C)$ such that*

$$(\hat{\alpha}_\chi(f))(x) = \chi(x) f(x) \quad \text{for all } x \in G,\ f \in C_c(G; C).$$

*This automorphism extends to an automorphism of the $C^*$-algebra $C \rtimes_{\alpha,r} G$. The assignment $\chi \mapsto \hat{\alpha}_\chi$ defines a strongly continuous action of $\widehat{G}$ on $C \rtimes_\alpha G$.*

*Proof.* See [121], [191] □

**Definition 9.1.9.** Let $\alpha$ be an action of $G$ on a $C^*$-algebra $C$, where $G$ is abelian. The action $\hat{\alpha}$ of $\widehat{G}$ on $C \rtimes_{\alpha,r} G$ constructed above is called the *dual action* of $\alpha$.

The construction of the dual action evidently is functorial in the following sense:

**Proposition 9.1.10.** *Let $\alpha$ and $\beta$ be actions of $G$ on $C^*$-algebras $C$ and $D$, respectively, where $G$ is abelian. Let $\rho\colon C \to M(D)$ be a non-degenerate covariant $*$-homomorphism. Then the $*$-homomorphism $\rho \rtimes_r \mathrm{id}\colon C \rtimes_{\alpha,r} G \to M(D \rtimes_{\beta,r} G)$ is covariant with respect to $\hat{\alpha}$ and $\hat{\beta}$.*

The main result in this section is the Takesaki–Takai duality theorem. Roughly, this theorem extends Pontrjagin duality from groups to actions: Given a locally compact abelian group $G$, we constructed a functor from the category of $G$-$C^*$-algebras to the category of $\widehat{G}$-$C^*$-algebras. Replacing $G$ by $\widehat{G}$, we get a functor from $\widehat{G}$-$C^*$-algebras to $\widehat{\widehat{G}}$-$C^*$-algebras, and by Pontrjagin duality, we can identify $\widehat{\widehat{G}}$ with $G$. Thus we obtain two functors

$$\begin{matrix}\text{category of} \\ G\text{-}C^*\text{-algebras}\end{matrix} \quad \underset{(D\rtimes_{\beta,r}\widehat{G},\hat{\beta})\leftarrow\!\shortmid(D,\beta)}{\overset{(C,\alpha)\mapsto(C\rtimes_{\alpha,r}G,\hat{\alpha})}{\rightleftarrows}} \quad \begin{matrix}\text{category of} \\ \widehat{G}\text{-}C^*\text{-algebras.}\end{matrix}$$

The Takesaki–Takai duality theorem says that these functors induce an equivalence of categories up to equivariant Morita equivalence. We shall only give a less abstract and more concrete formulation of the duality theorem: For every $G$-$C^*$-algebra $C$, the associated bidual $C \rtimes_{\alpha,r} G \rtimes_{\hat{\alpha},r} \widehat{G}$ is naturally covariantly isomorphic to a stabilization of $C$. This stabilization is a tensor product of $C$ with $\mathcal{K}(L^2(G,\lambda))$, where $G$ acts on $\mathcal{K}(L^2(G,\lambda))$ via the right regular representation $\rho\colon G \to \mathcal{L}(L^2(G))$, given by $(\rho(x)\xi)(y) = \xi(yx)$ for all $y \in G$, $\xi \in L^2(G,\lambda)$, $x \in G$.

**Theorem 9.1.11** (Takesaki–Takai duality). *Let $\alpha$ be an action of $G$ on a $C^*$-algebra $C$, where $G$ is abelian. Then there exists an isomorphism*

$$C \rtimes_{\alpha,r} G \rtimes_{\hat{\alpha},r} \widehat{G} \cong C \otimes \mathcal{K}(L^2(G,\lambda))$$

*that is covariant with respect to $\hat{\hat{\alpha}}$ and $\alpha \boxtimes \mathrm{Ad}_\rho$, respectively, where the action $\alpha \boxtimes \mathrm{Ad}_\rho$ is given by $(\alpha \boxtimes \mathrm{Ad}_\rho)_x(c \otimes T) = \alpha_x(c) \otimes \rho_x^{-1} T \rho_x$ for all $c \in C$, $T \in \mathcal{K}(L^2(G,\lambda))$.*

A proof is given in [121, Theorem 7.9.3], see also [150, Chapter X]. We shall prove a more general result – the Baaj–Skandalis duality theorem – in Section 9.5.2.

## 9.2 Coactions of $C^*$-bialgebras on $C^*$-algebras

The central concept in this chapter is that of a coaction:

**Definition 9.2.1.** A (*right*) *coaction* of a $C^*$-bialgebra $(A,\Delta)$ on a $C^*$-algebra $C$ is a non-degenerate $*$-homomorphism $\delta\colon C \to M(C \otimes A)$ which satisfies the following two conditions:

i) $\delta(C)(1 \otimes A) \subseteq C \otimes A$,

ii) $(\delta \otimes \mathrm{id}_A) \circ \delta = (\mathrm{id}_C \otimes \Delta) \circ \delta$, that is, the following diagram commutes:

$$\begin{array}{ccc} C & \xrightarrow{\delta} & M(C \otimes A) \\ {\scriptstyle\delta}\downarrow & & \downarrow{\scriptstyle\delta\otimes\mathrm{id}_A} \\ M(C \otimes A) & \xrightarrow[\mathrm{id}_C \otimes \Delta]{} & M(C \otimes A \otimes A). \end{array}$$

We also refer to the pair $(C, \delta)$ as a (right) coaction of $(A, \Delta)$. A coaction $(C, \delta)$ is an $(A, \Delta)$-*$C^*$-algebra* if

iii) $\delta$ is injective and

iv) $\delta(C)(1 \otimes A)$ is linearly dense in $C \otimes A$.

Let $(C, \delta_C)$ and $(D, \delta_D)$ be coactions of $(A, \Delta)$. A non-degenerate $*$-homomorphism $\rho\colon C \to M(D)$ is *covariant* with respect to $\delta_C$ and $\delta_D$ if $\delta_D \circ \rho = (\rho \otimes \mathrm{id}_A) \circ \delta_C$, that is, if the following diagram commutes:

$$\begin{array}{ccc} C & \xrightarrow{\rho} & M(D) \\ {\scriptstyle\delta_C}\downarrow & & \downarrow{\scriptstyle\delta_D} \\ M(C \otimes A) & \xrightarrow[\rho\otimes\mathrm{id}_A]{} & M(D \otimes A). \end{array}$$

**Remarks 9.2.2.** i) In condition ii) of the definition above, the $*$-homomorphisms $\delta \otimes \mathrm{id}_A$ and $\mathrm{id}_C \otimes \Delta$ are extended to the multiplier $C^*$-algebra $M(C \otimes A)$. Likewise, in the last paragraph, $\delta_D$ is extended to $M(D)$ and $\rho \otimes \mathrm{id}_A$ is extended to $M(C \otimes A)$.

ii) Evidently, the class of all coactions of a fixed $C^*$-bialgebra $(A, \Delta)$ forms a category with respect to covariant $*$-homomorphisms.

iii) *Left coactions* can be defined similarly as right coactions, and left coactions of a $C^*$-bialgebra $(A, \Delta)$ correspond bijectively with right coactions of the coopposite $C^*$-bialgebra $(A, \Delta)^{\mathrm{cop}}$ defined in Remark 4.1.2 ii). Since left coactions will not play a rôle in this book, we shall refer to right coactions simply as coactions.

iv) For every $C^*$-algebra $C$ and $C^*$-bialgebra $(A, \Delta)$, one can define a *trivial coaction* of $(A, \Delta)$ on $C$ by the formula $c \mapsto c \otimes 1$ for all $c \in C$. Clearly, the trivial coaction turns $C$ into an $(A, \Delta)$-$C^*$-algebra.

v) For every $C^*$-bialgebra $(A, \Delta)$, the map $\Delta\colon A \to M(A \otimes A)$ is a right coaction (and a left coaction) of $(A, \Delta)$ on $A$ itself. This is called the *right* (or *left*) *regular coaction*.

There exist several standard constructions to produce new coactions out of given ones:

**Remark 9.2.3.** Let $(A, \Delta_A)$ and $(B, \Delta_B)$ be $C^*$-bialgebras. Recall that $A \otimes B$ is a $C^*$-bialgebra with respect to the $*$-homomorphism $\Delta_{A\otimes B}$ given by

$$\begin{aligned} A \otimes B \xrightarrow{\Delta_A \otimes \Delta_B} & M(A \otimes A) \otimes M(B \otimes B) \\ & \hookrightarrow M(A \otimes A \otimes B \otimes B) \xrightarrow{\mathrm{id} \otimes \Sigma \otimes \mathrm{id}} M(A \otimes B \otimes A \otimes B); \end{aligned}$$

here, $\Sigma$ denotes the flip $A \otimes B \to B \otimes A$, $a \otimes b \mapsto b \otimes a$.

Given a coaction $(C, \delta_C)$ of $(A, \Delta_A)$ and a coaction $(D, \delta_D)$ of $(B, \Delta_B)$, we can form the following new coactions:

*External direct sum.* The composition of the map

$$\delta_C \oplus \delta_D \colon C \oplus D \to M(C \otimes A) \oplus M(D \otimes B)$$

with the extension of the non-degenerate $*$-homomorphism

$$\begin{aligned} (C \otimes A) \oplus (D \otimes B) &\to M((C \oplus D) \otimes A \otimes B), \\ ((c \otimes a), (d \otimes b)) &\mapsto (c, 0) \otimes a \otimes 1 + (0, d) \otimes 1 \otimes b, \end{aligned}$$

is a coaction of $(A \otimes B, \Delta_{A\otimes B})$ on $C \oplus D$, which we denote by $\delta_{C\oplus D}$.

*External tensor product.* The composition

$$\begin{aligned} C \otimes D \xrightarrow{\delta_C \otimes \delta_D} & M(C \otimes A) \otimes M(D \otimes B) \\ & \hookrightarrow M(C \otimes A \otimes D \otimes B) \xrightarrow{\mathrm{id} \otimes \Sigma \otimes \mathrm{id}} M(C \otimes D \otimes A \otimes B) \end{aligned}$$

is a coaction of $(A \otimes B, \Delta_{A\otimes B})$ on $C \otimes D$, which we denote by $\delta_{C\otimes D}$. Here, $\Sigma\colon A \otimes D \to D \otimes A$ denotes the flip $a \otimes d \mapsto d \otimes a$ again.

*Push-forward.* If $\phi\colon A \to M(B)$ is a morphism of $C^*$-bialgebras, then the pair $(C, (\mathrm{id}_C \otimes \phi) \circ \delta_C)$ is a coaction of $(B, \Delta_B)$.

*Internal direct sum.* Assume that $(A, \Delta_A) = (B, \Delta_B)$. Then the composition of the map

$$C \oplus D \xrightarrow{\delta_C \oplus \delta_D} M(C \otimes A) \oplus M(D \otimes A) \hookrightarrow M((C \otimes A) \oplus (D \otimes A))$$

with the extension of the isomorphism $(C \otimes A) \oplus (D \otimes A) \xrightarrow{\cong} (C \oplus D) \otimes A$ is a coaction of $(A, \Delta_A)$ on $C \oplus D$. If $A$ is commutative, this internal direct sum coincides with the push-forward of the external direct sum $(C \oplus D, \delta_{C\oplus D})$ along the multiplication map $m_A\colon A \otimes A \to A$, $a_1 \otimes a_2 \mapsto a_1 a_2$.

*Internal tensor product.* Assume that $(A, \Delta_A) = (B, \Delta_B)$ and that $A$ is commutative. Then the push-forward of the external tensor product $(C \otimes D, \delta_{C\otimes D})$ along the multiplication map $m_A : A \otimes A \to A$ is a coaction of $(A, \Delta_A)$ on $C \otimes D$. If $A$ is not commutative, then in general, such an internal tensor product can not be defined.

**Actions of a group $G$ and coactions of $C_0(G)$.** Coactions generalize actions in the following sense: For every locally compact group $G$, there exists a bijective correspondence between $G$-$C^*$-algebras and $(C_0(G), \Delta)$-$C^*$-algebras. To explain this correspondence, we need to recall a few facts and some notation.

Let $X$ be a locally compact space and $C$ a $C^*$-algebra. Then the $C^*$-algebra $C_b(X)$ of bounded continuous functions on $X$ can be identified with the multiplier algebra $M(C_0(X))$, and the $C^*$-algebra $C_b(X; C)$ of bounded continuous $C$-valued functions on $X$ can be identified with a $C^*$-subalgebra of $M(C \otimes C_0(X))$ [2, Corollary 3.4]. For each $x \in X$, we denote by $\mathrm{ev}_x : C_b(X) \to \mathbb{C}$ the evaluation at $x$. The non-degenerate $*$-homomorphism $\mathrm{id} \otimes \mathrm{ev}_x : C \otimes C_0(X) \to C$ extends to a $*$-homomorphism $M(C \otimes C_0(X)) \to M(C)$, which we denote by $\mathrm{id} \otimes \mathrm{ev}_x$ again.

**Theorem 9.2.4.** *Let $G$ be a locally compact group and $C$ a $C^*$-algebra.*

i) *Let $\alpha$ be an action of $G$ on $C$. Then the map $\delta_0 : C \to C_b(G; C)$ given by*

$$(\delta_0(c))(x) := \alpha_x(c) \quad \textit{for all } c \in C,\ x \in G,$$

*is a $*$-homomorphism. Denote by $\delta$ the composition of $\delta_0$ with the canonical embedding $C_b(G; C) \hookrightarrow M(C \otimes C_0(G))$. Then $\delta$ is a coaction of $(C_0(G), \Delta)$ and the pair $(C, \delta)$ is a $(C_0(G), \Delta)$-$C^*$-algebra.*

ii) *Let $\delta$ be a coaction of $(C_0(G), \Delta)$ on $C$ such that*

(a) *$\delta$ is injective, or*

(b) *$\delta(C)(1 \otimes C_0(G))$ is linearly dense in $C \otimes C_0(G)$.*

*Then for each $x \in G$, the map*

$$\alpha_x := (\mathrm{id} \otimes \mathrm{ev}_x) \circ \delta : C \to M(C)$$

*is an automorphism of $C$, and the map $\alpha : x \mapsto \alpha_x$ is an action of $G$ on $C$.*

iii) *The constructions in* i) *and* ii) *define a bijective correspondence between the set of all actions of $G$ on $C$ and the set of all coactions of $(C_0(G), \Delta)$ on $C$ that turn $C$ into a $(C_0(G), \Delta)$-$C^*$-algebra.*

iv) *Let $(C,\alpha)$, $(D,\beta)$ be actions of $G$ and $(C,\delta_C)$, $(D,\delta_D)$ the corresponding coactions of $(C_0(G),\Delta)$. Then a non-degenerate $*$-homomorphism $\rho\colon C \to M(D)$ is covariant with respect to $\alpha$ and $\beta$ if and only if it is covariant with respect to $\delta_C$ and $\delta_D$.*

v) *The assignment $(C,\alpha) \mapsto (C,\delta)$ constructed in* i) *and the assignment $\phi \mapsto \phi$ define an equivalence of the category of $G$-$C^*$-algebras and the category of $(C_0(G),\Delta)$-$C^*$-algebras, where the morphisms are all non-degenerate covariant $*$-homomorphisms.*

Some of the arguments of the following proof were used already in Example 5.2.5, where we related representations of a locally compact group $G$ to corepresentations of the $C^*$-bialgebra $(C_0(G),\Delta)$.

*Proof.* i) Evidently, $\delta_0$ is a $*$-homomorphism. We claim that $(\delta\otimes\mathrm{id})\circ\delta = (\mathrm{id}\otimes\Delta)\circ\delta$. By construction,

$$\begin{aligned}(\mathrm{id}\otimes \mathrm{ev}_x\otimes \mathrm{ev}_y)\circ(\delta\otimes\mathrm{id})\circ\delta &= (\mathrm{id}\otimes\mathrm{ev}_x)\circ\delta\circ(\mathrm{id}\otimes\mathrm{ev}_y)\circ\delta = \alpha_x\circ\alpha_y,\\ (\mathrm{id}\otimes \mathrm{ev}_x\otimes \mathrm{ev}_y)\circ(\mathrm{id}\otimes\Delta)\circ\delta &= (\mathrm{id}\otimes\mathrm{ev}_{xy})\circ\delta = \alpha_{xy} \quad \text{for all } x,y\in G.\end{aligned} \tag{9.2}$$

Now the claim follows from the fact that $\alpha_x\circ\alpha_y = \alpha_{xy}$ for all $x,y\in G$ and that maps of the form $\mathrm{id}\otimes\mathrm{ev}_x\otimes\mathrm{ev}_y$ separate the elements of $M(C\otimes C_0(G)\otimes C_0(G))$. Clearly, $\delta(C)(1\otimes C_0(G))\subseteq C\otimes C_0(G)$, so $\delta$ is a coaction.

Next, we show that $(C,\delta)$ is a $(C_0(G),\Delta)$-$C^*$-algebra. Evidently, the map $\delta$ is injective; in fact, $(\mathrm{id}_C\otimes\mathrm{ev}_e)\circ\delta = \mathrm{id}_C$, where $e\in G$ denotes the unit. It remains to show that the set $\delta(C)(1\otimes C_0(G))$ is linearly dense in $C\otimes C_0(G)$, or, equivalently, that the set $\delta_0(C)C_0(G)$ is linearly dense in $C_0(G;C)$. The equation $(\delta_0(\alpha_{x^{-1}}(c)))(x) = \alpha_x(\alpha_{x^{-1}}(c)) = c$ shows that for every $x\in G$ and every $c\in C$, there exists an element in $\delta_0(C)$ that takes the value $c$ at the point $x$. Now the density that we need to prove follows from a standard argument.

ii) The inclusion $\delta(C)(1\otimes C_0(G))\subseteq C\otimes C_0(G)$ implies that $\delta(C)$ is contained in the subspace $C_b(G;C)$ of $M(C\otimes C_0(G))$. Thus, the image of $\alpha_x$ is contained in $C$ for every $x\in G$, and the map $G\to C$ given by $x\mapsto \alpha_x(c)$ is continuous for every $c\in C$.

The relation $(\delta\otimes\mathrm{id})\circ\delta = (\mathrm{id}\otimes\Delta)\circ\delta$ and equation (9.2) imply that $\alpha_x\circ\alpha_y = \alpha_{xy}$ for all $x,y\in G$. Let us show that for each $x\in G$, the $*$-homomorphism $\alpha_x$ is an automorphism. The relations $\alpha_y = \alpha_{yx^{-1}}\circ\alpha_x$ and $\alpha_y = \alpha_x\circ\alpha_{x^{-1}y}$, and similar relations with $x$ and $y$ exchanged, show that

$$\ker\alpha_x = \ker\alpha_y,\quad \mathrm{Im}\,\alpha_x = \mathrm{Im}\,\alpha_y \quad \text{for all } x,y\in G.$$

Thus it suffices to show that $\ker\alpha_e = 0$ and that $\mathrm{Im}\,\alpha_e = C$. Since the map $\alpha_e$ is idempotent, these two conditions are equivalent. If assumption (a) is satisfied, then

$\ker \alpha_e = \bigcap_{x \in G} \ker \alpha_x = \ker \delta = 0$; if assumption (b) is satisfied, then

$$\operatorname{Im} \alpha_e = \overline{\operatorname{span}}\,(\mathrm{id} \otimes \mathrm{ev}_e)(\delta(C)(1 \otimes C_0(G))) = C \cdot \mathrm{ev}_e(C_0(G)) = C.$$

Summarizing, we find that the map $\alpha\colon G \to \operatorname{Aut}(C)$, $x \mapsto \alpha_x$, is an action.

iii) This follows immediately from the definition of $\delta_0$ in i) and of $\alpha_x$ in ii).

iv) We have $(\phi \otimes \mathrm{id}) \circ \delta_C = \delta_D \circ \phi$ if and only if

$$(\mathrm{id} \otimes \mathrm{ev}_x) \circ (\phi \otimes \mathrm{id}) \circ \delta_C = (\mathrm{id} \otimes \mathrm{ev}_x) \circ \delta_D \circ \phi \quad \text{for all } x \in G,$$

that is, if and only if $\phi \circ \alpha_x = \beta_x \circ \phi$ for all $x \in G$.

v) Obvious from iii) and iv). □

**Remark 9.2.5.** The previous result implies that every coaction $(C, \delta)$ of the $C^*$-bialgebra $(C_0(G), \Delta)$ that satisfies one of the conditions ii)(a) and ii)(b) automatically satisfies both conditions and is a $(C_0(G), \Delta)$-$C^*$-algebra.

The preceding result can be extended to groupoids, see Theorem 11.5.7.

**Fell bundles on a group $G$ and coactions of $C_r^*(G)$.** The concept of a coaction generalizes not only the concept of an action, but also the concept of a Fell bundle on a group: To every Fell bundle $F$ on a locally compact group $G$, one can associate a coaction of $C_r^*(G)$ on the reduced $C^*$-algebra of that Fell bundle. Let us briefly outline this construction. Standard references on Fell bundles are [53], [54].

Let $G$ be a locally compact group with left Haar measure $\lambda$ and associated right Haar measure $\lambda^{-1}$.

Recall that a *continuous Banach bundle* on $G$ is a topological space $F$ with an open continuous surjection $p\colon F \to G$ such that

i) for each point $x \in G$, the fiber $F_x := p^{-1}(x)$ is endowed with a Banach space structure;

ii) the maps $\mathbb{C} \times F \to F$ and $F^{(2)} \to F$ given in each fiber by scalar multiplication and addition, respectively, are continuous; here, $F^{(2)}$ denotes the set $\{(x, y) \in F \times F \mid p(x) = p(y)\}$;

iii) the map $F \to \mathbb{R}$ given by the norm on each fiber is continuous;

iv) for each point $x \in G$ and each neighborhood $W$ of the zero $0_x$ over $x$ in $F$, there exist a number $\epsilon > 0$ and a neighborhood $U$ of $x$ such that the set $\{e \in p^{-1}(U) \mid \|e\| < \epsilon\}$ is contained in $W$.

A *Fell bundle* on $G$ is a continuous Banach bundle $p\colon F \to G$ together with a continuous multiplication map $F \times F \to F$ and an involution $*\colon F \to F$ such that for all $a, a_1, a_2, a_3 \in F$,

i) $p(a_1a_2) = p(a_1)p(a_2)$ and $p(a^*) = p(a)^{-1}$;

ii) for all $x, y \in G$, the map $F_x \times F_y \to F_{xy}$ induced by the multiplication is bilinear, and the map $F_x \to F_{x^{-1}}$ induced by the involution is conjugate-linear;

iii) $(a_1a_2)a_3 = a_1(a_2a_3)$, $(a_1a_2)^* = a_2^*a_1^*$, and $(a^*)^* = a$;

iv) $\|a_1a_2\| \le \|a_1\|\|a_2\|$, $\|a^*a\| = \|a\|^2 = \|a^*\|^2$, and $a^*a \ge 0$.

Here, $F_x = p^{-1}(x) \subseteq F$ for each $x \in G$. Evidently, the fiber $F_e$ over the unit $e \in G$ is endowed with the structure of a $C^*$-algebra.

To the Fell bundle $F$, one can associate a reduced $C^*$-algebra $C_r^*(F)$ as follows. Denote by $\Gamma_c(F)$ the space of continuous sections of $F$ with compact support. This space is a pre-$C^*$-module over the $C^*$-algebra $F_e$, where

$$\langle \eta|\xi\rangle = \int_G \eta(x)^*\xi(x)d\lambda^{-1}(x), \quad (\xi a)(x) = \xi(x)a \quad \text{for all } \eta, \xi \in \Gamma_c(F),\ a \in F_e.$$

We denote the associated $C^*$-module by $L^2(G; F)$. Each section $f \in \Gamma_c(F)$ defines a left convolution operator $L_F(f) \in \mathcal{L}_{F_e}(L^2(G; F))$ via

$$(L_F(f)\xi)(x) = \int_G f(y)\xi(y^{-1}x)d\lambda(y) \quad \text{for all } \xi \in \Gamma_c(F),\ x \in G,$$

and each $a \in F$ defines a left convolution operator $L_F(a) \in \mathcal{L}_{F_e}(L^2(G; F))$ via

$$(L_F(a)\xi)(x) = a\xi(p(a)^{-1}x) \quad \text{for all } \xi \in \Gamma_c(F),\ x \in G.$$

The reduced $C^*$-algebra of $F$ is the $C^*$-subalgebra $C_r^*(F) \subseteq \mathcal{L}_{F_e}(L^2(G; F))$ generated by the operators $L_F(f)$, where $f \in \Gamma_c(F)$. It is easy to see that for each $a \in F$, the operator $L_F(a)$ belongs to $M(C_r^*(F))$.

Recall that for each $x \in G$, we defined a multiplier $L_x \in M(C_r^*(G))$ (see Example 4.2.2).

**Theorem 9.2.6.** i) *Let $F$ be a Fell bundle on a locally compact group $G$. There exists a unique coaction $\delta_F$ of $C_r^*(G)$ on $C_r^*(F)$ whose extension to $M(C_r^*(F))$ satisfies*

$$\delta_F(L_F(a)) = L_F(a) \otimes L_{p(a)} \quad \textit{for all } a \in F.$$

*This coaction turns $C_r^*(F)$ into a $C_r^*(G)$-$C^*$-algebra.*

ii) *Assume that $G$ is discrete and that $\delta$ is an injective coaction of $C_r^*(G)$ on a $C^*$-algebra $C$. Then $(C, \delta)$ is a $C_r^*(G)$-$C^*$-algebra. There exists a unique Fell bundle $F$ on $G$ such that*

$$F_x = \{c \in C \mid \delta(c) = c \otimes L_x\} \quad \textit{for all } x \in G,$$

*and such that the multiplication and involution on $F$ are induced by the multiplication and involution on $C$. Moreover, the map $\Gamma_c(F) \to C$ given by $f \mapsto \sum_{x \in G} f(x)$ extends to an isomorphism $C_r^*(F) \xrightarrow{\cong} C$. With respect to this isomorphism, the coaction $\delta_F$ corresponds to the coaction $\delta$.*

*Proof.* The coaction in i) is constructed by means of an auxiliary unitary, similarly as the comultiplication on $C_r^*(G)$; see, for example, [97]. For the proof of ii), see [6, Corollaire 15]. □

The preceding result can be extended to groupoids, see Theorem 11.5.7.

## 9.3 Weak Kac systems

The rest of this chapter is devoted to reduced crossed products for coactions of $C^*$-bialgebras. The proper framework for the construction of such crossed products are weak Kac systems. A weak Kac system consists of a well-behaved multiplicative unitary $V$ and an additional symmetry $U$. The multiplicative unitary $V$ gives rise to two $C^*$-bialgebras $(\widehat{A}(V), \widehat{\Delta})$ and $(A(V), \Delta)$, called the legs of $V$ (see Section 7.2), and for coactions of these $C^*$-bialgebras, we shall construct reduced crossed products. The additional symmetry $U$ serves two purposes:

- to overcome the asymmetry between the left leg $(\widehat{A}(V), \widehat{\Delta})$ and the right leg $(A(V), \Delta)$ with respect to right coactions (as opposed to both right and left coactions);
- to define a "right-regular" representation $x \mapsto \mathrm{Ad}_U(x)$ of each leg that commutes with the natural ("left-regular") representation on the Hilbert space underlying the multiplicative unitary.

These two purposes are addressed in Sections 9.3.1 and 9.3.2, where we introduce the notion of a balanced multiplicative unitary and a weak Kac system. Examples of weak Kac systems are discussed in Section 9.3.3.

We shall use the notation and concepts introduced in Chapter 7; in particular, the leg notation (see Notation 7.1.1) and slice maps (see Section 12.4).

### 9.3.1 Balanced multiplicative unitaries

A balanced multiplicative unitary is a multiplicative unitary $V$ equipped with an additional symmetry $U$ that allows us to identify the left and the right leg of $V$ with the right and the left leg of two auxiliary multiplicative unitaries $\check{V}$ and $\widehat{V}$. Then, the rôles of the right leg and the left leg get switched when we pass from $V$ to $\check{V}$ or $\widehat{V}$, so that we obtain a perfect symmetry between "right" and "left". This symmetry will be used in the construction of reduced crossed products for coactions of the legs of $V$ in Section 9.4. Let us turn to the precise definition.

Let $H$ be a Hilbert space and $U \in \mathcal{L}(H)$ a symmetry, that is, a self-adjoint unitary. As before, we denote by $\Sigma \in \mathcal{L}(H \otimes H)$ the flip $\eta \otimes \xi \mapsto \xi \otimes \eta$.

For each $T \in \mathcal{L}(H \otimes H)$, put

$$\check{T} := \Sigma(1 \otimes U)T(1 \otimes U)\Sigma, \quad \hat{T} := \Sigma(U \otimes 1)T(U \otimes 1)\Sigma. \tag{9.3}$$

The maps $T \mapsto \check{T}$ and $T \mapsto \hat{T}$ are automorphisms of order four. Indeed, the squares of these maps are given by $T \mapsto (U \otimes U)T(U \otimes U) = \mathrm{Ad}_{(U \otimes U)}(T)$ because

$$\Sigma(1 \otimes U)\Sigma(1 \otimes U) = U \otimes U = \Sigma(U \otimes 1)\Sigma(U \otimes 1),$$

and the square of the map $T \mapsto \mathrm{Ad}_{(U \otimes U)}(T)$ is equal to the identity because $(U \otimes U)^2 = \mathrm{id}_{H \otimes H}$. Furthermore, the relation $\Sigma(1 \otimes U)\Sigma(U \otimes 1) = \mathrm{id}_{H \otimes H}$ shows that the maps $T \mapsto \check{T}$ and $T \mapsto \hat{T}$ are inverse to each other.

**Definition 9.3.1.** A *balanced multiplicative unitary* on a Hilbert space $H$ is a pair $(V, U)$ consisting of a multiplicative unitary $V$ on $H$ and a symmetry $U$ on $H$ such that the unitaries $\check{V}$ and $\hat{V}$ defined in equation (9.3) are multiplicative.

The preceding definition is due to Baaj [5].

**Remarks 9.3.2.** i) In succinct leg notation,

$$\check{V} = \Sigma U_{[2]} V U_{[2]} \Sigma = U_{[1]} \Sigma V \Sigma U_{[1]} = U_{[1]} V_{[21]} U_{[1]},$$
$$\hat{V} = \Sigma U_{[1]} V U_{[1]} \Sigma = U_{[2]} \Sigma V \Sigma U_{[2]} = U_{[2]} V_{[21]} U_{[2]}.$$

ii) Since $\hat{V} = \mathrm{Ad}_{(U \otimes U)}(\check{V})$, the unitary $\hat{V}$ is multiplicative if and only if the unitary $\check{V}$ is multiplicative.

iii) The remarks preceding the definition above imply that iterated applications of the automorphisms $T \mapsto \check{T}$ and $T \mapsto \hat{T}$ to $V$ yield the operators

$$\check{\check{\check{V}}} = \hat{V}, \quad \check{\check{V}} = \mathrm{Ad}_{(U \otimes U)}(V) = \hat{\hat{V}}, \quad \check{V} = \hat{\hat{\hat{V}}}.$$

iv) If $(V, U)$ is a balanced multiplicative unitary, then also $(\check{V}, U)$, $(\hat{V}, U)$, and $(\mathrm{Ad}_{(U \otimes U)}(V),\ U)$ are balanced multiplicative unitaries. This follows immediately from iii).

v) If $(V, U)$ is a balanced multiplicative unitary, then also $(V^{\mathrm{op}}, U)$ is a balanced multiplicative unitary. Indeed,

$$\widehat{(V^{\mathrm{op}})} = (\check{V})^{\mathrm{op}}, \quad \check{(V^{\mathrm{op}})} = (\hat{V})^{\mathrm{op}}.$$

These relations follow from straightforward calculations, for example,

$$\widehat{(V^{\mathrm{op}})} = U_{[2]} \Sigma \Sigma V^* \Sigma \Sigma U_{[2]} = U_{[2]} V^* U_{[2]} = \Sigma \Sigma U_{[2]} V^* U_{[2]} \Sigma \Sigma = (\check{V})^{\mathrm{op}}.$$

Let $(V, U)$ be a balanced multiplicative unitary and assume that $V$ is well-behaved (Definition 7.2.6). Then we can associate to $V$ two $C^*$-bialgebras $(\widehat{A}(V), \widehat{\Delta})$ and $(A(V), \Delta)$, called the left and the right leg of $V$. Now we show that this left and right leg of $V$ can be identified with the right and the left legs of the auxiliary unitaries $\check{V}$ and $\widehat{V}$. Before we make this precise, let us consider the legs of the unitary $\check{\check{V}} = \mathrm{Ad}_{(U \otimes U)}(V) = \widehat{\widehat{V}}$:

**Lemma 9.3.3.** *Let $V$ be a well-behaved multiplicative unitary on a Hilbert space $H$ and let $U$ be a symmetry on $H$.*

i) *The $C^*$-algebras $\mathrm{Ad}_U(\widehat{A}(V))$ and $\mathrm{Ad}_U(A(V))$ are $C^*$-bialgebras with respect to the comultiplication given by*
$$\mathrm{Ad}_U(\hat{a}) \mapsto \mathrm{Ad}_{(U \otimes U)}(\widehat{\Delta}(\hat{a})) \quad \text{and} \quad \mathrm{Ad}_U(a) \mapsto \mathrm{Ad}_{(U \otimes U)}(\Delta(a)),$$
*where $\hat{a} \in \widehat{A}(V)$ and $a \in A(V)$, respectively.*

ii) *The operator $\check{\check{V}} = \mathrm{Ad}_{(U \otimes U)}(V) = \widehat{\widehat{V}}$ is a well-behaved multiplicative unitary, and the following $C^*$-bialgebras are isomorphic or equal, respectively:*
$$\widehat{A}(V) \cong \mathrm{Ad}_U(\widehat{A}(V)) = \widehat{A}(\check{\check{V}}), \quad A(V) \cong \mathrm{Ad}_U(A(V)) = A(\check{\check{V}}).$$

*Proof.* Straightforward. □

Let us turn to the legs of the unitaries $\check{V}$ and $\widehat{V}$:

**Proposition 9.3.4.** *Let $(V, U)$ be a balanced multiplicative unitary and assume that $V$ is well-behaved. Then also the multiplicative unitaries $\check{V}$ and $\widehat{V}$ are well-behaved, and the following $C^*$-bialgebras are equal:*
$$\widehat{A}(\check{V}) = \mathrm{Ad}_U(A(V)), \quad A(\check{V}) = \widehat{A}(V), \quad \widehat{A}(\widehat{V}) = A(V), \quad A(\widehat{V}) = \mathrm{Ad}_U(\widehat{A}(V)).$$

The proof of this proposition uses the following lemma.

**Lemma 9.3.5.** $V_{[13]} V_{[23]} \check{V}_{[12]} = \check{V}_{[12]} V_{[13]}$ *and* $\widehat{V}_{[23]} V_{[12]} V_{[13]} = V_{[13]} \widehat{V}_{[23]}$.

*Proof.* Let us prove the first equation. We insert the relation $\check{V} = U_{[1]} V_{[21]} U_{[1]}$ (see Remark 9.3.2 i)) into the pentagon equation $\check{V}_{[12]} \check{V}_{[13]} \check{V}_{[23]} = \check{V}_{[23]} \check{V}_{[12]}$ and obtain
$$U_{[1]} V_{[21]} U_{[1]} \cdot U_{[1]} V_{[31]} U_{[1]} \cdot \check{V}_{[23]} = \check{V}_{[23]} \cdot U_{[1]} V_{[21]} U_{[1]}.$$
Since $U_{[1]}$ commutes with $\check{V}_{[23]}$, we can cancel $U_{[1]}$ everywhere, and find
$$V_{[21]} V_{[31]} \check{V}_{[23]} = \check{V}_{[23]} V_{[21]}.$$

Now we conjugate both sides of this equation by the automorphism $\Sigma_{[23]}\Sigma_{[12]}\colon \eta\otimes\xi\otimes\zeta\mapsto\xi\otimes\zeta\otimes\eta$ of $H\otimes H\otimes H$, that is, we renumber the legs of the operators according to the permutation $(1,2,3)\mapsto(3,1,2)$, and obtain the first equation stated in the lemma. The proof of the second equation is similar. □

*Proof of Proposition* 9.3.4. We only prove the equation $A(\check{V}) = \hat{A}(V)$; the remaining equations can be proved by similar calculations or can be deduced from the former one using Lemma 9.3.3, Remark 9.3.2 v), and Proposition 7.2.11. Denote by $H$ the Hilbert space underlying $V$. Let $\omega\in\mathcal{L}(H)_*$ and put $\omega' := \omega\circ\mathrm{Ad}_U$. Then

$$\begin{aligned}(\omega\mathbin{\bar{\otimes}}\mathrm{id})(\check{V}) &= (\omega\mathbin{\bar{\otimes}}\mathrm{id})(\Sigma U_{[2]}VU_{[2]}\Sigma)\\ &= (\mathrm{id}\mathbin{\bar{\otimes}}\omega)(U_{[2]}VU_{[2]}) = (\mathrm{id}\mathbin{\bar{\otimes}}\omega')(V).\end{aligned}$$

Since the map $\mathcal{L}(H)_*\to\mathcal{L}(H)_*$ given by $\omega\mapsto\omega\circ\mathrm{Ad}_U$ is bijective, it follows that $A_0(\check{V}) = \hat{A}_0(V)$, and hence also $A(\check{V}) = \hat{A}(V)$. In particular, $A(\check{V})$ is a $C^*$-algebra.

Let us determine the action of the maps $\Delta_{\check{V}}$ and $\hat{\Delta}_V$ on an element of the form $\hat{a} := (\mathrm{id}\mathbin{\bar{\otimes}}\omega')(V)$, where $\omega'\in\mathcal{L}(H)_*$. The previous lemma implies

$$\begin{aligned}\Delta_{\check{V}}(\hat{a}) = \check{V}(\hat{a}\otimes 1)\check{V}^* &= (\mathrm{id}\mathbin{\bar{\otimes}}\mathrm{id}\mathbin{\bar{\otimes}}\omega')(\check{V}_{[12]}V_{[13]}\check{V}^*_{[12]})\\ &= (\mathrm{id}\mathbin{\bar{\otimes}}\mathrm{id}\mathbin{\bar{\otimes}}\omega')(V_{[13]}V_{[23]}),\end{aligned}$$

and the pentagon equation for $V$ implies

$$\begin{aligned}\hat{\Delta}_V(\hat{a}) = V^*(1\otimes\hat{a})V &= (\mathrm{id}\mathbin{\bar{\otimes}}\mathrm{id}\mathbin{\bar{\otimes}}\omega')(V^*_{[12]}V_{[23]}V_{[12]})\\ &= (\mathrm{id}\mathbin{\bar{\otimes}}\mathrm{id}\mathbin{\bar{\otimes}}\omega')(V_{[13]}V_{[23]}),\end{aligned}$$

that is, $\Delta_{\check{V}}(\hat{a}) = \hat{\Delta}_V(\hat{a})$. Since $A_0(\check{V})$ and $\hat{A}_0(V)$ are dense in $A(\check{V})$ and $\hat{A}(V)$, respectively, it follows that $\Delta_{\check{V}}(\hat{a}) = \hat{\Delta}_V(\hat{a})$ for all $\hat{a}\in\hat{A}(V) = A(\check{V})$. Therefore, $(A(\check{V}),\Delta_{\check{V}})$ is a bisimplifiable $C^*$-bialgebra, and equal to $(\hat{A}(V),\hat{\Delta}_V)$.

To prove that $\check{V}$ and $\hat{V}$ are well-behaved, it only remains to show that $\check{V}\in M(\hat{A}(\check{V})\otimes A(\check{V}))$ and $\hat{V}\in M(\hat{A}(\hat{V})\otimes A(\hat{V}))$. But since $V\in M(\hat{A}(V)\otimes A(V))$,

$$\begin{aligned}\check{V} = \Sigma U_2VU_2\Sigma &\in \Sigma U_2 M(\hat{A}(V)\otimes A(V))U_2\Sigma\\ &= M(\mathrm{Ad}_U(A(V))\otimes\hat{A}(V)) = M(\hat{A}(\check{V})\otimes A(\check{V})),\end{aligned}$$

and a similar calculation shows that $\hat{V}\in M(\hat{A}(\hat{V})\otimes A(\hat{V}))$. □

### 9.3.2 Weak Kac systems

In typical examples of balanced multiplicative unitaries, conjugation by the symmetry $U$ facilitates the transition between the left-regular and the right-regular representation of the legs of $V$. A natural condition to impose is that for each leg, these two representations commute. This condition leads to the following definition:

**Definition 9.3.6.** We call a balanced multiplicative unitary $(V, U)$ a *weak Kac system* if $V$ is well-behaved and $V_{[23]}\widehat{V}_{[12]} = \widehat{V}_{[12]}V_{[23]}$ and $V_{[12]}\check{V}_{[23]} = \check{V}_{[23]}V_{[12]}$.

**Remark 9.3.7.** In [182], Vergnioux introduced a notion of a weak Kac system that includes an additional condition which is needed for the construction of full crossed products. For our purposes, the weaker form given above is sufficient.

The preceding definition is motivated by the following observation:

**Lemma 9.3.8.** *Let $V$ be a multiplicative unitary on a Hilbert space $H$ and $U$ a symmetry on $H$. Then*

$$V_{[23]}\widehat{V}_{[12]} = \widehat{V}_{[12]}V_{[23]} \Leftrightarrow [\widehat{A}(V), \mathrm{Ad}_U(\widehat{A}(V))] = 0,$$
$$V_{[12]}\check{V}_{[23]} = \check{V}_{[23]}V_{[12]} \Leftrightarrow [A(V), \mathrm{Ad}_U(A(V))] = 0.$$

*Proof.* We only prove the first equivalence; the second one follows similarly. Since maps of the form $\omega \mathbin{\bar{\otimes}} \mathrm{id} \mathbin{\bar{\otimes}} \omega' \colon \mathcal{L}(H \otimes H \otimes H) \to \mathcal{L}(H)$, where $\omega, \omega' \in \mathcal{L}(H)_*$, separate the elements of $\mathcal{L}(H \otimes H \otimes H)$, the relation $V_{[23]}\widehat{V}_{[12]} = \widehat{V}_{[12]}V_{[23]}$ holds if and only if for all $\omega, \omega' \in \mathcal{L}(H)_*$, the operator

$$(\omega \mathbin{\bar{\otimes}} \mathrm{id} \mathbin{\bar{\otimes}} \omega')(V_{[23]}\widehat{V}_{[12]}) = (\mathrm{id} \mathbin{\bar{\otimes}} \omega')(V) \cdot (\omega \mathbin{\bar{\otimes}} \mathrm{id})(\widehat{V})$$

is equal to

$$(\omega \mathbin{\bar{\otimes}} \mathrm{id} \mathbin{\bar{\otimes}} \omega')(\widehat{V}_{[12]}V_{[23]}) = (\omega \mathbin{\bar{\otimes}} \mathrm{id})(\widehat{V}) \cdot (\mathrm{id} \mathbin{\bar{\otimes}} \omega')(V).$$

By definition and by a similar calculation as in the proof of Proposition 9.3.4, elements of the form $(\mathrm{id} \mathbin{\bar{\otimes}} \omega')(V)$ and $(\omega \mathbin{\bar{\otimes}} \mathrm{id})(\widehat{V})$, where $\omega, \omega' \in \mathcal{L}(H)_*$, are dense in $\widehat{A}(V)$ and $\mathrm{Ad}_U(\widehat{A}(V))$, respectively. Hence, the operators above are equal for all $\omega, \omega' \in \mathcal{L}(H)_*$ if and only if $\hat{a}\,\mathrm{Ad}_U(\hat{b}) = \mathrm{Ad}_U(\hat{b})\hat{a}$ for all $\hat{a}, \hat{b} \in \widehat{A}(V)$. □

The notion of a weak Kac system is symmetric in the following sense:

**Proposition 9.3.9.** *If $(V, U)$ is a weak Kac system, then also $(\check{V}, U)$, $(\widehat{V}, U)$, and $(V^{\mathrm{op}}, U)$ are weak Kac systems.*

*Proof.* If $(V, U)$ is a weak Kac system, then

- $\check{V}$, $\widehat{V}$, and $V^{\mathrm{op}}$ are well-behaved by Proposition 9.3.4 and 7.2.11;

- $(\check{V}, U)$, $(\widehat{V}, U)$, and $(V^{\mathrm{op}}, U)$ are balanced multiplicative unitaries by Remarks 9.3.2 iv) and v);
- $(\check{V}, U)$, $(\widehat{V}, U)$, and $(V^{\mathrm{op}}, U)$ satisfy the last condition in Definition 9.3.6 by Lemma 9.3.8 and Propositions 9.3.4, 7.2.11. □

**Definition 9.3.10.** Let $(V, U)$ be a weak Kac system. Then $(\check{V}, U)$, $(\widehat{V}, U)$, and $(V^{\mathrm{op}}, U)$ are called the *predual*, the *dual*, and the *opposite* weak Kac system of $(V, U)$, respectively.

### 9.3.3 Examples of weak Kac systems

To every locally compact group, to every compact quantum group, and, more generally, to every locally compact quantum group, one can associate a weak Kac system:

**Example 9.3.11.** Let $G$ be a locally compact group with left Haar measure $\lambda$. In Example 7.1.4, we defined a well-behaved multiplicative unitary $W_G$ on the Hilbert space $L^2(G, \lambda)$ via

$$(W_G \zeta)(x, y) = \zeta(x, x^{-1}y) \quad \text{for all } x, y \in G,\ \zeta \in L^2(G \times G, \lambda \times \lambda);$$

here, we identified $L^2(G, \lambda) \otimes L^2(G, \lambda)$ with $L^2(G \times G, \lambda \times \lambda)$. Now we use the group inversion on $G$ to construct a symmetry $U$ on $L^2(G, \lambda)$ such that $(W_G, U)$ is a weak Kac system.

Denote by $\lambda^{-1}$ the right Haar measure associated to the left Haar measure $\lambda$, and by $\delta = d\lambda/d\lambda^{-1}$ the modular function of $G$ (see Section 2.2.3). Then, it is easy to check that the formula

$$(U\xi)(x) := \xi(x^{-1})\delta(x)^{-1/2} \quad \text{for all } x \in G,\ \xi \in L^2(G, \lambda),$$

defines a symmetry $U$ on $L^2(G, \lambda)$ – use equation (2.8) from page 55 and the fact that $\delta$ is a homomorphism.

To prove that $(W_G, U)$ is a balanced multiplicative unitary, we determine the unitary $\widehat{W}_G = \Sigma U_{[1]} W_G U_{[1]} \Sigma$. For all $x, y \in G$ and $\zeta \in L^2(G \times G, \lambda \times \lambda)$,

$$\begin{aligned}(\widehat{W}_G \zeta)(x, y) &= (W_G U_{[1]} \Sigma \zeta)(y^{-1}, x) \cdot \delta(y)^{-1/2} \\ &= (U_{[1]} \Sigma \zeta)(y^{-1}, yx) \cdot \delta(y)^{-1/2} \\ &= \zeta(yx, y) \cdot \delta(y)^{-1/2} \delta(y^{-1})^{-1/2} = \zeta(yx, y).\end{aligned}$$

A comparison with the formula for $W_G^{\mathrm{op}}$ given at the end of Example 7.1.4 shows that $\widehat{W}_G = W_G^{\mathrm{op}}$. Therefore, $\widehat{W}_G$ is a multiplicative unitary and $(W_G, U)$ is a balanced multiplicative unitary.

A similar calculation as above shows that the unitary $\check{W}_G$ is given by

$$(\check{W}_G\zeta)(x,y) = \zeta(xy,y)\cdot\delta(y)^{1/2} \quad \text{for all } x,y \in G,\ \zeta \in L^2(G\times G,\lambda\times\lambda),$$

and that $\check{W}_G$ is equivalent to the multiplicative unitary $V_G$ on $L^2(G,\lambda^{-1})$ (see Example 7.1.4) with respect to the identification $L^2(G,\lambda) \xrightarrow{\cong} L^2(G,\lambda^{-1})$ given by $\xi \mapsto \xi\delta^{1/2}$.

Let us check that the pair $(W_G, U)$ forms a weak Kac system. Straightforward calculations show that for all $\zeta \in L^2(G\times G\times G, \lambda\times\lambda\times\lambda)$ and $x,y,z \in G$,

$$(\widehat{W}_{G[12]}W_{G[23]}\zeta)(x,y,z) = \zeta(yx,y,y^{-1}z) = (W_{G[23]}\widehat{W}_{G[12]}\zeta)(x,y,z),$$
$$(W_{G[12]}\check{W}_{G[23]}\zeta)(x,y,z) = \zeta(x,x^{-1}yz,z)\delta(z)^{1/2} = (\check{W}_{G[23]}W_{G[12]}\zeta)(x,y,z).$$

For completeness, we also determine the representation $\mathrm{Ad}_U$ of $\widehat{A}(W_G)$ and of $A(W_G)$ on $L^2(G,\lambda)$. By Example 7.2.13,

- $\widehat{A}(W_G) = \pi_M(C_0(G))$, where $\pi_M\colon C_0(G) \to \mathcal{L}(L^2(G,\lambda))$ denotes the representation via multiplication operators, and
- $A(W_G) = L(C^*(G)) = C^*_r(G)$, where $L\colon C^*(G) \to \mathcal{L}(L^2(G,\lambda))$ denotes the left regular representation.

Simple calculations show that

- $\mathrm{Ad}_U \circ \pi_M = \pi_M \circ \theta$, where $\theta$ denotes the automorphism of $C_0(G)$ given by $(\theta(g))(x) = g(x^{-1})$ for all $x \in G$, $g \in C_0(G)$;
- $\mathrm{Ad}_U \circ L$ is the right regular representation of $C^*(G)$ on $L^2(G,\lambda)$ given by

  $$((\mathrm{Ad}_U \circ L)(f)\xi)(x) = \int_G f(y)\delta(y)^{1/2}\xi(xy)d\lambda(y)$$

  for all $\xi \in L^2(G,\lambda)$, $x \in G$, $f \in L^1(G,\lambda)$.

**Example 9.3.12.** To every compact quantum group, one can associate a weak Kac system as follows. Let us start from an algebraic compact quantum group $(A_0,\Delta_0)$; given a $C^*$-algebraic compact quantum group $(A,\Delta)$, simply choose $(A_0,\Delta_0)$ to be the associated Hopf $*$-algebra of matrix coefficients of finite-dimensional corepresentations (see Theorem 5.4.1).

Denote by $h_0$ the Haar state of $(A_0,\Delta_0)$, and by $H$ the completion of $A_0$ with respect to the inner product given by $\langle a|b\rangle := h_0(a^*b)$ for all $a,b \in A_0$. In Example 7.1.6, we constructed a multiplicative unitary $V = V_{A_0}$ on $H$ such that

$$V(a \odot b) = \Delta_0(a)(1 \odot b) \quad \text{for all } a,b \in A_0 \subseteq H.$$

Now we use the antipode $S_0$ of $A_0$ and the characters $(f_z)_{z\in\mathbb{C}}$ introduced in Theorem 3.2.19 to construct a symmetry $U$ on $H$ such that $(V, U)$ is a weak Kac system.

Given $z, z' \in \mathbb{C}$, denote by $\rho_{z,z'}$ the automorphism of $A_0$ given by $a \mapsto f_z * a * f_{z'} = \sum f_{z'}(a_{(1)})a_{(2)} f_z(a_{(3)})$. We claim that the bijection

$$U_0 \colon A_0 \to A_0, \quad a \mapsto \rho_{1,0}(S_0(a)),$$

extends to a symmetry $U$ on $H$. By Corollary 3.2.20, $U_0 = \rho_{1,0} \circ S_0 = S_0 \circ \rho_{0,-1}$ and

$$U_0 \circ U_0 = \rho_{1,0} \circ S_0 \circ S_0 \circ \rho_{0,-1} = \rho_{1,0} \circ \rho_{-1,1} \circ \rho_{0,-1} = \mathrm{id}_{A_0}.$$

It remains to show that $h_0(U_0(a)^* U_0(b)) = h_0(a^* b)$ for all $a, b \in A_0$. Using Corollary 3.2.20 and the relation $* \circ S_0 = S_0^{-1} \circ *$ from Proposition 1.3.28, we find

$$* \circ U_0 = * \circ \rho_{1,0} \circ S_0 = \rho_{-1,0} \circ * \circ S_0 = \rho_{-1,0} \circ S_0^{-1} \circ * = \rho_{0,-1} \circ S_0 \circ *.$$

We insert this relation into the expression $h_0(U_0(a)^* U_0(b))$, use Theorem 3.2.19 v), Corollary 3.2.20, and the identity $h_0 \circ S_0 = h_0$ from Proposition 2.2.6, and get

$$\begin{aligned} h_0(U_0(a)^* U_0(b)) &= h_0(\rho_{0,-1}(S_0(a^*))\rho_{1,0}(S_0(b))) \\ &= h_0(\rho_{1,0}(S_0(b)S_0(a^*))) = h_0(S_0(a^* b)) = h_0(a^* b). \end{aligned}$$

Consider the unitary $\widehat{V} = \Sigma U_{[1]} V U_{[1]} \Sigma$. We claim that $\widehat{V}$ is equal to the multiplicative unitary $W = W_{A_0}$ constructed in Example 7.1.6, whose adjoint is given by

$$W^*(a \odot b) = \Delta_0(b)(a \odot 1) \quad \text{for all } a, b \in A_0 \subseteq H.$$

In particular, this claim implies that $(V, U)$ is a balanced multiplicative unitary. To prove this claim, we need to compute the composition $\Delta_0 \circ U_0$. By Corollary 3.2.20 and Proposition 1.3.12,

$$\begin{aligned} \Delta_0 \circ U_0 &= \Delta_0 \circ \rho_{1,0} \circ S_0 = (\mathrm{id} \odot \rho_{1,0}) \circ \Delta_0 \circ S_0 \\ &= \Sigma \circ (\rho_{1,0} \odot \mathrm{id}) \circ (S_0 \odot S_0) \circ \Delta_0 = \Sigma \circ (U_0 \odot S_0) \circ \Delta_0. \end{aligned}$$

The relation $W = \Sigma U_{[1]} V U_{[1]} \Sigma$ is equivalent to $W^* \Sigma U_{[1]} V = \Sigma U_{[1]}$. The operator $W^* \Sigma U_{[1]} V$ acts on an element $a \odot b \in A_0 \odot A_0 \subseteq H \otimes H$ as follows:

$$\begin{aligned} a \odot b \xmapsto{V} \sum a_{(1)} \odot a_{(2)} b &\xmapsto{\Sigma(U_0 \odot 1)} \sum a_{(2)} b \odot U_0(a_{(1)}) \\ &\xmapsto{W^*} \sum \Delta_0(U_0(a_{(1)}))(a_{(2)} b \odot 1). \end{aligned}$$

We insert the formula for $\Delta_0 \circ U_0$ derived above, use the defining properties of the antipode and counit (see also Example 1.3.4 ii)), and find that for all $a, b \in A_0$,

$$\begin{aligned} W^*\Sigma U_{[1]}V(a \odot b) &= \sum S_0(a_{(2)})a_{(3)}b \odot U_0(a_{(1)}) \\ &= b \odot U_0(a) = \Sigma U_{[1]}(a \odot b). \end{aligned}$$

Thus $\widehat{V} = W$, and the claim is proved.

The operator $\check{V} = \Sigma U_{[2]} V U_{[2]} \Sigma$ acts on an element $a \odot b \in A_0 \odot A_0$ as follows:

$$\begin{aligned} a \odot b &\xmapsto{(1 \odot U_0)\Sigma} b \odot U_0(a) \xmapsto{V} \sum b_{(1)} \odot b_{(2)}U_0(a) \\ &\xmapsto{\Sigma(1 \odot U_0)} \sum U_0(b_{(2)}U_0(a)) \odot b_{(1)}. \end{aligned}$$

Since $S_0$ is an antihomomorphism and $\rho_{1,0}$ is a homomorphism, the composition $U_0 = \rho_{1,0} \circ S_0$ is an antihomomorphism. Hence,

$$\check{V}(a \odot b) = \sum U_0^2(a)U_0(b_{(2)}) \odot b_{(1)} = \sum aU_0(b_{(2)}) \odot b_{(1)}$$

for all $a, b \in A_0 \subseteq H$.

Straightforward calculations show that the pair $(V, U)$ is a weak Kac system:

$$\begin{aligned} \widehat{V}^*_{[12]} V_{[23]}(a \odot b \odot c) &= \sum b_{(1)}a \odot b_{(2)} \odot b_{(3)}c \\ &= V_{[23]}\widehat{V}^*_{[12]}(a \odot b \odot c), \\ V_{[12]}\check{V}_{[23]}(a \odot b \odot c) &= \sum a_{(1)} \odot a_{(2)}bU_0(c_{(2)}) \odot c_{(1)} \\ &= \check{V}_{[23]}V_{[12]}(a \odot b \odot c) \end{aligned}$$

for all $a, b, c \in A_0 \subseteq H$.

For completeness, let us determine the representation $\mathrm{Ad}_U$ of $\widehat{A}(V)$ and of $A(V)$ on $H$. From Theorem 7.2.14, we know that

- $\widehat{A}(V) = \overline{\operatorname{span}}\, \rho(\widehat{A}_0)$, where $\widehat{A}_0$ is the dual of $A_0$ defined in Section 2.3, and $\rho\colon \widehat{A}_0 \to \mathcal{L}(H)$ is the $*$-homomorphism given by $\rho(\hat{a})b = \sum b_{(1)}\hat{a}(b_{(2)})$ for all $\hat{a} \in \widehat{A}_0$ and $b \in A_0 \subseteq H$;

- $A(V) = \overline{\operatorname{span}}\, \pi(A_0)$, where $\pi\colon A_0 \to \mathcal{L}(H)$ is the $*$-homomorphism given by $\pi(a)b = ab$ for all $a, b \in A_0$.

The representation $\mathrm{Ad}_U \circ \pi$ is easily computed: since $U_0$ is an antihomomorphism and $U_0^2 = \mathrm{id}$,

$$U_0\pi(a)U_0 b = U_0(aU_0(b)) = bU_0(a) \quad \text{for all } a, b \in A_0.$$

Next, we consider the representation $\mathrm{Ad}_U \circ \rho$. Combining the formula for $\Delta_0 \circ U_0$ derived above with the relations $U_0 \circ S_0 = (U_0)^2 \circ \rho_{0,1} = \rho_{0,1}$, $U_0 = S_0 \circ \rho_{0,-1}$, and the identity

$$\sum b_{(1)} \odot \rho_{0,1}(b_{(2)}) = \sum b_{(1)} \odot f_1(b_{(2)}) b_{(3)} = \sum \rho_{1,0}(b_{(1)}) \odot b_{(2)},$$

we find that for all $\hat{a} \in \widehat{A}_0$ and $b \in A_0$,

$$\begin{aligned} U_0 \rho(\hat{a}) U_0 b &= \sum U_0(U_0(b)_{(1)}) \hat{a}(U_0(b)_{(2)}) \\ &= \sum U_0(S_0(b_{(2)})) \hat{a}(U_0(b_{(1)})) \\ &= \sum \rho_{0,1}(b_{(2)}) \hat{a}(S_0(\rho_{0,-1}(b_{(1)}))) \\ &= \sum (\hat{a} \circ S_0 \circ \rho_{1,-1})(b_{(1)}) b_{(2)}. \end{aligned}$$

**Example 9.3.13.** To every locally compact quantum group, one can associate a weak Kac system as follows. Let us start from a reduced $C^*$-algebraic quantum group $(A, \Delta)$ with left Haar weight $\phi$ and right Haar weight $\psi$; for a locally compact quantum group in the setting of von Neumann algebras or a universal $C^*$-algebraic quantum group, the construction is completely analogous.

Let $(H, \pi, \Lambda)$ be a GNS-construction for the left Haar weight $\phi$. In Section 8.3.1, we introduced a multiplicative unitary $W_A := W$ on $H$, given by

$$W_A^*(\Lambda(a) \otimes \Lambda(b)) = (\Lambda \otimes \Lambda)(\Delta(b)(a \otimes 1)) \quad \text{for all } a, b \in \mathcal{N}_\phi.$$

Consider the operator $U := \nu^{i/8} J I$ on $H$, where

- $\nu$ is the scaling constant of $(A, \Delta)$, see Proposition 8.3.8;
- $J$ is the modular conjugation of $\phi$ on $H$, see Theorem 8.2.4;
- $I$ denotes the conjugate-linear isometry on $H$ constructed in Theorem 8.3.4.

We show that $(W_A, U)$ is a weak Kac system.

Before we begin the proof, let us note that $I$ is equal to the modular conjugation $\widehat{J}$ associated to the left Haar weight $\hat{\phi}$ on the dual $(\widehat{A}, \widehat{\Delta})$ and the GNS-map $\widehat{\Lambda} \colon \mathcal{N}_{\hat{\phi}} \to H$ constructed in Theorem 8.3.12, see [93, Corollary 2.9] or [158, Corollary 1.13.12].

Let us prove that $U$ is a symmetry. By construction, the maps $J$ and $I$ are conjugate-linear, isometric, and satisfy $J^* = J$, $I^* = I$, $J^2 = \mathrm{id}_H = I^2$. Moreover, by [93, Corollary 2.12] or [158, Corollary 1.13.15], $IJ = \widehat{J}J = \nu^{i/4} J \widehat{J} = \nu^{i/4} J I$. Combining these relations, we find that the operator $U$ is linear, isometric, and satisfies $U^2 = \nu^{i/4} J I J I = I J^2 I = \mathrm{id}_H$.

The unitary $\widetilde{W_A} = \Sigma U_{[2]} W_A U_{[2]} \Sigma$ can be described in terms of a certain GNS-map $\Gamma\colon \mathcal{N}_\psi \to H$ for the right Haar weight $\psi$ of $(A, \Delta)$, defined in [91, Notation 7.13] and [158, Notation 1.9.12], as follows. By [158, Section 1.13], there exists a multiplicative unitary $V_A$ on $H$ such that

$$V_A(\Gamma(a) \otimes \Gamma(b)) = (\Gamma \otimes \Gamma)(\Delta(a)(1 \otimes b)) \quad \text{for all } a, b \in \mathcal{N}_\psi,$$

and by [93, Corollary 2.2, Proposition 2.15] or [158, Corollary 1.13.3, Proposition 1.13.18],

$$(I \otimes J) W_A (I \otimes J) = W_A^*, \quad (I \otimes I) \Sigma W_A^* \Sigma (I \otimes I) = V_A.$$

Combining these relations, we find

$$\begin{aligned} \widetilde{W_A} &= \Sigma(1 \otimes IJ) W_A (1 \otimes JI) \Sigma = (I \otimes I) \Sigma (I \otimes J) W_A (I \otimes J) \Sigma (I \otimes I) \\ &= (I \otimes I) \Sigma W_A^* \Sigma (I \otimes I) = V_A. \end{aligned}$$

In particular, $\widetilde{W_A}$ is multiplicative and $(W_A, U)$ is a balanced multiplicative unitary.

The unitary $\widehat{W_A} = \Sigma U_{[1]} W_A U_{[1]} \Sigma$ does not coincide with the multiplicative unitary of the reduced dual $(\widehat{A}, \widehat{\Delta})$, which is usually denoted by $\widehat{W}$.[1] Using [93, Proposition 2.15] or [158, Proposition 1.13.18], one can show that the opposite $(\widehat{W_A})^{\mathrm{op}}$ of $\widehat{W_A}$ is equal to the unitary $V_{\widehat{A}}$ given by

$$V_{\widehat{A}}(\widehat{\Gamma}(\hat{a}) \otimes \widehat{\Gamma}(\hat{b})) = (\widehat{\Gamma} \otimes \widehat{\Gamma})(\widehat{\Delta}(\hat{a})(1 \otimes \hat{b})) \quad \text{for all } \hat{a}, \hat{b} \in \mathcal{N}_{\hat{\psi}};$$

here, $\widehat{\Gamma}\colon \mathcal{N}_{\hat{\psi}} \to H$ is a certain GNS-map for the right Haar weight $\hat{\psi}$ of $(\widehat{A}, \widehat{\Delta})$.

Next, consider the $C^*$-algebras $\widehat{A}(W_A)$ and $\mathrm{Ad}_U(\widehat{A}(W_A))$. By Proposition 8.3.2, $\widehat{A}(W_A) = \pi(A)$, and by Theorem 8.3.4 iv) and Theorem 8.2.4, the $C^*$-algebra

$$\mathrm{Ad}_U(\widehat{A}(W_A)) = U\pi(A)U = JI\pi(A)IJ = J\pi(R(A))^*J = J\pi(A)J \tag{9.4}$$

commutes with $\pi(A) = \widehat{A}(W_A)$.

A similar argument shows that the $C^*$-algebras $A(W_A)$ and $\mathrm{Ad}_U(A(W_A))$ commute – use the relation $A(W_A) = \widehat{A}$ (see Section 8.3.3), the symmetry in the definition of $U$ apparent from the equation $U = \nu^{i/4} J\widehat{J}$, and the biduality Theorem 8.3.15.

Summarizing, we find that $(W_A, U)$ is a weak Kac system as claimed.

---

[1] Here, notation does not match nicely; in the present context, the notation $W_{\widehat{A}}$ instead of $\widehat{W}$ for the multiplicative unitary of $(\widehat{A}, \widehat{\Delta})$ may be preferable.

## 9.4 Reduced crossed products and dual coactions

The concept of a weak Kac system introduced in the previous subsection provides the right framework for the construction of reduced crossed products, which is the topic of this subsection. We proceed as follows. Throughout this subsection, we fix a weak Kac system $(V, U)$ on a Hilbert space $H$ and consider coactions of the right leg $(A(V), \Delta)$ and the left leg $(\widehat{A}(V), \widehat{\Delta})$.

Given a coaction $(C, \delta)$ of $(A(V), \Delta)$, we define the reduced crossed product $C \rtimes_{\delta,r} \widehat{A}(V)$ and construct a dual coaction $\hat{\delta}$ of $(\widehat{A}(V), \widehat{\Delta})$ on $C \rtimes_{\delta,r} \widehat{A}(V)$, using the auxiliary multiplicative unitary $\check{V}$ introduced in the previous subsection. We show that $\hat{\delta}$ turns $C \rtimes_{\delta,r} \widehat{A}(V)$ into an $(\widehat{A}(V), \widehat{\Delta})$-$C^*$-algebra and that the assignment $(C, \delta) \mapsto (C \rtimes_{\delta,r} \widehat{A}(V), \hat{\delta})$ is functorial.

Next, we carry over these constructions and results to coactions of $(\widehat{A}(V), \widehat{\Delta})$. It is tempting to simply exchange $(A(V), \Delta)$ and $(\widehat{A}(V), \widehat{\Delta})$ in the preceding definitions, but we have to replace the weak Kac system $(V, U)$ by its dual $(\widehat{V}, U)$ or predual $(\check{V}, U)$.

Finally, we explain how these constructions for coactions relate to the corresponding constructions for group actions presented in Section 9.1.

### 9.4.1 The reduced crossed product of a coaction of $A(V)$

We define the reduced crossed product $C \rtimes_{\delta,r} \widehat{A}(V)$ of a coaction $(C, \delta)$ of $(A(V), \Delta)$ using the language of $C^*$-modules. The $C^*$-algebra $C$ of the coaction can be considered as a $C^*$-module over itself, and the Hilbert space $H$ of the weak Kac system can be considered as a $C^*$-module over $\mathbb{C}$. Thus we can form a tensor product $C \otimes H$, which is a $C^*$-module over $C$ (see Section 12.2).

**Definition 9.4.1.** Let $(V, U)$ be a weak Kac system on a Hilbert space $H$ and $(C, \delta)$ a coaction of $(A(V), \Delta)$. The *reduced crossed product* of $(C, \delta)$ is the $C^*$-subalgebra $C \rtimes_{\delta,r} \widehat{A}(V) \subseteq \mathcal{L}_C(C \otimes H)$ generated by the subset

$$\delta(C)(1 \otimes \widehat{A}(V)) \subseteq \mathcal{L}_C(C \otimes H). \tag{9.5}$$

If the coaction $\delta$ is understood, we shortly write $C \rtimes_r \widehat{A}(V)$ for $C \rtimes_{\delta,r} \widehat{A}(V)$.

**Remarks 9.4.2.** i) In formula (9.5), we use the extension of the non-degenerate embedding $C \otimes A(V) \hookrightarrow \mathcal{L}_C(C \otimes H)$ to define the composition $C \xrightarrow{\delta} M(C \otimes A(V)) \hookrightarrow \mathcal{L}_C(C \otimes H)$.

i) The $C^*$-algebra $C \rtimes_{\delta,r} \widehat{A}(V)$ can also be defined without reference to $C^*$-modules, using a non-degenerate faithful representation of $C$ on some Hilbert space $K$: If $\pi : C \to \mathcal{L}(K)$ is such a representation, then $C \rtimes_{\delta,r} \widehat{A}(V)$ is isomorphic to the $C^*$-subalgebra of $\mathcal{L}(K \otimes H)$ generated by the set $(\pi \otimes \mathrm{id})(\delta(C)) \cdot (1 \otimes \widehat{A}(V))$.

To see this, consider $K$ as a $C^*$-module over $\mathbb{C}$. Then the internal tensor product $(C \otimes H) \otimes_\pi K$ is a Hilbert space that is isomorphic to $K \otimes H$ via $(c \otimes \xi) \otimes_\pi \eta \equiv \pi(c)\eta \otimes \xi$ (see Section 12.2). Now the map $\mathrm{Ind}_\pi : \mathcal{L}_C(C \otimes H) \to \mathcal{L}((C \otimes H) \otimes_\pi K) \cong \mathcal{L}(K \otimes H)$ is an injective $*$-homomorphism (see Section 12.2) that maps $\delta(C)(1 \otimes \widehat{A}(V))$ onto $(\pi \otimes \mathrm{id})(\delta(C)) \cdot (1 \otimes \widehat{A}(V))$, and $C \rtimes_{\delta,r} \widehat{A}(V)$ onto the $C^*$-algebra generated by that subset.

Like the crossed product of a group action, the reduced crossed product of a coaction can be considered as a twisted tensor product. In the present setting, it is difficult to make this precise, but see the following example:

**Example 9.4.3.** Consider the weak Kac system $(V, U)$ associated to an algebraic compact quantum group $(A_0, \Delta_0)$ (see Example 9.3.12). Recall that $A_0$ embeds in $A(V)$ as a dense $*$-subalgebra and that the comultiplication $\Delta$ of $A(V)$ extends the comultiplication $\Delta_0$ of $A_0$ (see Theorem 7.2.14).

Assume that $(C, \delta)$ is a coaction of $(A(V), \Delta)$ and that there exists a dense $*$-subalgebra $C_0 \subseteq C$ such that $\delta(C_0)$ is contained in the algebraic tensor product $C_0 \odot A_0$. Denote by $\delta_0 \colon C_0 \to C_0 \odot A_0$ the restriction of $\delta$ to $C_0$.

The reduced crossed product $C \rtimes_{\delta,r} \widehat{A}(V)$ can be considered as the completion of a twisted tensor product $C_0 \rtimes \widehat{A}_0$ that is spanned by products of two $*$-algebras $C_0 \rtimes 1 \cong \delta(C_0)$ and $1 \rtimes \widehat{A}_0 \cong \widehat{A}_0$. Here, $(\widehat{A}_0, \widehat{\Delta}_0)$ denotes the dual of $(A_0, \Delta_0)$ as usual. Moreover, the twist governing the commutation relations between $C_0 \rtimes 1$ and $1 \rtimes \widehat{A}_0$ can be described in terms of $\delta_0$ as follows. Recall from Example 9.3.12 that

- the underlying Hilbert space $H$ of $(V, U)$ is the completion of $A_0$ with respect to the inner product induced by the Haar state;
- $\widehat{A}(V) = \overline{\operatorname{span}}\, \rho(\widehat{A}_0)$, where $\rho \colon \widehat{A}_0 \to \mathcal{L}(H)$ is the $*$-homomorphism given by $\rho(\hat{a})b = \sum b_{(1)}\hat{a}(b_{(2)})$ for all $\hat{a} \in \widehat{A}_0$ and $b \in A_0 \subseteq H$;
- $A(V) = \overline{\operatorname{span}}\, \pi(A_0)$, where $\pi : A_0 \to \mathcal{L}(H)$ is the $*$-homomorphism given by $\pi(a)b = ab$ for all $a, b \in A_0$.

By definition, the $C^*$-algebra $C \rtimes_{\delta,r} \widehat{A}(V) \subseteq \mathcal{L}_C(C \otimes H)$ is generated by products of the form $(c \rtimes 1)(1 \rtimes \hat{a})$, where

$$c \rtimes 1 := (\mathrm{id} \odot \pi)(\delta(c)), \quad 1 \rtimes \hat{a} := \mathrm{id}_C \odot \rho(\hat{a}),$$

and $c \in C_0$, $\hat{a} \in \widehat{A}_0$. Let us compute the product

$$(1 \rtimes \hat{a})(c \rtimes 1) = (\mathrm{id}_C \odot \rho(\hat{a})) \cdot (\mathrm{id} \odot \pi)(\delta(c)) \in \mathcal{L}_C(C \otimes H).$$

We apply this operator to an element $d \odot b \in C_0 \odot A_0 \subseteq C \otimes H$, and obtain

$$(1 \rtimes \hat{a})(c \rtimes 1)(d \odot b) = \sum c_{(0)}d \odot \rho(\hat{a})\pi(c_{(1)})b.$$

Here, we use the extended Sweedler notation for $\delta_0$ similarly as it was defined in Notation 3.1.3 for corepresentations, writing $\delta_0(c) = \sum c_{(0)} \odot c_{(1)} \in C_0 \odot \widehat{A}_0$. By definition of the representations $\pi$, $\rho$ and of the comultiplication of $\widehat{A}_0$,

$$\begin{aligned}\rho(\hat{a})\pi(c_{(1)})b &= \sum c_{(1)}b_{(1)}\hat{a}(c_{(2)}b_{(2)}) \\ &= \sum \pi(c_{(1)})\rho(\hat{a}(c_{(2)}\cdot))b.\end{aligned}$$

Therefore,

$$(1 \rtimes \hat{a})(c \rtimes 1)(d \odot b) = \sum ((c_{(0)} \rtimes 1)(1 \rtimes \hat{a}(c_{(1)}\cdot)))(d \odot b).$$

Thus, the reduced crossed product $C \rtimes_{\delta,r} \widehat{A}(V)$ is the completion of the space $C_0 \rtimes \widehat{A}_0 = \operatorname{span}(C_0 \rtimes 1)(1 \rtimes \widehat{A}_0)$, and this space can be considered as a twisted tensor product of the $*$-algebras $C_0 \rtimes 1$ and $1 \rtimes \widehat{A}_0$, where the twist is given by

$$(1 \rtimes \hat{a})(c \rtimes 1) = \sum (c_{(0)} \rtimes 1)(1 \rtimes \hat{a}(c_{(1)}\cdot)) \quad \text{for all } c \in C_0,\ \hat{a} \in \widehat{A}_0.$$

In Definition 9.4.1, the reduced crossed product of a coaction was defined in terms of a generating subset. Frequently, it is useful to know that this subset is linearly dense in the crossed product:

**Proposition 9.4.4.** *Let $(V, U)$ be a weak Kac system and $(C, \delta)$ a coaction of $(A(V), \Delta)$. Then $C \rtimes_{\delta,r} \widehat{A}(V) = \overline{\operatorname{span}}(\delta(C)(1 \otimes \widehat{A}(V)))$.*

*Proof.* Let us begin with some preliminaries. We denote by $H$ the Hilbert space underlying the weak Kac system $(V, U)$. For every $\xi \in H$, we define an operator

$$|\xi\rangle_{[3]} := \mathrm{id}_C \otimes \mathrm{id}_H \otimes |\xi\rangle \in \mathcal{L}_C(C \otimes H, C \otimes H \otimes H),\ c \otimes \zeta \mapsto c \otimes \zeta \otimes \xi,$$

and put $\langle\xi|_{[3]} := |\xi\rangle_{[3]}^*$. Moreover, we put $|H\rangle_{[3]} := \{|\xi\rangle_{[3]} \mid \xi \in H\}$, $\langle H|_{[3]} := |H\rangle_{[3]}^*$, and $\widehat{A} := \widehat{A}(V)$, $A := A(V)$. By Lemma 7.2.7,

$$1 \otimes \widehat{A} = \big[\langle H|_{[3]}(1 \otimes V)|H\rangle_{[3]}\big] \subseteq \mathcal{L}_C(C \otimes H).$$

To prove the assertion, it suffices to show that the space $[(1 \otimes \widehat{A})\delta(C)]$ is contained in $[\delta(C)(1 \otimes \widehat{A})]$. By the equation above,

$$\begin{aligned}[(1 \otimes \widehat{A})\delta(C)] &= [\langle H|_{[3]}(1 \otimes V)|H\rangle_{[3]}\,\delta(C)] \\ &= \big[\langle H|_{[3]}(1 \otimes V)(\delta(C) \otimes 1)|H\rangle_{[3]}\big].\end{aligned}$$

As $V(a \otimes 1)V^* = \Delta(a)$ for all $a \in A$, we have $(1 \otimes V)(\delta(c) \otimes 1) = \delta^{(2)}(c)(1 \otimes V)$ for all $c \in C$, where $\delta^{(2)} := (\mathrm{id} \otimes \Delta) \circ \delta = (\delta \otimes \mathrm{id}) \circ \delta$. Therefore,

$$[(1 \otimes \widehat{A})\delta(C)] = \big[\langle H|_{[3]}(\delta^{(2)}(C))(1 \otimes V)|H\rangle_{[3]}\big].$$

Since $A$ acts non-degenerately on $H$, we can replace $\langle H|_{[3]}$ by $\langle H|_{[3]}\, A_{[3]}$. Inserting the relation

$$[A_{[3]}\,\delta^{(2)}(C)] = [(\delta \otimes \mathrm{id})((1 \otimes A)\delta(C))] \subseteq [(\delta \otimes \mathrm{id})(C \otimes A)] = [\delta(C) \otimes A],$$

we find

$$[(1 \otimes \widehat{A})\delta(C)] \subseteq \big[\langle H|_{[3]}(\delta(C) \otimes A)(1 \otimes V)|H\rangle_{[3]}\big].$$

Now we move $A_{[3]}$ to the left, replace $\langle H|_{[3]}\, A_{[3]}$ by $\langle H|_{[3]}$ again, and find

$$[(1 \otimes \widehat{A})\delta(C)] \subseteq \big[\langle H|_{[3]}\,(\delta(C) \otimes 1)(1 \otimes V)|H\rangle_{[3]}\big] = [\delta(C)(1 \otimes \widehat{A})]. \qquad \square$$

**Corollary 9.4.5.** *Let $(V, U)$ be a weak Kac system and let $(C, \delta)$ be a coaction of $(A(V), \Delta)$. Then the maps*

$$C \to \mathcal{L}_C(C \otimes H),\quad c \mapsto \delta(c), \quad \textit{and} \quad \widehat{A}(V) \to \mathcal{L}_C(C \otimes H),\ \hat{a} \mapsto 1 \otimes \hat{a},$$

*send $C$ and $\widehat{A}(V)$ onto non-degenerate $C^*$-subalgebras of $M(C \rtimes_{\delta,r} \widehat{A}(V))$ and extend to $*$-homomorphisms*

$$M(C) \to M(C \rtimes_{\delta,r} \widehat{A}(V)) \quad \textit{and} \quad M(\widehat{A}(V)) \to M(C \rtimes_{\delta,r} \widehat{A}(V)). \qquad \square$$

### 9.4.2 The dual coaction of a coaction of $A(V)$

Given a coaction $(C, \delta)$ of $(A(V), \Delta)$, we construct a coaction of $(\widehat{A}(V), \widehat{\Delta})$ on the reduced crossed product $C \rtimes_{\delta,r} \widehat{A}(V)$, combining the trivial coaction on $\delta(C)$ with the right regular coaction on $\widehat{A}(V)$:

**Theorem 9.4.6.** *Let $(V, U)$ be a weak Kac system and $(C, \delta)$ a coaction of $(A(V), \Delta)$. Then the formula*

$$\delta(c)(1 \otimes \hat{a}) \mapsto (\delta(c) \otimes 1)(1 \otimes \widehat{\Delta}(\hat{a}))$$

*defines a coaction $\hat{\delta}$ of $(\widehat{A}(V), \widehat{\Delta})$ on $C \rtimes_{\delta,r} \widehat{A}(V)$ that turns $C \rtimes_{\delta,r} \widehat{A}(V)$ into an $(\widehat{A}(V), \widehat{\Delta})$-$C^*$-algebra.*

*Proof.* Denote by $H$ the Hilbert space underlying $(V, U)$ and put $A := A(V)$, $\widehat{A} := \widehat{A}(V)$. First, let us show that the map $\hat{\delta}$ is well defined. We identify

$$(C \rtimes_{\delta,r} \widehat{A}) \otimes \widehat{A} \subseteq \mathcal{L}_C(C \otimes H) \otimes \mathcal{L}(H)$$

with a non-degenerate $C^*$-subalgebra of $\mathcal{L}_C(C \otimes H \otimes H)$ and extend this identification to multiplier algebras. Consider the unitary $\check{V} = \Sigma U_{[2]} V U_{[2]} \Sigma$. By the proof of Lemma 9.3.8 and by Proposition 9.3.4,

$$\check{V}(a \otimes 1)\check{V}^* = a \otimes 1 \ \text{ for all } a \in A, \quad \check{V}(\hat{a} \otimes 1)\check{V}^* = \widehat{\Delta}(\hat{a}) \ \text{ for all } \hat{a} \in \widehat{A},$$

and therefore,

$$\hat{\delta}(x) = (1 \otimes \check{V})(x \otimes 1)(1 \otimes \check{V}^*) \quad \text{for all } x \in [\delta(C)(1 \otimes \widehat{A})] = C \rtimes_{\delta,r} \widehat{A}.$$

It is easy to see that the map $\hat{\delta}$ is a coaction. Let us prove that it turns $C \rtimes_{\delta,r} \widehat{A}$ into an $(\widehat{A}, \widehat{\Delta})$-$C^*$-algebra. The equation above shows that $\hat{\delta}$ is injective, and since $\widehat{\Delta}(\widehat{A}) \cdot (1 \otimes \widehat{A})$ is linearly dense in $\widehat{A} \otimes \widehat{A}$, the product $\hat{\delta}(C \rtimes_{\delta,r} \widehat{A}) \cdot (\mathrm{id}_{C \rtimes_r \widehat{A}} \otimes \widehat{A})$ is linearly dense in

$$\begin{aligned}[(\delta(C) \otimes 1)(1 \otimes \widehat{\Delta}(\widehat{A}))(1 \otimes 1 \otimes \widehat{A})] \\ = [(\delta(C) \otimes 1)(1 \otimes \widehat{A} \otimes \widehat{A})] = (C \rtimes_{\delta,r} \widehat{A}) \otimes \widehat{A}.\end{aligned} \qquad \square$$

**Definition 9.4.7.** Let $(V, U)$ be a weak Kac system. The *dual* of a coaction $(C, \delta)$ of $(A(V), \Delta)$ is the coaction $(C \rtimes_{\delta,r} \widehat{A}(V), \hat{\delta})$ of $(\widehat{A}(V), \widehat{\Delta})$ defined above.

The construction of the dual coaction defines an assignment from the class of all coactions of $(A(V), \Delta)$ to the class of all $(\widehat{A}(V), \widehat{\Delta})$-$C^*$-algebras. This assignment extends to a functor:

**Theorem 9.4.8.** *Let $(V, U)$ be a weak Kac system.*

i) *Let $(C, \delta_C)$ and $(D, \delta_D)$ be coactions of $(A(V), \Delta)$ and $\phi\colon C \to M(D)$ a non-degenerate covariant $*$-homomorphism. Then the formula*

$$\delta_C(c)(1 \otimes \hat{a}) \mapsto \delta_D(\phi(c))(1 \otimes \hat{a})$$

*defines a non-degenerate $*$-homomorphism*

$$\phi \rtimes_r \mathrm{id}\colon C \rtimes_r \widehat{A}(V) \to M(D \rtimes_r \widehat{A}(V))$$

*which is covariant with respect to the dual coactions $\hat{\delta}_C$ and $\hat{\delta}_D$.*

ii) *The assignment $(C, \delta) \mapsto (C \rtimes_{\delta,r} \widehat{A}(V), \hat{\delta})$ and $\phi \mapsto \phi \rtimes_r \mathrm{id}$ defines a functor from the category of coactions of $(A(V), \Delta)$ to the category of $(\widehat{A}(V), \widehat{\Delta})$-$C^*$-algebras.*

*Proof.* i) Denote by $H$ the Hilbert space underlying $(V, U)$. We consider $D$ as a $C^*$-module over itself and identify the internal tensor product $(C \otimes H) \otimes_\phi D$ with the $C^*$-module $D \otimes H$ via the map $(c \otimes \xi) \otimes_\phi d \mapsto \phi(c)d \otimes \xi$. The $*$-homomorphism

$$\begin{aligned}\mathrm{Ind}_\phi\colon \mathcal{L}_C(C \otimes H) &\to \mathcal{L}_D((C \otimes H) \otimes_\phi D) \cong \mathcal{L}_D(D \otimes H), \\ T &\mapsto T \otimes_\phi \mathrm{id}_D \equiv \mathrm{Ind}_\phi(T),\end{aligned}$$

(see Section 12.2) acts on the $C^*$-subalgebra $C \rtimes_r \widehat{A}(V) \subseteq \mathcal{L}_C(C \otimes H)$ as follows:

$$\begin{aligned}\operatorname{Ind}_\phi(\delta_C(c)(1 \otimes \hat{a})) &= (\phi \otimes \mathrm{id})(\delta_C(c)(1 \otimes \hat{a})) \\ &= \delta_D(\phi(c))(1 \otimes \hat{a}) \quad \text{for all } c \in C,\ \hat{a} \in \widehat{A}(V).\end{aligned}$$

Consequently, $\phi \rtimes_r \mathrm{id} = \operatorname{Ind}_\phi|_{C \rtimes_r \widehat{A}(V)}$ is well defined. This $*$-homomorphism evidently is covariant with respect to the dual coactions on $C \rtimes_r \widehat{A}(V)$ and $D \rtimes_r \widehat{A}(V)$.

ii) The proof of this statement is straightforward. □

### 9.4.3 The dual coaction of a coaction of $\widehat{A}(V)$

The constructions presented above carry over to coactions of the left leg $(\widehat{A}(V), \widehat{\Delta})$ of a weak Kac system $(V, U)$ as follows. It is not appropriate to simply exchange the $C^*$-bialgebras $(A(V), \Delta)$ and $(\widehat{A}(V), \widehat{\Delta})$ in all definitions. Rather, we have to replace the weak Kac system $(V, U)$ by its predual $(\check{V}, U)$ and use the identifications $A(\check{V}) = \widehat{A}(V)$ and $\widehat{A}(\check{V}) = \mathrm{Ad}_U(A(V)) \cong A(V)$ established in Proposition 9.3.4; alternatively, we could use the dual $(\widehat{V}, U)$ in a similar way.

**Definition 9.4.9.** Let $(V, U)$ be a weak Kac system on a Hilbert space $H$ and $(C, \delta)$ a coaction of $(\widehat{A}(V), \widehat{\Delta})$. The *reduced crossed product* of $(C, \delta)$ is the $C^*$-algebra $C \rtimes_{\delta,r} A(V) \subseteq \mathcal{L}_C(C \otimes H)$ generated by the subset

$$\delta(C)(1 \otimes \mathrm{Ad}_U(A(V))) \subseteq \mathcal{L}_C(C \otimes H).$$

If the coaction $\delta$ is understood, we shortly write $C \rtimes_r A(V)$ for $C \rtimes_{\delta,r} A(V)$.

**Theorem 9.4.10.** *Let $(V, U)$ be a weak Kac system.*

i) *Let $(C, \delta)$ be a coaction of $(\widehat{A}(V), \widehat{\Delta})$.*

(a) $C \rtimes_{\delta,r} A(V) = \overline{\operatorname{span}}(\delta(C)(1 \otimes \mathrm{Ad}_U(A(V))))$.

(b) *The formula*

$$\delta(c)(1 \otimes \mathrm{Ad}_U(a)) \mapsto (\delta(c) \otimes 1) \cdot (1 \otimes \mathrm{Ad}_{(U \otimes 1)}(\Delta(a)))$$

*defines a coaction $\hat{\delta}$ of $(A(V), \Delta)$ on $C \rtimes_{\delta,r} A(V)$ that turns $C \rtimes_{\delta,r} A(V)$ into an $(A(V), \Delta)$-$C^*$-algebra.*

ii) *Let $(C, \delta_C)$ and $(D, \delta_D)$ be coactions of $(\widehat{A}(V), \widehat{\Delta})$ and $\phi\colon C \to M(D)$ a non-degenerate covariant $*$-homomorphism. Then the formula*

$$\delta_C(c)(1 \otimes \mathrm{Ad}_U(a)) \mapsto \delta_D(\phi(c))(1 \otimes \mathrm{Ad}_U(a))$$

*defines a non-degenerate $*$-homomorphism*

$$\phi \rtimes_r \mathrm{id}\colon C \rtimes_r A(V) \to M(D \rtimes_r A(V))$$

*which is covariant with respect to the coactions $\hat{\delta}_C$ and $\hat{\delta}_D$.*

iii) *The assignment $(C, \delta) \mapsto (C \rtimes_{\delta,r} A(V), \hat{\delta})$ and $\phi \mapsto \phi \rtimes_r \mathrm{id}$ defines a functor from the category of coactions of $(\hat{A}(V), \hat{\Delta})$ to the category of $(A(V), \Delta)$-$C^*$-algebras.*

*Proof.* All assertions follow easily from Proposition 9.4.4, Theorem 9.4.6, and Theorem 9.4.8, if one replaces the weak Kac system $(V, U)$ by its predual $(\check{V}, U)$ and uses the identifications of the legs of these weak Kac systems listed in Proposition 9.3.4. □

**Definition 9.4.11.** Let $(V, U)$ be a weak Kac system. The *dual* of a coaction $(C, \delta)$ of $(\hat{A}(V), \hat{\Delta})$ is the coaction $(C \rtimes_{\delta,r} A(V), \hat{\delta})$ of $(A(V), \Delta)$ defined above.

### 9.4.4 Comparison with the reduced crossed product of a group action

The reduced crossed product and the dual coaction introduced above are related to the corresponding constructions for a group action as follows.

Let $G$ be a locally compact group. By Theorem 9.2.4, every action $(C, \alpha)$ of $G$ corresponds to a unique coaction $(C, \delta)$ of the $C^*$-bialgebra $C_0(G)$. The reduced crossed product $C \rtimes_{\alpha,r} G$ for an action $(C, \alpha)$ was defined in Definition 9.1.5. To construct the reduced crossed product for the corresponding coaction $(C, \delta)$, we use the weak Kac system $(W_G, U)$ introduced in Example 9.3.11. Since the legs of this weak Kac system are given by $\hat{A}(W_G) = \pi_M(C_0(G)) \cong C_0(G)$ and $A(W_G) = C_r^*(G)$, we can consider $\delta$ as a coaction of $\hat{A}(W_G)$ and form the reduced crossed product $C \rtimes_{\delta,r} C_r^*(G)$ as in Definition 9.4.9. We shall see that the crossed products $C \rtimes_{\alpha,r} G$ and $C \rtimes_{\delta,r} C_r^*(G)$ are naturally isomorphic.

If the group $G$ is abelian, we can also compare the dual action and the dual coaction on the respective crossed products: In this case, the $C^*$-bialgebra $C_r^*(G)$ is isomorphic to $C_0(\hat{G})$ (Proposition 4.2.3), so that the dual coaction of $C_r^*(G) \cong C_0(\hat{G})$ on $C \rtimes_{\delta,r} C_r^*(G)$ corresponds to an action of the dual group $\hat{G}$ (Theorem 9.2.4). We will see that the isomorphism $C \rtimes_{\delta,r} C_r^*(G) \cong C \rtimes_{\alpha,r} G$ referred to above identifies this action with the dual action of $\hat{G}$ on $C \rtimes_{\alpha,r} G$.

In the next theorem, we identify the $C^*$-bialgebra $C_0(G)$ with $\hat{A}(W_G)$ via the isomorphism $\pi_M$.

**Theorem 9.4.12.** *Let $\alpha$ be an action of a locally compact group $G$ on a $C^*$-algebra $C$. Denote by $\delta\colon C \to M(C \otimes C_0(G))$ the coaction of $C_0(G)$ that corresponds to the action $\alpha$.*

i) *Conjugation by the isomorphism* $C \otimes L^2(G,\lambda) \xrightarrow{\mathrm{id}\otimes U} C \otimes L^2(G,\lambda) \cong L^2(G;C)$ *induces an isomorphism*

$$C \rtimes_{\delta,r} C_r^*(G) \xrightarrow{\cong} C \rtimes_{\alpha,r} G, \quad \delta(c)(1 \otimes \mathrm{Ad}_U(g)) \mapsto c \rtimes_r g. \tag{9.6}$$

ii) *Assume that* $G$ *is abelian. With respect to the isomorphism* (9.6)*, the dual action of* $\widehat{G}$ *on* $C \rtimes_{\alpha,r} G$ *corresponds to the dual coaction of the* $C^*$*-bialgebra* $C_r^*(G) \cong C_0(\widehat{G})$ *on* $C \rtimes_{\delta,r} C_r^*(G)$.

*Proof.* i) Denote the canonical isomorphism $C \otimes L^2(G,\lambda) \cong L^2(G;C)$ by $\Upsilon$. By Proposition 9.1.6 and 9.4.4,

$$C \rtimes_{\delta,r} C_r^*(G) = [\delta(C)(1 \otimes C_r^*(G))], \quad C \rtimes_{\alpha,r} G = [(C \rtimes_r 1)(1 \rtimes_r C_r^*(G))].$$

Therefore it suffices to show that for all $c \in C$ and $g \in C_r^*(G)$,

$$\mathrm{Ad}_{\Upsilon(\mathrm{id}\otimes U)}(\delta(c)) = c \rtimes_r 1, \quad \mathrm{Ad}_{\Upsilon(\mathrm{id}\otimes U)}(1 \otimes \mathrm{Ad}_U(g)) = 1 \rtimes_r g. \tag{9.7}$$

The second equation is evident; let us prove the first one. The final remarks in Example 9.3.11 imply

$$\mathrm{Ad}_{(\mathrm{id}\otimes U)}(\delta(c)) = (\mathrm{id} \otimes (\pi_M \circ \theta))(\delta(c)),$$

where $\theta$ denotes the $*$-automorphism of $C_0(G)$ given by $(\theta f)(x) = f(x^{-1})$ for all $f \in C_0(G)$, $x \in G$. Let $h \in L^2(G;C)$, $x \in G$. Then

$$(\mathrm{Ad}_{\Upsilon(\mathrm{id}\otimes U)}(\delta(c))h)(x) = (\mathrm{id} \otimes \mathrm{ev}_x \circ \theta)(\delta(c)) \cdot h(x) = (\mathrm{id} \otimes \mathrm{ev}_{x^{-1}})(\delta(c)) \cdot h(x),$$

and by Theorem 9.2.4, $(\mathrm{Ad}_{\Upsilon(\mathrm{id}\otimes U)}(\delta(c))h)(x) = \alpha_{x^{-1}}(c)h(x)$. Comparing with the definition of $c \rtimes_r 1$ (see Proposition 9.1.4), we obtain the first equation in (9.7).

ii) First, note that $C_r^*(G) = C^*(G)$ because $G$ is abelian. Let us identify $C^*(G)$ with $C_0(\widehat{G})$ as in Proposition 4.2.3. Denote by $\beta$ the action of $\widehat{G}$ on $C \rtimes_{\delta,r} C^*(G)$ that corresponds to the dual coaction $\hat{\delta}$ of $C^*(G) \cong C_0(\widehat{G})$. Let $\chi \in \widehat{G}$. By Theorem 9.2.4, $\beta_\chi = (\mathrm{id} \otimes \mathrm{ev}_\chi) \circ \hat{\delta}$. The extension of this automorphism to the multiplier algebra acts as follows:

$$\begin{aligned}\beta_\chi \colon \delta(c)(1 \otimes \mathrm{Ad}_U(U_x)) &\overset{\hat{\delta}}{\longmapsto} (\delta(c)(1 \otimes \mathrm{Ad}_U(U_x))) \otimes U_x \\ &\overset{\mathrm{id}\otimes \mathrm{ev}_\chi}{\longmapsto} \delta(c)(1 \otimes \mathrm{Ad}_U(U_x))\chi(x) \quad \text{for all } c \in C,\ x \in G.\end{aligned}$$

On the other hand, the extension of the automorphism $\hat{\alpha}_\chi$ to the multiplier algebra $M(C \rtimes_\alpha G)$ is given by

$$\hat{\alpha}_\chi \colon c \rtimes U_x \mapsto c \rtimes U_x \chi(x) \quad \text{for all } c \in C,\ x \in G.$$

Comparing the formulas above, we find that with respect to the isomorphism (9.6), the action $\beta$ and the coaction $\hat{\delta}$ correspond to the action $\hat{\alpha}$. □

## 9.5 Kac systems and the Baaj–Skandalis duality theorem

We now come to the main result of this chapter – the Baaj–Skandalis duality theorem. In the preceding section, we associated to every coaction of the right leg of a weak Kac system $(V, U)$ a reduced crossed product and a dual coaction of the left leg of $(V, U)$. Conversely, to every coaction of the left leg, we associated a reduced crossed product and a dual coaction of the right leg. Summarizing, we constructed a pair of functors as follows:

$$\begin{array}{ccc} \begin{array}{c}\text{category of}\\ \text{coactions of } (A(V), \Delta)\end{array} & \xrightarrow[\phi \mapsto \phi \rtimes_r \mathrm{id}]{(C,\delta) \mapsto (C \rtimes_r \widehat{A}(V), \hat{\delta})} & \begin{array}{c}\text{category of}\\ (\widehat{A}(V), \widehat{\Delta})\text{-algebras}\end{array} \\ \cup & & \cap \\ \begin{array}{c}\text{category of}\\ (A(V), \Delta)\text{-algebras}\end{array} & \xleftarrow[\phi \rtimes_r \mathrm{id} \leftarrow\!\shortmid\, \phi]{(C \rtimes_r A(V), \hat{\delta}) \leftarrow\!\shortmid\, (C,\delta)} & \begin{array}{c}\text{category of}\\ \text{coactions of } (\widehat{A}(V), \widehat{\Delta}).\end{array} \end{array}$$

Naturally, we would like to know how these functors are related. The Baaj–Skandalis theorem says that under favorable conditions – if $(V, U)$ is a Kac system – these two functors induce an equivalence of the category of $(A(V), \Delta)$-$C^*$-algebras with the category of $(\widehat{A}(V), \widehat{\Delta})$-$C^*$-algebras up to equivariant Morita equivalence. To prove this assertion, one studies the composition of these two functors, that is, the iteration of the reduced crossed product construction.

We proceed as follows. First, we introduce the notion of a Kac system and consider some examples. Then, we determine the composition of the two functors above. We do not discuss equivariant Morita equivalence; for related concepts, see [6].

### 9.5.1 Kac systems

A Kac system is a special balanced multiplicative unitary. Recall that to every balanced multiplicative unitary $(V, U)$, we associated two auxiliary multiplicative unitaries

$$\check{V} = \Sigma(1 \otimes U)V(1 \otimes U)\Sigma \quad \text{and} \quad \hat{V} = \Sigma(U \otimes 1)V(U \otimes 1)\Sigma.$$

**Definition 9.5.1.** A balanced multiplicative unitary $(V, U)$ is called a *Kac system* if

i) the multiplicative unitaries $V$, $\check{V}$, $\hat{V}$ are regular, and

ii) $(\Sigma(1 \otimes U)V)^3 = 1$.

**Remarks 9.5.2.** i) The unitary $\hat{V}$ is regular if and only if $\check{V}$ is regular because $\hat{V} = \mathrm{Ad}_{(U \otimes U)}(\check{V})$.

ii) In leg notation, the equation $(\Sigma(1 \otimes U)V)^3 = 1$ takes the form $(\Sigma U_{[2]} V)^3 = 1$. Conjugating by $\Sigma$ or $V$, we find that this condition is equivalent to the relation $(U_{[2]} V \Sigma)^3 = 1$ and to the relation $(V \Sigma U_{[2]})^3 = 1$.

iii) A multiplicative unitary $V$ on a Hilbert space $H$ is called *irreducible* [7, Définition 6.2] if there exists a symmetry $U$ on $H$ such that $(V, U)$ is a balanced multiplicative unitary and $(\Sigma(1 \otimes U)V)^3 = 1$.

Condition ii) in Definition 9.5.1 may appear mysterious at first sight. However, we shall see in (the proof of) Proposition 9.5.3 and 9.5.5 that this condition is closely related to the Duality Theorem 9.5.11.

**Proposition 9.5.3.** *Every Kac system is a weak Kac system.*

The proof of this result uses the following lemma.

**Lemma 9.5.4.** *Let $(V, U)$ be a balanced multiplicative unitary.*

i) $[V_{[23]}, V_{[12]} \check{V}_{[12]} \Sigma_{[12]}] = 0$ *and* $[V_{[12]}, \Sigma_{[23]} \hat{V}_{[23]} V_{[23]}] = 0$.

ii) $(\Sigma U_{[2]} V)^3 = 1$ *if and only if* $\hat{V} V \check{V} = U_{[1]} \Sigma$.

*Proof.* i) We only prove the first equation, the second one follows similarly. By Lemma 9.3.5, $\check{V}_{[12]} V_{[13]} = V_{[13]} V_{[23]} \check{V}_{[12]}$. We multiply this equation by $V_{[12]}$ on the left and by $\Sigma_{[12]}$ on the right, use the pentagon equation for $V$, and obtain

$$\begin{aligned} V_{[12]} \check{V}_{[12]} \Sigma_{[12]} V_{[23]} &= V_{[12]} \check{V}_{[12]} V_{[13]} \Sigma_{[12]} \\ &= V_{[12]} V_{[13]} V_{[23]} \check{V}_{[12]} \Sigma_{[12]} = V_{[23]} V_{[12]} \check{V}_{[12]} \Sigma_{[12]}. \end{aligned}$$

ii) Rearranging the factors in the product $U_{[1]} U_{[2]} (\Sigma U_{[2]} V)^3 U_{[2]} \Sigma$, we find

$$U_{[1]} U_{[2]} (\Sigma U_{[2]} V)^3 U_{[2]} \Sigma = \Sigma U_{[1]} V U_{[1]} \Sigma \cdot V \cdot \Sigma U_{[2]} V U_{[2]} \Sigma = \hat{V} \cdot V \cdot \check{V}.$$

Thus, $(\Sigma U_{[2]} V)^3$ is equal to 1 if and only if $U_{[1]} \Sigma = U_{[1]} U_{[2]} U_{[2]} \Sigma$ is equal to $\hat{V} V \check{V}$. □

*Proof of Proposition* 9.5.3. By part i) of the previous lemma, $V_{[23]}$ commutes with $V_{[12]} \check{V}_{[12]} \Sigma_{[12]}$, and by part ii), $V_{[12]} \check{V}_{[12]} \Sigma_{[12]} = \hat{V}^*_{[12]} U_{[1]}$. Hence $V_{[23]}$ commutes with $\hat{V}^*_{[12]}$, and since $\hat{V}_{[12]}$ is unitary, also with $\hat{V}_{[12]}$. A similar argument shows that $V_{[12]}$ commutes with $\check{V}^*_{[23]}$ and hence also with $\check{V}_{[23]}$. □

The second main property of a Kac system – in addition to being a weak Kac system – is stated in the following proposition:

**Proposition 9.5.5.** *Let $(V, U)$ be a Kac system on a Hilbert space $H$. Then* $\overline{\operatorname{span}}\, A(V) \hat{A}(V) = \mathcal{K}(H)$.

*Proof.* We combine the leg notation and ket-bra notation as in Sections 7.2 and 7.3; see formulas (7.6)–(7.8) on page 176. By conditions i) and ii) of Definition 9.5.1,

$$\mathcal{K}(H) = \big[\langle H|_{[2]} V^* |H\rangle_{[1]}\big] \quad \text{and} \quad V^* = \Sigma U_{[2]} V \Sigma U_{[2]} V \Sigma U_{[2]}.$$

We insert the second equation into the first one, use $U\mathcal{K}(H)U = \mathcal{K}(H)$, and obtain

$$\mathcal{K}(H) = \big[\langle H|_{[2]} \Sigma U_{[2]} V \Sigma U_{[2]} V \Sigma U_{[2]} |H\rangle_{[1]}\big] = \big[\langle H|_{[1]} V \Sigma U_{[2]} V |H\rangle_{[2]}\big].$$

We multiply this equation on the right by $\widehat{A}(V)$. On the left-hand side we obtain $[\mathcal{K}(H)\widehat{A}(V)] = \mathcal{K}(H)$ because $\widehat{A}(V)$ acts non-degenerately on $H$. On the right-hand side, $[V|H\rangle_{[2]}\widehat{A}(V)] = [|H\rangle_{[2]}\widehat{A}(V)]$ by Lemma 7.3.13. Hence, by Lemma 7.2.7,

$$\begin{aligned}\mathcal{K}(H) &= [\langle H|_{[1]} V \Sigma U_{[2]} |H\rangle_{[2]} \widehat{A}(V)] \\ &= [\langle H|_{[1]} V |H\rangle_{[1]} \widehat{A}(V)] = [A(V)\widehat{A}(V)]. \qquad \square\end{aligned}$$

The preceding proposition has the following partial converse:

**Proposition 9.5.6.** *Let $V$ be a multiplicative unitary on a Hilbert space $H$ and $U$ a symmetry on $H$ such that $V_{[23]}\widehat{V}_{[12]} = \widehat{V}_{[12]}V_{[23]}$ and $V_{[12]}\check{V}_{[23]} = \check{V}_{[23]}V_{[12]}$.*

i) *If $\overline{\operatorname{span}}\, \widehat{A}(V)A(V) = \mathcal{K}(H)$, then $V$ is regular.*

ii) *If $\widehat{V}$, $\check{V}$ are multiplicative and the commutants of $A(V) \cup \widehat{A}(V)$ and of $A(V) \cup \operatorname{Ad}_U(\widehat{A}(V))$ both are only scalars, then $U_{[2]}\Sigma\widehat{V}V\check{V}$ is scalar.*

*Proof.* The proof uses similar techniques as presented in this section and in Sections 9.3 and 7.3; for details, see [7, Proposition 6.9]. □

We end this section with some examples and standard constructions:

**Example 9.5.7.** For every locally compact group $G$, the weak Kac system $(W_G, U)$ constructed in Example 9.3.11 is a Kac system. Indeed, the multiplicative unitaries $W_G$ and $\widehat{W}_G = (W_G)^{\mathrm{op}}$ are regular by Example 7.3.4 iii) and ii), and from the calculations in Example 9.3.11, we find

$$\begin{aligned}(\widehat{W}_G W_G \check{W}_G \zeta)(x,y) &= (W_G \check{W}_G \zeta)(yx, y) \\ &= (\check{W}_G \zeta)(yx, \underbrace{(yx)^{-1}y}_{=x^{-1}}) = \zeta(\underbrace{yx \cdot x^{-1}}_{=y}, x^{-1}) \cdot \delta(x^{-1})^{1/2}\end{aligned}$$

and

$$(U_{[1]}\Sigma\zeta)(x,y) = (\Sigma\zeta)(x^{-1}, y) \cdot \delta(x)^{-1/2} = \zeta(y, x^{-1}) \cdot \delta(x^{-1})^{1/2}$$

for all $\zeta \in L^2(G \times G, \lambda \times \lambda)$ and $x, y \in G$. Thus $\widehat{W}_G W_G \check{W}_G = U_{[1]}\Sigma$, and by Lemma 9.5.4, $(W_G, U)$ is a Kac system.

**Example 9.5.8.** For every algebraic compact quantum group $(A_0, \Delta_0)$, the weak Kac system $(V, U)$ constructed in Example 9.3.12 is a Kac system. Let us prove this claim. The multiplicative unitaries $V = V_{A_0}$ and $\widehat{V} = W_{A_0}$ are regular by Example 7.3.4 iv). We show that $V\check{V} = \widehat{V}^* U_{[1]}\Sigma$, and by Lemma 9.5.4, this equation implies that $(V, U)$ is a Kac system. By Proposition 1.3.12 and Corollary 3.2.20,

$$\begin{aligned}\Delta_0 \circ U_0 &= \Delta_0 \circ S_0 \circ \rho_{0,-1} = \Sigma \circ (S_0 \odot S_0) \circ \Delta_0 \circ \rho_{0,-1} \\ &= \Sigma \circ (S_0 \odot S_0) \circ (\rho_{1,-1} \odot \rho_{0,-1}) \circ \Delta_0 = \Sigma \circ (S_0^{-1} \odot U_0) \circ \Delta_0.\end{aligned}$$

Combining this equation with the calculations in Example 9.3.12, we find

$$V\check{V}(a \odot b) = \sum V(aU_0(b_{(2)}) \odot b_{(1)}) = \sum a_{(1)}U_0(b_{(3)}) \odot a_{(2)}S_0^{-1}(b_{(2)})b_{(1)}$$

for all $a, b \in A_0$. Since $S_0^{-1}$ is the antipode of $(A_0, \Delta_0)^{\mathrm{op}}$ (see Proposition 1.3.14), we can simplify this expression and find that for all $a, b \in A_0$,

$$V\check{V}(a \odot b) = \sum a_{(1)}U_0(b) \odot a_{(2)} = \widehat{V}^*(U_0(b) \odot a) = \widehat{V}^* U_{[1]}\Sigma(a \odot b).$$

Like the notion of a weak Kac system, the notion of a Kac system is highly symmetric:

**Proposition 9.5.9.** *If $(V, U)$ is a Kac system, then also $(\check{V}, U)$, $(\widehat{V}, U)$, and $(V^{\mathrm{op}}, U)$ are Kac systems.*

*Proof.* If $(V, U)$ is a Kac system, then

- $(\check{V}, U)$ and $(\widehat{V}, U)$ are balanced multiplicative unitaries by Remark 9.3.2 iv);
- the multiplicative unitaries $\check{V}$, $\widehat{V}$, and $\check{\check{V}} = \mathrm{Ad}_{U \otimes U}(V) = \widehat{\widehat{V}}$ are regular by assumption;
- $(\check{V}\Sigma U_{[2]})^3 = (\Sigma U_{[2]} V)^3 = 1$ and $(\Sigma U_{[2]}\widehat{V})^3 = (V\Sigma U_{[2]})^3 = 1$ by Remark 9.5.2 ii).

Remark 9.5.2 ii) implies that $(\check{V}, U)$ and $(\widehat{V}, U)$ are Kac systems. Moreover,

- $(V^{\mathrm{op}}, U)$ is a balanced multiplicative unitary by Remark 9.3.2 v);
- $\widecheck{V^{\mathrm{op}}}$ and $\widehat{V^{\mathrm{op}}}$ are equal to $\widehat{V}^{\mathrm{op}}$ and $\check{V}^{\mathrm{op}}$, respectively, by the same remark, and by Example 7.3.4 ii), the latter unitaries and $V^{\mathrm{op}}$ are regular;
- $(U_{[2]}V^{\mathrm{op}}\Sigma)^3 = (U_{[2]}\Sigma V^*)^3 = ((V\Sigma U_{[2]})^*)^3 = 1$ by Remark 9.5.2 ii).

Again, Remark 9.5.2 ii) implies that $(V^{\mathrm{op}}, U)$ is a Kac system. □

**Definition 9.5.10.** Let $(V, U)$ be a Kac system. Then $(\check{V}, U)$, $(\widehat{V}, U)$, and $(V^{\mathrm{op}}, U)$ are called the *predual*, the *dual*, and the *opposite* Kac system of $(V, U)$, respectively.

### 9.5.2 The Baaj–Skandalis duality theorem

The Baaj–Skandalis duality theorem is the main result of this chapter. Its content was described already in the introduction to this section. Given the results collected in the previous subsections, the proof is not difficult.

Our formulation of the duality theorem involves the following notation. Given a Hilbert space $H$ and a coaction $(C, \delta)$ of a $C^*$-bialgebra $(A, \Delta)$, we define a coaction $\delta_{[13]}$ of $(A, \Delta)$ on $C \otimes \mathcal{K}(H)$ by the formula

$$\delta_{[13]}(c \otimes T) = \mathrm{Ad}_{(1 \otimes \Sigma)}(\delta(c) \otimes \mathrm{id}_H) \cdot (1 \otimes T \otimes 1) \quad \text{for all } c \in C,\ T \in \mathcal{K}(H).$$

Here, $\mathrm{Ad}_{(1\otimes\Sigma)} \colon M(C \otimes A \otimes \mathcal{K}(H)) \to M(C \otimes \mathcal{K}(H) \otimes A)$ denotes the isomorphism induced by $c \otimes a \otimes T \mapsto c \otimes T \otimes a$.

**Theorem 9.5.11** (Baaj–Skandalis duality). *Let $(V, U)$ be a Kac system on a Hilbert space $H$.*

i) *Let $(C, \delta)$ be an $(A(V), \Delta)$-$C^*$-algebra and $\hat{\delta}$ the dual coaction of $(\widehat{A}(V), \widehat{\Delta})$ on $C \rtimes_{\delta,r} \widehat{A}(V)$. There exists a natural isomorphism*

$$C \rtimes_{\delta,r} \widehat{A}(V) \rtimes_{\hat{\delta},r} A(V) \cong C \otimes \mathcal{K}(H)$$

*which identifies the bidual coaction of $(A(V), \Delta)$ on $C \rtimes_{\delta,r} \widehat{A}(V) \rtimes_{\hat{\delta},r} A(V)$ with the map $\mathrm{Ad}_{(1\otimes V)} \circ \delta_{[13]} \colon C \otimes \mathcal{K}(H) \to M(C \otimes \mathcal{K}(H) \otimes A(V))$.*

ii) *Let $(C, \delta)$ be an $(\widehat{A}(V), \widehat{\Delta})$-$C^*$-algebra and $\hat{\delta}$ the dual coaction of $(A(V), \Delta)$ on $C \rtimes_{\delta,r} A(V)$. There exists a natural isomorphism*

$$C \rtimes_{\delta,r} A(V) \rtimes_{\hat{\delta},r} \widehat{A}(V) \cong C \otimes \mathcal{K}(H)$$

*which identifies the bidual coaction of $(\widehat{A}(V), \widehat{\Delta})$ on $C \rtimes_{\delta,r} A(V) \rtimes_{\hat{\delta},r} \widehat{A}(V)$ with the map $\mathrm{Ad}_{(1\otimes \check{V})} \circ \delta_{[13]} \colon C \otimes \mathcal{K}(H) \to M(C \otimes \mathcal{K}(H))$.*

*Proof.* i) To simplify notation, put $\widehat{A} := \widehat{A}(V)$ and $A := A(V)$. By Definitions 9.4.1, 9.4.7, 9.4.9, Proposition 9.4.4, and Theorem 9.4.10 i)(a), the iterated crossed product $C \rtimes_r \widehat{A} \rtimes_r A$ is equal to

$$[(\delta(C) \otimes 1)(1 \otimes \widehat{\Delta}(\widehat{A}))(1 \otimes 1 \otimes \mathrm{Ad}_U(A))] \subseteq \mathcal{L}_C(C \otimes H \otimes H).$$

By definition of $\Delta$ and $\widehat{\Delta}$ and by the proof of Lemma 9.3.8, conjugation by $1 \otimes V$ maps this $C^*$-algebra isomorphically onto

$$[\delta^{(2)}(C)(1 \otimes 1 \otimes \widehat{A})(1 \otimes 1 \otimes \mathrm{Ad}_U(A))],$$

where $\delta^{(2)} = (\mathrm{id} \otimes \Delta) \circ \delta = (\delta \otimes \mathrm{id}) \circ \delta$. Since $\delta$ is injective, the formula

$$\delta^{(2)}(c)(1 \otimes 1 \otimes T) \mapsto \delta(c)(1 \otimes T), \quad \text{where } c \in C,\ T \in \mathcal{L}(H),$$

defines an isomorphism of this $C^*$-algebra onto

$$[\delta(C)(1 \otimes \widehat{A} \cdot \mathrm{Ad}_U(A))] \subseteq \mathcal{L}_C(C \otimes H).$$

Now $[\widehat{A} \cdot \mathrm{Ad}_U(A)] = [A(\check{V}) \cdot \widehat{A}(\check{V})]$ by Proposition 9.3.4, and since $(\check{V}, U)$ is a Kac system (Proposition 9.5.9), $[A(\check{V}) \cdot \widehat{A}(\check{V})] = \mathcal{K}(H)$ (Proposition 9.5.5). Thus

$$C \rtimes_r \widehat{A} \rtimes_r A \cong [\delta(C)(1 \otimes \mathcal{K}(H))].$$

We make use of the fact that $A \subseteq \mathcal{L}(H)$ is non-degenerate and the assumption $[\delta(C)(1 \otimes A)] = C \otimes A$, and find

$$\begin{aligned} C \rtimes_r \widehat{A} \rtimes_r A &\cong [\delta(C)(1 \otimes A \cdot \mathcal{K}(H))] \\ &= [(C \otimes A)(1 \otimes \mathcal{K}(H))] = C \otimes \mathcal{K}(H). \end{aligned}$$

This isomorphism identifies an element of the form

$$(\delta(c) \otimes 1)(1 \otimes \widehat{\Delta}(\hat{a}))(1 \otimes 1 \otimes \mathrm{Ad}_U(a)) \in C \rtimes_r \widehat{A} \rtimes_r A \tag{9.8}$$

with

$$\delta(c)(1 \otimes \hat{a} \cdot \mathrm{Ad}_U(a)) \in C \otimes \mathcal{K}(H). \tag{9.9}$$

The bidual coaction of $(A, \Delta)$ on $C \rtimes_r \widehat{A} \rtimes_r A$ maps the element (9.8) to

$$(\delta(c) \otimes 1 \otimes 1)(1 \otimes \widehat{\Delta}(\hat{a}) \otimes 1)(1 \otimes 1 \otimes \mathrm{Ad}_{(U \otimes 1)}(\Delta(a))).$$

Put $W := U_{[1]} V U_{[1]} = \Sigma \widehat{V} \Sigma$. The map $\mathrm{Ad}_{(1 \otimes W)} \circ \delta_{[13]}$ sends the element (9.9) to

$$\begin{aligned} (1 \otimes W)&(\mathrm{Ad}_{(1 \otimes \Sigma)}(\delta^{(2)}(c)) \cdot (1 \otimes \hat{a} \cdot \mathrm{Ad}_U(a) \otimes 1))(1 \otimes W^*) \\ &= (\delta(c) \otimes 1)(1 \otimes \hat{a} \otimes 1)(1 \otimes \mathrm{Ad}_{(U \otimes 1)}(\Delta(a))). \end{aligned}$$

In this equation we used the following relations:

- $\mathrm{Ad}_{\Sigma \widehat{V} \Sigma}(\mathrm{Ad}_\Sigma(\Delta(b))) = b \otimes 1$ for all $b \in A$ by Proposition 9.3.4;
- $U_{[1]} V U_{[1]}$ and $\widehat{A} \otimes 1$ commute by the proof of Lemma 9.3.8;
- $\mathrm{Ad}_{U_{[1]} V U_{[1]}}(\mathrm{Ad}_U(a) \otimes 1) = \mathrm{Ad}_{(U \otimes 1)}(\Delta(a))$ by definition of $\Delta$.

Therefore, the bidual coaction of $(A, \Delta)$ on $C \rtimes_r \widehat{A} \rtimes_r A$ corresponds to the map $\mathrm{Ad}_{(1\otimes W)} \circ \delta_{[13]}$. With respect to the isomorphism

$$C \rtimes_r \widehat{A} \rtimes_r A \cong C \otimes \mathcal{K}(H) \xrightarrow{\mathrm{Ad}_{(1\otimes U)}} C \otimes \mathcal{K}(H),$$

the bidual coaction corresponds to $\mathrm{Ad}_{(1\otimes V)} \circ \delta_{[13]}$, because

$$\mathrm{Ad}_{(1\otimes U\otimes 1)} \circ \mathrm{Ad}_{(1\otimes W)} \circ \delta_{[13]} \circ \mathrm{Ad}_{(1\otimes U)} = \mathrm{Ad}_{(1\otimes V)} \circ \delta_{[13]}.$$

ii) The proof is similar as in i), simply replace $(V, U)$ by the predual $(\check{V}, U)$. □

# Chapter 10
# Pseudo-multiplicative unitaries on Hilbert spaces

In parts I and II of this book, we discussed several approaches to quantum groups. Most of these approaches can be extended to quantum groupoids, which simultaneously generalize quantum groups on one side and classical groupoids on the other side. The next two chapters provide an introduction to some techniques and basic concepts related to quantum groupoids in the setting of von Neumann algebras and $C^*$-algebras.

First steps towards a theory of quantum groupoids in the setting of von Neumann algebras were taken by Vallin and Enock, who generalized the concept of a von Neumann bialgebra and of a multiplicative unitary [50], [51], [171], [172]. Building on their work and on the theory of locally compact quantum groups of Kustermans and Vaes [91], Lesieur developed a general theory of measurable quantum groupoids [99]. A main feature of that theory is a generalization of Pontrjagin duality: to every measurable quantum groupoid, Lesieur associates a dual such that the bidual is naturally isomorphic to the initial quantum groupoid.

A major motivation for the introduction of quantum groupoids in the theory of von Neumann algebras comes from the theory of subfactors. A central problem, raised by Ocneanu, is to describe the symmetries of a given subfactor, that is, of an inclusion of factors, in terms of an associated generalized Galois group and to formulate a Galois theory for inclusions of factors [114], [115]. Let us briefly explain the problem. Given a factor $N_1$ with a suitable action of a group $G$, we obtain two inclusions of factors $N_0 \hookrightarrow N_1 \hookrightarrow N_2$, where $N_0 := N_1^G$ is the fixed point algebra and $N_2 := N_1 \rtimes G$ is the crossed product. The question is whether each inclusion of factors $N_0 \hookrightarrow N_1$ can be described in a similar way. Building on earlier results by various authors [30], [101], [146], Enock and Nest showed in [45] that for every irreducible depth 2 inclusion $N_0 \hookrightarrow N_1$, there exists a locally compact quantum group $M$ and an action of $M$ on $N_1$ such that $N_0$ is equal to the fixed point algebra $N_1^M$ and the inclusion $N_1 \hookrightarrow N_2$ given by the basic construction [72] is isomorphic to the inclusion $N_1 \hookrightarrow N_1 \rtimes M$. If the irreducibility assumption is dropped, then $M$ is no longer a quantum group but a quantum groupoid [43], [50], [51], [99]. Thus, quantum groupoids arise naturally in the study of subfactors.

Like the theory of locally compact quantum groups (or even more so), the theory of measurable quantum groupoids is technically demanding and involved. Let us briefly outline the main concepts and their rôle in the theory. For background on groupoids, see Section 10.3.3.

**Hopf–von Neumann bimodules.** A Hopf–von Neumann bimodule is a von Neumann algebra $M$ equipped with a comultiplication that is modeled on the multiplication map of a groupoid.

Recall that a groupoid consists of a set of arrows $G$, a set of units $G^0$, a range and a source map $r, s\colon G \to G^0$, and a multiplication $G\,{}_s\!\times_r^{}{}_{G^0}\, G \to G$ – the composition of arrows. The multiplication is only defined on the fiber product $G\,{}_s\!\underset{G^0}{\times}{}_r\, G = \{(x,y) \in G \times G \mid s(x) = r(y)\} \subseteq G \times G$ because two arrows can be composed only if the source of the left one matches up with the range of the right one.

Likewise, a Hopf–von Neumann bimodule consists of a von Neumann algebra $M$, a von Neumann algebra $N$ with a representation and an antirepresentation $r, s\colon N \to M$, and a comultiplication $\Delta\colon M \to M\,{}_s\!\underset{N}{*}{}_r\, M$. Here, the target of the comultiplication is no longer the ordinary tensor product $M \mathbin{\bar{\otimes}} M$ as in the case of a von Neumann bialgebra, but a fiber product of von Neumann algebras that is defined in terms of a relative tensor product of Hilbert modules. The notion of a Hopf–von Neumann bimodule was introduced by Vallin [171].

**Operator-valued weights.** Similar to a locally compact quantum group in the setting of von Neumann algebras, a measured quantum groupoid is a Hopf–von Neumann bimodule equipped with a left and a right Haar weight. Recall that for a groupoid $G$, the proper analogue of a Haar measure is a Haar system, which is a family of measures indexed by the elements of $G^0$. Integration with respect to such a Haar measure does not define a linear functional $C_c(G) \to \mathbb{C}$ but a linear map $C_c(G) \to C_c(G^0)$. Likewise, the Haar weights of a measured quantum groupoid $(N, M, r, s, \Delta)$ are operator-valued weights that take values in the base $N$. Operator-valued weights on von Neumann algebras were introduced by Haagerup [59], [60]; further references are [144], [150].

**Pseudo-multiplicative unitaries.** The preceding chapters showed that multiplicative unitaries play a central rôle in the theory of locally compact quantum groups. In the theory of measurable quantum groupoids, this rôle is taken by pseudo-multiplicative unitaries. These unitaries do not act on an ordinary tensor product of Hilbert spaces but on a relative tensor product of Hilbert modules. The concept of a pseudo-multiplicative unitary was introduced by Vallin [172].

**The relative tensor product of Hilbert modules.** The relative tensor product of Hilbert modules is an analogue of the tensor product of modules over a ring, where the modules are Hilbert spaces, the ring is a von Neumann algebra, and the relative tensor product is a Hilbert space again. If the underlying von Neumann algebra is commutative, the relative tensor product corresponds to the fiberwise tensor product of fields of Hilbert spaces. The construction in the general case is much more involved; it was introduced by Connes and is frequently called Connes' fusion.

We shall not attempt to give an overview of the theory of measured quantum groupoids, but provide an introduction to some of the concepts listed above. First, we give a fairly detailed introduction to the relative tensor product of Hilbert modules (Section 10.1). This construction is then used to define the fiber product of von Neumann algebras and Hopf–von Neumann bimodules (Section 10.2). Finally, we discuss pseudo-multiplicative unitaries on Hilbert spaces (Section 10.3) and study examples related to groupoids (Section 10.3.3).

## 10.1 The relative tensor product of Hilbert modules

The relative tensor product of Hilbert modules is an analogue of the tensor product of modules over a ring and of the internal tensor product of $C^*$-modules over a $C^*$-algebra. Roughly, this construction starts from a von Neumann algebra $N$ and Hilbert spaces $H$ and $K$ with an antirepresentation and a representation of $N$, respectively, and produces a new Hilbert space $H \underset{N}{\otimes} K$ by factoring out the actions of $N$ on $H$ and $K$. The precise construction is fairly involved.

Before we turn to the details, let us add some bibliographical remarks. For Hilbert spaces over a commutative von Neumann algebra, the relative tensor product was defined by Sauvageout. Connes generalized the construction to the non-commutative case; unfortunately, his manuscript was never published. Original references for the construction are [27], [138]; alternative references are [52] and [150, Section IX.3].

Throughout this section, let $N$ be a von Neumann algebra.

### 10.1.1 Hilbert modules over von Neumann algebras

We adopt the following terminology:

**Definition 10.1.1.** A *left/right Hilbert $N$-module* is a Hilbert space $H$ equipped with a non-degenerate normal representation/antirepresentation of $N$. We write $x\xi$ / $\xi x$ for the action of an element $x \in N$ on a vector $\xi \in H$.

Let $N_1$ and $N_2$ be von Neumann algebras. A *Hilbert $N_1$-$N_2$-bimodule* is a Hilbert space $H$ equipped with the structure of a left Hilbert $N_1$-module and a right Hilbert $N_2$-module such that $(x_1\xi)x_2 = x_1(\xi x_2)$ for all $x_1 \in N_1$, $x_2 \in N_2, \xi \in H$.

A *morphism of left Hilbert $N$-modules* $H_1$ and $H_2$ is an operator $T \in \mathcal{L}(H_1, H_2)$ that satisfies $Tx\xi = xT\xi$ for all $x \in N$ and $\xi \in H_1$. We denote the set of all such morphisms by $\mathcal{L}_N(H_1, H_2)$. Similarly, we define morphisms of right Hilbert $N$-modules and of Hilbert $N_1$-$N_2$-bimodules.

We denote by $N^{\mathrm{op}}$ the opposite von Neumann algebra of $N$ and by $N \to N^{\mathrm{op}}$, $x \mapsto x^{\mathrm{op}}$, the canonical antiautomorphism (see Section 12.3). Clearly, left

Hilbert $N$-modules correspond bijectively with right Hilbert $N^{\mathrm{op}}$-modules, and right Hilbert $N$-modules correspond bijectively with left Hilbert $N^{\mathrm{op}}$-modules.

The fundamental example of a Hilbert $N$-$N$-bimodule is the GNS-space associated to an n.s.f. weight. This bimodule plays a central rôle in the construction of the relative tensor product.

**Example 10.1.2.** Let $\psi$ be a normal semi-finite faithful weight on $N$ with GNS-construction $(H_\psi, \Lambda_\psi, \pi_\psi)$. Then the space $H_\psi$ carries the following structure of a Hilbert $N$-$N$-bimodule:

The representation $\pi_\psi$ turns $H_\psi$ into a left Hilbert $N$-module. Using the modular conjugation $J_\psi$ for $\psi$ (see Theorem 8.2.4), we can also define an antirepresentation

$$N \to \mathcal{L}(H_\psi), \quad z \mapsto J_\psi \pi_\psi(z)^* J_\psi. \tag{10.1}$$

This map is linear because $J_\psi$ is conjugate-linear, an antihomomorphism because $J_\psi^2 = \mathrm{id}_{H_\psi}$, and normal and non-degenerate because $\pi_\psi$ is normal and non-degenerate. Furthermore, this antirepresentation commutes with $\pi_\psi$ (see Theorem 8.2.4). Thus, $H_\psi$ becomes a Hilbert $N$-$N$-bimodule.

While $\pi_\psi$ amounts to left multiplication, the antirepresentation (10.1) amounts to right multiplication up to a twist by the modular automorphism group $\sigma^\psi$ of $\psi$:

$$\pi_\psi(x)\Lambda_\psi(y) = \Lambda_\psi(xy) \quad \text{and} \quad J_\psi \pi_\psi(z)^* J_\psi \Lambda_\psi(y) = \Lambda_\psi(y\sigma^\psi_{-i/2}(z))$$

for all $x \in N$, $y \in \mathcal{N}_\psi$, and $z \in \mathrm{Dom}(\sigma^\psi_{-i/2})$ (see Theorem 8.2.4 vi)).

The preceding construction yields two Hilbert $N^{\mathrm{op}}$-$N^{\mathrm{op}}$-bimodules:

- we can consider $H_\psi$ as a Hilbert $N^{\mathrm{op}}$-$N^{\mathrm{op}}$-bimodule via $x^{\mathrm{op}}\zeta y^{\mathrm{op}} := y\zeta x$ for all $x, y \in N$ and $\zeta \in H_\psi$, or
- we can define an opposite weight $\psi^{\mathrm{op}}$ on $N^{\mathrm{op}}$ by $\psi^{\mathrm{op}}(x^{\mathrm{op}}) = \psi(x)$ for all positive elements $x \in N$, and consider the associated $N^{\mathrm{op}}$-$N^{\mathrm{op}}$-bimodule $H_{\psi^{\mathrm{op}}}$.

It turns out that these two Hilbert $N^{\mathrm{op}}$-$N^{\mathrm{op}}$-bimodules are isomorphic: the relation $\psi^{\mathrm{op}}((y^{\mathrm{op}})^* y^{\mathrm{op}}) = \psi(yy^*)$, $y \in N$, implies

$$\mathcal{N}_{\psi^{\mathrm{op}}} = \{y^{\mathrm{op}} \mid y^* \in \mathcal{N}_\psi\} = (\mathcal{N}_\psi{}^*)^{\mathrm{op}},$$

and straightforward calculations show that the space $H_\psi$ and the maps

$$\begin{aligned} \Lambda_{\psi^{\mathrm{op}}} \colon \mathcal{N}_{\psi^{\mathrm{op}}} &\to H_\psi, & y^{\mathrm{op}} &\mapsto J_\psi \Lambda_\psi(y^*), \\ \pi_{\psi^{\mathrm{op}}} \colon N^{\mathrm{op}} &\to \mathcal{L}(H_\psi), & x^{\mathrm{op}} &\mapsto J_\psi \pi_\psi(x^*) J_\psi, \end{aligned}$$

form a GNS-construction for $\psi^{\mathrm{op}}$. Evidently, the representation $\pi_{\psi^{\mathrm{op}}}$ corresponds to the antirepresentation (10.1), and by symmetry, it follows that the representation

$\pi_\psi$ corresponds to the antirepresentation $N^{\mathrm{op}} \to \mathcal{L}(H_{\psi^{\mathrm{op}}})$ given by $x^{\mathrm{op}} \mapsto J_{\psi^{\mathrm{op}}} \cdot \pi_{\psi^{\mathrm{op}}}(x^{\mathrm{op}})^* \cdot J_{\psi^{\mathrm{op}}}$.

The category of Hilbert modules over a von Neumann algebra has sufficiently many morphisms:

**Proposition 10.1.3.** *Let $\psi$ be an n.s.f. weight on $N$. Consider the associated GNS-space $H_\psi$ as a right Hilbert $N$-module via the antirepresentation* (10.1). *For every right Hilbert $N$-module $H$ and every element $\eta \in H$, there exists a partial isometry $u \in \mathcal{L}_N(H_\psi, H)$, such that $uu^*\eta = \eta$.*

*Proof.* Given $\eta \in H$, consider the positive normal linear functional $\omega$ on $N$ given by $\omega(y) := \langle \eta | \eta y \rangle$ for all $y \in N$. By [58, Lemma 2.10] or [150, IX, Theorem 1.2 iv)], there exists a vector $\xi \in H_\psi$ such that $\omega(y) = \langle \xi | \xi y \rangle$ for all $y \in N$. By the choice of $\xi$, the map $u \colon H_\psi \to H$ given by $u(\xi y) := \eta y$ for all $y \in N$ and $u(\zeta) := 0$ for all $\zeta \in (\xi N)^\perp$ is a well-defined partial isometry that satisfies $uu^*\eta = \eta$. □

In the situation above, the set $\mathcal{L}_N(H_\psi, H)H_\psi$ is equal to $H$. If we replace $H_\psi$ by an arbitrary right $N$-module, a similar result holds:

**Proposition 10.1.4.** *Let $H_1$ and $H_2$ be Hilbert spaces, considered as right $N$-modules via non-degenerate normal $*$-antihomomorphisms $\pi_1 \colon N \to \mathcal{L}(H_1)$ and $\pi_2 \colon N \to \mathcal{L}(H_2)$. If $\ker \pi_1 \subseteq \ker \pi_2$, then $\mathcal{L}_N(H_1, H_2)H_1$ is linearly dense in $H_2$.*

*Proof.* By [35, I.4.4, Théorème 3], [137, Proposition 2.7.4], or [149, IV, Theorem 5.5], the $*$-antihomomorphism $\pi_2$ has the following form: there exist a Hilbert space $H$, a projection $p \in (\pi_1(N) \otimes 1)' \subseteq \mathcal{L}(H_1 \otimes H)$, and a unitary $U \in \mathcal{L}(p(H_1 \otimes H), H_2)$, such that

$$\pi_2(x) = \mathrm{Ad}_U(p(\pi_1(x) \otimes 1)) \quad \text{for all } x \in N.$$

Consider $H_1 \otimes H$ and $p(H_1 \otimes H)$ as right $N$-modules via the antirepresentation $x \mapsto \pi_1(x) \otimes 1$ and its restriction. Then $U \in \mathcal{L}_N(p(H_1 \otimes H), H_2)$ and $p \in \mathcal{L}_N(H_1 \otimes H, p(H_1 \otimes H))$, so that it suffices to show that $\mathcal{L}_N(H_1, H_1 \otimes H)H_1$ is linearly dense in $H_1 \otimes H$. But for every $\xi \in H$, the map $\eta \mapsto \eta \otimes \xi$ defines an element in $\mathcal{L}_N(H_1, H_1 \otimes H)$ with image $H_1 \otimes \mathbb{C}\xi$. The claim follows. □

## 10.1.2 Outline of the construction

The construction of the relative tensor product is quite involved. Therefore we first consider the special case where the underlying von Neumann algebra is commutative and outline the general approach before we turn to the details.

**The commutative case.** Let $N$ be a commutative von Neumann algebra on a separable Hilbert space. Then there exist a compact metrizable space $X$ and a positive Borel measure $\mu$ on $X$ with support $X$ such that $N$ is normally isomorphic to $L^\infty(X,\mu)$ [35, 1.7.3, Théorème 1]. Every Hilbert module over $N$ corresponds to a measurable field of Hilbert spaces over $(X,\mu)$, and the relative tensor product of two such modules is formed by taking the fiberwise tensor product of the corresponding fields. Let us explain the details; standard references are [35, Chapitre II], [149, Section IV.8].

**Definition 10.1.5.** A *measurable field of Hilbert spaces* on $(X,\mu)$ is a pair $\mathfrak{H} = ((\mathfrak{H}_x)_x, \Gamma(\mathfrak{H}))$ consisting of a family of Hilbert spaces $(\mathfrak{H}_x)_{x\in X}$ and a subset $\Gamma(\mathfrak{H}) \subseteq \prod_{x\in X} \mathfrak{H}_x$ subject to the following conditions:

i) for all $\eta, \xi \in \Gamma(\mathfrak{H})$, the function $x \mapsto \langle \eta(x)|\xi(x)\rangle$ is measurable;

ii) if $\eta \in \prod_{x\in X} \mathfrak{H}_x$ and for each $\xi \in \Gamma(\mathfrak{H})$, the function $x \mapsto \langle \eta(x)|\xi(x)\rangle$ is measurable, then $\eta \in \Gamma(\mathfrak{H})$;

iii) there exists a sequence $(\xi_n)_n$ in $\Gamma(\mathfrak{H})$ such that the sequence $(\xi_n(x))_n$ is linearly dense in $\mathfrak{H}_x$ for each $x \in X$.

The elements of $\Gamma(\mathfrak{H})$ are called the *measurable sections* of $\mathfrak{H}$. A measurable section $\xi$ is *square-integrable* if $\int_X \|\xi(x)\|^2 d\mu(x) < \infty$. The *direct integral* of $\mathfrak{H}$, denoted by $\int_X^\oplus \mathfrak{H}_x d\mu(x)$, is the space of all square-integrable sections of $\mathfrak{H}$, where two sections are identified if they coincide $\mu$-almost everywhere.

The direct integral $\int_X^\oplus \mathfrak{H}_x d\mu(x)$ above is a Hilbert space with respect to the inner product $\langle \eta|\xi\rangle := \int_X \langle \eta(x)|\xi(x)\rangle d\mu(x)$, and a Hilbert module over $L^\infty(X,\mu)$ with respect to pointwise multiplication of sections by functions.

**Theorem 10.1.6.** *For every separable Hilbert module $H$ over $L^\infty(X,\mu)$, there exists a measurable field of Hilbert spaces $\mathfrak{H}$ on $(X,\mu)$ such that $H \cong \int_X^\oplus \mathfrak{H}_x d\mu(x)$ as Hilbert $L^\infty(X,\mu)$-modules.*

*Proof.* See [35, II.6.2, Théorème 2]. □

The tensor product of fields of Hilbert spaces is easily constructed:

**Proposition 10.1.7.** *Let $\mathfrak{H}$ and $\mathfrak{K}$ be measurable fields of Hilbert spaces on $(X,\mu)$. Then there exists a unique measurable field of Hilbert spaces $\mathfrak{H} \otimes \mathfrak{K}$ on $(X,\mu)$ such that*

i) $(\mathfrak{H} \otimes \mathfrak{K})_x = \mathfrak{H}_x \otimes \mathfrak{K}_x$ *for all $x \in X$, and*

ii) *for all $\eta \in \Gamma(\mathfrak{H})$, $\xi \in \Gamma(\mathfrak{K})$, the section $x \mapsto \eta(x) \otimes \xi(x)$ belongs to* $\Gamma(\mathfrak{H} \otimes \mathfrak{K})$.

*Proof.* See [35, II.1.8, Proposition 10]. □

The relative tensor product of $L^\infty(X,\mu)$-modules can now be defined as follows: for each pair of measurable fields of Hilbert spaces $\mathfrak{H}$ and $\mathfrak{K}$, we put

$$\Big(\int_X^\oplus \mathfrak{H}_x d\mu(x)\Big) \underset{L^\infty(X,\mu)}{\otimes} \Big(\int_X^\oplus \mathfrak{K}_x d\mu(x)\Big) := \int_X^\oplus (\mathfrak{H}\otimes\mathfrak{K})_x d\mu(x).$$

Straightforward but tedious verifications show that the Hilbert space on the right-hand side does not depend on the precise choice of the fields $\mathfrak{H}$ and $\mathfrak{K}$ but only on the direct integrals that appear on the left-hand side.

**The general case.** Let us consider the general case where $N$ is an arbitrary von Neumann algebra and $H$ is a right and $K$ a left $N$-module. A first attempt at the definition of the relative tensor product $H \underset{N}{\otimes} K$ might be to look for an inner product on the algebraic tensor product $H \underset{N}{\odot} K$. The analogy with the internal tensor product of $C^*$-modules suggests to define an $N$-valued inner product $\langle\cdot|\cdot\rangle_N$ on $H$ and consider the sesquilinear form on the algebraic tensor product $H \odot K$ given by

$$\langle \eta'\odot\xi'|\eta\odot\xi\rangle := \langle \xi'|\langle\eta'|\eta\rangle_N\xi\rangle, \quad \text{where } \eta,\eta'\in H,\ \xi,\xi'\in K.$$

Then, $H \underset{N}{\otimes} K$ should be the Hilbert space obtained from that sesquilinear form by the standard procedure, that is, one factorizes out the null space of the form and completes the quotient with respect to the induced norm.

The approach sketched above does not work in general and has to be modified. Indeed, the spaces $H \underset{N}{\odot} K$ and $H$ turn out to be too large to carry the desired inner products: the space $H$ has to be replaced by a dense subspace of elements that are *bounded* with respect to the action of $N$. The precise definition involves the choice of a weight on $N$ but turns out to be essentially independent of that choice.

For illustration, let us return to the commutative case considered above: there, $H$ and $K$ correspond to measurable fields of Hilbert spaces $\mathfrak{H}$ and $\mathfrak{K}$ on $(X,\mu)$, and the elements of $H$ and $K$ correspond to square-integrable sections of $\mathfrak{H}$ and $\mathfrak{K}$, respectively. In general, the fiberwise tensor product of such sections is not square-integrable but integrable. However, the fiberwise tensor product of a section that is essentially bounded with a square-integrable section is always square-integrable. If $\eta,\eta'\in\int_X^\oplus \mathfrak{H}_x d\mu(x)\cong H$ are essentially bounded in the sense that the functions $x\mapsto\|\eta(x)\|$ and $x\mapsto\|\eta'(x)\|$ belong to $L^\infty(X,\mu)$, then also the function $x\mapsto\langle\eta'(x)|\eta(x)\rangle$ belongs to $L^\infty(X,\mu)\cong N$, and one can define $\langle\eta'|\eta\rangle_N$ to be that function.

### 10.1.3 Bounded elements of a Hilbert module

In this subsection, we introduce the space of bounded elements of a right Hilbert $N$-module and construct an $N$-valued inner product on that space. As explained above, the definition of that inner product is the crucial step in the construction of the relative tensor product. To define the space of bounded elements of a right Hilbert $N$-module $H$ and the $N$-valued inner product thereon, we choose an n.s.f. weight $\psi$ on $N$ and identify certain elements of $H$ – the $\psi^{\mathrm{op}}$-bounded elements – with morphisms between the GNS-space $H_\psi$ and $H$. Before we turn to the details, let us note that throughout this subsection, we use the $N$-$N$-bimodule structure on $H_\psi$ and the notation introduced in Example 10.1.2.

**Definition 10.1.8.** Let $\psi$ be an n.s.f. weight on $N$ and $K$ a left Hilbert $N$-module. An element $\xi \in K$ is *$\psi$-bounded* if the map

$$R_\psi(\xi)\colon \Lambda_\psi(\mathcal{N}_\psi) \to K, \quad \Lambda_\psi(y) \mapsto y\xi,$$

is bounded. In this case, we denote the extension to a bounded operator $H_\psi \to K$ by $R_\psi(\xi)$ again. We denote the set of $\psi$-bounded elements of $K$ by $D(K,\psi)$.

Let $H$ be a right Hilbert $N$-module. An element $\eta \in H$ is *$\psi^{\mathrm{op}}$-bounded* if the map

$$L_\psi(\eta)\colon \Lambda_{\psi^{\mathrm{op}}}(\mathcal{N}_{\psi^{\mathrm{op}}}) \to H, \quad \Lambda_{\psi^{\mathrm{op}}}(y^{\mathrm{op}}) \mapsto \eta y,$$

is bounded. In this case, we denote the extension to a bounded operator $H_\psi \to H$ by $L_\psi(\xi)$ again. We denote the set of $\psi^{\mathrm{op}}$-bounded elements of $H$ by $D(H,\psi^{\mathrm{op}})$.

**Remarks 10.1.9.** i) Evidently, an element $\xi$ of a left Hilbert $N$-module $K$ is $\psi$-bounded if and only if there exists a constant $C \geq 0$ such that $\|y\xi\|^2 \leq C\|\Lambda_\psi(y)\|^2 = C\psi(y^*y)$ for all $y \in \mathcal{N}_\psi$. Likewise, an element $\eta$ of a right Hilbert $N$-module $H$ is $\psi^{\mathrm{op}}$-bounded if and only if there exists a constant $C \geq 0$ such that $\|\eta y\|^2 \leq C\|\Lambda_{\psi^{\mathrm{op}}}(y^{\mathrm{op}})\|^2 = C\psi(yy^*)$ for all $y \in \mathcal{N}_\psi{}^*$.

ii) If we consider a left Hilbert $N$-module as a right Hilbert $N^{\mathrm{op}}$-module, we obtain two notions of $\psi^{\mathrm{op}}$-bounded elements; however, it is easy to see that these two notions coincide. A similar remark applies to right Hilbert $N$-modules, which can also be considered as left Hilbert $N^{\mathrm{op}}$-modules.

iii) In [150], Takesaki writes $D(H,\psi)$ instead of $D(H,\psi^{\mathrm{op}})$, and $D'(K,\psi)$ instead of $D(K,\psi)$.

We shall focus on right Hilbert $N$-modules, on $\psi^{\mathrm{op}}$-bounded elements, and on the maps $L_\psi(\eta)$ introduced above. These maps are morphisms of right $N$-modules, and this fact can be used to construct an $N$-valued inner product on $D(H,\psi^{\mathrm{op}})$:

**Lemma 10.1.10.** *Let $H$ be a right Hilbert $N$-module and $\psi$ an n.s.f. weight on $N$.*

i) *Consider* $H_\psi$ *as a right Hilbert* $N$*-module. Then* $L_\psi(\eta) \in \mathcal{L}_N(H_\psi, H)$ *and* $L_\psi(\eta')^* L_\psi(\eta) \in \pi_\psi(N)$ *for all* $\eta, \eta' \in D(H, \psi^{\mathrm{op}})$.

ii) *The map* $\langle \cdot | \cdot \rangle_\psi \colon D(H, \psi^{\mathrm{op}}) \times D(H, \psi^{\mathrm{op}}) \to N$ *given by*

$$\langle \eta' | \eta \rangle_\psi := x \ \Leftrightarrow\ L_\psi(\eta')^* L_\psi(\eta) = \pi_\psi(x)$$

*is well defined, sesquilinear, and positive definite in the sense that for each non-zero* $\eta \in D(H, \psi^{\mathrm{op}})$*, the element* $\langle \eta | \eta \rangle_\psi \in N$ *is positive and non-zero.*

*Proof.* i) Let $\eta, \eta' \in D(H, \psi^{\mathrm{op}})$, let $y \in \mathcal{N}_\psi{}^*$, and let $x \in N$. By definition, $\Lambda_{\psi^{\mathrm{op}}}(y^{\mathrm{op}})x = \pi_{\psi^{\mathrm{op}}}(x^{\mathrm{op}})\Lambda_{\psi^{\mathrm{op}}}(y^{\mathrm{op}}) = \Lambda_{\psi^{\mathrm{op}}}((yx)^{\mathrm{op}})$, and hence

$$L_\psi(\eta)(\Lambda_{\psi^{\mathrm{op}}}(y^{\mathrm{op}})x) = \eta(yx) = (\eta y)x = (L_\psi(\eta)\Lambda_{\psi^{\mathrm{op}}}(y^{\mathrm{op}}))x.$$

Consequently, $L_\psi(\eta) \in \mathcal{L}_N(H_\psi, H)$ and $L_\psi(\eta')^*\, L_\psi(\eta) \in \mathcal{L}_N(H_\psi)$. By Theorem 8.2.4, $\mathcal{L}_N(H_\psi) = (J_\psi \pi_\psi(N) J_\psi)'$ is equal to $\pi_\psi(N)$.

ii) The map $\langle \cdot | \cdot \rangle_\psi$ is well defined because $\pi_\psi$ is injective. Let $\eta \in D(H, \psi^{\mathrm{op}})$. Evidently, $\langle \eta | \eta \rangle_\psi = L_\psi(\eta)^* L_\psi(\eta)$ is positive. Assume that $\langle \eta | \eta \rangle_\psi = 0$. Then also $L_\psi(\eta) = 0$ and

$$\langle \xi y^* | \eta \rangle = \langle \xi | \eta y \rangle = \langle \xi | L_\psi(\eta)\Lambda_{\psi^{\mathrm{op}}}(y^{\mathrm{op}}) \rangle = 0 \quad \text{for all } \xi \in H,\ y \in \mathcal{N}_\psi{}^*.$$

Since $\mathcal{N}_\psi$ is weakly dense in $N$ and $HN = H$, we must have $\eta = 0$. □

**Remark 10.1.11.** Let $K$ be a left Hilbert $N$-module and $\psi$ an n.s.f. weight on $N$. Considering $K$ as a right Hilbert $N^{\mathrm{op}}$-module, we obtain an $N^{\mathrm{op}}$-valued inner product $\langle \cdot | \cdot \rangle_{\psi^{\mathrm{op}}}$ on $D(K, \psi)$. Since $L_{\psi^{\mathrm{op}}}(\xi) = R_\psi(\xi)$ for all $\xi \in K$,

$$\langle \xi' | \xi \rangle_{\psi^{\mathrm{op}}} = x^{\mathrm{op}} \ \Leftrightarrow\ R_\psi(\xi')^* R_\psi(\xi) = \pi_{\psi^{\mathrm{op}}}(x^{\mathrm{op}}) \quad \text{for all } \xi, \xi' \in D(K, \psi).$$

For later applications to the relative tensor product of Hilbert $N$-modules, we collect several properties of the maps $L_\psi(\eta)$ and of the $N$-valued product $\langle \cdot | \cdot \rangle_\psi$ constructed above:

**Proposition 10.1.12.** *Let* $H$ *be a right Hilbert* $N$*-module and* $\psi$ *an n.s.f. weight on* $N$.

i) *For each* $\eta \in D(H, \psi^{\mathrm{op}})$ *and* $x \in \mathrm{Dom}(\sigma^\psi_{-i/2})$*, we have* $\eta x \in D(H, \psi^{\mathrm{op}})$ *and* $L_\psi(\eta x) = L_\psi(\eta)\pi_\psi(\sigma^\psi_{-i/2}(x))$.

ii) *For every* $\eta \in D(H, \psi^{\mathrm{op}})$*, every right Hilbert* $N$*-module* $K$*, and every* $T \in \mathcal{L}_N(H, K)$*, we have* $T\eta \in D(K, \psi^{\mathrm{op}})$ *and* $L_\psi(T\eta) = TL_\psi(\eta)$*. Furthermore,* $\langle \eta' | T\eta \rangle_\psi = \langle T^*\eta' | \eta \rangle_\psi$ *for every* $\eta' \in D(K, \psi^{\mathrm{op}})$.

iii) $D(H_\psi, \psi^{\mathrm{op}}) = \Lambda_\psi(\mathcal{N}_\psi)$ *and* $L_\psi(\Lambda_\psi(y)) = \pi_\psi(y)$ *for all* $y \in \mathcal{N}_\psi$.

iv) $D(H, \psi^{\mathrm{op}}) = \mathcal{L}_N(H_\psi, H)\Lambda_\psi(\mathcal{N}_\psi)$ *is dense in* $H$.

v) *For all* $\eta, \eta' \in D(H, \psi^{\mathrm{op}})$,

$$\langle \eta'|\eta\rangle_\psi \in \mathcal{N}_\psi, \quad \Lambda_\psi(\langle \eta'|\eta\rangle_\psi) = L_\psi(\eta')^*\eta, \quad \psi(\langle \eta'|\eta\rangle_\psi) = \langle \eta'|\eta\rangle.$$

*Proof.* i) Let $y \in \mathcal{N}_\psi{}^*$. By Theorem 8.2.4 vi) and Proposition 8.1.11 iii),

$$\begin{aligned}\pi_\psi(\sigma^\psi_{-i/2}(x))\Lambda_{\psi^{\mathrm{op}}}(y^{\mathrm{op}}) &= \pi_\psi(\sigma^\psi_{i/2}(x^*))^* J_\psi \Lambda_\psi(y^*) \\ &= J_\psi\Lambda_\psi(y^*x^*) = \Lambda_{\psi^{\mathrm{op}}}((xy)^{\mathrm{op}}),\end{aligned}$$

and therefore

$$L_\psi(\eta x)\Lambda_{\psi^{\mathrm{op}}}(y^{\mathrm{op}}) = \eta x y = L_\psi(\eta)\pi_\psi(\sigma^\psi_{-i/2}(x))\Lambda_{\psi^{\mathrm{op}}}(y^{\mathrm{op}}).$$

ii) The first assertions follow from the relation

$$TL_\psi(\eta)\Lambda_{\psi^{\mathrm{op}}}(y^{\mathrm{op}}) = T(\eta y) = (T\eta)y = L_\psi(T\eta)\Lambda_{\psi^{\mathrm{op}}}(y^{\mathrm{op}}) \quad \text{for all } y \in \mathcal{N}_\psi{}^*,$$

and the last one from the relation

$$L_\psi(\eta')^*TL_\psi(\eta) = (T^*L_\psi(\eta'))^*L_\psi(\eta) = L_\psi(T^*\eta')^*L_\psi(\eta).$$

iii) This is a standard result from the theory of Hilbert algebras, see [150, Proof of Theorem VII.2.6].

iv) Statements ii) and iii) imply $\mathcal{L}_N(H_\psi, H)\Lambda_\psi(\mathcal{N}_\psi) \subseteq D(H, \psi^{\mathrm{op}})$. Let us prove the reverse inclusion. Given $\eta \in D(H, \psi^{\mathrm{op}})$, we choose a partial isometry $u \in \mathcal{L}_N(H_\psi, H)$ such that $uu^*\eta = \eta$ (see Proposition 10.1.3). Then $u^*\eta \in D(H_\psi, \psi^{\mathrm{op}}) = \Lambda_\psi(\mathcal{N}_\psi)$ by iii) and $\eta = u(u^*\eta) \in \mathcal{L}_N(H_\psi, H)\Lambda_\psi(\mathcal{N}_\psi)$. Thus, $D(H, \psi^{\mathrm{op}}) = \mathcal{L}_N(H_\psi, H)\Lambda_\psi(\mathcal{N}_\psi)$. By definition of $H_\psi$, this space is dense in $\mathcal{L}_N(H_\psi, H)H_\psi$, and by Proposition 10.1.3, also in $H$.

v) Let $\eta, \eta' \in D(H, \psi^{\mathrm{op}})$. By Lemma 10.1.10 i) and by ii), $L_\psi(\eta')^*\eta \in D(H_\psi, \psi^{\mathrm{op}})$, and by iii), $L_\psi(\eta')^*\eta = \Lambda_\psi(y)$ for some $y \in \mathcal{N}_\psi$. Therefore,

$$L_\psi(\eta')^*L_\psi(\eta) = L_\psi(L_\psi(\eta')^*\eta) = L_\psi(\Lambda_\psi(y)) = \pi_\psi(y) \in \pi_\psi(\mathcal{N}_\psi)$$

and $\Lambda_\psi(\langle \eta'|\eta\rangle_\psi) = \Lambda_\psi(y) = L_\psi(\eta')^*\eta$. Thus we have proved the first and second assertion. Let us prove $\psi(\langle \eta|\eta\rangle_\psi) = \langle \eta|\eta\rangle$. Choose a partial isometry $u \in \mathcal{L}_N(H_\psi, H)$ such that $uu^*\eta = \eta$ (see Proposition 10.1.3). By iii), $u^*\eta = \Lambda_\psi(x)$ for some $x \in \mathcal{N}_\psi$, and $\pi_\psi(x) = L_\psi(u^*\eta) = u^*L_\psi(\eta)$. Then $x^*x = \langle \eta|\eta\rangle_\psi$ because

$$\pi_\psi(x^*x) = L_\psi(\eta)^*uu^*L_\psi(\eta) = L_\psi(\eta)^*L_\psi(uu^*\eta) = L_\psi(\eta)^*L_\psi(\eta),$$

and

$$\psi(\langle \eta|\eta\rangle_\psi) = \psi(x^*x) = \|\Lambda_\psi(x)\|^2 = \|\eta\|^2 = \langle \eta|\eta\rangle.$$

Using polarization, we get $\psi(\langle \eta'|\eta\rangle_\psi) = \langle \eta'|\eta\rangle_\psi$ for all $\eta, \eta' \in D(H, \psi^{\mathrm{op}})$. □

**Remark 10.1.13.** In the situation of the preceding proposition, the linear span of the set $\{L_\psi(\eta)L_\psi(\eta')^* \mid \eta, \eta' \in D(H, \psi^{\mathrm{op}})\}$ is a weakly dense ideal in $\mathcal{L}_N(H)$, see [26, Proposition 3] or [150, IX.3, Lemma 3.9].

The preceding definitions and results can be rewritten in a suggestive symbolic calculus. We briefly summarize this calculus, but restrict to a symbolic level and do not make anything mathematically precise. Further details can be found in [27, V, Appendix B].

**Remark 10.1.14.** Before we begin, let us stress that *all calculations in this remark are purely formal symbolic manipulations*. We shall use the sign "$\equiv$" to indicate that two expressions are equal on a symbolic level.

Consider the Hilbert $N$-$N$-bimodule $H_\psi$ associated to an n.s.f. weight $\psi$ on $N$ (see Example 10.1.2). We put

$$\begin{aligned} y\psi^{1/2} &:\equiv \Lambda_\psi(y) && \text{for } y \in \mathcal{N}_\psi, \\ \psi^{1/2}x\psi^{-1/2} &:\equiv \sigma^\psi_{-i/2}(x) && \text{for } x \in \mathrm{Dom}(\sigma^\psi_{-i/2}), \end{aligned}$$

and treat the symbols $\psi^{1/2}$ and $\psi^{-1/2}$ like invertible self-adjoint operators. Then the $N$-$N$-bimodule structure of $H_\psi$ takes the form

$$x(y\psi^{1/2})z \equiv xy(\psi^{1/2}z\psi^{-1/2})\psi^{1/2} \equiv (xy\sigma^\psi_{-i/2}(z))\psi^{1/2}$$

for all $x \in N$, $y \in \mathcal{N}_\psi$, $z \in \mathrm{Dom}(\sigma^\psi_{-i/2})$. A comparison with Example 10.1.2 shows that this symbolic calculation correctly describes the actions of $N$.

For the GNS-map of the opposite weight $\psi^{\mathrm{op}}$, we put

$$\psi^{1/2}y :\equiv y^{\mathrm{op}}(\psi^{\mathrm{op}})^{1/2} \equiv \Lambda_{\psi^{\mathrm{op}}}(y^{\mathrm{op}}) \qquad \text{for } y \in \mathcal{N}_\psi^*.$$

Then the $N$-$N$-bimodule structure of $H_\psi$ takes the form

$$x(\psi^{1/2}y)z \equiv \psi^{1/2}(\psi^{-1/2}x\psi^{1/2})yz \equiv \psi^{1/2}\sigma^\psi_{i/2}(x)yz$$

for all $x \in \mathrm{Dom}(\sigma^\psi_{i/2})$, $y \in \mathcal{N}_\psi^*$, $z \in N$. Again, one can show that this symbolic calculation correctly describes the actions of $N$.

Given a Hilbert space $H$, we put

$$\eta'^*\eta :\equiv \langle \eta' | \eta \rangle \qquad \text{for } \eta, \eta' \in H.$$

Then the defining property of the modular automorphism group $\sigma^\psi$ (see Theorem 8.1.13 ii)) can be read off from the following symbolic calculations. Given $x \in \mathcal{N}_\psi \cap \mathrm{Dom}(\sigma^\psi_{i/2})$, put $z := \psi^{-1/2}x\psi^{1/2} \equiv \sigma^\psi_{i/2}(x)$. Then

$$\begin{aligned} \psi(x^*x) &\equiv \langle x\psi^{1/2} | x\psi^{1/2} \rangle \equiv \psi^{1/2}x^*x\psi^{1/2} \equiv z^*\psi^{1/2}\psi^{1/2}z \\ &\equiv \langle \psi^{1/2}z | \psi^{1/2}z \rangle \equiv \psi^{\mathrm{op}}((z^{\mathrm{op}})^*(z^{\mathrm{op}})) \equiv \psi(zz^*) \equiv \psi(\sigma^\psi_{i/2}(x)\sigma^\psi_{i/2}(x)^*). \end{aligned}$$

Moreover, in this notation, $\psi(x) \equiv \psi^{1/2}x\psi^{1/2}$ for each $x \in \mathcal{M}_\psi$; this relation will be used below.

Now consider a right Hilbert $N$-module $H$. Let us write

$$\eta\psi^{-1/2} :\equiv L_\psi(\eta), \quad \psi^{-1/2}\eta^* :\equiv (\eta\psi^{-1/2})^* \equiv L_\psi(\eta)^* \quad \text{for } \eta \in D(H, \psi^{\mathrm{op}}).$$

Then the action of $L_\psi(\eta)$ takes the simple form

$$L_\psi(\eta)\Lambda_{\psi^{\mathrm{op}}}(y^{\mathrm{op}}) \equiv \eta\psi^{-1/2} \cdot \psi^{1/2}y \equiv \eta y \quad \text{for all } \eta \in D(H, \psi^{\mathrm{op}}),\ y \in \mathcal{N}_\psi{}^*.$$

If we identify $N$ with $\pi_\psi(N)$, the inner product $\langle \cdot | \cdot \rangle_\psi$ can be written as follows:

$$\langle \eta' | \eta \rangle_\psi \equiv L_\psi(\eta')^* L_\psi(\eta) \equiv \psi^{-1/2}\eta'^*\eta\psi^{-1/2} \quad \text{for } \eta, \eta' \in D(H, \psi^{\mathrm{op}}).$$

Now we can "prove" some of the results of Proposition 10.1.12 by the following symbolic calculations:

$$\begin{aligned}
L_\psi(T\eta x) &\equiv (T\eta x)\psi^{-1/2} \equiv T\eta\psi^{-1/2}(\psi^{1/2}x\psi^{-1/2}) \equiv TL_\psi(\eta)\sigma^\psi_{-i/2}(x),\\
L_\psi(\Lambda_\psi(y)) &\equiv (y\psi^{1/2}) \cdot \psi^{-1/2} \equiv y \equiv \pi_\psi(y),\\
\Lambda_\psi(\langle \eta' | \eta \rangle_\psi) &\equiv (\psi^{-1/2}\eta'^*\eta\psi^{-1/2})\psi^{1/2} \equiv (\psi^{-1/2}\eta'^*)\eta \equiv L_\psi(\eta')^*\eta,\\
\psi(\langle \eta' | \eta \rangle_\psi) &\equiv \psi^{1/2}(\psi^{-1/2}\eta'^*\eta\psi^{-1/2})\psi^{1/2} \equiv \eta'^*\eta \equiv \langle \eta' | \eta \rangle,
\end{aligned}$$

where $\eta, \eta' \in D(H, \psi^{\mathrm{op}})$, $T \in \mathcal{L}_N(H, K)$, $x \in \mathrm{Dom}(\sigma^\psi_{-i/2})$, and $y \in \mathcal{N}_\psi$.

### 10.1.4 Construction of the relative tensor product

Equipped with the concepts introduced above, we are ready to define the relative tensor product of Hilbert modules.

Recall that to every sesquilinear form on a complex vector space, one can associate a Hilbert space by a standard procedure: one factorizes out the null space of the form and completes the quotient with respect to the induced norm.

**Lemma 10.1.15.** *Let $H$ be a right and $K$ a left Hilbert $N$-module, and let $\psi$ be an n.s.f. weight on $N$. Consider $H$ as a left and $K$ as a right Hilbert $N^{\mathrm{op}}$-module.*

i) *Formula* (10.2) *defines a positive sesquilinear form on $D(H, \psi^{\mathrm{op}}) \odot K$:*

$$\langle \eta' \odot \xi' | \eta \odot \xi \rangle := \langle \xi' | \langle \eta' | \eta \rangle_\psi \xi \rangle. \tag{10.2}$$

ii) *Formula* (10.3) *defines a positive sesquilinear form on $H \odot D(K, \psi)$:*

$$\langle \eta' \odot \xi' | \eta \odot \xi \rangle_{\mathrm{op}} := \langle \eta' | \langle \xi' | \xi \rangle_{\psi^{\mathrm{op}}} \eta \rangle. \tag{10.3}$$

iii) *The restrictions of* $\langle\cdot|\cdot\rangle$ *and* $\langle\cdot|\cdot\rangle_{\mathrm{op}}$ *to* $D(H,\psi^{\mathrm{op}})\odot D(K,\psi)$ *coincide, and for all* $\xi,\xi'\in D(H,\psi^{\mathrm{op}})$ *and* $\eta,\eta'\in D(K,\psi)$,

$$\langle\eta'\odot\xi'|\eta\odot\xi\rangle=\langle\Lambda_{\psi^{\mathrm{op}}}(\langle\xi|\xi'\rangle_{\psi^{\mathrm{op}}})|\Lambda_{\psi}(\langle\eta'|\eta\rangle_{\psi})\rangle=\langle\eta'\odot\xi'|\eta\odot\xi\rangle_{\mathrm{op}}.$$

iv) *The Hilbert spaces associated to the positive sesquilinear forms considered in* i)–iii) *are naturally isomorphic.*

*Proof.* i), ii) This follows from a standard argument (see [95, Chapter 4]).

iii) It suffices to prove the two equalities in the formula above. We only treat the first one; the proof of the second one is similar. By definition,

$$\langle\eta'\odot\xi'|\eta\odot\xi\rangle=\langle\xi'|\langle\eta'|\eta\rangle_{\psi}\,\xi\rangle=\langle\xi'|\xi((\langle\eta'|\eta\rangle_{\psi})^{\mathrm{op}}\rangle.$$

Now $\xi((\langle\eta'|\eta\rangle_{\psi})^{\mathrm{op}}=L_{\psi^{\mathrm{op}}}(\xi)\Lambda_{\psi}(\langle\eta'|\eta\rangle_{\psi})$ by definition of $L_{\psi^{\mathrm{op}}}(\xi)$, and hence

$$\langle\eta'\odot\xi'|\eta\odot\xi\rangle=\langle L_{\psi^{\mathrm{op}}}(\xi)^{*}\xi'|\Lambda_{\psi}(\langle\eta'|\eta\rangle_{\psi})\rangle.$$

Inserting the relation $L_{\psi^{\mathrm{op}}}(\xi)^{*}\xi'=\Lambda_{\psi^{\mathrm{op}}}(\langle\xi|\xi'\rangle_{\psi^{\mathrm{op}}})$ from Proposition 10.1.12 v), we obtain the desired equality.

iv) This follows immediately from iii) and the fact that $D(H,\psi^{\mathrm{op}})$ and $D(K,\psi)$ are dense in $H$ and $K$, respectively (see Proposition 10.1.12 iv)). □

**Definition 10.1.16.** The *relative tensor product* of a right Hilbert $N$-module $H$ and a left Hilbert $N$-module $K$ relative to an n.s.f. weight $\psi$ on $N$ is the Hilbert space $H\underset{\psi}{\otimes}K$ associated by the standard procedure to the sesquilinear forms defined in Lemma 10.1.15 i)–iii). Given elements $\eta\in D(H,\psi^{\mathrm{op}})$ and $\xi\in D(K,\psi)$, we write $\eta\underset{\psi}{\otimes}\xi$ for the image of $\eta\odot\xi$ under the natural map $D(H,\psi^{\mathrm{op}})\odot D(K,\psi)\to H\underset{\psi}{\otimes}K$. Likewise, we define the element $\eta\underset{\psi}{\otimes}\xi$ if either $\eta\in H$ or $\xi\in K$.

If $N$ is commutative, the relative tensor product of Hilbert $N$-modules corresponds to the fiberwise tensor product of measurable fields of Hilbert spaces:

**Proposition 10.1.17.** *Let* $N=L^{\infty}(X,\mu)$, *where* $X$ *is a compact metrizable space and* $\mu$ *is a positive Borel measure on* $X$. *Denote also by* $\psi$ *the weight on* $N$ *given by* $f\mapsto\int_X f\,d\mu$. *Let* $\mathfrak{H}$ *and* $\mathfrak{K}$ *be measurable fields of Hilbert spaces on* $(X,\mu)$ *and consider* $H=\int_X^{\oplus}\mathfrak{H}_x\,d\mu(x)$ *and* $K=\int_X^{\oplus}\mathfrak{K}_x\,d\mu(x)$ *as Hilbert* $N$*-modules via pointwise multiplication of sections by functions.*

i) *An element* $\eta\in H$ *belongs to* $D(H,\psi^{\mathrm{op}})$ *if and only if the function* $x\mapsto\|\eta(x)\|$ *belongs to* $L^2(X,\mu)\cap L^{\infty}(X,\mu)$.

ii) *For all* $\eta,\eta'\in D(H,\psi^{\mathrm{op}})$, *the inner product* $\langle\eta'|\eta\rangle_{\psi}$ *is equal to the function* $x\mapsto\langle\eta'(x)|\eta(x)\rangle$.

iii) $H \underset{\psi}{\otimes} K \cong \int_X^{\oplus} \mathfrak{H}_x \otimes \mathfrak{K}_x d\mu(x)$.

*Proof.* Straightforward. □

In the general case, the actions of $N$ on the factors of the relative tensor product do not behave as one might expect. Unlike the usual transformation rule $\eta x \underset{N}{\odot} \xi = \eta \underset{N}{\odot} x\xi$ that holds in an algebraic tensor product $H \underset{N}{\odot} K$ for all $\eta \in H$, $\xi \in K$, $x \in N$, we have the following relation:

**Proposition 10.1.18.** *Let $H$ be a right and $K$ a left Hilbert $N$-module, and let $\psi$ be an n.s.f. weight on $N$. Then for all $\eta \in D(H, \psi^{\mathrm{op}})$, $\xi \in K$, $x \in \mathrm{Dom}(\sigma^{\psi}_{-i/2})$,*

$$\eta x \underset{\psi}{\otimes} \xi = \eta \underset{\psi}{\otimes} \sigma^{\psi}_{-i/2}(x)\xi. \tag{10.4}$$

*Proof.* By Proposition 10.1.12 i), $\langle \eta' | \eta x\rangle_{\psi} = \langle \eta' | \eta\rangle_{\psi}\, \sigma^{\psi}_{-i/2}(x)$, and hence

$$\langle \eta' \underset{\psi}{\otimes} \xi' | \eta x \underset{\psi}{\otimes} \xi\rangle = \langle \xi' | \langle \eta' | \eta\rangle_{\psi}\, \sigma^{\psi}_{-i/2}(x)\xi\rangle = \langle \eta' \underset{\psi}{\otimes} \xi' | \eta \underset{\psi}{\otimes} \sigma^{\psi}_{-i/2}(x)\xi\rangle$$

for all $\eta' \in D(H, \psi^{\mathrm{op}})$ and $\xi' \in K$. □

**Remark 10.1.19.** We extend the notation introduced in Remark 10.1.14 and put

$$\eta_0 \psi^{-1/2} \underset{N}{\otimes} \xi :\equiv \eta_0 \underset{\psi}{\otimes} \xi \quad \text{and} \quad \eta \underset{N}{\otimes} \psi^{-1/2}\xi_0 :\equiv \eta \underset{\psi}{\otimes} \xi_0$$

for $\eta_0 \in D(H, \psi^{\mathrm{op}})$, $\xi \in K$ and $\eta \in H$, $\xi_0 \in D(K, \psi)$. Then the following symbolic calculation "proves" Proposition 10.1.18:

$$\begin{aligned}\eta_0 x \psi^{-1/2} \underset{N}{\otimes} \xi_0 &\equiv \eta_0 \psi^{-1/2}(\psi^{1/2} x \psi^{-1/2}) \underset{N}{\otimes} \xi_0 \\ &\equiv \eta_0 \psi^{-1/2} \sigma^{\psi}_{-i/2}(x) \underset{N}{\otimes} \xi_0 \equiv \eta_0 \psi^{-1/2} \underset{N}{\otimes} \sigma^{\psi}_{-i/2}(x)\xi_0.\end{aligned}$$

### 10.1.5 Properties of the relative tensor product

The relative tensor product has all functorial properties that one would naturally expect.

Lemma 10.1.15 implies that the construction of the relative tensor product is symmetric in the following sense:

**Proposition 10.1.20.** *Let $H$ be a right and $K$ a left Hilbert $N$-module, and let $\psi$ be an n.s.f. weight on $N$. Consider $H$ as a left and $K$ as a right Hilbert $N^{\mathrm{op}}$-module. Then the map*

$$D(H, \psi^{\mathrm{op}}) \underset{\psi}{\otimes} D(K, \psi) \to D(K, \psi) \underset{\psi^{\mathrm{op}}}{\otimes} D(H, \psi^{\mathrm{op}}), \quad \eta \underset{\psi}{\otimes} \xi \mapsto \xi \underset{\psi^{\mathrm{op}}}{\otimes} \eta,$$

*is well defined and extends to a unitary* $\Sigma_\psi \colon H \underset{\psi}{\otimes} K \xrightarrow{\cong} K \underset{\psi^{\mathrm{op}}}{\otimes} H$. □

Evidently, the relative tensor product commutes with direct sums:

**Proposition 10.1.21.** *Let* $(H_\nu)_\nu$ *be a family of right and* $(K_\mu)_\mu$ *a family of left Hilbert* $N$*-modules, and let* $\psi$ *be an n.s.f. weight on* $N$*. Consider the direct sums* $H := \bigoplus_\nu H_\nu$ *and* $K := \bigoplus_\mu K_\mu$ *as Hilbert* $N$*-modules in the obvious way. Then the family of natural inclusions* $H_\nu \underset{\psi}{\otimes} K_\mu \hookrightarrow H \underset{\psi}{\otimes} K$ *induces a unitary* $\bigoplus_{\nu,\mu}(H_\nu \underset{\psi}{\otimes} K_\mu) \xrightarrow{\cong} H \underset{\psi}{\otimes} K$.

*Proof.* Straightforward. □

The relative tensor product admits the following natural maps:

**Proposition 10.1.22.** *Let* $H$ *be a right and* $K$ *a left* $N$*-module, and let* $\psi$ *be an n.s.f. weight on* $N$*. For every* $\eta_0 \in D(H, \psi^{\mathrm{op}})$ *and* $\xi_0 \in D(K, \psi)$*, the maps*

$$|\eta_0\rangle_{\psi[1]} \colon K \to H \underset{\psi}{\otimes} K,\ \xi \mapsto \eta_0 \underset{\psi}{\otimes} \xi, \quad |\xi_0\rangle_{\psi[2]} \colon H \to H \underset{\psi}{\otimes} K,\ \eta \mapsto \eta \underset{\psi}{\otimes} \xi_0,$$

*are linear operators of norm*

$$\||\eta_0\rangle_{\psi[1]}\| \leq \|\langle\eta_0|\eta_0\rangle_\psi\|^{1/2}, \quad \||\xi_0\rangle_{\psi[2]}\| \leq \|\langle\xi_0|\xi_0\rangle_{\psi^{\mathrm{op}}}\|^{1/2}.$$

*If the representation of* $N$ *on* $K$*/on* $H$ *is injective, the left/right inequality above is an equality. The adjoints* $\langle\eta_0|_{\psi[1]} := |\eta_0\rangle^*_{\psi[1]}$ *and* $\langle\xi_0|_{\psi[2]} := |\xi_0\rangle^*_{\psi[2]}$ *are given by*

$$\langle\eta_0|_{\psi[1]}(\eta' \underset{\psi}{\otimes} \xi') = \langle\eta_0|\eta'\rangle_\psi\, \xi', \quad \langle\xi_0|_{\psi[2]}(\eta'' \underset{\psi}{\otimes} \xi'') = \langle\xi_0|\xi''\rangle_{\psi^{\mathrm{op}}}\, \eta'',$$

*where* $\eta' \in D(H, \psi^{\mathrm{op}})$*,* $\xi' \in K$ *and* $\eta'' \in H$*,* $\xi'' \in D(K, \psi)$*.*

*Proof.* Straightforward. □

We shall show that the construction of the relative tensor product is bifunctorial. First, we consider a special case.

**Proposition 10.1.23.** *Let* $H$ *be a right and* $K$ *a left* $N$*-module, and let* $\psi$ *be an n.s.f. weight on* $N$*.*

i) *For each* $S \in \mathcal{L}_N(H)$*, there exists an operator* $S \underset{\psi}{\otimes} 1 \in \mathcal{L}(H \underset{\psi}{\otimes} K)$ *such that* $(S \underset{\psi}{\otimes} 1)(\eta \underset{\psi}{\otimes} \xi) = S\eta \underset{\psi}{\otimes} \xi$ *for all* $\eta \in H$ *and* $\xi \in D(K, \psi)$*. The map* $\mathcal{L}_N(H) \to \mathcal{L}(H \underset{\psi}{\otimes} K)$ *given by* $S \mapsto S \underset{\psi}{\otimes} 1$ *is a normal non-degenerate* $*$*-homomorphism. If the representation of* $N$ *on* $K$ *is injective, this* $*$*-homomorphism is injective.*

ii) *For each* $T \in \mathcal{L}_N(K)$, *there exists an operator* $1 \underset{\psi}{\otimes} T \in \mathcal{L}(H \underset{\psi}{\otimes} K)$ *such that* $(1 \underset{\psi}{\otimes} T)(\eta \underset{\psi}{\otimes} \xi) = \eta \underset{\psi}{\otimes} T\xi$ *for all* $\eta \in D(H, \psi^{\mathrm{op}})$ *and* $\xi \in K$. *The map* $\mathcal{L}_N(K) \to \mathcal{L}(H \underset{\psi}{\otimes} K)$ *given by* $T \mapsto 1 \underset{\psi}{\otimes} T$ *is a normal non-degenerate* $*$*-homomorphism. If the antirepresentation of* $N$ *on* $H$ *is injective, this* $*$*-homomorphism is injective.*

*Proof.* By Proposition 10.1.20, it suffices to prove i). First, we show that for each $S \in \mathcal{L}_N(H)$, the map $S \underset{\psi}{\otimes} 1$ is well defined and bounded. Let $\eta_i \in D(H, \psi^{\mathrm{op}})$ and $\xi_i \in K$, where $i = 1, \dots, n$. By Proposition 10.1.12 ii), $S\eta_i \in D(H, \psi^{\mathrm{op}})$ for all $i$, and by definition, $\|\sum_i S\eta_i \underset{\psi}{\otimes} \xi_i\|^2 = \sum_{i,j} \langle \xi_i | \langle S\eta_i | S\eta_j \rangle_\psi \, \xi_j \rangle$. Now $(\langle S\eta_i | S\eta_j \rangle_\psi)_{i,j} \le \|S\|^2 (\langle \eta_i | \eta_j \rangle_\psi)_{i,j}$ in $M_n(N)$ because

$$(L_\psi(\eta_i)^* S^* S L_\psi(\eta_j))_{i,j} \le \|S\|^2 (L_\psi(\eta_i)^* L_\psi(\eta_j))_{i,j} \quad \text{in } M_n(\pi_\psi(N)),$$

and hence

$$\Big\| \sum_i S\eta_i \underset{\psi}{\otimes} \xi_i \Big\|^2 \le \|S\|^2 \cdot \sum_{i,j} \langle \xi_i | \langle \eta_i | \eta_j \rangle_\psi \, \xi_j \rangle = \|S\|^2 \cdot \Big\| \sum_i \eta_i \underset{\psi}{\otimes} \xi_i \Big\|^2.$$

Consequently, $S \underset{\psi}{\otimes} 1$ is well defined and bounded.

Using Proposition 10.1.12 ii), we find that the map $\pi : \mathcal{L}_N(H) \to \mathcal{L}(H \underset{\psi}{\otimes} K)$ given by $S \mapsto S \underset{\psi}{\otimes} 1$ is a $*$-homomorphism; in particular, it is norm-continuous. To prove that $\pi$ is normal, we show that for every normal functional $\omega \in \mathcal{L}(H \underset{\psi}{\otimes} K)_*$, the composition $\omega \circ \pi$ is normal. By [149, II.3, Theorem 2.6], functionals of the form $\omega_{\zeta',\zeta} = \langle \zeta' | \cdot \zeta \rangle$, where $\zeta, \zeta' \in H \underset{\psi}{\otimes} K$, are linearly dense in $\mathcal{L}(H \underset{\psi}{\otimes} K)_*$. Evidently, each such functional can be approximated in norm by linear combinations of functionals of the form $\omega = \langle \eta' \underset{\psi}{\otimes} \xi' | \cdot (\eta \underset{\psi}{\otimes} \xi) \rangle$, where $\eta, \eta' \in H$, $\xi, \xi' \in D(K, \psi)$. Since $\pi$ is norm-continuous and $\mathcal{L}(H)_*$ is closed in $\mathcal{L}(H)^*$ [149, II.3, Theorem 2.6], it is enough to show that for each such functional $\omega$, the composition $\omega \circ \pi$ is normal. But

$$(\omega \circ \pi)(S) = \langle \eta' \underset{\psi}{\otimes} \xi' | S\eta \underset{\psi}{\otimes} \xi \rangle = \langle \eta' | \langle \xi' | \xi \rangle_{\psi^{\mathrm{op}}} S\eta \rangle = \langle \eta' | S\eta'' \rangle$$

for all $S \in \mathcal{L}_N(H)$, where $\eta'' := \langle \xi' | \xi \rangle_{\psi^{\mathrm{op}}} \eta$. Hence $\omega \circ \pi$ is normal.

Finally, we assume that the representation of $N$ on $K$ is injective, and show that $\pi$ is injective. Let $S \in \mathcal{L}_N(H)$, $S \neq 0$. Since $D(H, \psi^{\mathrm{op}})$ is dense in $H$ (Proposition 10.1.12 iv)), there exists an $\eta \in D(H, \psi^{\mathrm{op}})$ such that $S\eta \neq 0$.

By Lemma 10.1.10 ii), $\langle S\eta|S\eta\rangle_\psi \neq 0$, and since the representation of $N$ on $K$ is injective, there exists a $\xi \in K$ such that $\|S\eta \underset{\psi}{\otimes} \xi\|^2 = \langle \xi|\langle S\eta|S\eta\rangle_\psi \xi\rangle \neq 0$. Therefore, $\pi(S) \neq 0$. □

Combining the previous result with Proposition 10.1.21, we obtain:

**Corollary 10.1.24.** *Let $\psi$ be an n.s.f. weight on $N$.*

i) *Let $H_1$, $H_2$ be right and $K_1$, $K_2$ left Hilbert $N$-modules. For all $S \in \mathcal{L}_N(H_1, H_2)$ and $T \in \mathcal{L}_N(K_1, K_2)$, there exists an operator*
$$S \underset{\psi}{\otimes} T \in \mathcal{L}(H_1 \underset{\psi}{\otimes} K_1, H_2 \underset{\psi}{\otimes} K_2), \ \eta \underset{\psi}{\otimes} \xi \mapsto S\eta \underset{\psi}{\otimes} T\xi.$$

ii) *The assignments $(H, K) \mapsto H \underset{\psi}{\otimes} K$ and $(S, T) \mapsto S \underset{\psi}{\otimes} T$ define a bifunctor from the categories of right and left Hilbert $N$-modules to the category of Hilbert spaces.*

*Proof.* Assertion i) follows easily from Proposition 10.1.21 and 10.1.23; simply consider $H := H_1 \oplus H_2$ as a right and $K := K_1 \oplus K_2$ as a left Hilbert $N$-module, and $S$ and $T$ as elements of $\mathcal{L}_N(H)$ and $\mathcal{L}_N(K)$, respectively. Statement ii) is an immediate consequence. □

The relative tensor product is associative:

**Proposition 10.1.25.** *Let $N_1$ and $N_2$ be von Neumann algebras with n.s.f. weights $\psi_1$ and $\psi_2$, respectively, and let $K_1$ be a right Hilbert $N_1$-module, $H$ a Hilbert $N_1$-$N_2$-bimodule, and $K_2$ a left Hilbert $N_2$-module. Then*

i) *$K_1 \underset{\psi_1}{\otimes} H$ is a right Hilbert $N_2$-module via $(\eta_1 \underset{\psi_1}{\otimes} \xi)x := \eta_1 \underset{\psi_1}{\otimes} \xi x$;*

ii) *$H \underset{\psi_2}{\otimes} K_2$ is a left Hilbert $N_1$-module via $y(\xi \underset{\psi_2}{\otimes} \eta_2) := y\xi \underset{\psi_2}{\otimes} \eta_2$;*

iii) *$(K_1 \underset{\psi_1}{\otimes} H) \underset{\psi_2}{\otimes} K_2 \cong K_1 \underset{\psi_1}{\otimes} (H \underset{\psi_2}{\otimes} K_2)$ via $(\eta_1 \underset{\psi_1}{\otimes} \xi) \underset{\psi_2}{\otimes} \eta_2 \leftrightarrow \eta_1 \underset{\psi_1}{\otimes} (\xi \underset{\psi_2}{\otimes} \eta_2)$ for all $\eta_1 \in D(K_1, \psi_1^{\mathrm{op}})$, $\eta_2 \in D(K_2, \psi_2)$, $\xi \in H$.*

*Proof.* i), ii) This follows immediately from Corollary 10.1.24.

iii) By Lemma 10.1.15, $(K_1 \underset{\psi_1}{\otimes} H) \underset{\psi_2}{\otimes} K_2$ and $K_1 \underset{\psi_1}{\otimes} (H \underset{\psi_2}{\otimes} K_2)$ are equal to
$$[(D(K_1, \psi_1^{\mathrm{op}}) \underset{\psi_1}{\otimes} H) \underset{\psi_2}{\otimes} D(K_2, \psi_2)] \quad \text{and} \quad [D(K_1, \psi_1^{\mathrm{op}}) \underset{\psi_1}{\otimes} (H \underset{\psi_2}{\otimes} D(K_2, \psi_2))],$$
respectively. Let $\eta_1 \in D(K_1, \psi_1^{\mathrm{op}})$, $\eta_2 \in D(K_2, \psi_2)$, $\xi \in H$. Several applications of Lemma 10.1.15 show that in $(K_1 \underset{\psi_1}{\otimes} H) \underset{\psi_2}{\otimes} K_2$ and in $K_1 \underset{\psi_1}{\otimes} (H \underset{\psi_2}{\otimes} K_2)$,
$$\|(\eta_1 \underset{\psi_1}{\otimes} \xi) \underset{\psi_2}{\otimes} \eta_2\|^2 = \langle \eta_1 \underset{\psi_1}{\otimes} \xi | \eta_1 \underset{\psi_1}{\otimes} \langle \eta_2|\eta_2\rangle_{\psi_2^{\mathrm{op}}} \xi\rangle = \langle \xi | \langle \eta_1|\eta_1\rangle_{\psi_1} \langle \eta_2|\eta_2\rangle_{\psi_2^{\mathrm{op}}} \xi\rangle$$

and

$$\|\eta_1 \underset{\psi_1}{\otimes} (\xi \underset{\psi_2}{\otimes} \eta_2)\|^2 = \langle \xi \underset{\psi_2}{\otimes} \eta_2 | \langle \eta_1|\eta_1\rangle_{\psi_1} \xi \underset{\psi_2}{\otimes} \eta_2\rangle = \langle \xi | \langle \eta_2|\eta_2\rangle_{\psi_2^{\mathrm{op}}} \langle \eta_1|\eta_1\rangle_{\psi_1} \xi\rangle.$$

Since $H$ is an $N_1$-$N_2$-bimodule, these expressions are equal, so that the identification given in statement ii) is a well-defined isometric isomorphism. □

The relative tensor product absorbs the standard Hilbert $N$-$N$-bimodule $H_\psi$ introduced in Example 10.1.2, that is, $H_\psi$ is a unit for the relative tensor product:

**Proposition 10.1.26.** *Let $\psi$ be an n.s.f. weight on $N$.*

i) *For every right Hilbert $N$-module $H$, we have $H \underset{\psi}{\otimes} H_\psi \cong H$ as right Hilbert $N$-modules via $\eta \underset{\psi}{\otimes} \Lambda_{\psi^{\mathrm{op}}}(y^{\mathrm{op}}) \leftrightarrow \eta y$ for all $\eta \in D(H, \psi^{\mathrm{op}})$, $y \in \mathcal{N}_\psi{}^*$.*

ii) *For every left Hilbert $N$-module $K$, we have $H_\psi \underset{\psi}{\otimes} K \cong K$ as left Hilbert $N$-modules via $\Lambda_\psi(y) \underset{\psi}{\otimes} \xi \leftrightarrow y\xi$ for all $\xi \in D(K, \psi)$, $y \in \mathcal{N}_\psi$.*

*Proof.* i) The map $\eta \underset{\psi}{\otimes} \Lambda_{\psi^{\mathrm{op}}}(y^{\mathrm{op}}) \mapsto \eta y$ is well defined and isometric because for all $\eta \in D(H, \psi^{\mathrm{op}})$ and $y \in \mathcal{N}_\psi{}^*$,

$$\begin{aligned}\|\eta \underset{\psi}{\otimes} \Lambda_{\psi^{\mathrm{op}}}(y^{\mathrm{op}})\|^2 &= \langle \Lambda_{\psi^{\mathrm{op}}}(y) | \langle \eta|\eta\rangle_\psi \Lambda_{\psi^{\mathrm{op}}}(y)\rangle \\ &= \langle L_\psi(\eta)\Lambda_{\psi^{\mathrm{op}}}(y) | L_\psi(\eta)\Lambda_{\psi^{\mathrm{op}}}(y)\rangle = \langle \eta y|\eta y\rangle = \|\eta y\|^2.\end{aligned}$$

The map is surjective because $(D(H, \psi^{\mathrm{op}})\mathcal{N}_\psi{}^*)^\perp = (H\mathcal{N}_\psi{}^*)^\perp = (HN)^\perp = 0$; here, we used that $D(H, \psi^{\mathrm{op}})$ is dense in $H$ (Proposition 10.1.12 iv)), that $\mathcal{N}_\psi{}^*$ is weakly dense in $N$, and that $N$ acts non-degenerately and normal on $H$.

Finally, the map is a morphism of right $N$-modules because $\Lambda_{\psi^{\mathrm{op}}}(y^{\mathrm{op}})x = \pi_{\psi^{\mathrm{op}}}(x^{\mathrm{op}})\Lambda_{\psi^{\mathrm{op}}}(y^{\mathrm{op}}) = \Lambda_{\psi^{\mathrm{op}}}((yx)^{\mathrm{op}})$ for all $y \in \mathcal{N}_\psi{}^*$ and $x \in N$.

ii) Consider $K$ as a right $N^{\mathrm{op}}$-module and use i) and Proposition 10.1.20. □

Finally, the relative tensor product is essentially independent of the choice of the weight on $N$:

**Theorem 10.1.27.** *Let $\psi$ and $\phi$ be n.s.f. weights on $N$. Then the bifunctors $(H, K) \mapsto H \underset{\psi}{\otimes} K$, $(S, T) \mapsto S \underset{\psi}{\otimes} T$ and $(H, K) \mapsto H \underset{\phi}{\otimes} K$, $(S, T) \mapsto S \underset{\phi}{\otimes} T$ from the categories of right and left Hilbert $N$-modules to the category of Hilbert spaces are naturally isomorphic.*

The proof involves the Radon–Nikodym derivative of n.s.f. weights and can be found in [150, IX.3, Theorem 3.21].

**Remarks 10.1.28.** i) The natural isomorphism $H \underset{\phi}{\otimes} K \cong H \underset{\psi}{\otimes} K$ is *not* given by $\eta \underset{\phi}{\otimes} \xi \leftrightarrow \eta \underset{\psi}{\otimes} \xi$. In the notation of Remark 10.1.19, it has the form $\eta\phi^{-1/2} \underset{N}{\otimes} \xi \leftrightarrow \eta(d\psi/d\phi)^{1/2}\psi^{-1/2} \underset{N}{\otimes} \xi$, where $d\psi/d\phi$ denotes the Radon–Nikodym derivative of $\psi$ with respect to $\phi$.

ii) The natural isomorphisms $H \underset{\phi}{\otimes} K \cong H \underset{\psi}{\otimes} K$ and $K \underset{\phi^{\mathrm{op}}}{\otimes} H \cong K \underset{\psi^{\mathrm{op}}}{\otimes} H$ are compatible with the flip maps $\Sigma_\psi$ and $\Sigma_\phi$ defined in Proposition 10.1.20.

**Notation 10.1.29.** In the remainder of this chapter, we neglect the precise choice of the n.s.f. weight $\psi$ and denote the bifunctor $(\cdot) \underset{\psi}{\otimes} (\cdot)$ by $(\cdot) \underset{N}{\otimes} (\cdot)$.

## 10.2 Hopf–von Neumann bimodules

Hopf–von Neumann bimodules generalize von Neumann bialgebras just like groupoids generalize groups, and groupoids just like von Neumann bialgebras generalize groups, see the introduction to this chapter. The definition is due to Vallin [171] and involves the fiber product of von Neumann algebras which was introduced by Sauvageot [139].

### 10.2.1 The fiber product of von Neumann algebras

The fiber product of von Neumann algebras is an analogue of the algebraic tensor product of algebras over a common subalgebra. Its definition involves the relative tensor product of Hilbert modules.

We fix a von Neumann algebra $N$ and adopt the following terminology:

**Definition 10.2.1.** A *left/right von Neumann $N$-module* is a von Neumann algebra $M$ equipped with a unital normal $*$-homomorphism/$*$-antihomomorphism $N \to M$.

Let $N_1$ and $N_2$ be von Neumann algebras. A *von Neumann $N_1$-$N_2$-bimodule* is a von Neumann algebra $M$ equipped with the structure of a left $N_1$-module and a right $N_2$-module such that the images of $N_1$ and $N_2$ in $M$ commute.

A *morphism of left von Neumann $N$-modules $M_1, M_2$* is a non-degenerate normal $*$-homomorphism $\pi : M_1 \to M_2$ that satisfies $x\pi(y) = \pi(xy)$ for all $x \in N$ and $y \in M_1$. Similarly, we define morphisms of right von Neumann $N$-modules and of von Neumann $N_1$-$N_2$-bimodules.

**Remarks 10.2.2.** i) Evidently, left von Neumann $N$-modules correspond bijectively with right von Neumann $N^{\mathrm{op}}$-modules, and right von Neumann $N$-modules correspond bijectively with left von Neumann $N^{\mathrm{op}}$-modules.

i) If $M$ is a von Neumann algebra on a Hilbert space $H$ and $M$ is a left/right $N$-module, then the composition $N \to M \hookrightarrow \mathcal{L}(H)$ turns $H$ into a left/right Hilbert $N$-module.

Throughout this section, we apply Notation 10.1.29 and write "$\underset{N}{\otimes}$" instead of "$\underset{\psi}{\otimes}$" in all constructions that involve the relative tensor product. As usual, we denote the commutant of a set $X$ of operators on some Hilbert space by $X'$.

The definition of the fiber product of von Neumann algebras involves the following auxiliary construction:

**Definition 10.2.3.** Let $H_1$ be a right and $H_2$ a left Hilbert $N$-module. The *relative tensor product* of von Neumann algebras $M_1 \subseteq \mathcal{L}_N(H_1)$ and $M_2 \subseteq \mathcal{L}_N(H_2)$ is the von Neumann algebra

$$M_1 \underset{N}{\bar{\otimes}} M_2 := \{x \underset{N}{\otimes} y \mid x \in M_1,\ y \in M_2\}'' \subseteq \mathcal{L}(H_1 \underset{N}{\otimes} H_2).$$

The fiber product of von Neumann modules is defined as follows:

**Definition 10.2.4.** Let $M_1$ and $M_2$ be von Neumann algebras on Hilbert spaces $H_1$ and $H_2$, respectively. Assume that $M_1$ is a right and $M_2$ a left $N$-module, and consider $H_1$ and $H_2$ as $N$-modules as in Remark 10.2.2 ii). Then the *fiber product* of $M_1$ and $M_2$ over $N$ is the von Neumann algebra

$$M_1 \underset{N}{*} M_2 := ({M_1}' \underset{N}{\bar{\otimes}} {M_2}')' \subseteq \mathcal{L}(H_1 \underset{N}{\otimes} H_2).$$

**Remarks 10.2.5.** i) Strictly speaking, the von Neumann algebras constructed in Definition 10.2.3 and 10.2.4 are well defined up to spatial isomorphism only, see Theorem 10.1.27.

ii) For $i = 1, 2$, consider $H_i$ as a left Hilbert $M_i$-module. Evidently, ${M_i}' = \mathcal{L}_{M_i}(H_i) \subseteq \mathcal{L}_N(H_i)$ for $i = 1, 2$, so that the relative tensor product ${M_1}' \underset{N}{\bar{\otimes}} {M_2}'$ is well defined. Moreover,

$$M_1 \underset{N}{*} M_2 = (\mathcal{L}_{M_1}(H_1) \underset{N}{\bar{\otimes}} \mathcal{L}_{M_2}(H_2))' = \mathcal{L}_{({M_1}' \underset{N}{\bar{\otimes}} {M_2}')}(H_1 \underset{N}{\otimes} H_2).$$

iii) The flip map $\Sigma_N \colon H_1 \underset{N}{\otimes} H_2 \xrightarrow{\cong} H_2 \underset{N^{op}}{\otimes} H_1$ (see Proposition 10.1.20) induces an isomorphism $\mathrm{Ad}_{\Sigma_N} \colon \mathcal{L}(H_1 \underset{N}{\otimes} H_2) \xrightarrow{\cong} \mathcal{L}(H_2 \underset{N^{\mathrm{op}}}{\otimes} H_1)$ which restricts to an isomorphism $\mathrm{Ad}_\Sigma \colon M_1 \underset{N}{*} M_2 \xrightarrow{\cong} M_2 \underset{N^{\mathrm{op}}}{*} M_1$.

iv) If $N$ is commutative and the maps $N \to M_i$, $i = 1, 2$, take values only in the center $Z(M_i)$, then $M_1$ and $M_2$ are direct integrals of measurable fields of von Neumann algebras, and their fiber product $M_1 \underset{N}{*} M_2$ is the direct integral of the fiberwise tensor product of these fields. For details on fields of von Neumann algebras and their fiberwise tensor product, see [35, Chapitre II].

In general, the fiber product $M_1 \underset{N}{*} M_2$ of von Neumann algebras can not be described in terms of elementary tensor products $x_1 \underset{N}{\otimes} x_2$, where $x_1 \in M_1$ and $x_2 \in M_2$. More precisely, let $H_1$, $H_2$ and $M_1$, $M_2$ be as in the definition above. Then an elementary tensor $x_1 \underset{N}{\otimes} x_2$ as above is well defined if and only if $x_i \in \mathcal{L}_N(H_i)$ for $i = 1, 2$ (see Proposition 10.1.23). In that case, $x_1 \underset{N}{\otimes} x_2$ belongs to $M_1 \underset{N}{*} M_2$:

**Lemma 10.2.6.** *Let $H_1$, $H_2$ and $M_1$, $M_2$ be as in Definition* 10.2.4. *Then*

$$(M_1 \cap \mathcal{L}_N(H_1)) \underset{N}{\overline{\otimes}} (M_2 \cap \mathcal{L}_N(H_2)) \subseteq M_1 \underset{N}{*} M_2.$$

*Proof.* For $i = 1, 2$, put $M_i^N := M_i \cap \mathcal{L}_N(H_i)$. Then $M_i{}' \subseteq (M_i^N)'$ and

$$M_1^N \underset{N}{\overline{\otimes}} M_2^N \subseteq (M_1{}' \underset{N}{\overline{\otimes}} M_2{}')' = M_1 \underset{N}{*} M_2. \qquad \square$$

As one should expect, the fiber product of bimodules is a bimodule:

**Proposition 10.2.7.** *Let $M_1$ and $M_2$ be von Neumann algebras on Hilbert spaces $H_1$ and $H_2$, respectively. Assume that $M_1$ is an $N_1$-$N$-bimodule and $M_2$ an $N$-$N_2$-bimodule, where $N_1$ and $N_2$ are some von Neumann algebras. Then $M_1 \underset{N}{*} M_2$ is a von Neumann $N_1$-$N_2$-bimodule via $x_1 \mapsto x_1 \underset{N}{\otimes} 1$ for all $x_1 \in N_1$ and $x_2 \mapsto 1 \underset{N}{\otimes} x_2$ for all $x_2 \in N_2$.*

*Proof.* For $i = 1, 2$, denote by $\rho_i$ the fixed $*$-(anti)homomorphism $N_i \to M_i$. From Proposition 10.1.23, we obtain normal unital $*$-(anti)homomorphisms

$$N_1 \to \mathcal{L}_N(H_1) \cap M_1 \to \mathcal{L}_N(H_1 \underset{N}{\otimes} H_2), \quad x_1 \mapsto \rho_1(x_1) \underset{N}{\otimes} 1,$$
$$N_2 \to \mathcal{L}_N(H_2) \cap M_2 \to \mathcal{L}_N(H_1 \underset{N}{\otimes} H_2), \quad x_2 \mapsto 1 \underset{N}{\otimes} \rho_2(x_2),$$

and by Lemma 10.2.6, the images of these maps are contained in $M_1 \underset{N}{*} M_2$. $\square$

The fiber product is associative:

**Proposition 10.2.8.** *Let $M_1$, $M_2$, $M_3$ be von Neumann algebras on Hilbert spaces $H_1$, $H_2$, $H_3$, respectively. Assume that $M_1$ is a right $N_1$-module, $M_2$ an $N_1$-$N_2$-bimodule, and $M_3$ a left $N_2$-module, where $N_1$ and $N_2$ are some von Neumann* *algebras. Then the natural isomorphism* $(H_1 \underset{N_1}{\otimes} H_2) \underset{N_2}{\otimes} H_3 \cong H_1 \underset{N_1}{\otimes} (H_2 \underset{N_2}{\otimes} H_3)$ *induces an isomorphism*

$$(M_1 \underset{N_1}{*} M_2) \underset{N_2}{*} M_3 \cong M_1 \underset{N_1}{*} (M_2 \underset{N_2}{*} M_3).$$

*Proof.* The isomorphism of the Hilbert spaces identifies $(M_1' \underset{N_1}{\bar{\otimes}} M_2') \underset{N_2}{\bar{\otimes}} M_3'$ with $M_1' \underset{N_1}{\bar{\otimes}} (M_2' \underset{N_2}{\bar{\otimes}} M_3')$, and the iterated fiber products are the respective commutants of these von Neumann algebras. □

As one should expect, the fiber product is bifunctorial:

**Proposition 10.2.9.** *Let* $H_1$, $H_2$ *and* $M_1$, $M_2$ *be as in Definition* 10.2.4. *Let* $\pi_i$ $(i = 1, 2)$ *be a non-degenerate normal representation of* $M_i$ *on a Hilbert space* $K_i$ *and consider* $K_i$ *as a Hilbert* $N$*-module via the map* $N \to M_i \xrightarrow{\pi_i} \mathcal{L}(K_i)$.

i) *There exists a unique normal unital* $*$*-homomorphism*

$$\pi_1 \underset{N}{*} \pi_2 \colon M_1 \underset{N}{*} M_2 \to \pi_1(M_1) \underset{N}{*} \pi_2(M_2)$$

*such that for all* $T_1 \in \mathcal{L}_{M_1}(H_1, K_1)$, $T_2 \in \mathcal{L}_{M_2}(H_2, K_2)$, $x \in M_1 \underset{N}{*} M_2$,

$$(\pi_1 \underset{N}{*} \pi_2)(x) \circ (T_1 \underset{N}{\otimes} T_2) = (T_1 \underset{N}{\otimes} T_2) \circ x. \tag{10.5}$$

ii) $(\pi_1 \underset{N}{*} \pi_2)(M_1 \underset{N}{*} M_2) = \pi_1(M_1) \underset{N}{*} \pi_2(M_2)$.

iii) *If* $\pi_1$ *and* $\pi_2$ *are injective, so is* $\pi_1 \underset{N}{*} \pi_2$.

*Proof.* i) By Proposition 10.1.4, elements of the form $(T_1 \underset{N}{\otimes} T_2)\zeta$, where $T_i \in \mathcal{L}_{M_i}(H_i, K_i)$ and $\zeta \in H_1 \underset{N}{\otimes} H_2$, are linearly dense in $K_1 \underset{N}{\otimes} K_2$. Therefore, $\pi_1 \underset{N}{*} \pi_2$ is uniquely determined by condition (10.5). To prove existence, we consider the special cases $K_1 = H_1$, $\pi_1 = \mathrm{id}_{M_1}$ and $K_2 = H_2$, $\pi_2 = \mathrm{id}_{M_2}$ separately; then it is easy to see that the composition $\pi_1 \underset{N}{*} \pi_2 := (\mathrm{id}_{\pi(M_1)} \underset{N}{*} \pi_2) \circ (\pi_1 \underset{N}{*} \mathrm{id}_{M_2})$ satisfies condition (10.5). We treat the case $K_2 = H_2$, $\pi_2 = \mathrm{id}_{M_2}$; the other case is similar.

By [35, I.4.4, Théorème 3], [137, Proposition 2.7.4], or [149, IV, Theorem 5.5], $\pi_1$ can be written as follows: there exist a Hilbert space $H_0$, a projection $p \in (1 \bar{\otimes} M_1)' \subseteq \mathcal{L}(H_0 \otimes H_1)$, and a unitary $U \in \mathcal{L}(p(H_0 \otimes H_1), K_1)$, such that

$$\pi_1(z) = \mathrm{Ad}_U(p(1 \otimes z)) \quad \text{for all } z \in M_1.$$

We consider $H_0 \otimes H_1$ and $p(H_0 \otimes H_1)$ as right Hilbert $N$-modules via the right $N$-module structure of $M_1$ and the action of $M_1$ on $H_1$; then $p$ and $U$ are morphisms of right $N$-modules. Furthermore, we use the identification $(H_0 \otimes H_1) \underset{N}{\otimes} H_2 \cong H_0 \otimes (H_1 \underset{N}{\otimes} H_2)$ and omit the parentheses. We define $\pi_1 \underset{N}{*} \mathrm{id}_{M_2}$ as the composition of the following three normal $*$-homomorphisms:

- amplification by $H_0$, given by $M_1 \underset{N}{*} M_2 \to \mathcal{L}(H_0 \otimes H_1 \underset{N}{\otimes} H_2)$, $x \mapsto \mathrm{id}_{H_0} \otimes x$,
- induction by the projection $p \underset{N}{\otimes} \mathrm{id}_{H_2}$, given by

$$\begin{aligned}\mathbb{C}\,\mathrm{id}_{H_0} \mathbin{\bar{\otimes}} (M_1 \underset{N}{*} M_2) &\to \mathcal{L}(p(H_0 \otimes H_1) \underset{N}{\otimes} H_2),\\ \mathrm{id}_{H_0} \otimes x &\mapsto (p \underset{N}{\otimes} \mathrm{id}_{H_2})(\mathrm{id}_{H_0} \otimes x);\end{aligned}$$

  note that this is a normal $*$-homomorphism because $p \underset{N}{\otimes} \mathrm{id}_{H_2}$ is contained in

$$\mathcal{L}(H_0) \mathbin{\bar{\otimes}} M_1{}' \underset{N}{\bar{\otimes}} M_2{}' = \mathcal{L}(H_0) \mathbin{\bar{\otimes}} (M_1 \underset{N}{*} M_2)' = (1 \mathbin{\bar{\otimes}} (M_1 \underset{N}{*} M_2))';$$

- conjugation by $U \underset{N}{\otimes} \mathrm{id}_{H_2} \in \mathcal{L}(p(H_0 \otimes H_1) \underset{N}{\otimes} H_2, K_1 \underset{N}{\otimes} H_2)$.

Thus we put

$$(\pi_1 \underset{N}{*} \mathrm{id}_{M_2})(x) := \mathrm{Ad}_{(U \underset{N}{\otimes} 1)}((p \underset{N}{\otimes} 1)(1 \otimes x)) \quad \text{for all } x \in M_1 \underset{N}{*} M_2.$$

Evidently, this composition is non-degenerate and normal, and a straightforward calculation shows that it satisfies condition (10.5).

ii) It suffices to prove $((\pi_1 \underset{N}{*} \pi_2)(M_1 \underset{N}{*} M_2))' = \pi_1(M_1)' \underset{N}{\bar{\otimes}} \pi_2(M_2)'$. Let us show that the right-hand side is contained in the left-hand side. By condition (10.5), we have for all $x \in M_1 \underset{N}{*} M_2$, $T_i \in \mathcal{L}_{M_i}(H_i, K_i)$, $S_i \in \pi_i(M_i)'$

$$\begin{aligned}(S_1 \underset{N}{\otimes} S_2) \circ (\pi_1 \underset{N}{*} \pi_2)(x) \circ (T_1 \underset{N}{\otimes} T_2) &= (S_1 T_1 \underset{N}{\otimes} S_2 T_2) \circ x\\ &= (\pi_1 \underset{N}{*} \pi_2)(x) \circ (S_1 \underset{N}{\otimes} S_2) \circ (T_1 \underset{N}{\otimes} T_2).\end{aligned}$$

Since elements of the form $(T_1 \underset{N}{\otimes} T_2)\zeta$, where $T_i \in \mathcal{L}_{M_i}(H_i, K_i)$ and $\zeta \in H_1 \underset{N}{\otimes} H_2$, are linearly dense in $K_1 \underset{N}{\otimes} K_2$ (Proposition 10.1.4), the equation above implies that $S_1 \underset{N}{\otimes} S_2$ commutes with $(\pi_1 \underset{N}{*} \pi_2)(x)$. Thus, $\pi_1(M_1)' \underset{N}{\bar{\otimes}} \pi_2(M_2)'$ is contained in $((\pi_1 \underset{N}{*} \pi_2)(M_1 \underset{N}{*} M_2))'$.

Let us prove the reverse inclusion. For $i = 1, 2$, put $Y_i = \mathcal{L}_{M_i}(H_i, K_i)$. Consider the von Neumann algebra

$$A := \begin{pmatrix} M_1' \underset{N}{\otimes} M_2' & Y_1^* \underset{N}{\otimes} Y_2^* \\ Y_1 \underset{N}{\otimes} Y_2 & \pi_1(M_1)' \underset{N}{\otimes} \pi_2(M_2)' \end{pmatrix}'' \subseteq \mathcal{L}((H_1 \underset{N}{\otimes} H_2) \oplus (K_1 \underset{N}{\otimes} K_2)).$$

Here, we think of elements of the Hilbert space $(H_1 \underset{N}{\otimes} H_2) \oplus (K_1 \underset{N}{\otimes} K_2)$ as column vectors with upper entry in $H_1 \underset{N}{\otimes} H_2$ and lower entry in $K_1 \underset{N}{\otimes} K_2$, and of elements of $A$ as 2-by-2-matrices in the obvious way. It is easy to see that

$$A' = \left\{ \begin{pmatrix} x & \\ & (\pi_1 \underset{N}{*} \pi_2)(x) \end{pmatrix} \;\middle|\; x \in M_1 \underset{N}{*} M_2 \right\},$$

and that

$$\begin{pmatrix} 0 & \\ & ((\pi_1 \underset{N}{*} \pi_2)(M_1 \underset{N}{*} M_2))' \end{pmatrix} \subseteq A'' = A.$$

Therefore, $((\pi_1 \underset{N}{*} \pi_2)(M_1 \underset{N}{*} M_2))' \subseteq \pi_1(M_1)' \underset{N}{\bar{\otimes}} \pi_2(M_2)'$.

iii) For $i = 1, 2$, assume that $\pi_i$ is injective, and denote by $\pi_i^{-1} \colon \pi_i(M_i) \to M_i$ its inverse. Then $\pi_1^{-1} \underset{N}{*} \pi_2^{-1}$ is inverse to $\pi_1 \underset{N}{*} \pi_2$. □

Using the previous proposition, we can extend the fiber product to a bifunctor: Given a morphism $\pi_1 \colon M_1 \to N_1$ of right von Neumann $N$-modules and a morphism $\pi_2 \colon M_2 \to N_2$ of left von Neumann $N$-modules, we can define a normal $*$-homomorphism $\pi_1 \underset{N}{*} \pi_2 \colon M_1 \underset{N}{*} M_2 \to N_1 \underset{N}{*} N_2$. Then, we find:

**Corollary 10.2.10.** *The assignments*

$$(M_1, M_2) \mapsto M_1 \underset{N}{*} M_2 \quad \textit{and} \quad (\pi_1, \pi_2) \mapsto \pi_1 \underset{N}{*} \pi_2$$

*define a bifunctor from the categories of right and left von Neumann $N$-modules to the category of von Neumann algebras.* □

### 10.2.2 Hopf–von Neumann bimodules

Informally, the concept of a Hopf–von Neumann bimodule has already been described in the introduction to this chapter. Let us now give the precise definition.

**Notation 10.2.11.** Let $M$ and $N$ be von Neumann algebras. If we want to emphasize that we consider $M$ as a left/right $N$-module via a specific $*$-homomorphism/$*$-antihomomorphism $\pi$, we write ${}_{\pi}M$ or $M_{\pi}$ for that $N$-module, respectively.

**Definition 10.2.12.** A *Hopf–von Neumann bimodule* is a tuple $(N, M, r, s, \Delta)$ consisting of a von Neumann algebra $N$ called the *base*, a von Neumann algebra $M$, a $*$-homomorphisms $r\colon N \to M$ called the *range map*, a $*$-antihomomorphism $s\colon N \to M$ called the *source map*, and a $*$-homomorphism $\Delta\colon M \to M_s \underset{N}{*} {}_r M$ called the *comultiplication*, such that

i) $r$, $s$, $\Delta$ are normal, injective and unital;

ii) $\Delta(s(x)) = 1 \underset{N}{\otimes} s(x)$ and $\Delta(r(x)) = r(x) \underset{N}{\otimes} 1$ for all $x \in N$;

iii) $\Delta$ is *coassociative* in the sense that $(\Delta \underset{N}{*} \mathrm{id}_M) \circ \Delta = (\mathrm{id}_M \underset{N}{*} \Delta) \circ \Delta$ as maps $M \to M_s \underset{N}{*} {}_r M_s \underset{N}{*} {}_r M$, that is, the following diagram commutes:

$$\begin{array}{ccc} M & \xrightarrow{\quad\Delta\quad} & M_s \underset{N}{*} {}_r M \\ {\scriptstyle\Delta}\big\downarrow & & \big\downarrow{\scriptstyle\Delta \underset{N}{*} \mathrm{id}} \\ M_s \underset{N}{*} {}_r M & \xrightarrow[\mathrm{id} \underset{N}{*} \Delta]{} & M_s \underset{N}{*} {}_r M_s \underset{N}{*} {}_r M. \end{array}$$

**Remarks 10.2.13.** Let $(N, M, r, s, \Delta)$ be a Hopf–von Neumann bimodule.

i) The maps $r$ and $s$ turn $M$ into an $N$-$N$-bimodule. Since no other (anti)representations of $N$ on $M$ occur at the moment, we may omit the indices $s$ and $r$ in the fiber product and write $M \underset{N}{*} M$ for $M_s \underset{N}{*} {}_r M$.

ii) The maps $N \to M \underset{N}{*} M$ given by $x \mapsto r(x) \underset{N}{\otimes} 1$ and $x \mapsto 1 \underset{N}{\otimes} s(x)$ turn $M \underset{N}{*} M$ into an $N$-$N$-bimodule (Proposition 10.2.7). Condition 10.2.12 ii) says that the map $\Delta\colon M \to M \underset{N}{*} M$ is a morphism of $N$-$N$-bimodules. Therefore the maps $\Delta \underset{N}{*} \mathrm{id}$ and $\mathrm{id} \underset{N}{*} \Delta$ in condition iii) are well defined (Proposition 10.2.9).

iii) In condition 10.2.12 i), the assumption that $\Delta$ is unital can be omitted, because this follows from the assumption that $r$ and $s$ are unital and from condition 10.2.12 ii).

iv) The tuple $(N^{\mathrm{op}}, M, s, r, \mathrm{Ad}_{\Sigma} \circ \Delta)$ is a Hopf–von Neumann bimodule again, called the *coopposite* of $(N, M, r, s, \Delta)$. Here, $\mathrm{Ad}_{\Sigma_N}$ denotes the natural isomorphism $M_s \underset{N}{*} {}_r M \to M_r \underset{N^{\mathrm{op}}}{*} {}_s M$ (see Remark 10.2.5 iii)).

**Examples 10.2.14.** The main examples of Hopf–von Neumann bimodules arise from

- locally compact groupoids, see Section 10.3.3 or [171];
- inclusions of factors, see [42], [43], [44], [50], [51].

Further examples of Hopf–von Neumann bimodules are discussed in the thesis of Lesieur [99]. In most examples, the construction of a Hopf–von Neumann bimodule proceeds via a pseudo-multiplicative unitary, see the next section.

## 10.3 Pseudo-multiplicative unitaries on Hilbert spaces

The concept of a pseudo-multiplicative unitary on a Hilbert space, introduced by Vallin [172], plays a similar fundamental rôle in the theory of measurable quantum groupoids like the concept of a multiplicative unitary in the theory of locally compact quantum groups.

### 10.3.1 Definition

A pseudo-multiplicative unitary acts on relative tensor products of a Hilbert space, taken with respect to several representations:

**Assumption 10.3.1.** Throughout this section, let $N$ be a von Neumann algebra with an n.s.f. weight $\psi$, and let $H$ be a Hilbert space with

- a non-degenerate injective normal representation $\alpha$ of $N$, and
- non-degenerate injective normal antirepresentations $\beta$, $\hat{\beta}$ of $N$,

such that $\alpha(N)$, $\beta(N)$, $\hat{\beta}(N)$ commute pairwise.

The Hilbert space $H$ can be considered as a module in various ways. To distinguish between the different possibilities, we use the following notation:

**Notation 10.3.2.** We denote by ${}_{\alpha}H$ the left Hilbert $N$-module given by $H$ and $\alpha$, and by $H_{\beta}$, $H_{\hat{\beta}}$ the right Hilbert $N$-modules given by $H$ and $\beta$ or $\hat{\beta}$, respectively. If we consider ${}_{\alpha}H$, $H_{\beta}$, $H_{\hat{\beta}}$ as Hilbert $N^{\mathrm{op}}$-modules, we write $H_{\alpha}$, ${}_{\beta}H$, ${}_{\hat{\beta}}H$, respectively. We combine these notations to denote bimodule structures on $H$, and use a similar notation to denote module or bimodule structures on an arbitrary Hilbert space with a given (anti)representation.

Given $H$ and $\alpha$, $\beta$, $\hat{\beta}$, we can form the following relative tensor products:

$$H_{\hat{\beta}} \underset{\psi}{\otimes} {}_{\alpha}H, \quad H_{\beta} \underset{\psi}{\otimes} {}_{\alpha}H, \quad H_{\alpha} \underset{\psi^{\mathrm{op}}}{\otimes} {}_{\hat{\beta}}H, \quad H_{\alpha} \underset{\psi^{\mathrm{op}}}{\otimes} {}_{\beta}H.$$

A pseudo-multiplicative unitary on $H$ is a unitary

$$V \colon H_{\hat{\beta}} \underset{\psi}{\otimes} {}_{\alpha}H \to H_{\alpha} \underset{\psi^{\mathrm{op}}}{\otimes} {}_{\beta}H$$

that satisfies the pentagon equation $V_{[12]}V_{[13]}V_{[23]} = V_{[23]}V_{[12]}$ known from the definition of a multiplicative unitary. In the present context, this equation requires some explanation, for example, the individual factors $V_{[ij]}$ do not act on one fixed space but on several intermediate relative tensor products. Therefore, several intertwining conditions have to be imposed on $V$ and $\alpha$, $\beta$, $\hat{\beta}$ to ensure that each factor $V_{[ij]}$ occurring in the pentagon equation is well defined.

To describe these intertwining relations, we use the following leg notation for (anti)representations: Since $\alpha(N)$, $\beta(N)$, $\hat{\beta}(N)$ commute, we can define (anti)representations

$$\begin{aligned} &\alpha_{[1]}, \beta_{[1]}, \beta_{[2]}, \hat{\beta}_{[2]} \text{ on } H_{\hat{\beta}} \underset{\psi}{\otimes}{}_{\alpha} H, && \alpha_{[1]}, \hat{\beta}_{[1]}, \hat{\beta}_{[2]}, \beta_{[2]} \text{ on } H_{\beta} \underset{\psi}{\otimes}{}_{\alpha} H, \\ &\hat{\beta}_{[1]}, \beta_{[1]}, \beta_{[2]}, \alpha_{[2]} \text{ on } H_{\alpha} \underset{\psi^{\mathrm{op}}}{\otimes}{}_{\hat{\beta}} H, && \beta_{[1]}, \hat{\beta}_{[1]}, \hat{\beta}_{[2]}, \alpha_{[2]} \text{ on } H_{\alpha} \underset{\psi^{\mathrm{op}}}{\otimes}{}_{\beta} H; \end{aligned}$$

for example, the first four (anti)representations are given by

$$\begin{aligned} \alpha_{[1]}(x) &= \alpha(x) \underset{\psi}{\otimes} 1, & \beta_{[1]}(x) &= \beta(x) \underset{\psi}{\otimes} 1, \\ \beta_{[2]}(x) &= 1 \underset{\psi}{\otimes} \beta(x), & \hat{\beta}_{[2]}(x) &= 1 \underset{\psi}{\otimes} \hat{\beta}(x). \end{aligned}$$

**Lemma 10.3.3.** *If a bounded linear operator* $V\colon H_{\hat{\beta}} \underset{\psi}{\otimes}{}_{\alpha} H \to H_{\alpha} \underset{\psi^{\mathrm{op}}}{\otimes}{}_{\beta} H$ *satisfies*

$$\begin{aligned} V\alpha_{[1]}(x) &= \alpha_{[2]}(x)V, & V\beta_{[2]}(x) &= \hat{\beta}_{[1]}(x)V, \\ V\beta_{[1]}(x) &= \beta_{[1]}(x)V, & V\hat{\beta}_{[2]}(x) &= \hat{\beta}_{[2]}(x)V \quad \textit{for all } x \in N, \end{aligned} \tag{10.6}$$

*then all operators in the following diagram are well defined:*

$$\begin{array}{ccccc} & & H_{\alpha} \underset{\psi^{\mathrm{op}}}{\otimes}{}_{\beta} H_{\hat{\beta}} \underset{\psi}{\otimes}{}_{\alpha} H & & \\ & \nearrow^{V \underset{\psi}{\otimes} 1} & & \searrow^{1 \underset{\psi^{\mathrm{op}}}{\otimes} V} & \\ H_{\hat{\beta}} \underset{\psi}{\otimes}{}_{\alpha} H_{\hat{\beta}} \underset{\psi}{\otimes}{}_{\alpha} H & & & & H_{\alpha} \underset{\psi^{\mathrm{op}}}{\otimes}{}_{\beta} H_{\alpha} \underset{\psi^{\mathrm{op}}}{\otimes}{}_{\beta} H \\ \downarrow{\scriptstyle 1 \underset{\psi}{\otimes} V} & & & & \uparrow{\scriptstyle V \underset{\psi^{\mathrm{op}}}{\otimes} 1} \\ H_{\hat{\beta}} \underset{\psi}{\otimes}{}_{\alpha_{[2]}} (H_{\alpha} \underset{\psi^{\mathrm{op}}}{\otimes}{}_{\beta} H) & & & & (H_{\hat{\beta}} \underset{\psi}{\otimes}{}_{\alpha} H)_{\alpha_{[1]}} \underset{\psi^{\mathrm{op}}}{\otimes}{}_{\beta} H \\ \downarrow{\scriptstyle 1 \underset{\psi}{\otimes} \Sigma} & & & & \uparrow{\scriptstyle \Sigma_{[23]}} \\ H_{\hat{\beta}} \underset{\psi}{\otimes}{}_{\alpha} H_{\beta} \underset{\psi}{\otimes}{}_{\alpha} H & & \xrightarrow{V \underset{\psi}{\otimes} 1} & & (H_{\alpha} \underset{\psi^{\mathrm{op}}}{\otimes}{}_{\beta} H)_{\hat{\beta}_{[1]}} \underset{\psi}{\otimes}{}_{\alpha} H. \end{array} \tag{10.7}$$

*Here,* $\Sigma_{[23]}$ *denotes the isomorphism given by* $(\xi_1 \underset{\psi^{\mathrm{op}}}{\otimes} \xi_2) \underset{\psi}{\otimes} \xi_3 \mapsto (\xi_1 \underset{\psi}{\otimes} \xi_3) \underset{\psi^{\mathrm{op}}}{\otimes} \xi_2$.

*Proof.* This follows immediately from the fact that the relative tensor product is symmetric (Proposition 10.1.20), associative (Proposition 10.1.25), and bifunctorial (Corollary 10.1.24). □

We extend the leg notation to the operators that occur in diagram (10.7) and write

$$V_{[12]} \text{ for } V \underset{\psi}{\otimes} 1 \text{ and } V \underset{\psi^{\mathrm{op}}}{\otimes} 1; \quad V_{[23]} \text{ for } 1 \underset{\psi}{\otimes} V \text{ and } 1 \underset{\psi^{\mathrm{op}}}{\otimes} V;$$
$$V_{[13]} \text{ for } \Sigma_{[23]}(V \underset{\psi}{\otimes} 1)(1 \underset{\psi}{\otimes} \Sigma).$$

Thus each of the symbols $V_{[12]}$ and $V_{[23]}$ simultaneously denotes two different operators, which have in common that they act like $V$ on the first and second or on the second and third factor of different relative tensor products. Now the diagram (10.7) commutes if and only if $V$ satisfies the pentagon equation $V_{[12]}V_{[13]}V_{[23]} = V_{[23]}V_{[12]}$.

**Definition 10.3.4.** A unitary $V\colon H_{\hat{\beta}} \underset{\psi}{\otimes} {}_{\alpha}H \to H_{\alpha} \underset{\psi^{\mathrm{op}}}{\otimes} {}_{\beta}H$ is *pseudo-multiplicative* if it satisfies the intertwining conditions (10.6) and if diagram (10.7) commutes.

**Remark 10.3.5.** Let $V\colon H_{\hat{\beta}} \underset{\psi}{\otimes} {}_{\alpha}H \to H_{\alpha} \underset{\psi^{\mathrm{op}}}{\otimes} {}_{\beta}H$ be a pseudo-multiplicative unitary. Then also the composition

$$V^{\mathrm{op}} := \Sigma V^{*}\Sigma\colon H_{\beta} \underset{\psi}{\otimes} {}_{\alpha}H \xrightarrow{\Sigma} H_{\alpha} \underset{\psi^{\mathrm{op}}}{\otimes} {}_{\beta}H \xrightarrow{V^{*}} H_{\hat{\beta}} \underset{\psi}{\otimes} {}_{\alpha}H \xrightarrow{\Sigma} H_{\alpha} \underset{\psi^{\mathrm{op}}}{\otimes} {}_{\hat{\beta}}H$$

is a pseudo-multiplicative unitary, as one can easily check; here, the rôles of $\beta$ and $\hat{\beta}$ get reversed. This unitary is called the *opposite* of $V$.

### 10.3.2 The legs of a pseudo-multiplicative unitary

To every pseudo-multiplicative unitary, one can associate two Hopf–von Neumann bimodules – the left and the right leg of that unitary. This construction generalizes the construction presented in Section 7.2, where we associated to every multiplicative unitary two von Neumann bialgebras. Throughout this subsection, we consider a fixed pseudo-multiplicative unitary

$$V\colon H_{\hat{\beta}} \underset{\psi}{\otimes} {}_{\alpha}H \to H_{\alpha} \underset{\psi^{\mathrm{op}}}{\otimes} {}_{\beta}H,$$

where $N$, $\psi$, $H$, and $\alpha, \beta, \hat{\beta}$ are as in Assumption 10.3.1.

**The von Neumann bimodules $\beta(N)'$ and $\hat{\beta}(N)'$.** Consider the von Neumann algebras

$$\beta(N)' = \mathcal{L}_{\beta(N)}(H) \subseteq \mathcal{L}(H) \quad \text{and} \quad \hat{\beta}(N)' = \mathcal{L}_{\hat{\beta}(N)}(H) \subseteq \mathcal{L}(H).$$

Since $\alpha, \beta, \hat{\beta}$ commute pairwise, we can regard

- $\beta(N)'$ as a von Neumann $N$-$N$-bimodule via $\alpha, \hat{\beta}\colon N \to \beta(N)'$, and
- $\hat{\beta}(N)'$ as a von Neumann $N^{\mathrm{op}}$-$N^{\mathrm{op}}$-bimodule via $\beta, \alpha\colon N \to \hat{\beta}(N)'$.

In particular, we can form the fiber products

$$\beta(N)' \underset{\psi}{*} \beta(N)' = (\beta(N) \underset{\psi}{\bar{\otimes}} \beta(N))' \subseteq \mathcal{L}(H_{\hat{\beta}} \underset{\psi}{\otimes} {}_{\alpha}H)$$

and

$$\hat{\beta}(N)' \underset{\psi^{\mathrm{op}}}{*} \hat{\beta}(N)' = (\hat{\beta}(N) \underset{\psi^{\mathrm{op}}}{\bar{\otimes}} \hat{\beta}(N))' \subseteq \mathcal{L}(H_{\alpha} \underset{\psi^{\mathrm{op}}}{\otimes} {}_{\beta}H),$$

and consider these fiber products as a von Neumann $N$-$N$-bimodule and a von Neumann $N^{\mathrm{op}}$-$N^{\mathrm{op}}$-bimodule, respectively, via the maps

$$\alpha_{[1]}, \hat{\beta}_{[2]}\colon N \to \beta(N)' \underset{\psi}{*} \beta(N)'$$

and

$$\beta_{[1]}, \alpha_{[2]}\colon N \to \hat{\beta}(N)' \underset{\psi^{\mathrm{op}}}{*} \hat{\beta}(N)'.$$

**The comultiplications $\widehat{\Delta}$ and $\Delta$.** Like a multiplicative unitary, the pseudo-multiplicative unitary $V$ gives rise to two maps $\widehat{\Delta}$ and $\Delta$ which define the comultiplication on the left and on the right leg of $V$, respectively.

**Proposition 10.3.6.** *The maps*

$$\widehat{\Delta} = \widehat{\Delta}_V\colon \beta(N)' \to \mathcal{L}(H_{\hat{\beta}} \underset{\psi}{\otimes} {}_{\alpha}H), \quad y \mapsto V^*(1 \underset{\psi^{\mathrm{op}}}{\otimes} y)V,$$

*and*

$$\Delta = \Delta_V\colon \hat{\beta}(N)' \to \mathcal{L}(H_{\alpha} \underset{\psi^{\mathrm{op}}}{\otimes} {}_{\beta}H), \quad z \mapsto V(z \underset{\psi}{\otimes} 1)V^*,$$

*define injective morphisms of von Neumann bimodules*

$$\widehat{\Delta}\colon \beta(N)' \to \beta(N)' \underset{\psi}{*} \beta(N)' \quad \textit{and} \quad \Delta\colon \hat{\beta}(N)' \to \hat{\beta}(N)' \underset{\psi^{\mathrm{op}}}{*} \hat{\beta}(N)' \tag{10.8}$$

*and satisfy*

$$(\widehat{\Delta} \underset{\psi}{*} \mathrm{id}) \circ \widehat{\Delta} = (\mathrm{id} \underset{\psi}{*} \widehat{\Delta}) \circ \widehat{\Delta}, \quad (\Delta \underset{\psi^{\mathrm{op}}}{*} \mathrm{id}) \circ \Delta = (\mathrm{id} \underset{\psi^{\mathrm{op}}}{*} \Delta) \circ \Delta.$$

*Proof.* The maps $\widehat{\Delta}$ and $\Delta$ are injective normal $*$-homomorphisms because the maps $\beta(N)' \to \mathcal{L}(H_\alpha \underset{\psi^{\mathrm{op}}}{\otimes} {}_\beta H)$ and $\hat{\beta}(N)' \to \mathcal{L}(H_{\hat{\beta}} \underset{\psi}{\otimes} {}_\alpha H)$ given by $y \mapsto 1 \underset{\psi^{\mathrm{op}}}{\otimes} y$ and $z \mapsto z \underset{\psi}{\otimes} 1$, respectively, are injective normal $*$-homomorphisms (Proposition 10.1.23) and because $V$ is unitary.

Let us show that $\widehat{\Delta}(\beta(N)')$ is contained in $\beta(N)' \underset{\psi}{*} \beta(N)'$. Using the intertwining relations (10.6), we find that for all $x_1, x_2 \in N$ and $y \in \beta(N)'$, the operators $\widehat{\Delta}(y)$ and $\beta(x_1) \underset{\psi}{\otimes} \beta(x_2)$ commute:

$$\begin{aligned}(\beta(x_1) \underset{\psi}{\otimes} \beta(x_2)) V^*(1 \underset{\psi^{\mathrm{op}}}{\otimes} y) V &= V^*(\beta(x_1)\hat{\beta}(x_2) \underset{\psi^{\mathrm{op}}}{\otimes} y) V \\ &= V^*(1 \underset{\psi^{\mathrm{op}}}{\otimes} y) V (\beta(x_1) \underset{\psi}{\otimes} \beta(x_2)).\end{aligned}$$

Thus, $\widehat{\Delta}(\beta(N)') \subseteq (\beta(N) \underset{\psi}{\bar{\otimes}} \beta(N))' = \beta(N)' \underset{\psi}{*} \beta(N)'$. A similar argument shows that $\Delta(\hat{\beta}(N)')$ is contained in $\hat{\beta}(N)' \underset{\psi^{\mathrm{op}}}{*} \hat{\beta}(N)'$.

The intertwining relations (10.6) immediately imply that for all $x \in N$,

$$\begin{aligned}\widehat{\Delta}(\alpha(x)) &= \alpha_{[1]}(x), & \widehat{\Delta}(\hat{\beta}(x)) &= \hat{\beta}_{[2]}(x), \\ \Delta(\alpha(x)) &= \alpha_{[2]}(x), & \Delta(\beta(x)) &= \beta_{[1]}(x).\end{aligned}$$

Therefore, $\widehat{\Delta}$ and $\Delta$ are morphisms of von Neumann bimodules as claimed.

The coassociativity equations follow from the pentagon equation for $V$; the proof is similar as in the case of a multiplicative unitary (see Lemma 7.2.1). □

**The spaces $\widehat{A}_{\mathrm{alg}}(V)$ and $A_{\mathrm{alg}}(V)$.** The legs of $V$ are defined in terms of operators on $H$ that are constructed similarly as in the case of a multiplicative unitary in Lemma 7.2.7. Let us begin with the left leg. By Proposition 10.1.22, we can define for each $\xi \in D({}_\alpha H, \psi)$ and $\xi' \in D({}_\beta H, \psi^{\mathrm{op}})$ the following operators:

$$\begin{aligned}|\xi\rangle^{\hat{\beta},\alpha}_{\psi[2]} \colon H &\to H_{\hat{\beta}} \underset{\psi}{\otimes} {}_\alpha H, & |\xi'\rangle^{\alpha,\beta}_{\psi^{\mathrm{op}}[2]} \colon H &\to H_\alpha \underset{\psi^{\mathrm{op}}}{\otimes} {}_\beta H, \\ \zeta &\mapsto \zeta \underset{\psi}{\otimes} \xi, & \zeta &\mapsto \zeta \underset{\psi^{\mathrm{op}}}{\otimes} \xi',\end{aligned}$$

and

$$\hat{a}_{(\xi',\xi)} := \langle \xi'|^{\alpha,\beta}_{\psi^{\mathrm{op}}[2]} V |\xi\rangle^{\hat{\beta},\alpha}_{\psi[2]} \colon H \to H_{\hat{\beta}} \underset{\psi}{\otimes} {}_\alpha H \to H_\alpha \underset{\psi^{\mathrm{op}}}{\otimes} {}_\beta H \to H,$$

where $\langle\xi'|^{\alpha,\beta}_{\psi^{\mathrm{op}}[2]} = (|\xi'\rangle^{\alpha,\beta}_{\psi^{\mathrm{op}}[2]})^*$. The operator $\hat{a}_{(\xi',\xi)}$ is characterized by the relation

$$\langle\zeta'|\hat{a}_{(\xi',\xi)}\zeta\rangle = \langle\zeta' \underset{\psi^{\mathrm{op}}}{\otimes} \xi'|V(\zeta \underset{\psi}{\otimes} \xi)\rangle \quad \text{for all } \zeta,\zeta' \in H, \tag{10.9}$$

and can be written in the form $\hat{a}_{(\xi',\xi)} = (\mathrm{id} * \omega_{\xi',\xi})(V)$, where $\mathrm{id} * \omega_{\xi',\xi}$ denotes the generalized slice map

$$\mathrm{id} * \omega_{\xi',\xi} \colon \mathcal{L}(H_{\hat{\beta}} \underset{\psi}{\otimes} {}_{\alpha}H, H_{\alpha} \underset{\psi^{\mathrm{op}}}{\otimes} {}_{\beta}H) \to \mathcal{L}(H), \quad X \mapsto \langle\xi'|^{\alpha,\beta}_{\psi^{\mathrm{op}}[2]} X |\xi\rangle^{\hat{\beta},\alpha}_{\psi[2]}.$$

Symmetrically, we define for each $\eta \in D(H_{\hat{\beta}}, \psi^{\mathrm{op}})$ and $\eta' \in D(H_{\alpha}, \psi)$

$$\begin{aligned} |\eta\rangle^{\hat{\beta},\alpha}_{\psi[1]} \colon H &\to H_{\hat{\beta}} \underset{\psi}{\otimes} {}_{\alpha}H, & |\eta'\rangle^{\alpha,\beta}_{\psi^{\mathrm{op}}[2]} \colon H &\to H_{\alpha} \underset{\psi^{\mathrm{op}}}{\otimes} {}_{\beta}H, \\ \zeta &\mapsto \eta \underset{\psi}{\otimes} \zeta, & \zeta &\mapsto \eta' \underset{\psi^{\mathrm{op}}}{\otimes} \zeta, \end{aligned}$$

and

$$a_{(\eta',\eta)} := \langle\eta'|^{\alpha,\beta}_{\psi^{\mathrm{op}}[2]} V |\eta\rangle^{\hat{\beta},\alpha}_{\psi[1]} \colon H \to H_{\hat{\beta}} \underset{\psi}{\otimes} {}_{\alpha}H \to H_{\alpha} \underset{\psi^{\mathrm{op}}}{\otimes} {}_{\beta}H \to H,$$

where $\langle\eta'|^{\alpha,\beta}_{\psi^{\mathrm{op}}[2]} = (|\eta'\rangle^{\alpha,\beta}_{\psi^{\mathrm{op}}[2]})^*$. Similarly as $\hat{a}_{(\xi',\xi)}$, the operator $a_{(\eta',\eta)}$ is characterized by the equation

$$\langle\zeta'|a_{(\eta',\eta)}\zeta\rangle = \langle\eta' \underset{\psi^{\mathrm{op}}}{\otimes} \zeta'|V(\eta \underset{\psi}{\otimes} \zeta)\rangle \quad \text{for all } \zeta,\zeta' \in H,$$

and can be written in the form $a_{(\eta',\eta)} = (\omega_{\eta',\eta} * \mathrm{id})(V)$, where

$$\omega_{\eta',\eta} * \mathrm{id} \colon \mathcal{L}(H_{\hat{\beta}} \underset{\psi}{\otimes} {}_{\alpha}H, H_{\alpha} \underset{\psi^{\mathrm{op}}}{\otimes} {}_{\beta}H) \to \mathcal{L}(H), \quad X \mapsto \langle\eta'|^{\alpha,\beta}_{\psi^{\mathrm{op}}[2]} X |\eta\rangle^{\hat{\beta},\alpha}_{\psi[1]}.$$

We consider the spaces

$$\widehat{A}_{\mathrm{alg}}(V) := \mathrm{span}\{\hat{a}_{(\xi',\xi)} \mid \xi \in D({}_{\alpha}H, \psi),\ \xi' \in D({}_{\beta}H, \psi^{\mathrm{op}})\} \subseteq \mathcal{L}(H)$$

and

$$A_{\mathrm{alg}}(V) := \mathrm{span}\{a_{(\eta',\eta)} \mid \eta \in D(H_{\hat{\beta}}, \psi^{\mathrm{op}}),\ \eta' \in D(H_{\alpha}, \psi)\} \subseteq \mathcal{L}(H).$$

The intertwining relations of $V$ imply the following inclusions:

**Lemma 10.3.7.** i) $\hat{\beta}(N)\widehat{A}_{\mathrm{alg}}(V)\alpha(N) = \widehat{A}_{\mathrm{alg}}(V) \subseteq \beta(N)'$.

ii) $\alpha(N)A_{\mathrm{alg}}(V)\beta(N) = A_{\mathrm{alg}}(V) \subseteq \hat{\beta}(N)'$.

*Proof.* i) Let $\xi \in D({}_{\alpha}H, \psi)$, $\xi' \in D({}_{\beta}H, \psi^{\mathrm{op}})$, and $x, y, z \in N$. We show that

$$\beta(x)\hat{\beta}(y) \cdot \hat{a}_{(\xi',\xi)} \cdot \alpha(z) = \hat{a}_{(\vartheta',\vartheta)} \cdot \beta(x), \quad \text{where } \vartheta = \beta(y)\xi,\ \vartheta' = \alpha(z)^*\xi',$$

and this implies the assertion. Note that $\vartheta \in D({}_{\alpha}H, \psi)$ and $\vartheta' \in D({}_{\beta}H, \psi^{\mathrm{op}})$ by Proposition 10.1.12. Let $\zeta, \zeta' \in H$. By equation (10.9),

$$\begin{aligned} \langle \zeta' | \beta(x)\hat{\beta}(y) \cdot \hat{a}_{(\xi',\xi)} \cdot \alpha(z)\zeta \rangle &= \langle \hat{\beta}(y^*)\beta(x^*)\zeta' | \hat{a}_{(\xi',\xi)}\alpha(z)\zeta \rangle \\ &= \langle \hat{\beta}(y^*)\beta(x^*)\zeta' \underset{\psi^{\mathrm{op}}}{\otimes} \xi' | V(\alpha(z)\zeta \underset{\psi}{\otimes} \xi) \rangle, \end{aligned}$$

and the intertwining relations (10.6) imply that this is equal to

$$\langle \zeta' \underset{\psi^{\mathrm{op}}}{\otimes} \alpha(z)^*\xi' | V(\beta(x)\zeta \underset{\psi}{\otimes} \beta(y)\xi) \rangle = \langle \zeta' | \hat{a}_{(\vartheta',\vartheta)}\beta(x)\zeta \rangle.$$

ii) The proof is similar to the proof of i). □

The subspaces $\widehat{A}_{\mathrm{alg}}(V)$ and $A_{\mathrm{alg}}(V)$ of $\mathcal{L}(H)$ are non-degenerate in the following sense:

**Lemma 10.3.8.** *The sets $\widehat{A}_{\mathrm{alg}}(V)H$, $\widehat{A}_{\mathrm{alg}}(V)^*H$ and $A_{\mathrm{alg}}(V)H$, $A_{\mathrm{alg}}(V)^*H$ are linearly dense in $H$.*

*Proof.* The proof is essentially the same as in the case of a multiplicative unitary, see Lemma 7.2.2. □

**The left-right symmetry.** As in the setting of multiplicative unitaries, the preceding definitions are symmetric in the following sense:

**Lemma 10.3.9.** *We have*

$$\begin{aligned} \widehat{A}_{\mathrm{alg}}(V^{\mathrm{op}}) &= A_{\mathrm{alg}}(V)^*, & A_{\mathrm{alg}}(V^{\mathrm{op}}) &= \widehat{A}_{\mathrm{alg}}(V)^*, \\ \widehat{\Delta}_{V^{\mathrm{op}}} &= \mathrm{Ad}_{\Sigma} \circ \Delta_V, & \Delta_{V^{\mathrm{op}}} &= \mathrm{Ad}_{\Sigma} \circ \widehat{\Delta}_V. \end{aligned}$$

*Proof.* The proof is essentially the same as in the case of a multiplicative unitary, see Lemma 7.2.5. □

**Hopf–von Neumann bimodules associated to $V$.** Let us study various closures of the spaces $\widehat{A}_{\mathrm{alg}}(V)$ and $A_{\mathrm{alg}}(V)$. First, consider the von Neumann algebras

$$\widehat{B}_w(V) := (\widehat{A}_{\mathrm{alg}}(V) + \widehat{A}_{\mathrm{alg}}(V)^*)'' \quad \text{and} \quad B_w(V) := (A_{\mathrm{alg}}(V) + A_{\mathrm{alg}}(V)^*)''.$$

**Proposition 10.3.10.** i) *We have* $(\alpha(N) \cup \hat{\beta}(N)) \subseteq \widehat{B}_w(V) \subseteq \beta(N)'$ *and* $(\beta(N) \cup \alpha(N)) \subseteq B_w(V) \subseteq \hat{\beta}(N)'$.

ii) *The commutant* $(\widehat{B}_w(V))'$ *is equal to the set of all* $T \in \alpha(N)' \cap \hat{\beta}(N)'$ *that satisfy* $(T \underset{\psi^{\mathrm{op}}}{\otimes} 1)V = V(T \underset{\psi}{\otimes} 1)$, *and the commutant* $(B_w(V))'$ *is equal to the set of all* $T \in \beta(N)' \cap \alpha(N)'$ *that satisfy* $(1 \underset{\psi^{\mathrm{op}}}{\otimes} T)V = V(1 \underset{\psi}{\otimes} T)$.

*Consider* $\widehat{B}_w(V)$ *as a von Neumann* $N$-$N$*-bimodule via* $\alpha$ *and* $\hat{\beta}$, *and* $B_w(V)$ *as a von Neumann* $N^{\mathrm{op}}$-$N^{\mathrm{op}}$*-bimodule via* $\beta$ *and* $\alpha$.

iii) *The maps* $\widehat{\Delta}$ *and* $\Delta$ *restrict to normal* $*$*-homomorphisms*

$$\widehat{\Delta}\colon \widehat{B}_w(V) \to \widehat{B}_w(V) \underset{\psi}{*} \widehat{B}_w(V) \quad \text{and} \quad \Delta\colon B_w(V) \to B_w(V) \underset{\psi^{\mathrm{op}}}{*} B_w(V).$$

*Proof.* Assertion i) follows easily from Lemma 10.3.7 and Lemma 10.3.8, and the proof of assertions ii) and iii) is essentially the same as in the case of a multiplicative unitary, see Lemma 7.2.9. □

The preceding proposition immediately implies the following main result:

**Theorem 10.3.11.** *Let* $V\colon H_{\hat{\beta}} \underset{\psi}{\otimes}{}_{\alpha} H \to H_{\alpha} \underset{\psi^{\mathrm{op}}}{\otimes}{}_{\beta} H$ *be a pseudo-multiplicative unitary. Then the tuples* $(N, \widehat{B}_w(V), \alpha, \hat{\beta}, \widehat{\Delta})$ *and* $(N^{\mathrm{op}}, B_w(V), \beta, \alpha, \Delta)$ *are Hopf–von Neumann bimodules.* □

The closures of the spaces $\widehat{A}_{\mathrm{alg}}(V)$ and $A_{\mathrm{alg}}(V)$ with respect to the norm or the weak operator topology are more difficult to analyze than in the setting of multiplicative unitaries. For the proofs of the following statements, we refer to the literature. Put

$$\widehat{A}(V) := \|\cdot\|\text{-closure of } \widehat{A}_{\mathrm{alg}}(V), \qquad A(V) := \|\cdot\|\text{-closure of } A_{\mathrm{alg}}(V),$$
$$\widehat{A}_w(V) := w\text{-closure of } \widehat{A}_{\mathrm{alg}}(V), \qquad A_w(V) := w\text{-closure of } A_{\mathrm{alg}}(V),$$

where "$w$-closure" denotes the closure with respect to the weak operator topology.

**Proposition 10.3.12** ([44, Proposition 3.6]). *The spaces* $\widehat{A}(V)$ *and* $A(V)$ *(and hence also* $\widehat{A}_w(V)$ *and* $A_w(V)$*) are non-degenerate subalgebras of* $\mathcal{L}(H)$.

As for multiplicative unitaries, we use the following terminology:

**Definition 10.3.13.** A pseudo-multiplicative unitary $V\colon H_{\hat{\beta}} \underset{\psi}{\otimes}{}_{\alpha} H \to H_{\alpha} \underset{\psi^{\mathrm{op}}}{\otimes}{}_{\beta} H$ is *weakly well-behaved* if $\widehat{A}_w(V)$ and $A_w(V)$ are von Neumann algebras.

By Proposition 10.3.12, a pseudo-multiplicative unitary $V$ is well-behaved as soon as $\widehat{A}_w(V)^* = \widehat{A}_w(V)$ and $A_w(V)^* = A_w(V)$. In that case, $\widehat{A}_w(V) = \widehat{B}_w(V)$ and $A_w(V) = B_w(V)$. Thus, Theorem 10.3.11 implies:

**Corollary 10.3.14.** *If* $V\colon H_{\hat{\beta}} \underset{\psi}{\otimes} {}_{\alpha}H \to H_{\alpha} \underset{\psi^{\mathrm{op}}}{\otimes} {}_{\beta}H$ *is a weakly well-behaved pseudo-multiplicative unitary, then* $(N, \widehat{A}_w(V), \alpha, \hat{\beta}, \widehat{\Delta})$, $(N^{\mathrm{op}}, A_w(V), \beta, \alpha, \Delta)$ *are Hopf–von Neumann bimodules.* □

Let us mention two classes of weakly well-behaved pseudo-multiplicative unitaries: the weakly regular ones, and the manageable ones. For a discussion of manageability, see [99].

**Weak regularity.** The concept of regularity was carried over from multiplicative unitaries to pseudo-multiplicative unitaries by Enock [44]. The definition involves operators of the following form. Given $\xi, \eta' \in D({}_{\alpha}H, \psi) = D(H_{\alpha}, \psi)$, we can consider the composition

$$c_{(\eta',\xi)} := \langle \eta' |^{\alpha,\beta}_{\psi^{\mathrm{op}}[1]} V |\xi\rangle^{\hat{\beta},\alpha}_{\psi[2]} \colon H \to H_{\hat{\beta}} \underset{\psi}{\otimes} {}_{\alpha}H \to H_{\alpha} \underset{\psi^{\mathrm{op}}}{\otimes} {}_{\beta}H \to H,$$

which is characterized by the equation $\langle \zeta' | c_{(\eta',\xi)}\zeta\rangle = \langle \eta' \underset{\psi^{\mathrm{op}}}{\otimes} \zeta' | V(\zeta \underset{\psi}{\otimes} \xi)\rangle$ for all $\zeta, \zeta' \in H$. Let

$$\mathcal{C}_{\mathrm{alg}}(V) := \operatorname{span}\{c_{(\eta',\xi)} \mid \eta', \xi \in D(H_{\alpha}, \psi)\} \subseteq \mathcal{L}(H),$$

and denote by $\mathcal{C}(V)$ and $\mathcal{C}_w(V)$ the closure of $\mathcal{C}_{\mathrm{alg}}(V)$ with respect to the norm topology and the weak operator topology, respectively.

**Lemma 10.3.15.** i) $\mathcal{C}_{\mathrm{alg}}(V) \subseteq \alpha(N)'$.

ii) *The sets* $\mathcal{C}_{\mathrm{alg}}(V)H$ *and* $\mathcal{C}_{\mathrm{alg}}(V)^*H$ *are linearly dense in* $H$.

*Proof.* The proof of i) is similar to the proof of Lemma 10.3.7, and the proof of ii) is similar to the proof of the last statement of Lemma 7.2.2. □

Recall that to each $\xi \in D({}_{\alpha}H, \psi) = D(H_{\alpha}, \psi)$, we associated an operator $R_{\psi}(\xi) = L_{\psi^{\mathrm{op}}}(\xi)\colon H_{\psi} \to H$ (Definition 10.1.8). This operator occurs in the following definition.

**Definition 10.3.16.** A pseudo-multiplicative unitary $V\colon H_{\hat{\beta}} \underset{\psi}{\otimes} {}_{\alpha}H \to H_{\alpha} \underset{\psi^{\mathrm{op}}}{\otimes} {}_{\beta}H$ is *weakly regular* if $\mathcal{C}_w(V) = \alpha(N)'$, and *regular* if

$$\mathcal{C}(V) = \overline{\operatorname{span}}\{L_{\psi^{\mathrm{op}}}(\xi)L_{\psi^{\mathrm{op}}}(\eta')^* \mid \eta', \xi \in D(H_{\alpha}, \psi)\}.$$

The following results generalize Lemma 7.3.5 and Theorem 7.3.10, however, the proofs involve new techniques.

**Proposition 10.3.17** ([44, Proposition 3.10]). *The space* $\mathcal{C}(V)$ *(and hence also* $\mathcal{C}_w(V)$*) is a subalgebra of* $\mathcal{L}(H)$.

**Theorem 10.3.18** ([44, Proposition 3.12]). *Let* $V : H_{\hat{\beta}} \underset{\psi}{\otimes} {}_{\alpha}H \to H_{\alpha} \underset{\psi^{\mathrm{op}}}{\otimes} {}_{\beta}H$ *be a pseudo-multiplicative unitary. If* $\mathcal{C}_w(V) = \mathcal{C}_w(V)^*$*, then* $V$ *is weakly well-behaved.*

### 10.3.3 The pseudo-multiplicative unitary of a groupoid

The prototypical example of a pseudo-multiplicative unitary is the unitary associated to a locally compact groupoid. After a review of some preliminaries, we construct this unitary and determine the associated Hopf–von Neumann bimodules. The original references are [171], [172].

**Preliminaries.** Let us briefly recall some concepts related to groupoids and integration over groupoids. Standard references are [120], [129].

A *groupoid* is a small category in which every morphism is invertible. Equivalently, a groupoid consists of a set of *morphisms* or *arrows* $G$, a set of *objects* or *units* $G^0 \subseteq G$, two maps $r, s : G \to G^0$ called the *range* and *source* map, and a composition map $G^2_{s,r} \to G$, where $G^2_{s,r} = \{(x, y) \subseteq G \times G \mid s(x) = r(y)\}$, subject to several conditions (see [129, Definition 1.1]). A *topological groupoid* is a groupoid equipped with a topology on its set of morphisms for which the inversion and the composition are continuous; then also the range and source map are continuous.

Let $G$ be a topological groupoid that is locally compact, Hausdorff, and second countable. To perform translation-invariant integration on $G$, we need an analogue of the Haar measure of a locally compact group. The precise definition of this analogue involves the fibers of the range and the source map, which we denote by $G^u := r^{-1}(u)$ and $G_u := s^{-1}(u)$ for each $u \in G^0$. A *left Haar system* on $G$ is a family of Borel measures $\lambda = (\lambda^u)_{u \in G^0}$ such that

i) for each $u \in G^0$, $\lambda^u$ is a regular Borel measure on $G^u$ with support $G^u$;

ii) for each $f \in C_c(G)$, the function $G^0 \to \mathbb{C}$, $u \mapsto \int_{G^u} f \, d\lambda^u$, is continuous;

iii) for each $x \in G$ and $f \in C_c(G^{r(x)})$,

$$\int_{G^{r(x)}} f(y) \, d\lambda^{r(x)}(y) = \int_{G^{s(x)}} f(xy') \, d\lambda^{s(x)}(y').$$

Denote by $i : G \to G$ the inversion map $x \mapsto x^{-1}$. For each $u \in G^0$, define the measure $\lambda_u^{-1}$ on $G_u$ as the push-forward of $\lambda^u$ along $i$, that is, $\lambda_u^{-1} = i_*(\lambda^u)$. The family $\lambda^{-1} = (\lambda_u^{-1})_{u \in G^0}$ is called the *right Haar system associated to* $\lambda$. Haar systems need not exist and need not be unique, but in many examples, they arise naturally.

Let $\mu$ be a Radon measure on the unit space $G^0$, and define a measure $\nu$ on $G$ via

$$\int_G f\,d\nu := \int_{G^0}\int_{G^u} f(x)\,d\lambda^u(x)\,d\mu(u) \quad \text{for all } f \in C_c(G).$$

The push-forward of $\nu$ via the inversion map $i$ is denoted by $\nu^{-1}$; evidently,

$$\int_G f d\nu^{-1} = \int_{G^0}\int_{G_u} f(x) d\lambda_u^{-1}(x)\,d\mu(u).$$

The measure $\mu$ is called *quasi-invariant* if $\nu$ and $\nu^{-1}$ are equivalent. There always exist sufficiently many quasi-invariant measures [129, I, Proposition 3.6].

**Hilbert modules associated to a groupoid.** Let $G$ be a topological groupoid that is locally compact, Hausdorff, and second countable. Furthermore, let $\lambda$ be a left Haar system on $G$ and $\mu$ a quasi-invariant Radon measure on $G^0$, and define $\nu$ and $\nu^{-1}$ as above. From now on, we denote the range and the source map of $G$ by $r_G$ and $s_G$, respectively, and reserve the letters $r$ and $s$ for certain representations constructed out of $r_G$ and $s_G$.

The measure $\mu$ on $G^0$ defines an n.s.f. weight on the von Neumann algebra $L^\infty(G^0, \mu)$ via $f \mapsto \int_{G^0} f\,d\mu$. Let us denote this weight by $\mu$ again.

The Hilbert space $L^2(G, \nu)$ can be considered as a Hilbert module over the Neumann algebra $L^\infty(G^0, \mu)$ via the $*$-homomorphisms

$$\begin{aligned} r\colon L^\infty(G^0,\mu) &\to L^\infty(G,\nu) &\quad\text{and}\quad s\colon L^\infty(G^0,\mu) &\to L^\infty(G,\nu),\\ (r(f))(x) &:= f(r_G(x)), & (s(f))(x) &:= f(s_G(x)), \end{aligned}$$

and the representation $L^\infty(G, \nu) \to \mathcal{L}(L^2(G, \nu))$ given by multiplication operators. We denote by $L^2(G, \nu)_r$ and $L^2(G, \nu)_s$ the Hilbert $L^\infty(G^0, \mu)$-module given by $L^2(G, \nu)$ and the representation $r$ or $s$, respectively.

By construction, the Hilbert module $L^2(G, \nu)_r$ is a direct integral:

$$L^2(G,\nu)_r = \int_{G^0}^{\oplus} L^2(G^u, \lambda^u) d\mu(u).$$

The Hilbert module $L^2(G, \nu)_s$ can be described similarly. Denote by $D := d\nu/d\nu^{-1}$ the Radon–Nikodym derivative of $\nu$ with respect to $\nu^{-1}$. Then

$$\int_G f(x) d\nu(x) = \int_{G^0}\int_{G_u} f(x) D(x) d\lambda_u^{-1}(x) d\mu(u) \quad \text{for all } f \in L^1(G,\nu)$$

and

$$L^2(G,\nu)_s = \int_{G^0}^{\oplus} L^2(G_u, D_u\lambda_u^{-1}) d\mu(u), \quad \text{where } D_u = D|_{G_u}.$$

For later use, we identify the bounded elements of the Hilbert modules $L^2(G,\nu)_r$ and $L^2(G,\nu)_s$. By direct calculations or an application of Proposition 10.1.17, we find:

**Lemma 10.3.19.** i) *An element $\xi \in L^2(G,\nu)_r$ is bounded with respect to $\mu$ if and only if the function on $G^0$ given by $u \mapsto \int_{G^u} |\xi(x)|^2 d\lambda^u(x)$ belongs to $L^\infty(G^0,\mu)$. For $\xi,\xi' \in D(L^2(G,\nu)_r,\mu)$, the product $\langle \xi'|\xi\rangle_\mu \in L^\infty(G^0,\mu)$ is given by $u \mapsto \int_{G^u} \overline{\xi'(x)}\xi(x)d\lambda^u(x)$.*

ii) *An element $\xi \in L^2(G,\nu)_s$ is bounded with respect to $\mu$ if and only if the function on $G^0$ given by $u \mapsto \int_{G_u} |\xi(x)|^2 D(x)d\lambda_u^{-1}(x)$ belongs to $L^\infty(G^0,\mu)$. For $\xi,\xi' \in D(L^2(G,\nu)_s,\mu)$, the product $\langle \xi'|\xi\rangle_\mu \in L^\infty(G^0,\mu)$ is given by $u \mapsto \int_{G_u} \overline{\xi'(x)}\xi(x)D(x)d\lambda_u^{-1}(x)$.* □

**The pseudo-multiplicative unitary.** We construct a pseudo-multiplicative unitary

$$W_G\colon L^2(G,\nu)_s \underset{\mu}{\otimes} {}_rL^2(G,\nu) \to L^2(G,\nu)_r \underset{\mu}{\otimes} {}_rL^2(G,\nu); \tag{10.10}$$

in the notation of Definition 10.3.4,

$$N = L^\infty(G^0,\mu) = N^{\mathrm{op}}, \quad H = L^2(G,\nu),$$
$$\psi = \mu = \psi^{\mathrm{op}}, \quad \alpha = r, \quad \beta = r, \quad \hat\beta = s.$$

The relative tensor products that appear in (10.10) can conveniently be described in terms of the measure $\nu^2_{s,r}$ on $G^2_{s,r}$ given by

$$\int_{G^2_{s,r}} f\, d\nu^2_{s,r} := \int_{G^0}\int_{G^u}\int_{G^{s_G(x)}} f(x,y)\, d\lambda^{s_G(x)}(y)\, d\lambda^u(x)\, d\mu(u)$$

and the measure $\nu^2_{r,r}$ on $G^2_{r,r} = \{(x,y) \in G^2 \mid r_G(x) = r_G(y)\}$ given by

$$\int_{G^2_{r,r}} g\, d\nu^2_{r,r} := \int_{G^0}\int_{G^u}\int_{G^u} g(x,y)\, d\lambda^u(y)\, d\lambda^u(x)\, d\mu(u),$$

where $f \in C_c(G^2_{s,r})$ and $g \in C_c(G^2_{r,r})$:

**Lemma 10.3.20.** *For $k = s, r$, respectively, the map*

$$\Phi_{k,r}\colon L^2(G,\nu)_k \underset{\mu}{\otimes} {}_rL^2(G,\nu) \to L^2(G^2_{k,r},\nu^2_{k,r})$$

*given by*

$$(\Phi_{k,r}(\eta \underset{\mu}{\otimes} \xi))(x,y) := \eta(x)\xi(y),$$

*where $\eta \in L^2(G,\nu)$ and $\xi \in D(L^2(G,\nu)_r,\mu)$, is an isometric isomorphism.*

*Proof.* We prove the assertion for $k = s$; the case $k = r$ is similar. Let $\eta \in L^2(G, \nu)$, $\xi \in D(L^2(G, \nu)_r, \mu)$, and put $\zeta := \eta \underset{\mu}{\otimes} \xi$. By Lemma 10.3.19 i),

$$\begin{aligned}
\|\zeta\|^2 &= \langle \eta | s(\langle \xi | \xi \rangle_\mu) \eta \rangle \\
&= \int_{G^0} \int_{G^u} \overline{\eta(x)} (\langle \xi | \xi \rangle_\mu (s_G(x))) \eta(x) d\lambda^u(x)\, d\mu(u) \\
&= \int_{G^0} \int_{G^u} \int_{G^{s_G(x)}} |\eta(x)|^2 |\xi(y)|^2 \, d\lambda^{s_G(x)}(y)\, d\lambda^u(x)\, d\mu(u) \\
&= \int_{G^2_{s,r}} |(\Phi_{s,r}(\zeta))(x, y)|^2 d\nu^2_{s,r}(x, y) \\
&= \|\Phi_{s,r}(\zeta)\|^2.
\end{aligned}$$

This calculation implies that $\Phi_{s,r}$ is well defined and isometric, and it is easy to see that this map has dense image. □

Alternatively, the preceding result can also be deduced from Proposition 10.1.17. From now on, we identify $L^2(G, \nu)_k \underset{\mu}{\otimes}{}_r L^2(G, \nu)$ with $L^2(G^2_{k,r}, \nu^2_{k,r})$ $(k = r, s)$ via the map $\Phi_{k,r}$.

**Proposition 10.3.21.** *The map*

$$W_G \colon L^2(G, \nu)_s \underset{\mu}{\otimes}{}_r L^2(G, \nu) \to L^2(G, \nu)_r \underset{\mu}{\otimes}{}_r L^2(G, \nu)$$

*given by*

$$(W_G\zeta)(x, y) = \zeta(x, x^{-1}y) \quad \textit{for all } \zeta \in L^2(G^2_{s,r}, \nu^2_{s,r}),\ (x, y) \in G^2_{r,r},$$

*is a pseudo-multiplicative unitary.*

*Proof.* The fact that $W_G$ is a unitary follows easily from Lemma 10.3.20 and from the left-invariance of the Haar system $\lambda$. To verify the pentagon equation, one describes the iterated relative tensor products of $L^2(G, \nu)$ that occur in the pentagon diagram (10.7) in terms of measures on iterated fiber products of $G$, similarly as it was done in Lemma 10.3.20 for simple relative tensor products, and uses a similar calculation as in Example 7.1.4. □

Let us identify the legs of $W_G$. Recall that the von Neumann algebra of $G$ is the von Neumann algebra $LG \subseteq \mathcal{L}(L^2(G, \nu))$ generated by operators of the form $L(g)$, where $g \in L^1(G, \nu)$,

$$(L(g)\xi)(x) = \int_{G^{r_G(x)}} g(y)\xi(y^{-1}x) d\lambda^{r_G(x)}(y) \quad \text{for all } x \in G,\ \xi \in L^2(G, \nu).$$

**Proposition 10.3.22.** *$\widehat{A}_w(W_G) = L^\infty(G, \nu)$ and $A_w(W_G) = LG$; in particular, the pseudo-multiplicative unitary $W_G$ is weakly well-behaved.*

*Proof.* This follows from similar calculations and arguments as in Example 7.2.13, where $G$ was a group. □

By Theorem 10.3.11, we obtain Hopf–von Neumann bimodules

$$(L^\infty(G^0, \mu), L^\infty(G, \nu), r, s, \widehat{\Delta}) \quad \text{and} \quad (L^\infty(G^0, \mu), LG, r, r, \Delta).$$

Let us describe the comultiplications

$$\widehat{\Delta}: L^\infty(G, \nu) \to L^\infty(G, \nu)_s \underset{\mu}{*}{}_r L^\infty(G, \nu), \quad f \mapsto W_G^*(1 \underset{\mu}{\otimes} f)W_G,$$

and

$$\Delta: LG \to LG_r \underset{\mu}{*}{}_r LG, \quad L(g) \mapsto W_G(L(g) \underset{\mu}{\otimes} 1)W_G^*.$$

Using the isomorphisms of Lemma 10.3.20, we identify $L^\infty(G, \nu)_s \underset{\mu}{*}{}_r L^\infty(G, \nu)$ and $LG_r \underset{\mu}{*}{}_r LG$ with von Neumann algebras on the Hilbert space $L^2(G^2_{k,r}, \nu^2_{k,r}) \cong L^2(G, \nu)_k \underset{\mu}{\otimes}{}_r L^2(G, \nu)$, where $k = s$ or $k = r$, respectively.

**Proposition 10.3.23.** i) *For all $f \in L^\infty(G, \nu)$, $\zeta \in L^2(G^2_{s,r}, \nu^2_{s,r})$, $(x, y) \in G^2_{s,r}$,*

$$(\widehat{\Delta}(f)\zeta)(x, y) = f(xy)\zeta(x, y).$$

ii) *For all $g \in L^1(G, \nu)$, $\zeta \in L^2(G^2_{r,r}, \nu^2_{r,r})$, $(x, y) \in G^2_{r,r}$,*

$$(\Delta(L(g))\zeta)(x, y) = \int_{G^{r_G(x)}} g(z)\zeta(z^{-1}x, z^{-1}y)\, d\lambda^{r_G(x)}(z).$$

*Proof.* Again, this follows from similar calculations as in Example 7.2.13, where $G$ was a group. □

Finally, we note:

**Proposition 10.3.24.** *$W_G$ is regular.*

*Proof.* This follows from similar calculations as in Example 7.3.4 iii); a detailed proof can be found in [44, Proposition 4.8]. □

# Chapter 11
# Pseudo-multiplicative unitaries on $C^*$-modules

In contrast to the theory of measurable quantum groupoids, a general theory of locally compact quantum groupoids in the setting of $C^*$-algebras does not exist yet. Such a theory should be based on $C^*$-algebraic analogues of the following von Neumann-algebraic concepts:

- pseudo-multiplicative unitaries on Hilbert spaces,
- Hopf–von Neumann bimodules, and
- operator-valued weights.

Such $C^*$-algebraic analogues of the first and second concepts have recently been introduced and studied in [154], [155], [156].

In this book we restrict ourselves to a special class of "decomposable" locally compact quantum groupoids or, more precisely, to a special class of "decomposable" pseudo-multiplicative unitaries on $C^*$-modules. For classical groupoids, this implies a restriction to groupoids that are $r$-discrete or extensions of $r$-discrete groupoids by group bundles.

We proceed as follows. First, we introduce pseudo-multiplicative unitaries on $C^*$-modules (Section 11.1). When we try to define the legs of such a pseudo-multiplicative unitary, we immediately encounter several problems that we can only solve under a certain decomposability assumption.

Tailored to this decomposability assumption, we develop a general calculus of homogeneous operators on $C^*$-bimodules (Section 11.2) and introduce $C^*$-families of such operators, which generalize $C^*$-algebras. Next, we define internal tensor products and morphisms of $C^*$-families, which enter the definition of a Hopf $C^*$-family (Section 11.3).

The legs of a decomposable regular pseudo-multiplicative unitary can then be constructed in the form of Hopf $C^*$-families (Section 11.4). Moreover, one can define coactions of Hopf $C^*$-families, and relate coactions of the Hopf $C^*$-families associated to a decomposable groupoid to actions and Fell bundles of that groupoid (Section 11.5).

The results and concepts presented in this chapter are part of the author's thesis [152]. For further details, we refer to this thesis, the article [153], and forthcoming publications.

## 11.1 Pseudo-multiplicative unitaries on $C^*$-modules

Pseudo-multiplicative unitaries on $C^*$-modules are natural generalizations of multiplicative unitaries and analogues of pseudo-multiplicative unitaries on Hilbert spaces. The definition and the main example – the pseudo-multiplicative unitary of a locally compact groupoid – are given in the first sections. Thereafter, we discuss the question how to define the legs of such a pseudo-multiplicative unitary, and outline a strategy for a partial solution that will be implemented in the following sections. Let us note that a general solution based on a better notion of a pseudo-multiplicative unitary was recently proposed in [154].

### 11.1.1 The flipped internal tensor product of $C^*$-modules

The definition of a pseudo-multiplicative unitary on a $C^*$-module involves the internal tensor product of $C^*$-modules, which is reviewed in Appendix 12.2, and the flipped internal tensor product of $C^*$-modules, which is defined as follows.

Let $A$ and $B$ be $C^*$-algebras, let $E$ be a $C^*$-module over $B$ with a fixed representation $A \to \mathcal{L}_B(E)$, and let $F$ be a $C^*$-module over $A$. Then one can form the internal tensor product $F \otimes_A E$, whose structure maps are given by

$$\langle \eta' \otimes_A \xi' | \eta \otimes_A \xi \rangle_{(F \otimes_A E)} = \langle \xi' | \langle \eta' | \eta \rangle_F \xi \rangle_E \quad \text{and} \quad (\eta \otimes_A \xi) b = \eta \otimes_A \xi b$$

for all $\eta, \eta' \in F$, $\xi, \xi' \in E$, and $b \in B$. It will be convenient to denote the internal tensor product by "$\circleddash$"; thus, for example, $F \circleddash E = F \otimes_A E$.

Similarly, we can construct a flipped internal tensor product $E \oslash F$: We equip the algebraic tensor product $E \odot F$ with the structure maps $\langle \xi' \odot \eta' | \xi \odot \eta \rangle := \langle \xi' | \langle \eta' | \eta \rangle_F \xi \rangle_E$ and $(\xi' \odot \eta') b := \xi' b \odot \eta'$, factor out the null space of the seminorm $\zeta \mapsto \| \langle \zeta | \zeta \rangle \|$, form the completion with respect to the induced norm on the quotient, and obtain a $C^*$-module over $B$ which we call the *flipped internal tensor product* $E \oslash F$. Thus, $E \oslash F$ is the closed linear span of elements $\xi \oslash \eta$, where $\xi \in E$ and $\eta \in F$, and

$$\langle \xi' \oslash \eta' | \xi \oslash \eta \rangle = \langle \xi' | \langle \eta' | \eta \rangle_F \xi \rangle_E \quad \text{and} \quad (\xi \oslash \eta) b = \xi b \oslash \eta$$

for all $\xi, \xi' \in E$, $\eta, \eta' \in F$, and $b \in B$. Evidently, we have a unitary map

$$\Sigma \colon F \circleddash E \xrightarrow{\cong} E \oslash F, \quad \eta \circleddash \xi \mapsto \xi \oslash \eta.$$

The flipped internal tensor product $E \oslash F$ can be interpreted as follows. The right $C^*$-$B$-module $E$ and the right $C^*$-$A$-module $F$ can be considered in a canonical way as a left $C^*$-$B^{\mathrm{op}}$-module and a left $C^*$-$A^{\mathrm{op}}$-module, respectively, and the representation of $A$ on $E$ can be considered as an antirepresentation of $A^{\mathrm{op}}$ on $E$. Then, the flipped internal tensor product $E \oslash F$ is just the natural internal tensor product of the left $C^*$-modules $E$ and $F$. However, we consider this internal tensor product as a right $C^*$-module again.

If we want to emphasize that the (flipped) internal tensor product $F \ominus E$ or $E \oslash F$ is formed with respect to a fixed representation $\pi\colon A \to \mathcal{L}_B(E)$, we denote these tensor products by $F \ominus_\pi E$ or $E_\pi \oslash F$, respectively.

### 11.1.2 Definition and examples

Recall that a multiplicative unitary on a Hilbert space $H$ is a unitary $V\colon H \otimes H \to H \otimes H$ that satisfies the pentagon equation $V_{[12]}V_{[13]}V_{[23]} = V_{[23]}V_{[12]}$. We extend this concept, replacing $H$ by a $C^*$-module $E$ with representations $\hat{\beta}, \beta$.

**Definition 11.1.1.** A *$C^*$-trimodule* $(E, \hat{\beta}, \beta)$ over $B$ is a full $C^*$-$B$-module $E$ with two commuting nondegenerate faithful representations $\hat{\beta}, \beta$ of $B$ on $E$.

Let $(E, \hat{\beta}, \beta)$ be a $C^*$-trimodule over $B$. Then we can form (flipped) internal tensor products $E \ominus_{\hat{\beta}} E$, $E \ominus_\beta E$ and $E_{\hat{\beta}} \oslash E$, $E_\beta \oslash E$. Roughly, a pseudo-multiplicative unitary on $E$ is a unitary

$$V\colon E_{\hat{\beta}} \oslash E \to E \ominus_\beta E$$

that satisfies the pentagon equation $V_{[12]}V_{[13]}V_{[23]} = V_{[23]}V_{[13]}$. Similarly as in the case of a pseudo-multiplicative unitary on a Hilbert space, the individual factors $V_{[ij]}$ act on iterated internal tensor products. Therefore, several intertwining conditions on $V$ and $\beta$, $\hat{\beta}$ have to be assumed to ensure that each factor $V_{[ij]}$ is well defined. Since $\beta$ and $\hat{\beta}$ commute, we can define representations

$$\beta_{[1]}, \hat{\beta}_{[1]}, \beta_{[2]} \text{ on } E \ominus_{\hat{\beta}} E, \quad \beta_{[1]}, \hat{\beta}_{[1]}, \hat{\beta}_{[2]} \text{ on } E \ominus_\beta E,$$
$$\beta_{[1]}, \beta_{[2]}, \hat{\beta}_{[2]} \text{ on } E_{\hat{\beta}} \oslash E, \quad \hat{\beta}_{[1]}, \hat{\beta}_{[2]}, \beta_{[2]} \text{ on } E_\beta \oslash E;$$

for example, the first three representations are given by

$$\beta_{[1]}(b) := \beta(b) \ominus 1, \quad \hat{\beta}_{[1]}(b) := \hat{\beta}(b) \ominus 1, \quad \beta_{[2]}(b) := 1 \ominus \beta(b)$$

for all $b \in B$.

**Lemma 11.1.2.** *Let $V \in \mathcal{L}_B(E_{\hat{\beta}} \oslash E, E \ominus_\beta E)$ and assume that for all $b \in B$,*

$$V\beta_{[2]}(b) = \hat{\beta}_{[1]}(b)V, \quad V\beta_{[1]}(b) = \beta_{[1]}(b)V, \quad V\hat{\beta}_{[2]}(b) = \hat{\beta}_{[2]}(b)V. \tag{11.1}$$

*Then all operators in the following diagram are well defined:*

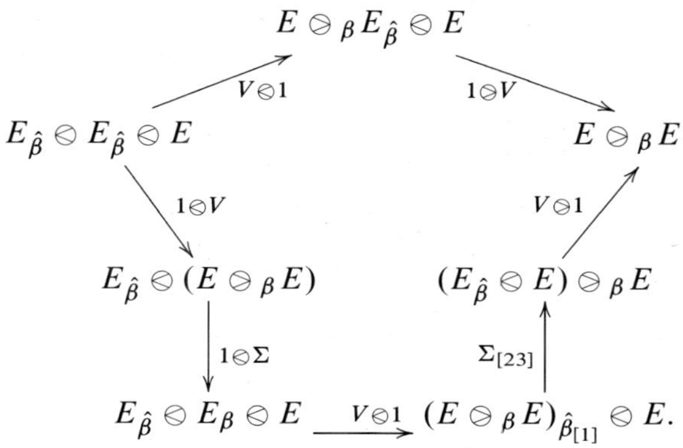


(11.2)

*Here,* $\Sigma_{[23]}$ *denotes the isomorphism given by* $(\xi_1 \otimes \xi_2) \otimes \xi_3 \mapsto (\xi_1 \otimes \xi_3) \otimes \xi_2$.

*Proof.* This follows from Proposition 12.2.1. □

We extend the leg notation to the operators occurring in diagram (11.2), and write

$$V_{[12]} \text{ for } V \otimes 1 \text{ and } V \otimes 1; \quad V_{[23]} \text{ for } 1 \otimes V \text{ and } 1 \otimes V;$$
$$V_{[13]} \text{ for } \Sigma_{[23]}(V \otimes 1)(1 \otimes \Sigma).$$

Then diagram (11.2) commutes if and only if $V_{[12]}V_{[13]}V_{[23]} = V_{[23]}V_{[12]}$.

**Definition 11.1.3.** Let $(E, \hat\beta, \beta)$ be a $C^*$-trimodule over $B$. We call a unitary $V \in \mathcal{L}_B(E_{\hat\beta} \otimes E,\ E \otimes_\beta E)$ *pseudo-multiplicative* if it satisfies equation (11.1) and if diagram (11.2) commutes.

Evidently, this definition generalizes the notion of a multiplicative unitary on Hilbert spaces and translates the concept of a pseudo-multiplicative unitary on a Hilbert space into the language of $C^*$-modules. For commutative $B$, Definition 11.1.3 subsumes the following special cases:

i) If $\beta(b)\xi = \xi b = \hat\beta(b)\xi$ for all $\xi \in E$ and $b \in B$, then $V$ is a continuous field of multiplicative unitaries as defined by Blanchard [17].

ii) If $\hat\beta(b)\xi = \xi b$ for all $\xi \in E$ and $b \in B$, then $V$ is a pseudo-multiplicative unitary in the sense of O'uchi [116].

**Remark 11.1.4.** Let $(E, \hat{\beta}, \beta)$ and $V: E_{\hat{\beta}} \otimes E \to E \otimes_{\beta} E$ be as in Definition 11.1.3. Then $(E, \beta, \hat{\beta})$ is a $C^*$-trimodule over $B$, and the unitary

$$V^{\mathrm{op}} := \Sigma V^* \Sigma \colon E_{\beta} \otimes E \xrightarrow{\Sigma} E \otimes_{\beta} E \xrightarrow{V^*} E_{\hat{\beta}} \otimes E \xrightarrow{\Sigma} E \otimes_{\hat{\beta}} E$$

is pseudo-multiplicative, as one can easily check. The unitary $V^{\mathrm{op}}$ is called the *opposite* of $V$.

Let us consider the fundamental example of a pseudo-multiplicative unitary, discussed already in [116] – the unitary associated to a locally compact groupoid. For background on groupoids and Haar systems, see Section 10.3.3.

**Example 11.1.5.** Let $G$ be a locally compact, second countable, Hausdorff groupoid with left Haar system $\lambda$. We denote its unit space by $G^0$, its range map by $r_G$, its source map by $s_G$, and put $G^u := r_G^{-1}(u)$ for each $u \in G^0$.

Let $B := C_0(G^0)$. Denote by $L^2(G, \lambda)$ the $C^*$-module over $B$ associated to $G$ and $\lambda$; this is the completion of the pre-$C^*$-module $C_c(G)$, whose structure maps are given by

$$\langle \xi' | \xi \rangle(u) = \int_{G^u} \overline{\xi'(x)} \xi(x) \, d\lambda^u(x) \quad \text{and} \quad (\xi f)(x) = \xi(x) f(r_G(x))$$

for all $u \in G^0$, $x \in G$, $\xi, \xi' \in C_c(G)$, $f \in B$. The $C^*$-module $L^2(G, \lambda)$ corresponds to a continuous field of Hilbert spaces [34], [38], [57] over $G^0$ with fiber $L^2(G^u, \lambda^u)$ at $u \in G^0$. Define representations $r, s \colon B \to \mathcal{L}_B(L^2(G, \lambda))$ by

$$(r(f)\xi)(x) := f(r_G(x))\xi(x) \quad \text{and} \quad (s(f)\xi)(x) := f(s_G(x))\xi(x)$$

for all $x \in G$, $\xi \in C_c(G)$, $f \in B$. Then $(E, \hat{\beta}, \beta) := (L^2(G, \lambda), s, r)$ is a $C^*$-trimodule over $B$.

The (flipped) internal tensor products $E_{\hat{\beta}} \otimes E$ and $E \otimes_{\beta} E$ can be described as follows. For $k = r, s$, put $G^2_{k,r} := \{(x, y) \in G \times G \mid k_G(x) = r_G(y)\}$. Consider $C_c(G^2_{s,r})$ and $C_c(G^2_{r,r})$ as pre-$C^*$-modules over $B$ via the structure maps

$$\langle \zeta' | \zeta \rangle(u) := \int_{G^u} \int_{G^{s_G(x)}} \overline{\zeta'(x, y)} \zeta(x, y) \, d\lambda^{s_G(x)}(y) d\lambda^u(x) \quad \text{for } C_c(G^2_{s,r}),$$

$$\langle \zeta' | \zeta \rangle(u) := \int_{G^u} \int_{G^u} \overline{\zeta'(x, y)} \zeta(x, y) \, d\lambda^u(y) d\lambda^u(x) \quad \text{for } C_c(G^2_{r,r}),$$

$$(\zeta f)(x, y) := \zeta(x, y) f(r_G(x)) \quad \text{for both,}$$

and denote by $L^2(G^2_{s,r})$ and $L^2(G^2_{r,r})$ the respective completions. Then we have natural isomorphisms $E_{\hat{\beta}} \otimes E \cong L^2(G^2_{s,r})$ and $E \otimes_{\beta} E \cong L^2(G^2_{r,r})$, similarly as in Lemma 10.3.20.

Now the map $W_0\colon C_c(G^2_{s,r}) \to C_c(G^2_{r,r})$ given by

$$(W_0\zeta)(x,y) := \zeta(x, x^{-1}y) \quad \text{for all } (x,y) \in G^2_{r,r},\ \zeta \in C_c(G^2_{s,r}),$$

extends to a pseudo-multiplicative unitary $W_G\colon E_{\hat{\beta}} \ominus E \to E \ominus_\beta E$. Indeed, the map $W_0$ is a linear isomorphism because it is the transpose of a homeomorphism $G^2_{r,r} \to G^2_{s,r}$, and it extends to a unitary because

$$\begin{aligned}\langle W_0\zeta | W_0\zeta\rangle(u) &= \int_{G^u}\int_{G^u} \overline{\zeta(x,x^{-1}y)}\zeta(x,x^{-1}y)\, d\lambda^u(y) d\lambda^u(x)\\ &= \int_{G^u}\int_{G^{s_G(x)}} \overline{\zeta(x,y')}\zeta(x,y')\, d\lambda^{s_G(x)}(y') d\lambda^u(x) = \langle\zeta|\zeta\rangle(u)\end{aligned}$$

for all $u \in G^0$ and $\zeta \in C_c(G^2_{s,r})$. To check that $W_G$ satisfies the pentagon equation, one can use a similar calculation as in the case where $G$ is a group, see Example 7.1.4.

The pseudo-multiplicative unitary $W_G$ is related to the pseudo-multiplicative unitary constructed in Section 10.3.3 as follows. Let $\mu$ be a quasi-invariant Radon measure on $G^0$. Denote also by $\mu$ the weight on $L^\infty(G^0,\mu)$ given by $f \mapsto \int_{G^0} f d\mu$, and consider the pseudo-multiplicative unitary

$$W_G^\mu\colon L^2(G,\nu)_s \underset{\mu}{\otimes}{}_r L^2(G,\nu) \to L^2(G,\nu)_r \underset{\mu}{\otimes}{}_r L^2(G,\nu)$$

constructed in Proposition 10.3.21. Let $B$ act on $L^2(G^0,\mu)$ by multiplication operators. Then

$$L^2(G,\nu)_k \underset{\mu}{\otimes}{}_r L^2(G,\nu) \cong L^2(G^2_{k,r}, \nu^2_{k,r}) \cong L^2(G^2_{k,r}) \ominus L^2(G^0,\mu)$$

for $k = s, r$, and these isomorphisms identify $W_G^\mu$ with $W_G \ominus \mathrm{id}_{L^2(G,\mu)}$, as one can easily check.

The following example is a $C^*$-algebraic analogue of a pseudo-multiplicative unitary on Hilbert spaces considered by Lesieur [99, Section 7.6].

**Example 11.1.6.** Let $B$ be a unital $C^*$-algebra, $C \subseteq Z(B)$ a $C^*$-subalgebra containing $1_B$, and $\tau\colon B \to C$ a faithful conditional expectation, that is, a faithful positive $C$-linear map such that $\tau|_C = \mathrm{id}_C$. We associate to $\tau$ a pseudo-multiplicative unitary $W_\tau$ as follows.

First, consider $B$ as a pre-$C^*$-module over $C$ via the inner product $\langle a'|a\rangle := \tau(a'^*a)$ and via right multiplication, and denote by $B_\tau$ the completion. Next, consider $B$ as a right $C^*$-$B$-$B$-bimodule in the natural way, and denote by $E := B_\tau \ominus B$ the internal tensor product over $C$. Thus $E$ is generated by elements $a \ominus b$, where $a, b \in B$, and $\langle a' \ominus b' | a \ominus b\rangle = b'^*\tau(a'^*a)b$, $(a \ominus b)b' = a \ominus bb'$ for all $a, b, a', b' \in B$.

Routine arguments show that there exist representations $\hat{\beta}, \beta \colon B \to \mathcal{L}_B(E)$ such that $\hat{\beta}(b')(a \ogreaterthan b) = b'a \ogreaterthan b$ and $\beta(b')(a \ogreaterthan b) = a \ogreaterthan b'b$ for all $a, b, b' \in B$; here, we use $\tau(B) \subseteq Z(B)$. Evidently, $(E, \hat{\beta}, \beta)$ is a $C^*$-trimodule.

We claim that there exist unitaries

$$X \colon E_{\hat{\beta}} \olessthan E \to B_\tau \ogreaterthan B_\tau \ogreaterthan B, \quad (a \ogreaterthan b) \olessthan (c \ogreaterthan d) \mapsto da \ogreaterthan c \ogreaterthan b,$$
$$Y \colon E \ogreaterthan_\beta E \to B_\tau \ogreaterthan B_\tau \ogreaterthan B, \quad (a \ogreaterthan b) \ogreaterthan (c \ogreaterthan d) \mapsto a \ogreaterthan c \ogreaterthan bd.$$

Indeed, for $x := (a \ogreaterthan b) \olessthan (c \ogreaterthan d)$ and $y := (a \ogreaterthan b) \ogreaterthan (c \ogreaterthan d)$ as above,

$$\begin{aligned} \|Xx\|^2 = \|\langle b|\langle c|\langle da|da\rangle c\rangle b\rangle\| &= \|b^*\tau(c^*\tau(a^*d^*da)c)b\| \\ &= \|b^*\tau(a^*d^*\tau(c^*c)da)b\| \\ &= \|\langle a \ogreaterthan b|\langle c \ogreaterthan d|c \ogreaterthan d\rangle a \ogreaterthan b\rangle\| = \|x\|^2 \end{aligned}$$

and

$$\begin{aligned} \|Yy\|^2 = \|\langle bd|\langle c|\langle a|a\rangle c\rangle bd\rangle\| &= \|d^*b^*\tau(c^*\tau(a^*a)c)bd\| \\ &= \|d^*\tau(c^*c)b^*\tau(a^*a)bd\| \\ &= \|\langle c \ogreaterthan d|c \ogreaterthan \langle a \ogreaterthan b|a \ogreaterthan b\rangle d\rangle\| = \|y\|^2; \end{aligned}$$

here, we use $\tau(B) \subseteq Z(B)$ and $\tau(e\tau(f)) = \tau(e)\tau(f)$ for $e, f \in B$. Now consider the unitary $W_\tau := Y^*X \colon E_{\hat{\beta}} \olessthan E \to E \ogreaterthan_\beta E$. Explicitly,

$$W_\tau((a \ogreaterthan b) \olessthan (c \ogreaterthan d)) = (da \ogreaterthan b) \ogreaterthan (c \ogreaterthan 1) \quad \text{for all } a, b, c, d \in B, \tag{11.3}$$

as can be seen from the relation

$$Y((da \ogreaterthan b) \ogreaterthan (c \ogreaterthan 1)) = da \ogreaterthan c \ogreaterthan b = X((a \ogreaterthan b) \olessthan (c \ogreaterthan d)).$$

The following calculations show that $W_\tau$ satisfies the intertwining conditions in equation (11.1) and that diagram (11.2) commutes: for $a, b, c, d, e, f, g \in B$,

$$\begin{array}{ccc} (a \ogreaterthan b) \olessthan (c \ogreaterthan d) & \xrightarrow{\beta_{[1]}(e)\beta_{[2]}(f)\hat{\beta}_{[2]}(g)} & (a \ogreaterthan eb) \olessthan (gc \ogreaterthan fd) \\ {\scriptstyle W_\tau}\downarrow & & \downarrow{\scriptstyle W_\tau} \\ (da \ogreaterthan b) \ogreaterthan (c \ogreaterthan 1) & \xrightarrow[\beta_{[1]}(e)\hat{\beta}_{[1]}(f)\hat{\beta}_{[2]}(g)]{} & (fda \ogreaterthan eb) \ogreaterthan (gc \ogreaterthan 1) \end{array}$$

and

$$\begin{array}{ccc} & (da \ogreaterthan b) \ogreaterthan (c \ogreaterthan 1) \olessthan (e \ogreaterthan f) & \\ {\scriptstyle W_{\tau[12]}} \nearrow & & \searrow {\scriptstyle W_{\tau[23]}} \\ (a \ogreaterthan b) \olessthan (c \ogreaterthan d) \olessthan (e \ogreaterthan f) & & (da \ogreaterthan b) \ogreaterthan (fc \ogreaterthan 1) \ogreaterthan (e \ogreaterthan 1) \\ {\scriptstyle W_{\tau[23]}}\downarrow & & \uparrow{\scriptstyle W_{\tau[12]}} \\ (a \ogreaterthan b) \olessthan ((fc \ogreaterthan d) \ogreaterthan (e \ogreaterthan 1)) & \xrightarrow{W_{\tau[13]}} & ((a \ogreaterthan b) \olessthan (fc \ogreaterthan d)) \ogreaterthan (e \ogreaterthan 1). \end{array}$$

### 11.1.3 Obstructions to the construction of the legs

Given a pseudo-multiplicative unitary $V: E_{\hat{\beta}} \otimes E \to E \otimes_{\beta} E$, it is natural to ask whether we can define a left and a right leg of $V$ in a similar way as it was done in Section 7.2 and Section 10.3.2 for every multiplicative unitary and every pseudo-multiplicative unitary on a Hilbert space. The constructions in Section 7.2 and Section 10.3.2 suggest the following plan:

1. Define $C^*$-algebras $\widehat{A}(V)$ and $A(V)$ of operators on $E$, spanned by compositions of the form $\hat{a}_{(\xi',\xi)} = [\xi'|_{[2]}V|\xi\rangle_{[2]}$ and $a_{(\eta',\eta)} = \langle\eta'|_{[1]}V|\eta]_{[1]}$, respectively, where $\xi$, $\xi'$ and $\eta$, $\eta'$ are suitable elements of $E$, and the maps $|\xi\rangle_{[2]}$, $|\xi']_{[2]} = ([\xi'|_{[2]})^*$ and $|\eta]_{[1]}$, $|\eta'\rangle_{[1]} = (\langle\eta'|_{[1]})^*$ have the form
$$|\xi\rangle_{[2]}: E \to E_{\hat{\beta}} \otimes E, \ \zeta \mapsto \zeta \otimes \xi, \quad |\xi']_{[2]}: E \to E \otimes_{\beta} E, \ \zeta \mapsto \zeta \otimes \xi',$$
$$|\eta]_{[1]}: E \to E_{\hat{\beta}} \otimes E, \ \zeta \mapsto \eta \otimes \zeta, \quad |\eta'\rangle_{[1]}: E \to E \otimes_{\beta} E, \ \zeta \mapsto \eta' \otimes \zeta.$$

2. Define fiber products $\widehat{A}(V) \otimes \widehat{A}(V)$ and $A(V) \otimes A(V)$ as $C^*$-algebras of operators on the $C^*$-module $E_{\hat{\beta}} \otimes E$ or $E \otimes_{\beta} E$, respectively.

3. Define comultiplications $\widehat{\Delta}: \widehat{A}(V) \to \widehat{A}(V) \otimes \widehat{A}(V)$ and $\Delta: A(V) \to A(V) \otimes A(V)$ via the formulas $\widehat{\Delta}(\hat{a}) = V^*(1 \otimes \hat{a})V$ and $\Delta(a) = V(a \otimes 1)V^*$, respectively.

If we try to carry out this plan, we encounter the following problems:

1. At a first attempt, we might try to define $\widehat{A}(V)$ and $A(V)$ as $C^*$-subalgebras of $\mathcal{L}_B(E)$. But then $\widehat{A}(V)$ and $A(V)$ would have to commute with the right multiplication of $B$ on $E$, and this excludes many examples:

   - For the pseudo-multiplicative unitary $W_G$ associated to a locally compact groupoid $G$ with left Haar system $\lambda$ (Example 11.1.5), the right leg $A(W_G)$ should correspond to the left regular representation of $G$ on $L^2(G, \lambda)$. But in general, this left regular representation, which is an integrated form of left multiplication on $G$, does not commute with the module multiplication, which is the transpose of the range map of $G$.
   - If $V$ were a pseudo-multiplicative unitary on a Hilbert space, then the right module multiplication would correspond to the representation $\alpha$, and Lemma 10.3.7 indicates that the legs of $V$ need not commute with $\alpha$.

   Therefore we need a substitute for the $C^*$-algebra $\mathcal{L}_B(E)$ that is large enough to accommodate $\widehat{A}(V)$ and $A(V)$.

   The fact that $\widehat{A}(V)$ and $A(V)$ can not be defined as $C^*$-subalgebras of $\mathcal{L}_B(E)$ is related to the problem that the maps $|\xi']_{[2]}$ and $|\eta]_{[1]}$ that we wanted to use for the definition of the operators $\hat{a}_{(\xi',\xi)}$ and $a_{(\eta',\eta)}$ are not well-behaved. Indeed, $|\xi']_{[2]}$ need not commute with the right module multiplication because

$$|\xi']_{[2]}(\zeta b) = \zeta b \ominus \xi' = \zeta \ominus \beta(b)\xi' \neq \zeta \ominus \xi' b = (\zeta \ominus \xi')b = (|\xi']_{[2]}\zeta)b$$

unless $\beta(b)\xi' = \xi' b$ for all $b \in B$, which is a severe restriction on the choice of $\xi'$. Likewise, $|\eta]_{[1]}$ is not adjointable unless $\hat{\beta}(b)\eta = \eta b$ for all $b \in B$.

2. It is not clear how to define $\widehat{A}(V) \ominus \widehat{A}(V)$ and $A(V) \ominus A(V)$. In general, these fiber products can not be spanned by operators of the form

$$\hat{a}_1 \ominus \hat{a}_2 \colon \eta \ominus \xi \mapsto \hat{a}_1\eta \ominus \hat{a}_2\xi \quad \text{and} \quad a_1 \ominus a_2 \colon \eta \ominus \xi \mapsto a_1\eta \ominus a_2\xi,$$

respectively, where $\hat{a}_1, \hat{a}_2 \in \widehat{A}(V)$ and $a_1, a_2 \in A(V)$, because such operators need not be well defined. If we put $\widehat{A}(V) \ominus \widehat{A}(V) := (\widehat{A}(V)' \ominus \widehat{A}(V)')'$ and $A(V) \ominus A(V) := (A(V)' \ominus A(V)')'$ as in the setting of von Neumann algebras (see Section 10.2.1), we obtain spaces that are too large. Note that it is also not clear in which ambient $C^*$-algebra the respective commutants should be taken.

**The strategy for a special case.** We can solve the problems outlined above and define the legs of a pseudo-multiplicative unitary on $C^*$-modules only in a special case, using the following strategy:

The problems listed in Step 1 originate from the fact that the operators involved are not necessarily adjointable. We consider operators on the $C^*$-module $E$ that are not strictly adjointable, but adjointable up to a twist by some partial automorphism $\sigma$ of $B$ (Section 11.2.1). We demand that these operators also commute with the representation $\beta$ or $\hat{\beta}$ up to a twist by some partial automorphism $\rho$ of $B$. Such operators form graded families $\mathcal{L}({}_{\hat{\beta}}E) = (\mathcal{L}^{\rho}_{\sigma}({}_{\hat{\beta}}E))_{\rho,\sigma}$ and $\mathcal{L}({}_{\beta}E) = (\mathcal{L}^{\rho}_{\sigma}({}_{\beta}E))_{\rho,\sigma}$, which carry an involution and are closed under multiplication and addition of homogeneous elements. Under suitable assumptions, we can then define the left and the right leg of $V$ in the form of subfamilies $\widehat{\mathcal{A}}(V) \subseteq \mathcal{L}({}_{\hat{\beta}}E)$ and $\mathcal{A}(V) \subseteq \mathcal{L}({}_{\beta}E)$ (Section 11.4).

The assumption that we need to impose is the following. The non-adjointability of the operators considered in Step 1 is related to the discrepancy between the right module multiplication on $E$ and the representations $\hat{\beta}$ and $\beta$. Similarly as for operators on $E$, we consider elements of $E$ that intertwine the representation $\hat{\beta}$ or $\beta$, respectively, and the right module multiplication up to a twist by some partial automorphism $\theta$ of $B$ (Section 11.2.2). These elements form families $\mathcal{H}({}_{\hat{\beta}}E) = (\mathcal{H}_{\theta}({}_{\hat{\beta}}E))_{\theta}$ and $\mathcal{H}({}_{\beta}E) = (\mathcal{H}_{\theta}({}_{\beta}E))_{\theta}$, and the assumption that we need to impose is that these families are linearly dense in $E$. If this condition is satisfied, we call the pseudo-multiplicative unitary $V$ *decomposable*.

## 11.2 Semigroup grading techniques on right $C^*$-bimodules

In this section, we develop a general calculus of homogeneous operators on $C^*$-bimodules and of homogeneous elements of $C^*$-bimodules. Moreover, we introduce the concept of a $C^*$-family, which can be thought of as a generalized $C^*$-algebra of homogeneous operators on a $C^*$-module. These concepts will be used in Section 11.4 to construct the legs of a decomposable pseudo-multiplicative unitary on a $C^*$-module.

Before we start, let us fix some terminology and notation.

Recall that a *partial automorphism* of a $C^*$-algebra $C$ is a $*$-isomorphism $\sigma\colon \mathrm{Dom}(\sigma) \to \mathrm{Im}(\sigma)$, where $\mathrm{Dom}(\sigma)$ and $\mathrm{Im}(\sigma)$ are closed ideals of $C$. Compositions and inverses of partial automorphisms are partial automorphisms again; thus the set $\mathrm{PAut}(C)$ of all partial automorphisms of $C$ forms an inverse semigroup. We denote the inverse of a partial automorphism $\sigma$ by $\sigma^*$. Let $\sigma, \sigma' \in \mathrm{PAut}(C)$. We say that $\sigma'$ extends $\sigma$ and write $\sigma' \geq \sigma$ if $\mathrm{Dom}(\sigma) \subseteq \mathrm{Dom}(\sigma')$ and $\sigma'|_{\mathrm{Dom}(\sigma)} = \sigma$. We put

$$\sigma \wedge \sigma' := \max\{\sigma'' \in \mathrm{PAut}(C) \mid \sigma'' \leq \sigma,\ \sigma'' \leq \sigma'\};$$

thus $\sigma \wedge \sigma' = \sigma|_I = \sigma'|_I$, where $I \subseteq \mathrm{Dom}(\sigma) \cap \mathrm{Dom}(\sigma')$ is the largest ideal on which $\sigma$ and $\sigma'$ coincide.

Throughout this section, let $A$ and $B$ be $C^*$-algebras.

**Definition 11.2.1.** A *right $C^*$-$A$-$B$-bimodule* is a $C^*$-module $E$ over $B$ with a fixed non-degenerate $*$-homomorphism $A \to \mathcal{L}_B(E)$. Given right $C^*$-$A$-$B$-bimodules $E$ and $F$, we put

$$\mathcal{L}_B^A(E, F) := \{T \in \mathcal{L}_B(E, F) \mid aT\xi = Ta\xi \text{ for all } a \in A,\ \xi \in E\}.$$

### 11.2.1 Homogeneous operators and $C^*$-families

We consider maps of right $C^*$-bimodules which almost preserve the bimodule structure:

**Definition 11.2.2.** Let $E, F$ be right $C^*$-$A$-$B$-bimodules and $\rho \in \mathrm{PAut}(A)$, $\sigma \in \mathrm{PAut}(B)$. We call a map $T\colon E \to F$ a *$(\rho, \sigma)$-homogeneous operator* if

i) $\mathrm{Im}(T) \subseteq \overline{\mathrm{span}}\, \mathrm{Im}(\rho) F$ and $Ta\xi = \rho(a)T\xi$ for all $a \in \mathrm{Dom}(\rho)$, $\xi \in E$, and

ii) there exists a map $S\colon F \to E$ such that $\langle SF|E\rangle \subseteq \mathrm{Dom}(\sigma)$ and $\langle \eta|T\xi\rangle = \sigma(\langle S\eta|\xi\rangle)$ for all $\xi \in E$, $\eta \in F$.

Let us collect some first properties of homogeneous operators:

**Proposition 11.2.3.** *Let $E$, $F$, $\rho$, $\sigma$ and $T$, $S$ be as above.*

i) *$T$ and $S$ are linear and bounded, and* $\|T\| = \|S\|$.

ii) $T(\xi b) = (T\xi)\sigma(b)$ *for all* $b \in \mathrm{Dom}(\sigma)$ *and* $\xi \in E$.

iii) *There exists a partial automorphism* $\sigma_0 \in \mathrm{PAut}(B)$ *such that whenever $T$ is $(\rho', \sigma')$-homogeneous for some* $\rho' \in \mathrm{PAut}(A)$ *and* $\sigma' \in \mathrm{PAut}(B)$, *then* $\sigma_0 \leq \sigma'$.

iv) *$S$ is uniquely determined by $T$.*

v) *Let $(u_\nu)_\nu$ and $(v_\mu)_\mu$ be approximate units of the ideals* $\mathrm{Dom}(\rho)$ *and* $\mathrm{Dom}(\sigma)$, *respectively. Then* $\lim_\nu T(u_\nu \xi) = T\xi = \lim_\mu T(\xi v_\mu)$ *for all* $\xi \in E$.

*Proof.* i) The proof is completely analogous to the case of ordinary adjointable operators on $C^*$-modules.

ii) This relation follows from the fact that for all $\eta, \xi \in E$ and $b \in \mathrm{Dom}(\sigma)$,

$$\langle \eta | T(\xi b)\rangle = \sigma(\langle S\eta|\xi b\rangle) = \sigma(\langle S\eta|\xi\rangle b) = \sigma(\langle S\eta|\xi\rangle)\sigma(b) = \langle \eta|(T\xi)\sigma(b)\rangle.$$

iii) Put $J := [\langle F|TE\rangle] \subseteq B$. Then $J$ is a subspace of $\mathrm{Im}(\sigma)$, and an ideal in $B$ because $BJ \subseteq [\langle FB|TE\rangle]$ and $J\,\mathrm{Im}(\sigma) \subseteq [\langle F|T(EB)\rangle]$ by ii). Denote by $\sigma_0$ the restriction of $\sigma$ to $\sigma^*(J)$. Assume that $T$ is also $(\rho', \sigma')$-homogeneous for some $\rho' \in \mathrm{PAut}(A)$ and $\sigma' \in \mathrm{PAut}(B)$, and that $S'$ satisfies condition ii) in Definition 11.2.2 for $T$ and $(\rho', \sigma')$. Then

$$\sigma(\langle S\eta|\xi\rangle b) = \langle \eta|T(\xi b)\rangle = \sigma'(\langle S'\eta|\xi\rangle b) \quad \text{for all } \eta, \xi \in E,\ b \in B,$$

and hence $\sigma(\sigma^*(a)b) = \sigma'(\sigma'^*(a)b)$ for all $a \in J$, $b \in B$. Let $(u_\nu)_\nu$ be an approximate unit for $J$ and $d \in J$. The last relation and the inclusion $J \subseteq \mathrm{Im}(\sigma')$ imply

$$d = \lim_\nu \sigma'(\sigma'^*(d)\sigma'^*(u_\nu)) = \lim_\nu \sigma(\sigma^*(d)\sigma'^*(u_\nu)).$$

Therefore, $\sigma^*(d) = \lim_\nu \sigma^*(d)\sigma'^*(u_\nu) \in \sigma'^*(J)$. Now

$$d = \lim_\nu \sigma(\sigma^*(u_\nu)\sigma^*(d)) = \lim_\nu \sigma'(\sigma'^*(u_\nu)\sigma^*(d)) = \sigma'(\sigma^*(d)).$$

Consequently, $\sigma_0 \leq \sigma'$.

iv) Similarly as in the case of ordinary adjointable operators, one finds that $S$ is uniquely determined by $T$ and $\sigma$. But by ii), $S$ is independent of $\sigma$.

v) This follows from standard arguments. □

The preceding proposition justifies the following definition:

**Definition 11.2.4.** Let $E$, $F$ be right $C^*$-$A$-$B$-bimodules and let $T\colon E \to F$ be a $(\rho, \sigma)$-homogeneous operator, where $\rho \in \mathrm{PAut}(A)$, $\sigma \in \mathrm{PAut}(B)$. Then the map $S\colon F \to E$ in Definition 11.2.2 ii) is called the *adjoint* of $T$ and denoted by $T^*$.

For later use, we note the following simple example:

**Example 11.2.5.** Let $\rho \in \mathrm{PAut}(B)$, and consider $\mathrm{Dom}(\rho), \mathrm{Im}(\rho) \subseteq B$ as right sub-$C^*$-bimodules of $B$. Then $\rho \in \mathcal{L}_\rho^\rho(\mathrm{Dom}(\rho), \mathrm{Im}(\rho))$. Indeed, condition i) in Definition 11.2.2 is easily checked, and for condition ii), note that for all $b \in \mathrm{Dom}(\rho)$ and $c \in \mathrm{Im}(\rho)$,

$$\langle c|\rho(b)\rangle = c^*\rho(b) = \rho(\rho^*(c^*)b) = \rho(\langle \rho^*(c)|b\rangle).$$

**Remark 11.2.6.** In the situation of Definition 11.2.2, one can consider

$$E_{(\rho,\sigma)} := [\mathrm{Dom}(\rho)E\,\mathrm{Dom}(\sigma)] \subseteq E \quad \text{and} \quad F^{(\rho,\sigma)} := [\mathrm{Im}(\rho)F\,\mathrm{Im}(\sigma)] \subseteq F$$

as right $C^*$-$\mathrm{Dom}(\rho)$-$\mathrm{Dom}(\sigma)$-bimodules, where the structure maps of $E_{(\rho,\sigma)}$ are inherited from $E$, and the structure maps of $F^{(\rho,\sigma)}$ are twisted by $\rho$ and $\sigma$ in a straightforward way. Then every $(\rho,\sigma)$-homogeneous operator $T\colon E \to F$ restricts to an operator $T_{(\rho,\sigma)} \in \mathcal{L}_{\mathrm{Dom}(\sigma)}^{\mathrm{Dom}(\rho)}(E_{(\rho,\sigma)}, F^{(\rho,\sigma)})$, whose adjoint is a restriction of $T^*$.

The preceding remark shows that homogeneous operators generalize ordinary operators on right $C^*$-bimodules only slightly. The point is that we shall consider entire families of homogeneous operators:

**Notation 11.2.7.** Let $E$, $F$ be right $C^*$-$A$-$B$-bimodules and $\rho \in \mathrm{PAut}(A)$, $\sigma \in \mathrm{PAut}(B)$. We denote the set of all $(\rho,\sigma)$-homogeneous operators from $E$ to $F$ by $\mathcal{L}_\sigma^\rho(E,F)$ and put $\mathcal{L}_\sigma^\rho(E) := \mathcal{L}_\sigma^\rho(E,E)$. The *strict topology* on $\mathcal{L}_\sigma^\rho(E,F)$ is the topology given by the family of seminorms $T \mapsto \|T\xi\|$ and $T \mapsto \|T^*\eta\|$, where $\xi \in E$ and $\eta \in F$. Finally, we put $\mathcal{L}(E,F) := (\mathcal{L}_\sigma^\rho(E,F))_{\rho,\sigma}$ and $\mathcal{L}(E) := (\mathcal{L}_\sigma^\rho(E))_{\rho,\sigma}$

The family of all homogeneous operators has the following properties:

**Proposition 11.2.8.** *Let $E$, $F$, $G$ be right $C^*$-$A$-$B$-bimodules and $\rho, \rho' \in \mathrm{PAut}(A)$, $\sigma, \sigma' \in \mathrm{PAut}(B)$.*

i) *$\mathcal{L}_\sigma^\rho(E,F)$ is a closed subspace of the space of all bounded linear maps from $E$ to $F$, and complete with respect to the strict topology.*

ii) *$\mathcal{L}_{\sigma'}^{\rho'}(F,G)\mathcal{L}_\sigma^\rho(E,F) \subseteq \mathcal{L}_{\sigma'\sigma}^{\rho'\rho}(E,G)$.*

iii) *$\mathcal{L}_\sigma^\rho(E,F)^* = \mathcal{L}_{\sigma^*}^{\rho^*}(F,E)$, and $(\lambda T)^* = \bar{\lambda}T^*$, $\|T^*\| = \|T\| = \|T^*T\|^{1/2}$, $(ST)^* = T^*S^*$ for all $\lambda \in \mathbb{C}$, $T \in \mathcal{L}_\sigma^\rho(E,F)$, $S \in \mathcal{L}_{\sigma'}^{\rho'}(F,G)$.*

iv) *$\mathcal{L}_{\mathrm{id}}^{\mathrm{id}}(E,F) = \mathcal{L}_B^A(E,F)$, and for each pair of partial identities $\epsilon' \in \mathrm{PAut}(A)$, $\epsilon \in \mathrm{PAut}(B)$, the space $\mathcal{L}_\epsilon^{\epsilon'}(E)$ is a $C^*$-subalgebra of $\mathcal{L}_B^A(E)$.*

v) $\mathcal{L}^{\rho}_{\sigma}(E,F)$ *is a right* $C^*$-$\mathcal{L}^{\rho\rho^*}_{\sigma\sigma^*}(F)$-$\mathcal{L}^{\rho^*\rho}_{\sigma^*\sigma}(E)$*-bimodule.*

vi) $\mathcal{L}^{\rho}_{\sigma}(E,F) \subseteq \mathcal{L}^{\rho'}_{\sigma'}(E,F)$ *if* $\rho \leq \rho'$ *and* $\sigma \leq \sigma'$.

*Proof.* Most of these assertions generalize facts about ordinary operators on right $C^*$-bimodules and can be proved in a similar way by the help of Proposition 11.2.3. Therefore we only prove ii). Let $T \in \mathcal{L}^{\rho}_{\sigma}(E,F)$, $T' \in \mathcal{L}^{\rho'}_{\sigma'}(F,G)$. By Definition 11.2.2 i) and Proposition 11.2.3 v),

$$[T'TE] \subseteq [T' \operatorname{Im}(\rho)F] \subseteq [\rho'(\operatorname{Dom}(\rho') \cap \operatorname{Im}(\rho))G] = [\operatorname{Im}(\rho'\rho)G]$$

and $T'Tb\xi = \rho'(\rho(b))T'T\xi$ for all $b \in \operatorname{Dom}(\rho'\rho)$, $\xi \in E$. Moreover, by Definition 11.2.2 ii) and Proposition 11.2.3 v), $\langle T'^*G|TE\rangle \subseteq \operatorname{Dom}(\sigma') \cap \operatorname{Im}(\sigma)$ and

$$\langle T^*T'^*G|E\rangle = \sigma^*(\langle T'^*G|TE\rangle) \subseteq \sigma^*(\operatorname{Dom}(\sigma') \cap \operatorname{Im}(\sigma)) = \operatorname{Dom}(\sigma'\sigma).$$

Finally, $\langle \eta|T'T\xi\rangle = \sigma'(\langle T'^*\eta|T\xi\rangle) = (\sigma'\sigma)(\langle T^*T'^*\eta|\xi\rangle)$ for all $\xi \in E, \eta \in G$. Therefore, $T'T \in \mathcal{L}^{\rho'\rho}_{\sigma'\sigma}(E,G)$ and $(T'T)^* = T^*T'^*$. □

We adopt the following notation:

**Notation 11.2.9.** Let $E$, $F$ be right $C^*$-$A$-$B$-bimodules, and let $\mathcal{C} = (\mathcal{C}^{\rho}_{\sigma})_{\rho,\sigma}$ be a family of closed subspaces $\mathcal{C}^{\rho}_{\sigma} \subseteq \mathcal{L}^{\rho}_{\sigma}(E,F)$, where $\rho \in \operatorname{PAut}(A)$, $\sigma \in \operatorname{PAut}(B)$.

- Given a family $\mathcal{D} = (\mathcal{D}^{\rho}_{\sigma})_{\rho,\sigma}$ of closed subspaces $\mathcal{D}^{\rho}_{\sigma} \subseteq \mathcal{L}^{\rho}_{\sigma}(E,F)$, we write $\mathcal{D} \subseteq \mathcal{C}$ if and only if $\mathcal{D}^{\rho}_{\sigma} \subseteq \mathcal{C}^{\rho}_{\sigma}$ for all $\rho \in \operatorname{PAut}(A)$, $\sigma \in \operatorname{PAut}(B)$.
- We define a family $\mathcal{C}^* \subseteq \mathcal{L}(F,E)$ by $(\mathcal{C}^*)^{\rho}_{\sigma} := (\mathcal{C}^{\rho^*}_{\sigma^*})^*$ for all $\rho, \sigma$.
- We put $[\mathcal{C}E] := \overline{\operatorname{span}}\{T\xi \mid T \in \mathcal{C}^{\rho}_{\sigma},\ \rho \in \operatorname{PAut}(A),\ \sigma \in \operatorname{PAut}(B),\ \xi \in E\}$.
- Let $G$ be a right $C^*$-$A$-$B$-bimodule and $\mathcal{D} \subseteq \mathcal{L}(F,G)$ a family of closed subspaces. The product $[\mathcal{D}\mathcal{C}] \subseteq \mathcal{L}(E,G)$ is the family given by

  $$[\mathcal{D}\mathcal{C}]^{\rho''}_{\sigma''} := \overline{\operatorname{span}}\{T'T \mid T' \in \mathcal{D}^{\rho'}_{\sigma'},\ T \in \mathcal{C}^{\rho}_{\sigma},\ \rho'\rho \leq \rho'',\ \sigma'\sigma \leq \sigma''\}$$

  for all $\rho'' \in \operatorname{PAut}(A)$, $\sigma'' \in \operatorname{PAut}(B)$. Clearly, the product $(\mathcal{D},\mathcal{C}) \mapsto [\mathcal{D}\mathcal{C}]$ is associative.

  Similarly, we define families $[\mathcal{D}T], [S\mathcal{C}] \subseteq \mathcal{L}(E,G)$ for operators $T \in \mathcal{L}^{\rho}_{\sigma}(E,F)$ and $S \in \mathcal{L}^{\rho'}_{\sigma'}(F,G)$, where $\rho, \rho' \in \operatorname{PAut}(A)$, $\sigma, \sigma' \in \operatorname{PAut}(B)$.
- By a slight abuse of notation, we define a family $\mathcal{L}^{\mathrm{id}}(E,F) \subseteq \mathcal{L}(E,F)$ by

  $$(\mathcal{L}^{\mathrm{id}}(E,F))^{\mathrm{id}}_{\sigma} := \mathcal{L}^{\mathrm{id}}_{\sigma}(E,F), \quad (\mathcal{L}^{\mathrm{id}}(E,F))^{\rho}_{\sigma} := 0 \text{ for } \rho \neq \mathrm{id}\,.$$

  Similarly, we define a family $\mathcal{L}_{\mathrm{id}}(E,F) \subseteq \mathcal{L}(E,F)$.

In the notation introduced above, we have $[\mathcal{L}(F,G)\mathcal{L}(E,F)] \subseteq \mathcal{L}(E,G)$ and $\mathcal{L}(E,F)^* = \mathcal{L}(F,E)$ for all right $C^*$-$A$-$B$-bimodules $E, F, G$.

**Definition 11.2.10.** Let $E$ be a right $C^*$-$A$-$B$-bimodule. We call a family of closed subspaces $\mathcal{C} \subseteq \mathcal{L}(E)$ a *$C^*$-family on $E$* if $[\mathcal{C}\mathcal{C}] \subseteq \mathcal{C}$, $\mathcal{C}^* \subseteq \mathcal{C}$, and $\mathcal{C}^{\rho_1}_{\sigma_1} \subseteq \mathcal{C}^{\rho_2}_{\sigma_2}$ whenever $\rho_1 \le \rho_2$ and $\sigma_1 \le \sigma_2$. We call a $C^*$-family $\mathcal{C}$ *non-degenerate* if $[\mathcal{C}E] = E$.

**Remarks 11.2.11.** Let $\mathcal{C}$ be a $C^*$-family on a right $C^*$-$A$-$B$-bimodule $E$.

i) For each pair of partial identities $\epsilon' \in \mathrm{PAut}(A)$ and $\epsilon \in \mathrm{PAut}(B)$, the space $\mathcal{C}^{\epsilon'}_{\epsilon} \subseteq \mathcal{L}^{\mathrm{id}}_{\mathrm{id}}(E) = \mathcal{L}^{A}_{B}(E)$ is a $C^*$-algebra because

$$(\mathcal{C}^{\epsilon'}_{\epsilon})^* = \mathcal{C}^{\epsilon'^*}_{\epsilon^*} = \mathcal{C}^{\epsilon'}_{\epsilon} \quad \text{and} \quad \mathcal{C}^{\epsilon'}_{\epsilon}\,\mathcal{C}^{\epsilon'}_{\epsilon} \subseteq \mathcal{C}^{\epsilon'\epsilon'}_{\epsilon\epsilon} = \mathcal{C}^{\epsilon'}_{\epsilon}.$$

ii) For each $\rho \in \mathrm{PAut}(A)$ and $\sigma \in \mathrm{PAut}(B)$, the space $\mathcal{C}^{\rho}_{\sigma}$ is a $C^*$-module over the $C^*$-algebra $\mathcal{C}^{\rho^*\rho}_{\sigma^*\sigma}$ because

$$(\mathcal{C}^{\rho}_{\sigma})^*\,\mathcal{C}^{\rho}_{\sigma} = \mathcal{C}^{\rho^*}_{\sigma^*}\,\mathcal{C}^{\rho}_{\sigma} \subseteq \mathcal{C}^{\rho^*\rho}_{\sigma^*\sigma} \quad \text{and} \quad \mathcal{C}^{\rho}_{\sigma}\,\mathcal{C}^{\rho^*\rho}_{\sigma^*\sigma} \subseteq \mathcal{C}^{\rho\rho^*\rho}_{\sigma\sigma^*\sigma} = \mathcal{C}^{\rho}_{\sigma}.$$

Likewise, $\mathcal{C}^{\rho}_{\sigma}$ is a left $C^*$-module over the $C^*$-algebra $\mathcal{C}^{\rho\rho^*}_{\sigma\sigma^*}$ and a $C^*$-bimodule over $\mathcal{C}^{\rho\rho^*}_{\sigma\sigma^*}$ and $\mathcal{C}^{\rho^*\rho}_{\sigma^*\sigma}$.

iii) $[\mathcal{C}^{\mathrm{id}}_{\mathrm{id}}\mathcal{C}^{\rho}_{\sigma}] = \mathcal{C}^{\rho}_{\sigma} = [\mathcal{C}^{\rho}_{\sigma}\,\mathcal{C}^{\mathrm{id}}_{\mathrm{id}}]$ for each $\rho, \sigma$; this follows from ii) (see Section 12.2).

iv) The $C^*$-family $\mathcal{C}$ is non-degenerate if and only if the $C^*$-algebra $\mathcal{C}^{\mathrm{id}}_{\mathrm{id}} \subseteq \mathcal{L}^{A}_{B}(E)$ is non-degenerate in the usual sense. This follows easily from iii).

To every $C^*$-family, one can associate a multiplier $C^*$-family:

**Definition 11.2.12.** Let $\mathcal{C}$ be a $C^*$-family on a right $C^*$-$A$-$B$-bimodule $E$. The *multiplier family* of $\mathcal{C}$ is the family $\mathcal{M}(\mathcal{C}) \subseteq \mathcal{L}(E)$ given by

$$\mathcal{M}(\mathcal{C})^{\rho}_{\sigma} = \{T \in \mathcal{L}^{\rho}_{\sigma}(E) \mid [T\mathcal{C}] \subseteq \mathcal{C} \text{ and } [\mathcal{C}T] \subseteq \mathcal{C}\}$$

for all $\rho \in \mathrm{PAut}(A)$ and $\sigma \in \mathrm{PAut}(B)$. Evidently, this is a $C^*$-family.

**Remark 11.2.13.** Remark 11.2.11 iii) implies that for all $\rho \in \mathrm{PAut}(A)$, $\sigma \in \mathrm{PAut}(B)$,

$$\mathcal{M}(\mathcal{C})^{\rho}_{\sigma} = \{T \in \mathcal{L}^{\rho}_{\sigma}(E) \mid T\mathcal{C}^{\mathrm{id}}_{\mathrm{id}} \subseteq \mathcal{C}^{\rho}_{\sigma} \text{ and } \mathcal{C}^{\mathrm{id}}_{\mathrm{id}}T \subseteq \mathcal{C}^{\rho}_{\sigma}\}.$$

In Section 11.4, we shall construct for every decomposable regular pseudo-multiplicative unitary a left and a right leg in the form of $C^*$-families. In principle, we could embed these $C^*$-families in some ambient $C^*$-algebras, which would bring us back into more familiar terrain. However, until recently, it was not clear how to define a fiber product, which is necessary to define the analogue of a Hopf–von Neumann bimodule, on the level of the ambient $C^*$-algebras. On the level of $C^*$-families, an internal tensor will be defined in Section 11.3 quite easily. For recent developments in the setting of $C^*$-algebras, see [154], [155], [156].

### 11.2.2 Homogeneous elements of right $C^*$-bimodules

We consider elements of right $C^*$-bimodules that almost intertwine left and right multiplication:

**Definition 11.2.14.** Let $E$ be a right $C^*$-$B$-$B$-bimodule and $\theta \in \mathrm{PAut}(B)$. An element $\xi \in E$ is *$\theta$-homogeneous* if $\xi \in [E\,\mathrm{Dom}(\theta)]$ and $\xi b = \theta(b)\xi$ for all $b \in \mathrm{Dom}(\theta)$. The set of all $\theta$-homogeneous elements of $E$ is denoted by $\mathcal{H}_\theta(E)$. We call $E$ *decomposable* if the family $\mathcal{H}(E) := (\mathcal{H}_\theta(E))_\theta$ is linearly dense in $E$.

Note that $B$ can be regarded as a $C^*$-module over $B$ in a natural way, and that left multiplication turns $B$ into a right $C^*$-$B$-$B$-bimodule. Thus we can speak of homogeneous elements of $B$; these will be studied later.

Let $E$ be a right $C^*$-$B$-$B$-bimodule. For each $\xi \in E$, we define maps

$$|\xi\rangle\colon B \to E,\ b \mapsto \xi b, \quad |\xi]\colon B \to E,\ b \mapsto b\xi.$$

Then $|\xi\rangle$ has an adjoint $\langle\xi| = |\xi\rangle^*\colon \eta \mapsto \langle\xi|\eta\rangle$, and $\||\xi\rangle\| = \|\xi\|$ [95, p. 12–13].

**Proposition 11.2.15.** *Let $\xi \in E$ and $\theta \in \mathrm{PAut}(B)$. Then the following conditions are equivalent:*

$$\text{i) } \xi \in \mathcal{H}_\theta(E), \qquad \text{ii) } |\xi\rangle \in \mathcal{L}^{\theta}_{\mathrm{id}}(B,E), \qquad \text{iii) } |\xi] \in \mathcal{L}^{\mathrm{id}}_{\theta*}(B,E).$$

*If* i)–iii) *hold, then* $\||\xi]\| = \|\xi\|$ *and* $[\xi| := |\xi]^*$ *is given by* $\eta \mapsto \theta(\langle\xi|\eta\rangle)$.

*Proof.* i) $\Rightarrow$ ii), iii): Assume that i) holds. To prove ii), we only need to show that $|\xi\rangle$ satisfies condition i) of Definition 11.2.2. But by assumption, $\mathrm{Im}\,|\xi\rangle \subseteq [\mathrm{Im}(\theta)E]$ and $|\xi\rangle(bb') = \xi bb' = \theta(b)|\xi\rangle b'$ for all $b \in \mathrm{Dom}(\theta)$, $b' \in B$. Let us prove iii). Evidently, $|\xi]$ commutes with left multiplication. By assumption, $\langle\xi|\eta\rangle \in \mathrm{Dom}(\theta)$ for all $\eta \in E$, so that the map $[\xi|\colon \eta \mapsto \theta(\langle\xi|\eta\rangle)$ is well defined. Let $(u_\nu)_\nu$ be an approximate unit of $\mathrm{Im}(\theta)$. Then

$$\langle\eta||\xi]b\rangle = \lim_\nu \langle\eta|u_\nu b\xi\rangle = \lim_\nu \langle\eta|\xi\rangle\theta^*(u_\nu b) = \theta^*(\theta(\langle\xi|\eta\rangle)^* b) = \theta^*(\langle[\xi|\eta|b\rangle)$$

for all $\eta \in E$, $b \in B$. Hence iii) holds. Moreover, we may assume $\|u_\nu\| \le 1$ for all $\nu$, and then $\|\xi\| = \lim_\nu \||\xi]u_\nu\| \le \||\xi]\|$. The reverse inequality is evident.

ii) $\Rightarrow$ i): If ii) holds, then $\xi \in [\xi B] = [\mathrm{Im}\,|\xi\rangle] \subseteq [\mathrm{Im}(\theta)E]$, and $\xi c = \theta(c)\xi$ for each $c \in \mathrm{Dom}(\theta)$ because $\xi cb = |\xi\rangle cb = \theta(c)(|\xi\rangle b) = \theta(c)\xi b$ for each $b \in B$.

iii) $\Rightarrow$ i): This follows from a similar argument like the implication "ii) $\Rightarrow$ i)". □

Let $E$ be a $C^*$-module over $A$ and $F$ a right $C^*$-$A$-$B$-bimodule. For each $\eta \in E$ and $\xi \in F$, we define maps

$$|\eta\rangle_{[1]} \colon F \to E \otimes F,\ \zeta \mapsto \eta \otimes \zeta, \quad |\xi]_{[2]} \colon E \to E \otimes F,\ \zeta \mapsto \zeta \otimes \xi.$$

Then $|\eta\rangle_{[1]}$ has an adjoint $\langle\eta|_{[1]} = |\eta\rangle_{[1]}^* \colon \zeta \otimes \zeta' \mapsto \langle\eta|\zeta\rangle\zeta'$, and $\||\xi\rangle\| = \|\xi\|$ if the representation $A \to \mathcal{L}_B(F)$ is injective [95, Lemma 4.6].

**Proposition 11.2.16.** *Let $E$, $F$ be right $C^*$-$B$-$B$-bimodules and $\theta \in \mathrm{PAut}(B)$.*

i) *If $\eta \in \mathcal{H}_\theta(E)$, then $|\eta\rangle_{[1]} \in \mathcal{L}_{\mathrm{id}}^\theta(F, E \otimes F)$.*

ii) *Let $\xi \in \mathcal{H}_\theta(F)$. Then $|\xi]_{[2]} \in \mathcal{L}_{\theta^*}^{\mathrm{id}}(E, E \otimes F)$, and $[\xi|_{[2]} := |\xi]_{[2]}{}^*$ is given by $\zeta \otimes \zeta' \mapsto \zeta\theta(\langle\xi|\zeta'\rangle)$. If $E$ is full, then $\||\xi]\| = \|\xi\|$.*

*Proof.* The proof is similar to that of Proposition 11.2.15; we only sketch the main steps for ii). Let $\xi \in \mathcal{H}_\theta(F)$. For all $\zeta, \zeta' \in E$ and $\xi' \in F$,

$$\langle\zeta' \otimes \xi'|\zeta \otimes \xi\rangle = \langle\xi'|\langle\zeta'|\zeta\rangle\xi\rangle = \theta^*(\theta(\langle\xi'|\xi\rangle)\langle\zeta'|\zeta\rangle) = \theta^*(\langle\zeta'\theta(\langle\xi|\xi'\rangle)|\zeta\rangle).$$

For $\zeta' = \zeta$, $\xi' = \xi$, this equation shows $\||\xi]_{[2]}\zeta\|^2 \leq \|\theta(\langle\xi|\xi\rangle)\|\|\zeta\|^2$, and hence $\||\xi]_{[2]}\| \leq \|\xi\|$. If $E$ is full, this inequality is an equality. Finally, the equation above shows that the formula for $[\xi|_{[2]}$ defines a bounded map $E \otimes F \to E$, and that $\langle\zeta' \otimes \xi'||\xi]_{[2]}\zeta\rangle = \theta^*(\langle[\xi|_{[2]}(\zeta' \otimes \xi')|\zeta\rangle)$ for all $\zeta, \zeta' \in E$ and $\xi' \in F$. □

Next, we collect several useful formulas concerning homogeneous elements. Let $E$ and $F$ be right $C^*$-$B$-$B$-bimodules. For $\theta, \theta' \in \mathrm{PAut}(B)$, put

$$\mathcal{H}_\theta(E) \otimes \mathcal{H}_{\theta'}(F) := \overline{\mathrm{span}}\{\eta \otimes \xi \mid \eta \in \mathcal{H}_\theta(E),\ \xi \in \mathcal{H}_{\theta'}(F)\} \subseteq E \otimes F.$$

**Proposition 11.2.17.** *Let $\theta, \theta', \sigma, \rho \in \mathrm{PAut}(B)$. Then:*

i) $\mathcal{H}_\theta(E) = [\mathcal{H}_{\theta'}(E)\,\mathrm{Dom}(\theta)] \subseteq \mathcal{H}_{\theta'}(E)$ *if $\theta \leq \theta'$.*

ii) $\langle\mathcal{H}_\theta(E)|\,\mathcal{H}_{\theta'}(E)\rangle \subseteq \mathcal{H}_{\theta^*\theta'}(B)$.

iii) *For each $\xi \in E$, the set $\{\theta' \in \mathrm{PAut}(B) \mid \xi \in \mathcal{H}_{\theta'}(E)\}$ either is empty or has a minimal element.*

iv) *$\mathcal{L}_\sigma^\rho(E, F)\,\mathcal{H}_\theta(E) \subseteq \mathcal{H}_{\rho\theta\sigma^*}(F)$ and $\mathcal{H}_\rho(A)\,\mathcal{H}_\theta(E)\,\mathcal{H}_\sigma(B) \subseteq \mathcal{H}_{\rho\theta\sigma}(E)$.*

v) *The space $I_\theta := [\langle\mathcal{H}_\theta(E)|\,\mathcal{H}_\theta(E)\rangle]$ is an ideal in $Z(B)$, and $\mathcal{H}_\theta(E)$ is a right $C^*$-$Z(B)$-$I_\theta$-bimodule. In particular, $\mathcal{H}_\theta(E)I_\theta = \mathcal{H}_\theta(E)$.*

vi) *If $E$ is full and decomposable, then $B$ is decomposable and the ideal of $Z(B)$ spanned by all $I_{\theta''}$, where $\theta'' \in \mathrm{PAut}(B)$, is non-degenerate in $B$.*

vii) $\mathcal{H}_\theta(E) \cap \mathcal{H}_{\theta'}(E) = \mathcal{H}_{(\theta\wedge\theta')}(E)$.

viii) $\mathcal{H}_\theta(E) \otimes \mathcal{H}_{\theta'}(F) \subseteq \mathcal{H}_{\theta\theta'}(E \otimes F)$.

*Proof.* We only prove assertions iii), iv), v), iiv); the others follow from straightforward calculations or can be deduced from Proposition 11.2.8 and 11.2.15.

iii) Given $\xi \in E$, apply Propositions 11.2.3 iii) and 11.2.15 to $|\xi]$.

iv) Let $T \in \mathcal{L}^\rho_\sigma(E, F)$ and $\xi \in \mathcal{H}_\theta(E)$. Choose approximate units $(u_\kappa)_\kappa$, $(v_\mu)_\mu$, $(w_\nu)_\nu$ of $\mathrm{Dom}(\rho)$, $\mathrm{Im}(\theta)$, $\mathrm{Dom}(\sigma)$, respectively. By Proposition 11.2.3,

$$T\xi = \lim_{\kappa,\mu,\nu} T(u_\kappa v_\mu \xi w_\nu) = \lim_{\kappa,\mu,\nu} T(\xi\theta^*(u_\kappa v_\mu)w_\nu) = \lim_{\kappa,\mu,\nu} (T\xi)\sigma(\theta^*(u_\kappa v_\mu)w_\nu).$$

Since $(\sigma(\theta^*(u_\kappa v_\mu)w_\nu))_{\kappa,\mu,\nu}$ is an approximate unit for $\mathrm{Dom}(\rho\theta\sigma^*)$, the equation above implies $T\xi \in [F\,\mathrm{Dom}(\rho\theta\sigma^*)]$. Moreover, for all $b \in \mathrm{Dom}(\rho\theta\sigma^*)$,

$$(T\xi)b = T(\xi\sigma^*(b)) = T\theta(\sigma^*(b))\xi = ((\rho\theta\sigma^*)(b))T\xi.$$

Therefore, $T\xi \in \mathcal{H}_{\rho\theta\sigma^*}(F)$. The second inclusion in assertion iv) follows similarly.

v) The assumptions imply that $B$ is contained in the closure of

$$\sum_{\theta,\theta'} \langle \mathcal{H}_{\theta'}(E) | \mathcal{H}_\theta(E)\rangle = \sum_{\theta,\theta'} I_{\theta'} \langle \mathcal{H}_{\theta'}(E) | \mathcal{H}_\theta(E)\rangle \subseteq \sum_{\theta,\theta'} I_{\theta'}\, \mathcal{H}_{\theta'^*\theta}(B);$$

here, we used ii) and iv). The claims follow.

vii) By i), $\mathcal{H}_{(\theta\wedge\theta')}(E) \subseteq \mathcal{H}_\theta(E) \cap \mathcal{H}_{\theta'}(E)$. Conversely, if $\xi \in \mathcal{H}_\theta(E) \cap \mathcal{H}_{\theta'}(E)$ and $\theta'' \in \mathrm{PAut}(B)$ is minimal with $\xi \in \mathcal{H}_{\theta''}(E)$ (see iii)), then $\theta'' \leq \theta$ and $\theta'' \leq \theta'$, whence $\theta'' \leq \theta \wedge \theta'$ and $\xi \in \mathcal{H}_{(\theta\wedge\theta')}(E)$. □

The preceding proposition suggests the following notation:

**Notation 11.2.18.** Let $E$ be a right $C^*$-$B$-$B$-bimodule, and let $\mathcal{E} = (\mathcal{E}_\theta)_\theta$ and $\mathcal{E}' = (\mathcal{E}'_\theta)_\theta$ be families of closed subspaces $\mathcal{E}_\theta \subseteq \mathcal{H}_\theta(E)$ and $\mathcal{E}'_\theta \subseteq \mathcal{H}_\theta(E)$, where $\theta \in \mathrm{PAut}(B)$.

- We write $\mathcal{E}' \subseteq \mathcal{E}$ if $\mathcal{E}'_\theta \subseteq \mathcal{E}_\theta$ for all $\theta \in \mathrm{PAut}(B)$.
- We define a family $[\langle \mathcal{E}' | \mathcal{E}\rangle] \subseteq \mathcal{H}(B)$ by

$$[\langle \mathcal{E}'|\mathcal{E}\rangle]_{\theta''} = \overline{\mathrm{span}}\{\langle \xi'|\xi\rangle \mid \xi \in \mathcal{E}_\theta,\ \xi' \in \mathcal{E}'_{\theta'},\ \theta,\theta' \in \mathrm{PAut}(B),\ \theta'^*\theta \leq \theta''\}.$$

- Given a family $\mathcal{C} \subseteq \mathcal{L}(E, F)$, where $F$ is a right $C^*$-$B$-$B$-bimodule, we define a family $[\mathcal{C}\mathcal{E}] \subseteq \mathcal{H}(F)$ by

$$[\mathcal{C}\mathcal{E}]_\theta = \overline{\mathrm{span}}\{S\xi \mid S \in \mathcal{C}^\rho_\sigma,\ \xi \in \mathcal{E}_{\theta'},\ \rho,\theta',\sigma \in \mathrm{PAut}(B),\ \rho\theta'\sigma^* \leq \theta\}.$$

  Similarly, we define a family $[S\mathcal{E}] \subseteq \mathcal{H}(F)$ for each homogeneous operator $S\colon E \to F$.

- Given a right $C^*$-$B$-$B$-bimodule $F$ and a family $\mathcal{F} \subseteq \mathcal{H}(F)$, we define a family $[\mathcal{E} \circledcirc \mathcal{F}] \subseteq \mathcal{H}(E \circledcirc F)$ by

$$[\mathcal{E} \circledcirc \mathcal{F}]_{\theta''} := \overline{\operatorname{span}}\{\eta \circledcirc \xi \mid \eta \in \mathcal{E}_\theta, \xi \in \mathcal{F}_{\theta'},\ \theta, \theta' \in \mathrm{PAut}(B),\ \ \theta\theta' \leq \theta''\}.$$

Let $E$, $F$ be right $C^*$-$B$-$B$-bimodules. For each $\theta'' \in \mathrm{PAut}(B)$, put

$$\mathcal{K}_{\mathrm{id}}^{\theta''}(E, F) := \overline{\operatorname{span}}\{|\xi\rangle\langle\xi'| \mid \xi \in \mathcal{H}_\theta(F), \xi' \in \mathcal{H}_{\theta'}(E), \theta\theta'^* \leq \theta''\}.$$

**Proposition 11.2.19.** *Let $E$, $F$ be right $C^*$-$B$-$B$-bimodules.*

i) *If $E$ or $F$ is decomposable, then $\mathcal{K}_{\mathrm{id}}^{\theta}(E, F) = \mathcal{K}_B(E, F) \cap \mathcal{L}_{\mathrm{id}}^{\theta}(E, F)$ for each $\theta \in \mathrm{PAut}(B)$.*

ii) *If $E$ and $F$ are decomposable, then $E \circledcirc F$ is decomposable and $\mathcal{H}(E \circledcirc F) = [\mathcal{H}(E) \circledcirc \mathcal{H}(F)]$.*

*Proof.* See [153, Propositions 3.16, 3.17]. □

Let $E$ be a $C^*$-$A$-module, $F$ a right $C^*$-$B$-$B$-bimodule, and $\pi : A \to \mathcal{L}_B^B(F)$ a $*$-homomorphism. Then $E \circledcirc_\pi F$ is a right $C^*$-$B$-$B$-bimodule via the representation $B \to \mathcal{L}_B(E \circledcirc_\pi F)$, $b \mapsto \mathrm{id} \circledcirc b$ (use Proposition 12.2.1). Given a family $\mathcal{F} \subseteq \mathcal{H}(F)$, we define a family $[E \circledcirc_\pi \mathcal{F}] \subseteq \mathcal{H}(E \circledcirc_\pi F)$ by

$$[E \circledcirc_\pi \mathcal{F}]_\theta := \overline{\operatorname{span}}\{\eta \circledcirc \xi \mid \eta \in E,\ \xi \in \mathcal{F}_\theta\}.$$

**Proposition 11.2.20.** *If $F$ is decomposable, then $E \circledcirc_\pi F$ is decomposable and $\mathcal{H}(E \circledcirc_\pi F) = [E \circledcirc_\pi \mathcal{H}(F)]$.*

*Proof.* See [153, Proposition 3.18]. □

**Homogeneous elements of $C^*$-algebras.** Let us collect some useful properties of homogeneous elements of $C^*$-algebras.

**Proposition 11.2.21.** *Let $b \in \mathcal{H}_\theta(B)$, $\theta \in \mathrm{PAut}(B)$, and denote by $I_b \subseteq B$ the ideal generated by $b^*b$. Then:*

i) *$b$ is normal and $b^*b$ is central.*

ii) *There exists a unitary $u \in M(I_b)$ such that $b = u(b^*b)^{1/2}$.*

iii) *With $u$ as in* ii)*, the map $\mathrm{Ad}_u : I_b \to I_b$ is the minimal partial automorphism of $B$ with respect to which $b$ is homogeneous.*

iv) *$\theta(b) = b$; in particular, $b \in \mathrm{Dom}(\theta^*)$ and $\theta^*(b) = b$.*

*Proof.* i) The positive elements $b^*b$ and $bb^*$ are central by Proposition 11.2.17 ii), whence $bb^* \cdot bb^* = b^*bbb^* = b^*b \cdot b^*b$. Consequently, $bb^* = b^*b$.

ii) Put $D := \operatorname{spec}(b) \setminus \{0\}$. For $n \geq 1$, define $f_n \in C_0(D)$ by $f_n(z) := z/|z|$ if $|z| \geq 1/n$, and $f_n(z) := nz$ if $|z| \leq 1/n$. Then $(f_n)_n$ converges in $M(D)$ strictly to a unitary, and functional calculus shows that the sequence $(f_n(b))_n$ converges in $M(I_b)$ strictly to some unitary $u$. Denote by $\mathrm{id}_D \in C_0(D)$ the identity map. Then $\lim_n f_n|\mathrm{id}_D| = \mathrm{id}_D$ in $C_0(D)$, and hence $u(b^*b)^{1/2} = \lim_n f_n(b)|\mathrm{id}_D(b)| = \mathrm{id}_D(b) = b$.

iii) Evidently, $b \in I_b$ and $bd = u(b^*b)^{1/2}d = udu^*u(b^*b)^{1/2} = \mathrm{Ad}_u(d)b$ for all $d \in I_b$, so $b \in \mathcal{H}_{\mathrm{Ad}_u}(B)$. If $b \in \mathcal{H}_{\theta'}(B)$ for some $\theta' \in \mathrm{PAut}(B)$, then $I_b \subseteq \mathrm{Dom}(\theta')$ because $b \in \mathrm{Dom}(\theta')$, and $\mathrm{Ad}_u \leq \theta'$ by Proposition 11.2.17 iii).

iv) $\theta(b) = \mathrm{Ad}_u(b) = u(u(b^*b)^{1/2})u^* = u(b^*b)^{1/2} = b$ by iii) and because $(b^*b)^{1/2}$ is central. The relations $b \in \mathrm{Dom}(\theta^*)$ and $b = \theta^*(b)$ follow. □

**Proposition 11.2.22.** *Let $\theta, \theta', \rho \in \mathrm{PAut}(B)$. Then:*

i) *$bc = \theta(cb)$ and $cb = \theta^*(bc)$ for all $b \in \mathcal{H}_\theta(B)$, $c \in B$.*

ii) *$\mathcal{H}_\theta(B) = \mathcal{H}_\theta(B) \cap \mathrm{Dom}(\theta \wedge \mathrm{id})$.*

iii) *$\rho(\mathcal{H}_\theta(B) \cap \mathrm{Dom}(\rho)) \subseteq \mathcal{H}_{\rho\theta\rho^*}(B)$.*

iv) *$\mathcal{H}_{\theta'}(B)\mathcal{H}_\theta(B) \subseteq \mathcal{H}_{\theta'\theta}(B)$ and $\mathcal{H}_\theta(B)^* = \mathcal{H}_{\theta^*}(B)$.*

v) *$B$ is decomposable if and only if the inclusion $Z(B) \subseteq B$ is non-degenerate. In particular, every unital $C^*$-algebra is decomposable.*

*Proof.* i) Let $b \in \mathcal{H}_\theta(B)$, $c \in B$, and let $(u_\nu)_\nu$ be an approximate unit of $\mathrm{Dom}(\theta)$. Then

$$bc = \lim_\nu bu_\nu c = \lim_\nu \theta(u_\nu c)b = \theta(c\theta^*(b)) = \theta(cb)$$

by Proposition 11.2.21 iv), and similarly $cb = \theta^*(bc)$.

ii) This follows from Proposition 11.2.21 iv).

iii) Combine Example 11.2.5 with Proposition 11.2.17 iv).

iv) Straightforward.

v) If $B$ is decomposable, then $[BZ(B)] = B$ by Proposition 11.2.17 vi). Conversely, assume $[Z(B)B] = B$. For each unitary $u \in M(B)$ and each $b \in Z(B)$, the product $bu$ is contained in $\mathcal{H}_{\mathrm{Ad}_u}(B)$ because $buc = (ucu^*)bu$ for all $c \in B$. By [113, Remark 2.2.2], each element of $B$ can be written as a sum of four unitaries in $M(B)$. Therefore $B$ is decomposable. □

To every $C^*$-bimodule $E$, we associate a $C^*$-family $\mathcal{O}(E)$ as follows:

**Proposition 11.2.23.** *Let $A$, $B$ be $C^*$-algebras and $E$ a right $C^*$-$A$-$B$-bimodule.*

i) *Let* $a \in \mathcal{H}_\rho(A)$, $\rho \in \mathrm{PAut}(A)$, *and* $b \in \mathcal{H}_\sigma(B)$, $\sigma \in \mathrm{PAut}(B)$. *Then the map* $o_{a,b}\colon E \to E$, $\xi \mapsto a\xi b$, *is a* $(\rho, \sigma^*)$*-homogeneous operator with adjoint* $(o_{a,b})^* = o_{a^*,b^*}$.

ii) *For all* $\rho \in \mathrm{PAut}(A)$, $\sigma \in \mathrm{PAut}(B)$, *put*

$$\mathcal{O}^\rho_\sigma(E) := \overline{\mathrm{span}}\{o_{a,b} \mid a \in \mathcal{H}_\rho(A),\ b \in \mathcal{H}_{\sigma^*}(B)\}.$$

*Then* $\mathcal{O}(E) \subseteq \mathcal{L}(E)$ *is a* $C^*$*-family.*

*Proof.* i) Let $a, b$ as above. Then $o_{a,b}$ satisfies condition i) of Definition 11.2.2 because $\mathrm{Im}(o_{a,b}) \subseteq aE \subseteq \mathrm{Im}(\rho)E$ and

$$o_{a,b}a'\xi = aa'\xi b = \rho(a')a\xi b = \rho(a')o_{a,b}\xi$$

for all $a' \in \mathrm{Dom}(\rho)$, $\xi \in E$. Further, by Proposition 11.2.22 i), iv) and 11.2.21 iv), $a^* \in \mathcal{H}_{\rho^*}(A)$, $\sigma(b^*) = b^* \in \mathcal{H}_{\sigma^*}(B)$, $\langle o_{a^*,b^*}E|E\rangle \subseteq b\langle E|E\rangle \subseteq \mathrm{Dom}(\sigma^*)$, and

$$\langle \eta|a\xi b\rangle = \langle a^*\eta|\xi\rangle b = \sigma^*(b\langle a^*\eta|\xi\rangle) = \sigma^*(\langle a^*\eta b^*|\xi\rangle)$$

for all $\eta, \xi \in E$. The claim follows.

ii) Obvious from i) and Proposition 11.2.22 iv). □

**Definition 11.2.24.** Let $E$ be a right $C^*$-$A$-$B$-bimodule, where $A$ and $B$ are decomposable. A family $\mathcal{C} \subseteq \mathcal{L}(E)$ is called an $\mathcal{O}(E)$*-module* if $[\mathcal{O}(E)\mathcal{C}] \subseteq \mathcal{C}$, and a *non-degenerate* $\mathcal{O}(E)$*-module* if additionally $\mathcal{C}^\rho_\sigma = [\mathcal{O}^{\rho\rho^*}_{\sigma\sigma^*}(E)\mathcal{C}^\rho_\sigma]$ for all $\rho \in \mathrm{PAut}(A)$ and $\sigma \in \mathrm{PAut}(B)$.

**Remark 11.2.25.** The $C^*$-family $\mathcal{O}(E)$ defined above is interesting primarily if $A$ and $B$ are decomposable. However, we can consider $E$ as a right $C^*$-$M(A)$-$M(B)$-bimodule via the identification $E \cong A \otimes E \otimes M(B)$, and $M(A)$ and $M(B)$ are decomposable by Proposition 11.2.22 v).

### 11.2.3 Examples related to groupoids

Let $G$ be a locally compact, second countable, Hausdorff groupoid with left Haar system $\lambda$, and consider the $C^*$-module $L^2(G,\lambda)$ over the $C^*$-algebra $B := C_0(G^0)$. Recall that the range and source map of $G$ induce representations $r, s\colon B \to \mathcal{L}_B(L^2(G,\lambda))$. To avoid confusion with these representations, we denote the range and the source map of $G$ by $r_G$ and $s_G$, respectively. As before, we write ${}_rL^2(G,\lambda)$ or ${}_sL^2(G,\lambda)$ to indicate whether we consider $L^2(G,\lambda)$ as a right $C^*$-bimodule via the representation $r$ or $s$.

**Homogeneous elements of $L^2(G,\lambda)$.** The right $C^*$-bimodule ${}_rL^2(G,\lambda)$ is decomposable: Evidently,

$$\mathcal{H}_{\mathrm{id}}({}_rL^2(G,\lambda)) = L^2(G,\lambda).$$

The right $C^*$-bimodule ${}_sL^2(G,\lambda)$ is decomposable if the groupoid $G$ satisfies the following decomposability condition:

**Definition 11.2.26.** We call an open subset $U \subseteq G$ *homogeneous* if for all $x, y \in U$, we have $r_G(x) = r_G(y)$ if and only if $s_G(x) = s_G(y)$. We call the groupoid $G$ *decomposable* if it is equal to the union of its open homogeneous subsets.

**Remarks 11.2.27.** i) If $U, V \subseteq G$ are open homogeneous subsets, then also $U^{-1}$ and $UV = \{xy \mid (x,y) \in G^2_{s,r} \cap (U \times V)\}$ are open and homogeneous.

ii) Recall that an open subset $U \subseteq G$ is called a *$G$-set* if the restrictions $r|_U : U \to r(U)$ and $s|_U : U \to s(U)$ are homeomorphisms and $r(U), s(U) \subseteq G^0$ are open. Moreover, recall that $G$ is $r$-discrete if and only if it is the union of open $G$-sets [129, Proposition 2.8]. Evidently, every $G$-set is homogeneous, and if $G$ is $r$-discrete, then it is decomposable.

iii) It is easy to see that every extension of an $r$-discrete groupoid by a bundle of groups is decomposable. In fact, it is not difficult to show that a groupoid is decomposable if and only if it is an extension of an $r$-discrete groupoid by a bundle of groups [152, Proposition 3.4].

Let us prove that ${}_sL^2(G,\lambda)$ is decomposable if $G$ is decomposable. Since the range and the source map of $G$ are open [129, I, Proposition 2.4], every open homogeneous subset $U \subseteq G$ defines a homeomorphism $q_U : s_G(U) \to r_G(U)$ by $s_G(x) \mapsto r_G(x)$, which induces partial automorphisms

$$q_{U*} : C_0(s_G(U)) \to C_0(r_G(U)) \quad \text{and} \quad q_U^* : C_0(r_G(U)) \to C_0(s_G(U))$$

of the $C^*$-algebra $B = C_0(G^0)$.

**Lemma 11.2.28.** *Let $U \subseteq G$ be open and homogeneous. Then*

$$C_c(U) \subseteq \mathcal{H}_{q_U^*}({}_sL^2(G,\lambda)).$$

*Proof.* Clearly, $C_c(U) \subseteq L^2(G,\lambda)C_0(r_G(U))$, and for all $\xi \in C_c(U)$, $f \in C_0(r_G(U))$, and $x \in U$,

$$(\xi f)(x) = \xi(x) f(r_G(x)) = \xi(x) f(q_U(s_G(x))) = (s(q_U^*(f))\xi)(x). \qquad \square$$

Combining this lemma with a partition of unity argument, we find:

**Proposition 11.2.29.** *If the groupoid $G$ is decomposable, then the right $C^*$-bimodule ${}_sL^2(G,\lambda)$ is decomposable.* □

**The left regular representation on $L^2(G,\lambda)$.** We shall see that if $G$ is decomposable, then the left regular representation on $L^2(G,\lambda)$ gives rise to a $C^*$-family. To simplify the discussion, we impose the following condition:

**Definition 11.2.30.** We call the left Haar system $\lambda$ *unimodular* if for every open homogeneous subset $U \subseteq G$, every function $f \in C_c(U)$, and every $z \in U$,

$$\int_{G^{r_G(z)}} f(y)d\lambda^{r_G(z)}(y) = \int_{G^{s_G(z)}} f(y^{-1})d\lambda^{s_G(z)}(y).$$

**Example 11.2.31.** If $G$ is $r$-discrete, then the family of counting measures is a unimodular left Haar system.

For a function $f \in C_c(G)$, we denote by $L(f)\colon L^2(G,\lambda) \to L^2(G,\lambda)$ the map given by left convolution with $f$,

$$(L(f)\xi)(x) := \int_{G^{r_G(x)}} f(y)\xi(y^{-1}x)d\lambda^{r_G(x)}(y) \quad \text{for all } x \in G,\ \xi \in C_c(G).$$

Moreover, we denote by $f^* \in C_c(G)$ the function given by $f^*(x) := \overline{f(x^{-1})}$ for all $x \in G$.

**Proposition 11.2.32.** *Assume that G is decomposable and that the left Haar system $\lambda$ is unimodular.*

i) *For every open homogeneous subset $U \subseteq G$ and every $f \in C_c(U)$, we have $L(f) \in \mathcal{L}^{\mathrm{id}}_{q_{U*}}({}_sL^2(G,\lambda))$, $L(f) \in \mathcal{L}^{q_{U*}}_{q_{U*}}({}_rL^2(G,\lambda))$, and $L(f)^* = L(f^*)$.*

ii) *Let $f \in C_c(U)$ and $g \in C_c(V)$, where $U, V \subseteq G$ are open homogeneous subsets. Then $L(f)L(g) = L(h)$, where $h \in C_c(UV)$ is given by*

$$h(x) = \int_{G^{r_G(x)}} f(y)g(y^{-1}x)d\lambda^{r_G(x)}(y) \quad \text{for all } x \in G.$$

iii) *The family of closed subspaces $\mathcal{C}^*_r(G) \subseteq \mathcal{L}({}_rL^2(G,\lambda))$ given by*

$$\mathcal{C}^*_r(G)^{\rho}_{\sigma} = \overline{\operatorname{span}}\{L(f) \mid f \in C_c(U),\ U \subseteq G \text{ open homogeneous}, \\ q_{U*} \leq \rho \text{ and } q_{U*} \leq \sigma\}$$

*for all $\rho, \sigma \in \mathrm{PAut}(B)$ is a non-degenerate $C^*$-family and a non-degenerate $\mathcal{O}({}_rL^2(G,\lambda))$-module.*

*Proof.* i) It is easy to see that $L(f)$ commutes with the representation $s$, that the image of $L(f)$ is contained in $r(C_0(r_G(U)))L^2(G,\lambda)$, and that $L(f)r(b) = r(q_{U*}(b))L(f)$ for each $b \in C_0(s_G(U))$.

Let $\xi, \eta \in C_c(G)$. Then the inner products $\langle \eta | L(f)\xi \rangle$ and $\langle L(f^*)\eta | \xi \rangle$, considered as functions on $G^0$, vanish outside $r_G(U)$ and $s_G(U)$, respectively. For each $z \in U$,

$$\langle \eta | L(f)\xi \rangle (r_G(z)) = \int_{G^{r_G(z)}} \int_{G^{r_G(z)}} \overline{\eta(x)} f(y) \xi(y^{-1}x) d\lambda^{r_G(z)}(y) d\lambda^{r_G(z)}(x).$$

Since $\lambda$ is left-invariant and unimodular, this iterated integral is equal to

$$\begin{aligned} &\int_{G^{r_G(z)}} \int_{G^{s_G(y)}} \overline{\eta(yx') f^*(y^{-1})} \xi(x') d\lambda^{s_G(y)}(x') d\lambda^{r_G(z)}(y) \\ &= \int_{G^{s_G(z)}} \int_{G^{s_G(z)}} \overline{\eta(y^{-1}x') f^*(y)} \xi(x') d\lambda^{s_G(z)}(x') d\lambda^{s_G(z)}(y) \\ &= \langle L(f^*)\eta | \xi \rangle (s_G(z)). \end{aligned}$$

Since $s_G(z) = q_U(r_G(z))$, it follows that $\langle \eta | L(f)\xi \rangle = q_U^*(\langle L(f^*)\eta | \xi \rangle)$. Thus assertion i) is proved.

ii), iii) Statement ii) follows from a routine calculation, and statement iii) follows from i), ii), and Remark 11.2.27 i). □

## 11.3 Hopf $C^*$-families

In this section, we introduce the internal tensor product of $C^*$-families and the notion of a morphism of $C^*$-families. These concepts are needed for the definition of a Hopf $C^*$-family, which is given afterwards. Throughout this subsection, let $A$, $B$, $C$ be $C^*$-algebras.

### 11.3.1 The internal tensor product of $C^*$-families

Let $E$ be a right $C^*$-$A$-$B$-bimodule and $F$ a right $C^*$-$B$-$C$-bimodule. Then we can form the internal tensor product $E \otimes F$, which is a right $C^*$-$A$-$C$-bimodule. Given $C^*$-families $\mathcal{C} \subseteq \mathcal{L}(E)$ and $\mathcal{D} \subseteq \mathcal{L}(F)$, we shall define an internal tensor product $\mathcal{C} \otimes \mathcal{D} \subseteq \mathcal{L}(E \otimes F)$, which is a $C^*$-family again. This internal tensor product will be generated by operators of the form

$$S \otimes T : \eta \otimes \xi \mapsto S\eta \otimes T\xi,$$

where $S$ and $T$ are suitable operators of the $C^*$-families $\mathcal{C}$ and $\mathcal{D}$, respectively. The relation

$$S(\eta b) \otimes T\xi = (S \otimes T)(\eta b \otimes \xi) = (S \otimes T)(\eta \otimes b\xi) = S\eta \otimes Tb\xi$$

shows that the map $S \circledcirc T$ is well defined only if $S$ intertwines the right multiplication on $E$ in a way that matches up with the way in which $T$ intertwines the left multiplication on $F$. In the next definition and lemma, we formulate and investigate this compatibility condition:

**Definition 11.3.1.** We call two partial automorphisms $\rho, \sigma \in \mathrm{PAut}(B)$ *compatible* and write $\rho \curlyvee \sigma$ if $\rho\sigma^* \leq \mathrm{id}$ and $\rho^*\sigma \leq \mathrm{id}$.

**Lemma 11.3.2.** *Let $\rho, \sigma \in \mathrm{PAut}(B)$ such that $\rho \curlyvee \sigma$.*

i) $\rho^* \curlyvee \sigma^*$;

ii) $\rho(a) = \sigma(a)$ *for all* $a, b \in \mathrm{Dom}(\rho) \cap \mathrm{Dom}(\sigma)$;

iii) $\rho(\mathrm{Dom}(\rho) \cap \mathrm{Dom}(\sigma)) = \mathrm{Im}(\rho) \cap \mathrm{Im}(\sigma) = \sigma(\mathrm{Dom}(\rho) \cap \mathrm{Dom}(\sigma))$;

iv) $\rho(ab) = \rho(a)\sigma(b) = \sigma(ab)$ *for all* $a \in \mathrm{Dom}(\rho)$, $b \in \mathrm{Dom}(\sigma)$;

v) *if* $\rho', \sigma' \in \mathrm{PAut}(B)$ *and* $\rho' \curlyvee \sigma'$, *then* $\rho\rho' \curlyvee \sigma\sigma'$.

*Proof.* Assertions i) and ii) follow immediately from the definition.

iii) By ii), $\rho(\mathrm{Dom}(\rho) \cap \mathrm{Dom}(\sigma)) = \sigma(\mathrm{Dom}(\rho) \cap \mathrm{Dom}(\sigma))$ is contained in $\mathrm{Im}(\rho) \cap \mathrm{Im}(\sigma)$. To obtain the reverse inclusion, replace $\rho, \sigma$ by $\rho^*, \sigma^*$.

iv) Let $(u_\nu)_\nu$ be an approximate unit for $\mathrm{Dom}(\rho) \cap \mathrm{Dom}(\sigma)$. By iii), $(\sigma(u_\nu))_\nu$ is an approximate unit for $\mathrm{Im}(\rho) \cap \mathrm{Im}(\sigma)$, and for all $a \in \mathrm{Dom}(\rho)$, $b \in \mathrm{Dom}(\sigma)$,

$$\begin{aligned}\rho(ab) = \lim_\nu \rho(abu_\nu) &= \lim_\nu \rho(a)\rho(bu_\nu) \\ &= \lim_\nu \rho(a)\sigma(bu_\nu) = \lim_\nu \rho(a)\sigma(b)\sigma(u_\nu) = \rho(a)\sigma(b).\end{aligned}$$

v) Since $\rho'\sigma'^* \leq \mathrm{id}$ and $\rho^*\sigma \leq \mathrm{id}$, the products $(\rho\rho')(\sigma\sigma')^* = \rho(\rho'\sigma'^*)\sigma^*$ and $(\rho\rho')^*(\sigma\sigma') = \rho'^*(\rho^*\sigma)\sigma'$ are restrictions of $\rho\sigma^*$ and $\rho'^*\sigma'$, respectively. ☐

In general, compatibility is not transitive: the automorphism of the ideal $\{0\}$ is compatible with every other partial automorphism of $B$.

**Proposition 11.3.3.** *Let $E_1, E_2$ be right $C^*$-$A$-$B$-bimodules and $F_1, F_2$ right $C^*$-$B$-$C$-bimodules. Furthermore, let*

- $S \in \mathcal{L}_{\sigma_S}^{\rho_S}(E_1, E_2)$, *where* $\rho_S \in \mathrm{PAut}(A)$, $\sigma_S \in \mathrm{PAut}(B)$,
- $T \in \mathcal{L}_{\sigma_T}^{\rho_T}(F_1, F_2)$, *where* $\rho_T \in \mathrm{PAut}(B)$, $\sigma_T \in \mathrm{PAut}(C)$.

*If $\sigma_S \curlyvee \rho_T$, then there exists an operator $S \circledcirc T \in \mathcal{L}_{\sigma_T}^{\rho_S}(E_1 \circledcirc F_1, E_2 \circledcirc F_2)$ such that*

$$(S \circledcirc T)(\eta \circledcirc \xi) = S\eta \circledcirc T\xi \quad \textit{for all } \eta \in E_1,\ \xi \in F_1. \tag{11.4}$$

*Moreover, $\|S \circledcirc T\| \leq \|S\|\|T\|$ and $(S \circledcirc T)^* = S^* \circledcirc T^*$.*

*Proof.* To simplify notation, we put $E := E_1 \oplus E_2$, $F := F_1 \oplus F_2$, and consider $S$ and $T$ as elements of $\mathcal{L}^{\rho_S}_{\sigma_S}(E)$ and $\mathcal{L}^{\rho_T}_{\sigma_T}(F)$, respectively, in the natural way. Let $\eta, \eta' \in E$ and $\xi, \xi' \in F$. Then

$$\langle \eta' \ominus \xi' | S\eta \ominus T\xi \rangle = \langle \xi' | \langle \eta' | S\eta \rangle T\xi \rangle = \langle \xi' | \sigma_S(\langle S^*\eta' | \eta \rangle) T\xi \rangle.$$

Let $(u_\nu)_\nu$ be an approximate unit for $\mathrm{Dom}(\rho_T)$. Then the homogeneity of $T$ and $S$, Proposition 11.2.3 v), and Lemma 11.3.2 iv) imply

$$\begin{aligned} \sigma_S(\langle S^*\eta' | \eta \rangle) T\xi &= \lim_\nu \rho_T(u_\nu) \sigma_S(\langle S^*\eta' | \eta \rangle) T\xi \\ &= \lim_\nu T u_\nu \langle S^*\eta' | \eta \rangle \xi = T \langle S^*\eta' | \eta \rangle \xi. \end{aligned}$$

We insert this relation into the equation above, and find

$$\begin{aligned} \langle \eta' \ominus \xi' | S\eta \ominus T\xi \rangle = \langle \xi' | T \langle S^*\eta' | \eta \rangle \xi \rangle &= \sigma_T(\langle T^*\xi' | \langle S^*\eta' | \eta \rangle \xi \rangle) \\ &= \sigma_T(\langle S^*\eta' \ominus T^*\xi' | \eta \ominus \xi \rangle). \end{aligned} \tag{11.5}$$

Let us show that formula (11.4) defines a bounded map $S \ominus T$. By equation (11.5), we have for all $\eta_i \in E$, $\xi_i \in F$, where $i = 1, \dots, n$,

$$\Big\| \sum_i S\eta_i \ominus T\xi_i \Big\|^2 = \Big\| \sum_{i,j} \langle S^*S\eta_i \ominus T^*T\xi_i | \eta_j \ominus \xi_j \rangle \Big\|$$

Now $S^*S \in \mathcal{L}_B(E)$, $T^*T \in \mathcal{L}^B_C(F)$, and by Proposition 12.2.1, the operators $S^*S \ominus 1$, $1 \ominus T^*T$, $S^*S \ominus T^*T \in \mathcal{L}_C(E \ominus F)$ are well defined. Since

$$S^*S \ominus T^*T = (S^*S \ominus 1)(1 \ominus T^*T) = (1 \ominus T^*T)(S^*S \ominus 1),$$

we have

$$\|S \ominus T\|^2 \le \|S^*S \ominus T^*T\| \le \|S^*S \ominus 1\| \|1 \ominus T^*T\| \le \|S\|^2 \|T\|^2.$$

Evidently, the image of $S \ominus T$ is contained in $\mathrm{Im}(\rho_S)(E \ominus F)$, and

$$\begin{aligned} (S \ominus T) a (\eta \ominus \xi) &= Sa\eta \ominus T\xi \\ &= \rho_S(a) S\eta \ominus T\xi = \rho_S(a)(S \ominus T)(\eta \ominus \xi) \end{aligned}$$

for all $\eta \in E$, $\xi \in F$, and $a \in \mathrm{Dom}(\rho_S)$. Replacing $S$ and $T$ by their adjoints, we obtain a bounded map $S^* \ominus T^* \colon E \ominus F \to E \ominus F$, and formula (11.5) shows that $S \ominus T$ is $(\rho_S, \sigma_T)$-homogeneous with adjoint $S^* \ominus T^*$ as claimed. □

Next, we introduce the internal tensor product of $C^*$-families. For later applications, we state the definition in a slightly wider generality.

**Notation 11.3.4.** Let $E_1, E_2$ be right $C^*$-$A$-$B$-bimodules and $F_1, F_2$ right $C^*$-$B$-$C$-bimodules. Furthermore, let $\mathcal{C} \subseteq \mathcal{L}(E_1, E_2)$ and $\mathcal{D} \subseteq \mathcal{L}(F_1, F_2)$ be families of closed subspaces. Then the *internal tensor product* $\mathcal{C} \ominus \mathcal{D} \subseteq \mathcal{L}(E_1 \ominus F_1, E_2 \ominus F_2)$ is the family given by

$$(\mathcal{C} \ominus \mathcal{D})^{\rho}_{\sigma} := \overline{\operatorname{span}}\{S \ominus T \mid S \in \mathcal{C}^{\rho}_{\sigma_S}, \ T \in \mathcal{D}^{\rho_T}_{\sigma}, \ \sigma_S, \rho_T \in \operatorname{PAut}(B), \sigma_S \curlyvee \rho_T\}$$

for all $\rho \in \operatorname{PAut}(A)$ and $\sigma \in \operatorname{PAut}(C)$.

The internal tensor product of families does not always behave as one might naïvely expect:

**Remark 11.3.5.** Let $E$ be a right $C^*$-$A$-$B$-bimodule, $F$ a right $C^*$-$B$-$C$-bimodule, and $\mathcal{A}, \mathcal{C} \subseteq \mathcal{L}(E)$, $\mathcal{B}, \mathcal{D} \subseteq \mathcal{L}(F)$ families of closed subspaces. Then

$$[(\mathcal{A} \ominus \mathcal{B})(\mathcal{C} \ominus \mathcal{D})] \subseteq [\mathcal{A}\mathcal{C}] \ominus [\mathcal{B}\mathcal{D}].$$

*This inclusion may be strict and fail to be an equality.* As a simple example, consider the case where all spaces comprising the families $\mathcal{C}$ and $\mathcal{D}$ are 0 except for $\mathcal{C}^{\rho_1}_{\sigma_1}$ and $\mathcal{D}^{\rho_2}_{\sigma_2}$, where $\rho_1 \in \operatorname{PAut}(A)$, $\sigma_2 \in \operatorname{PAut}(C)$, and $\sigma_1, \rho_2 \in \operatorname{PAut}(B)$ are fixed and not compatible. Then $\mathcal{C}^* \ominus \mathcal{D}^* = 0 = \mathcal{C} \ominus \mathcal{D}$, but $\mathcal{C}^*\mathcal{C} \ominus \mathcal{D}^*\mathcal{D}$ need not be 0.

The internal tensor product of $C^*$-families is a $C^*$-family again:

**Proposition 11.3.6.** *Let $E$ be a right $C^*$-$A$-$B$-bimodule, $F$ a right $C^*$-$B$-$C$-bimodule, and let $\mathcal{C} \subseteq \mathcal{L}(E)$ and $\mathcal{D} \subseteq \mathcal{L}(F)$ be $C^*$-families.*

i) *$\mathcal{C} \ominus \mathcal{D} \subseteq \mathcal{L}(E \ominus F)$ is a $C^*$-family.*

ii) *If $\mathcal{C}$ and $\mathcal{D}$ are non-degenerate, so is $\mathcal{C} \ominus \mathcal{D}$.*

iii) *If $A$, $B$, $C$ are decomposable, $\mathcal{C}$ is a (non-degenerate) $\mathcal{O}(E)$-module, and $\mathcal{D}$ is a (non-degenerate) $\mathcal{O}(F)$-module, then $\mathcal{C} \ominus \mathcal{D}$ is a (non-degenerate) $\mathcal{O}(E \ominus F)$-module.*

iv) *$\mathcal{M}(\mathcal{C}) \ominus \mathcal{M}(\mathcal{D}) \subseteq \mathcal{M}(\mathcal{C} \ominus \mathcal{D})$.*

*Proof.* This follows easily from the definitions and from Lemma 11.3.2. □

It is easy to see that the internal tensor product is associative:

**Proposition 11.3.7.** *Let $A$, $B$, $C$, $D$ be $C^*$-algebras, let $E$ be a right $C^*$-$A$-$B$-bimodule, $F$ a right $C^*$-$B$-$C$-bimodule, and $G$ a right $C^*$-$C$-$D$-bimodule. Furthermore, let $\mathcal{B} \subseteq \mathcal{L}(E)$, $\mathcal{C} \subseteq \mathcal{L}(F)$, $\mathcal{D} \subseteq \mathcal{L}(G)$ be $C^*$-families. Then the natural isomorphism $(E \ominus F) \ominus G \cong E \ominus (F \ominus G)$ induces an isomorphism of $C^*$-families $(\mathcal{B} \ominus \mathcal{C}) \ominus \mathcal{D} \cong \mathcal{B} \ominus (\mathcal{C} \ominus \mathcal{D})$.* □

**Remark 11.3.8.** The internal tensor product is not quite associative on the level of operators: It may happen that $R, S$, and $T$ are operators such that the internal tensor products $R \ominus S$ and $(R \ominus S) \ominus T$ are well defined while the internal tensor products $S \ominus T$ and $R \ominus (S \ominus T)$ are not defined. This phenomenon occurs, for example, if $R = 0$, but it is not restricted to such trivial cases.

The constructions introduced above can easily be adapted to the flipped internal tensor product of right $C^*$-bimodules and give rise to a flipped internal tensor product of homogeneous operators and $C^*$-families.

### 11.3.2 Morphisms of $C^*$-families

It seems to be difficult to find a natural notion of a morphism of $C^*$-families that makes the internal tensor product bifunctorial. Therefore we adopt a pragmatic approach:

**Definition 11.3.9.** Let $\mathcal{C}$ and $\mathcal{D}$ be $C^*$-families on right $C^*$-$A$-$B$-bimodules. A *family of linear maps* $\phi\colon \mathcal{C} \to \mathcal{D}$ is a family $\phi = (\phi^\rho_\sigma)_{\rho,\sigma}$ of linear maps $\phi^\rho_\sigma\colon \mathcal{C}^\rho_\sigma \to \mathcal{D}^\rho_\sigma$, defined for all $\rho \in \mathrm{PAut}(A)$, $\sigma \in \mathrm{PAut}(B)$. We call a family of linear maps $\phi\colon \mathcal{C} \to \mathcal{D}$

- *$A'$-$B'$-extendible*, where $A'$ and $B'$ are $C^*$-algebras, if for each right $C^*$-$A'$-$A$-bimodule $X$ and each right $C^*$-$B$-$B'$-bimodule $Y$, there exists a linear map
$$\phi^X_Y : (\mathcal{L}(X) \ominus \mathcal{C} \ominus \mathcal{L}(Y))^{\mathrm{id}}_{\mathrm{id}} \to (\mathcal{L}(X) \ominus \mathcal{D} \ominus \mathcal{L}(Y))^{\mathrm{id}}_{\mathrm{id}}$$
such that $\phi^X_Y(R \ominus S \ominus T) = R \ominus \phi^\rho_\sigma(S) \ominus T$ for all $R \in \mathcal{L}^{\mathrm{id}}_{\sigma'}(X)$, $S \in \mathcal{C}^\rho_\sigma$, $T \in \mathcal{L}^{\rho'}_{\mathrm{id}}(Y)$, where $\sigma', \rho \in \mathrm{PAut}(A)$, $\sigma, \rho' \in \mathrm{PAut}(B)$, $\sigma' \curlyvee \rho$, $\sigma \curlyvee \rho'$;
- *extendible* if $\phi$ is $A'$-$B'$-extendible for every $C^*$-algebra $A'$ and $B'$;
- *injective* if each component $\phi^\rho_\sigma$ is injective;
- a *morphism* if $\phi$ is extendible and $\phi^X_Y$ always is a $*$-homomorphism.

We call a morphism $\phi\colon \mathcal{C} \to \mathcal{M}(\mathcal{D})$ *non-degenerate* if $[\phi(\mathcal{C})\mathcal{D}] = \mathcal{D}$.

Let $\mathcal{B}$, $\mathcal{C}$, $\mathcal{D}$ be $C^*$-families on right $C^*$-$A$-$B$-bimodules. The *composition* of two families of linear maps $\phi\colon \mathcal{B} \to \mathcal{C}$ and $\psi\colon \mathcal{C} \to \mathcal{D}$ is the family $\psi \circ \phi\colon \mathcal{B} \to \mathcal{D}$ given by $(\psi \circ \phi)^\rho_\sigma := \psi^\rho_\sigma \circ \phi^\rho_\sigma$ for all $\rho, \sigma$.

**Remark 11.3.10.** i) $(\mathcal{L}(X) \ominus \mathcal{C} \ominus \mathcal{L}(Y))^{\mathrm{id}}_{\mathrm{id}}$ and $(\mathcal{L}(X) \ominus \mathcal{D} \ominus \mathcal{L}(Y))^{\mathrm{id}}_{\mathrm{id}}$ are $C^*$-subalgebras of $\mathscr{L}^{A'}_{B'}(X \ominus E \ominus Y)$ and $\mathscr{L}^{A'}_{B'}(X \ominus F \ominus Y)$, respectively.

ii) Clearly, the composition of (extendible) families of linear maps/of morphisms is a (extendible) family of linear maps/a morphism again, and the collection of all $C^*$-families on right $C^*$-$A$-$B$-bimodules and all (extendible) families of linear maps/all morphisms forms a category.

To determine whether a family of linear maps is extendible, it suffices to check that it is $\mathbb{C}$-$\mathbb{C}$-extendible:

**Proposition 11.3.11.** *Let $\phi\colon \mathcal{C} \to \mathcal{D}$ be a family of linear maps between $C^*$-families. Then $\phi$ is extendible if and only if it is $\mathbb{C}$-$\mathbb{C}$-extendible.*

*Proof.* Assume that $\phi$ is $\mathbb{C}$-$\mathbb{C}$-extendible, and let $A'$, $B'$ be $C^*$-algebras. We show that $\phi$ is $A'$-$B'$-extendible. Let $X'$ be a right $C^*$-$A'$-$A$-bimodule and $Y'$ a right $C^*$-$B$-$B'$-bimodule. Denote by $X$ the $C^*$-module $X'$, considered as a right $C^*$-$\mathbb{C}$-$A$-bimodule via multiplication by scalars. Choose a faithful representation of $B'$ on a Hilbert space $H$ and put $Y := Y' \ominus H$. For $G = E, F$ and $\mathcal{B} = \mathcal{C}, \mathcal{D}$, the embedding $\mathcal{L}^{A'}_{B'}(X' \ominus G \ominus Y') \hookrightarrow \mathcal{L}^{\mathbb{C}}_{\mathbb{C}}(X \ominus G \ominus Y' \ominus H)$, $T \mapsto T \ominus \mathrm{id}_H$, maps $(\mathcal{L}(X') \ominus \mathcal{B} \ominus \mathcal{L}(Y'))^{\mathrm{id}}_{\mathrm{id}}$ to $(\mathcal{L}(X) \ominus \mathcal{B} \ominus \mathcal{L}(Y))^{\mathrm{id}}_{\mathrm{id}}$. Restricting the map $\phi^X_Y$ (which exists by assumption), we obtain the desired map $\phi^{X'}_{Y'}$. □

For the study of morphisms of $C^*$-families, the following embedding result is useful:

**Theorem 11.3.12.** *There exist a right $C^*$-$A$-$A$-bimodule $\mathfrak{I}A$, a right $C^*$-$B$-$B$-bimodule $\mathfrak{I}B$, and operators $V_\sigma \in \mathcal{L}^{\mathrm{id}}_\sigma(\mathfrak{I}A)$, $W_\rho \in \mathcal{L}^{\rho}_{\mathrm{id}}(\mathfrak{I}B)$, defined for each $\sigma \in \mathrm{PAut}(A)$, $\rho \in \mathrm{PAut}(B)$, such that for each right $C^*$-$A$-$B$-bimodule $E$, the maps*

$$\iota^\rho_\sigma\colon \mathcal{L}^\rho_\sigma(E) \to \mathcal{L}^A_B(\mathfrak{I}A \ominus E \ominus \mathfrak{I}B), \quad T \mapsto V_\rho \ominus T \ominus W_\sigma,$$

*defined for each $\rho \in \mathrm{PAut}(A)$, $\sigma \in \mathrm{PAut}(B)$, satisfy*

$$\|\iota^\rho_\sigma(T)\| = \|T\|, \quad \iota^\rho_\sigma(T)^* = \iota^{\rho^*}_{\sigma^*}(T^*), \quad \iota^\rho_\sigma(T)\iota^{\rho'}_{\sigma'}(T') = \iota^{\rho\rho'}_{\sigma\sigma'}(TT')$$

*for all $T \in \mathcal{L}^\rho_\sigma(E)$, $T' \in \mathcal{L}^{\rho'}_{\sigma'}(E)$, $\rho, \rho' \in \mathrm{PAut}(A)$, $\sigma, \sigma' \in \mathrm{PAut}(B)$.*

*Proof.* See [153, Theorem 5.10]. □

Morphisms of $C^*$-families behave in many respects as one should expect:

**Proposition 11.3.13.** *Let $\phi\colon \mathcal{C} \to \mathcal{D}$ be a morphism of $C^*$-families. Then*

$$\phi^{\rho'}_{\sigma'}(c')\phi^\rho_\sigma(c) = \phi^{\rho'\rho}_{\sigma'\sigma}(c'c), \quad \phi^\rho_\sigma(c)^* = \phi^{\rho^*}_{\sigma^*}(c^*), \quad \|\phi^\rho_\sigma(c)\| \leq \|c\|,$$
$$\phi^\rho_\sigma(c) = \phi^{\rho'}_{\sigma'}(c) \quad \text{if } (\rho,\sigma) \leq (\rho',\sigma')$$

*for all $c \in \mathcal{C}^\rho_\sigma$, $c' \in \mathcal{C}^{\rho'}_{\sigma'}$, $\rho, \rho' \in \mathrm{PAut}(A)$, $\sigma, \sigma' \in \mathrm{PAut}(B)$. In particular, $\phi^{\mathrm{id}}_{\mathrm{id}}\colon \mathcal{C}^{\mathrm{id}}_{\mathrm{id}} \to \mathcal{D}^{\mathrm{id}}_{\mathrm{id}}$ is a $*$-homomorphism of $C^*$-algebras.*

*Proof.* This follows from the existence of a $*$-homomorphism $\phi^{\mathfrak{I}A}_{\mathfrak{I}B}$ that makes the diagram below commute for all $\rho \in \mathrm{PAut}(A)$ and $\sigma \in \mathrm{PAut}(B)$:

$$\begin{array}{ccc} \mathcal{C}^\rho_\sigma & \xhookrightarrow[\iota^\rho_\sigma]{} & [\iota(\mathcal{C})] \subseteq (\mathcal{L}(\mathfrak{I}A) \otimes \mathcal{C} \otimes \mathcal{L}(\mathfrak{I}B))^{\mathrm{id}}_{\mathrm{id}} \\ \phi^\rho_\sigma \downarrow & & \downarrow \phi^{\mathfrak{I}A}_{\mathfrak{I}B} \\ \mathcal{D}^\rho_\sigma & \xhookrightarrow{\iota^\rho_\sigma} & [\iota(\mathcal{D})] \subseteq (\mathcal{L}(\mathfrak{I}A) \otimes \mathcal{D} \otimes \mathcal{L}(\mathfrak{I}B))^{\mathrm{id}}_{\mathrm{id}}. \end{array}$$ □

**Remarks 11.3.14.** i) A morphism $\phi \colon \mathcal{C} \to \mathcal{D}$ of $C^*$-families is injective if and only if the component $\phi^{\mathrm{id}}_{\mathrm{id}}$ is injective because

$$\|\phi^\rho_\sigma(c)\|^2 = \|\phi^\rho_\sigma(c)^*\phi^\rho_\sigma(c)\| = \|\phi^{\rho^*\rho}_{\sigma^*\sigma}(c^*c)\| = \|\phi^{\mathrm{id}}_{\mathrm{id}}(c^*c)\|$$

for all $c \in \mathcal{C}^\rho_\sigma$ and all $\rho, \sigma$.

i) A morphism $\phi \colon \mathcal{C} \to \mathcal{M}(\mathcal{D})$ of $C^*$-families is non-degenerate if and only if the natural map $\phi^{\mathrm{id}}_{\mathrm{id}} \colon \mathcal{C}^{\mathrm{id}}_{\mathrm{id}} \to \mathcal{M}(\mathcal{D})^{\mathrm{id}}_{\mathrm{id}} \to M(\mathcal{D}^{\mathrm{id}}_{\mathrm{id}})$ is a non-degenerate $*$-homomorphism of $C^*$-algebras. This follows from Remark 11.2.11 iii).

The internal tensor product of $C^*$-families is bifunctorial:

**Proposition 11.3.15.** *Let $\phi \colon \mathcal{A} \to \mathcal{C}$ and $\psi \colon \mathcal{B} \to \mathcal{D}$ be extendible families of linear maps/(non-degenerate) morphisms of $C^*$-families on right $C^*$-$A$-$B$-bimodules and right $C^*$-$B$-$C$-bimodules, respectively. Then there exists an extendible family of linear maps/(non-degenerate) morphism*

$$\phi \otimes \psi \colon \mathcal{A} \otimes \mathcal{B} \to \mathcal{C} \otimes \mathcal{D}$$

*such that*

$$(\phi \otimes \psi)^\rho_{\sigma'}(a \otimes b) = \phi^\rho_\sigma(a) \otimes \psi^{\rho'}_{\sigma'}(b)$$

*for all $a \in \mathcal{A}^\rho_\sigma$, $b \in \mathcal{B}^{\rho'}_{\sigma'}$, $\rho \in \mathrm{PAut}(A)$, $\sigma, \rho' \in \mathrm{PAut}(B)$, $\sigma' \in \mathrm{PAut}(C)$, $\sigma \curlyvee \rho'$.*

*Proof.* If we can prove the assertion for the case that $\mathcal{B} = \mathcal{D}$, $\psi = \mathrm{id}_{\mathcal{B}}$ and for the case that $\mathcal{A} = \mathcal{C}$, $\phi = \mathrm{id}_{\mathcal{A}}$, then we can simply put $\phi \otimes \psi := (\phi \otimes \mathrm{id}) \circ (\mathrm{id} \otimes \psi)$. We treat the first case, the second one is similar.

Let $\rho \in \mathrm{PAut}(A)$ and $\sigma' \in \mathrm{PAut}(C)$. Denote by $F$ the right $C^*$-bimodule on which $\mathcal{B}$ acts. If $\sigma, \rho' \in \mathrm{PAut}(B)$ and $\sigma \curlyvee \rho'$, then the diagram

$$\begin{array}{ccc} \mathcal{A}^\rho_\sigma \otimes \mathcal{B}^{\rho'}_{\sigma'} & \xhookrightarrow[\iota^\rho_{\sigma'}]{} & \iota^\rho_{\sigma'}((\mathcal{A} \otimes \mathcal{B})^\rho_{\sigma'}) \subseteq (\mathcal{L}(\mathfrak{I}A) \otimes \mathcal{A} \otimes \mathcal{L}(F \otimes \mathfrak{I}C))^{\mathrm{id}}_{\mathrm{id}} \\ \phi^\rho_\sigma \otimes \mathrm{id} \downarrow & & \downarrow \phi^{\mathfrak{I}A}_{F \otimes \mathfrak{I}C} \\ \mathcal{C}^\rho_\sigma \otimes \mathcal{B}^{\rho'}_{\sigma'} & \xhookrightarrow{\iota^\rho_{\sigma'}} & \iota^\rho_{\sigma'}((\mathcal{C} \otimes \mathcal{B})^\rho_{\sigma'}) \subseteq (\mathcal{L}(\mathfrak{I}A) \otimes \mathcal{C} \otimes \mathcal{L}(F \otimes \mathfrak{I}C))^{\mathrm{id}}_{\mathrm{id}} \end{array}$$

commutes, and we can insert a unique linear map $(\phi \otimes \mathrm{id})^\rho_{\sigma'} \colon (\mathcal{A} \otimes \mathcal{B})^\rho_{\sigma'} \to (\mathcal{C} \otimes \mathcal{B})^\rho_{\sigma'}$ that does not depend on $\sigma, \rho'$ such that the diagram still commutes.

The family $((\phi \otimes \mathrm{id})^{\rho}_{\sigma'})_{\rho,\sigma'}$ thus defined is extendible. Indeed, let $X$ be a right $C^*$-$\mathbb{C}$-$A$-bimodule and $Y$ a right $C^*$-$C$-$\mathbb{C}$-bimodule. Then $F \otimes Y$ is a right $C^*$-$B$-$\mathbb{C}$-bimodule, and the linear map

$$\phi^X_{F\otimes Y} : (\mathcal{L}(X) \otimes \mathcal{A} \otimes \mathcal{L}(F \otimes Y))^{\mathrm{id}}_{\mathrm{id}} \to (\mathcal{L}(X) \otimes \mathcal{C} \otimes \mathcal{L}(F \otimes Y))^{\mathrm{id}}_{\mathrm{id}}$$

restricts to a linear map

$$(\phi \otimes \mathrm{id})^X_Y : (\mathcal{L}(X) \otimes \mathcal{A} \otimes \mathcal{B} \otimes \mathcal{L}(Y))^{\mathrm{id}}_{\mathrm{id}} \to (\mathcal{L}(X) \otimes \mathcal{C} \otimes \mathcal{B} \otimes \mathcal{L}(Y))^{\mathrm{id}}_{\mathrm{id}}$$

which has the desired properties. If $\phi$ is a morphism, then $\phi^X_{F\otimes Y}$ and hence also $(\phi \otimes \mathrm{id})^X_Y$ are $*$-homomorphisms, so that $\phi \otimes \mathrm{id}$ is a morphism. □

Non-degenerate morphisms of $C^*$-families can be extended to multipliers:

**Proposition 11.3.16.** *Let $\phi \colon \mathcal{C} \to \mathcal{M}(\mathcal{D})$ be a non-degenerate morphism of $C^*$-families. If the $C^*$-family $\mathcal{D}$ is non-degenerate, then $\phi$ extends uniquely to a morphism $\mathcal{M}(\mathcal{C}) \to \mathcal{M}(\mathcal{D})$.*

*Proof.* Uniqueness follows, once existence is proved, by a standard argument. Denote by $F$ the underlying right $C^*$-bimodule of $\mathcal{D}$. Choose an approximate unit $(u_\nu)_\nu$ for the $C^*$-algebra $\mathcal{C}^{\mathrm{id}}_{\mathrm{id}}$ such that $0 \le u_\nu \le 1$ for all $\nu$.

For each $\rho \in \mathrm{PAut}(A)$, $\sigma \in \mathrm{PAut}(B)$, we construct an extension $\bar{\phi}^{\rho}_{\sigma} : \mathcal{M}(\mathcal{C})^{\rho}_{\sigma} \to \mathcal{M}(\mathcal{D})^{\rho}_{\sigma}$ of $\phi^{\rho}_{\sigma}$ as follows. Let $c \in \mathcal{M}(\mathcal{C})^{\rho}_{\sigma}$. Since $\phi$ and $\mathcal{D}$ are non-degenerate, the net $(\phi^{\rho}_{\sigma}(cu_\nu))_\nu$ converges strictly to some $\bar{\phi}^{\rho}_{\sigma}(c) \in \mathcal{L}^{\rho}_{\sigma}(F)$ (see Proposition 11.2.8 i)). Now $\bar{\phi}^{\rho}_{\sigma}(c) \in \mathcal{M}(\mathcal{D})^{\rho}_{\sigma}$ because

$$\bar{\phi}^{\rho}_{\sigma}(c)\mathcal{D}^{\mathrm{id}}_{\mathrm{id}} = \bar{\phi}^{\rho}_{\sigma}(c)\big[\phi^{\mathrm{id}}_{\mathrm{id}}(\mathcal{C}^{\mathrm{id}}_{\mathrm{id}})\mathcal{D}^{\mathrm{id}}_{\mathrm{id}}\big] \subseteq \big[\phi^{\rho}_{\sigma}(c\mathcal{C}^{\mathrm{id}}_{\mathrm{id}})\mathcal{D}^{\mathrm{id}}_{\mathrm{id}}\big] \subseteq \mathcal{D}^{\rho}_{\sigma},$$

and likewise $\mathcal{D}^{\mathrm{id}}_{\mathrm{id}}\bar{\phi}^{\rho}_{\sigma}(c) \subseteq \mathcal{D}^{\rho}_{\sigma}$.

We show that the family $\bar{\phi} \colon \mathcal{M}(\mathcal{C}) \to \mathcal{M}(\mathcal{D})$ thus defined is a morphism. Let $X$ be a right $C^*$-$\mathbb{C}$-$A$-bimodule and $Y$ a right $C^*$-$B$-$\mathbb{C}$-bimodule. By assumption on $\phi$, the $*$-homomorphism $\phi^X_Y$ is non-degenerate and extends to a $*$-homomorphism

$$\overline{\phi^X_Y} : M((\mathcal{L}(X) \otimes \mathcal{C} \otimes \mathcal{L}(Y))^{\mathrm{id}}_{\mathrm{id}}) \to M((\mathcal{L}(X) \otimes \mathcal{M}(\mathcal{D}) \otimes \mathcal{L}(Y))^{\mathrm{id}}_{\mathrm{id}}).$$

For all $R \in \mathcal{L}^{\mathrm{id}}_{\sigma'}(X)$, $S \in \mathcal{M}(\mathcal{C})^{\rho}_{\sigma}$, $T \in \mathcal{L}^{\rho'}_{\mathrm{id}}(Y)$, where $\sigma', \rho \in \mathrm{PAut}(A)$, $\sigma, \rho' \in \mathrm{PAut}(B)$, and $\sigma' \curlyvee \rho$, $\sigma \curlyvee \rho'$, the operators $\overline{\phi^X_Y}(R \otimes S \otimes T)$ and $R \otimes \bar{\phi}^{\rho}_{\sigma}(S) \otimes T$ are equal because they coincide with the strict limit of the net $(R \otimes \phi^{\rho}_{\sigma}(Su_\nu) \otimes T)_\nu$. Hence, $\overline{\phi^X_Y}$ restricts to a $*$-homomorphism

$$\bar{\phi}^X_Y : (\mathcal{L}(X) \otimes \mathcal{M}(\mathcal{C}) \otimes \mathcal{L}(Y))^{\mathrm{id}}_{\mathrm{id}} \to (\mathcal{L}(X) \otimes \mathcal{M}(\mathcal{D}) \otimes \mathcal{L}(Y))^{\mathrm{id}}_{\mathrm{id}}.$$ □

We are primarily concerned with the following examples of morphisms:

**Examples 11.3.17.** i) An inclusion of $C^*$-families is a morphism.

Let $\mathcal{C}$ be a $C^*$-family on a right $C^*$-$A$-$B$-bimodule $E$.

ii) Let $F$ be a right $C^*$-$A$-$B$-bimodule and $V \in \mathcal{L}_B^A(E, F)$ an isometry. Then $\mathrm{Ad}_V(\mathcal{C}) := [V\mathcal{C}V^*] \subseteq \mathcal{L}(F)$ is a $C^*$-family, and the formula $c \mapsto VcV^*$ defines an isomorphism $\mathrm{Ad}_V : \mathcal{C} \to \mathrm{Ad}_V(\mathcal{C})$. If $\mathcal{C}$ is a (non-degenerate) $\mathcal{O}(E)$-module, then $\mathrm{Ad}_V(\mathcal{C})$ is a (non-degenerate) $\mathcal{O}(F)$-module; if $V$ is unitary and $\mathcal{C}$ non-degenerate, then $\mathrm{Ad}_V(\mathcal{C})$ is non-degenerate.

iiii) Let $F$ be a $C^*$-module over $C$ and $\pi : C \to \mathcal{L}_B(E)$ a $*$-homomorphism such that $\pi(C)$ commutes with each operator in $\mathcal{C}$. Consider $F \ogreaterthan_\pi E$ as a right $C^*$-$A$-$B$-bimodule via $a(\eta \ogreaterthan \xi) := \eta \ogreaterthan a\xi$ for all $a \in A, \eta \in F, \xi \in E$. By a slight abuse of notation, we denote by $1 \ogreaterthan \mathcal{C} \subseteq \mathcal{L}(F \ogreaterthan_\pi E)$ the internal tensor product of $\mathcal{C}$ with the $C^*$-family generated by the identity operator on $F$. Then $1 \ogreaterthan \mathcal{C}$ is a $C^*$-family, and the map $T \mapsto 1 \ogreaterthan T$ defines a non-degenerate morphism $\mathcal{C} \to 1 \ogreaterthan \mathcal{C}$. If $\pi(\langle F|F\rangle) \subseteq \mathcal{L}_B(E)$ is non-degenerate, then this morphism is injective. If the $C^*$-family $\mathcal{C}$ is non-degenerate, then $1 \ogreaterthan \mathcal{C}$ is non-degenerate.

### 11.3.3 Hopf $C^*$-families

The notion of a Hopf $C^*$-family is the straightforward generalization of a bisimplifiable $C^*$-bialgebra (Definition 4.1.1). For our applications to pseudo-multiplicative unitaries, we define two variants which differ only in the usage of the flipped or non-flipped internal tensor product:

**Definition 11.3.18.** A *Hopf $C^*$-family* over a $C^*$-algebra $B$ is a non-degenerate $C^*$-family $\mathcal{A}$ on a right $C^*$-$B$-$B$-bimodule equipped with a non-degenerate morphism $\Delta : \mathcal{A} \to \mathcal{M}(\mathcal{A} \ogreaterthan \mathcal{A})$ called the *comultiplication* such that

i) $[\Delta(\mathcal{A})(1 \ogreaterthan \mathcal{A})] = \mathcal{A} \ogreaterthan \mathcal{A} = [\Delta(\mathcal{A})(\mathcal{A} \ogreaterthan 1)]$,

ii) $(\mathrm{id} \ogreaterthan \Delta) \circ \Delta = (\Delta \ogreaterthan \mathrm{id}) \circ \Delta$, that is, the following diagram commutes:

$$\begin{array}{ccc}
\mathcal{A} & \xrightarrow{\quad\Delta\quad} & \mathcal{M}(\mathcal{A} \ogreaterthan \mathcal{A}) \\
{\scriptstyle\Delta}\downarrow & & \downarrow{\scriptstyle\mathrm{id} \ogreaterthan \Delta} \\
\mathcal{M}(\mathcal{A} \ogreaterthan \mathcal{A}) & \xrightarrow[\Delta \ogreaterthan \mathrm{id}]{} & \mathcal{M}(\mathcal{A} \ogreaterthan \mathcal{A} \ogreaterthan \mathcal{A}).
\end{array}$$

A *flipped Hopf $C^*$-family* over a $C^*$-algebra $B$ is a non-degenerate $C^*$-family $\widehat{\mathcal{A}}$ on a right $C^*$-$B$-$B$-bimodule equipped with a non-degenerate morphism $\widehat{\Delta} : \widehat{\mathcal{A}} \to \mathcal{M}(\widehat{\mathcal{A}} \ogreaterthan \widehat{\mathcal{A}})$ called the *comultiplication* such that

i) $[\widehat{\Delta}(\widehat{\mathcal{A}})(1 \otimes \widehat{\mathcal{A}})] = \widehat{\mathcal{A}} \otimes \widehat{\mathcal{A}} = [\widehat{\Delta}(\widehat{\mathcal{A}})(\widehat{\mathcal{A}} \otimes 1)]$,

ii) $(\mathrm{id} \otimes \widehat{\Delta}) \circ \widehat{\Delta} = (\widehat{\Delta} \otimes \mathrm{id}) \circ \widehat{\Delta}$.

**Remark 11.3.19.** Condition i) implies that the morphisms $\Delta$ and $\widehat{\Delta}$ are non-degenerate; therefore we can extend the morphisms $\mathrm{id} \otimes \Delta$, $\Delta \otimes \mathrm{id}$ and $\mathrm{id} \otimes \widehat{\Delta}$, $\widehat{\Delta} \otimes \mathrm{id}$ to $\mathcal{M}(\mathcal{A} \otimes \mathcal{A})$ and $\mathcal{M}(\widehat{\mathcal{A}} \otimes \widehat{\mathcal{A}})$, respectively.

## 11.4 The legs of a decomposable pseudo-multiplicative unitary

Let $(E, \hat{\beta}, \beta)$ be a $C^*$-trimodule over a $C^*$-algebra $B$ and $V \colon E_{\hat{\beta}} \otimes E \to E \otimes_{\beta} E$ a pseudo-multiplicative unitary. We shall construct legs $(\widehat{\mathcal{A}}(V), \widehat{\Delta})$ and $(\mathcal{A}(V), \Delta)$, similarly as for a multiplicative unitary (see Section 7.2.1) or a pseudo-multiplicative unitary on a Hilbert space (see Section 10.3.2). The construction will be interesting, however, only if the right $C^*$-bimodule $_{\beta}E$ or the right $C^*$-bimodule $_{\hat{\beta}}E$ is decomposable. If that is the case and if the unitary $V$ satisfies a certain regularity condition, then $(\widehat{\mathcal{A}}(V), \widehat{\Delta})$ and $(\mathcal{A}(V), \Delta)$ are (flipped) Hopf $C^*$-families.

For detailed proofs, we refer to the article [153].

**The families $\widehat{\mathcal{A}}(V)$ and $\mathcal{A}(V)$.** The families $\widehat{\mathcal{A}}(V)$ and $\mathcal{A}(V)$ are spanned by operators of the following form:

**Lemma 11.4.1.** *Let $\rho, \sigma \in \mathrm{PAut}(B)$.*

i) *Let $\xi \in \mathcal{H}_{\rho}(_{\beta}E)$, $\xi' \in \mathcal{H}_{\sigma}(_{\beta}E)$. Then we have homogeneous operators*

$$_{\hat{\beta}}E \xrightarrow[(\rho,\mathrm{id})\text{-}hmg.]{|\xi\rangle_{[2]}} {}_{\beta_{[2]}}(E_{\hat{\beta}} \otimes E) \xrightarrow[(\mathrm{id},\mathrm{id})\text{-}hmg.]{V} {}_{\hat{\beta}_{[1]}}(E \otimes_{\beta} E) \xrightarrow[(\mathrm{id},\sigma)\text{-}hmg.]{|\xi']^*_{[2]}} {}_{\hat{\beta}}E,$$

*where $|\xi\rangle_{[2]}\zeta = \zeta \otimes \xi$ and $|\xi']_{[2]}\zeta = \zeta \otimes \xi'$ for all $\zeta \in E$. Put $[\xi'|_{[2]} := |\xi']^*_{[2]}$. The composition $\hat{a}_{(\xi',\xi)} := [\xi'|_{[2]} V |\xi\rangle_{[2]}$ belongs to $\mathcal{L}^{\rho}_{\sigma}(_{\hat{\beta}}E)$ and satisfies*

$$\langle\zeta'|\hat{a}_{(\xi',\xi)}\zeta\rangle = \sigma(\langle\zeta' \otimes \xi'|V(\zeta \otimes \xi)\rangle) \quad \textit{for all } \zeta, \zeta' \in E.$$

ii) *Let $\eta \in \mathcal{H}_{\rho^*}(_{\hat{\beta}}E)$, $\eta' \in \mathcal{H}_{\sigma^*}(_{\hat{\beta}}E)$. Then we have homogeneous operators*

$$_{\beta}E \xrightarrow[(\mathrm{id},\sigma)\text{-}hmg.]{|\eta]_{[1]}} {}_{\beta_{[2]}}(E_{\hat{\beta}} \otimes E) \xrightarrow[(\mathrm{id},\mathrm{id})\text{-}hmg.]{V} {}_{\hat{\beta}_{[1]}}(E \otimes_{\beta} E) \xrightarrow[(\rho,\mathrm{id})\text{-}hmg.]{|\eta'\rangle^*_{[1]}} {}_{\beta}E,$$

*where $|\eta]_{[1]}\zeta = \eta \otimes \zeta$ and $|\eta'\rangle_{[1]}\zeta = \eta' \otimes \zeta$ for all $\zeta \in E$. Put $\langle\eta'|_{[1]} := |\eta'\rangle^*_{[1]}$. The composition $a_{(\eta',\eta)} := \langle\eta'|_{[1]} V |\eta]_{[1]}$ belongs to $\mathcal{L}^{\rho}_{\sigma}(_{\beta}E)$ and satisfies*

$$\langle\zeta'|a_{(\eta',\eta)}\zeta\rangle = \langle\eta' \otimes \zeta'|V(\eta \otimes \zeta)\rangle \quad \textit{for all } \zeta, \zeta' \in E.$$

*Proof.* All assertions follow easily from Proposition 11.2.16, equation (10.6), and Proposition 11.2.8. □

We define families $\widehat{\mathcal{A}}(V) \subseteq \mathcal{L}({}_{\hat{\beta}}E)$ and $\mathcal{A}(V) \subseteq \mathcal{L}({}_{\beta}E)$ by

$$\widehat{\mathcal{A}}(V)^{\rho}_{\sigma} := \overline{\operatorname{span}}\{\hat{a}_{(\xi',\xi)} \mid \xi \in \mathcal{H}_{\rho}({}_{\beta}E),\ \xi' \in \mathcal{H}_{\sigma}({}_{\beta}E)\} \subseteq \mathcal{L}^{\rho}_{\sigma}({}_{\hat{\beta}}E)$$

and

$$\mathcal{A}(V)^{\rho}_{\sigma} := \overline{\operatorname{span}}\{a_{(\eta',\eta)} \mid \eta \in \mathcal{H}_{\sigma^*}({}_{\hat{\beta}}E),\ \eta' \in \mathcal{H}_{\rho^*}({}_{\hat{\beta}}E)\} \subseteq \mathcal{L}^{\rho}_{\sigma}({}_{\beta}E)$$

for each $\rho, \sigma \in \mathrm{PAut}(B)$.

Applying the ket-bra notation to families of homogeneous elements, we can rewrite the definition of $\widehat{\mathcal{A}}(V)$ and $\mathcal{A}(V)$ as follows. Define $|{}_{\beta}\mathcal{E}\rangle \subseteq \mathcal{L}_{\mathrm{id}}(B, {}_{\beta}E)$ and $|{}_{\beta}\mathcal{E}] \subseteq \mathcal{L}^{\mathrm{id}}(B, {}_{\beta}E)$ by

$$|{}_{\beta}\mathcal{E}\rangle^{\rho}_{\mathrm{id}} := \{|\xi\rangle \mid \xi \in \mathcal{H}_{\rho}({}_{\beta}E)\}, \quad |{}_{\beta}\mathcal{E}]^{\mathrm{id}}_{\sigma} := \{|\xi'] \mid \xi' \in \mathcal{H}_{\sigma^*}({}_{\beta}E)\}$$

(see Proposition 11.2.15). Put $\langle{}_{\beta}\mathcal{E}| := |{}_{\beta}\mathcal{E}\rangle^*$ and $[{}_{\beta}\mathcal{E}| := |{}_{\beta}\mathcal{E}]^*$. Replacing ${}_{\beta}E$ by ${}_{\hat{\beta}}E$, we similarly define $|{}_{\hat{\beta}}\mathcal{E}\rangle$, $\langle{}_{\hat{\beta}}\mathcal{E}|$, $|{}_{\hat{\beta}}\mathcal{E}]$, $[{}_{\hat{\beta}}\mathcal{E}|$. To all of these families, we apply the leg notation just like to individual ket-bra operators. Then

$$\widehat{\mathcal{A}}(V) = \big[[{}_{\beta}\mathcal{E}|_{[2]} V |{}_{\beta}\mathcal{E}\rangle_{[2]}\big] \quad \text{and} \quad \mathcal{A}(V) = \big[\langle{}_{\hat{\beta}}\mathcal{E}|_{[1]} V |{}_{\hat{\beta}}\mathcal{E}]_{[1]}\big].$$

For each $\theta \in \mathrm{PAut}(B)$, $b \in \mathcal{H}_{\theta}(B)$, we have an $(\mathrm{id}, \theta^*)$-homogeneous operator (see the proof of Proposition 11.2.23)

$$\alpha(b)\colon E \to E, \quad \xi \mapsto \xi b.$$

**Lemma 11.4.2.** *Let $b \in B$, $\xi, \xi' \in {}_{\hat{\beta}}E$, $\eta, \eta' \in {}_{\beta}E$ be homogeneous. Then*

$$\hat{a}_{(\xi',\xi)}\hat{\beta}(b) = \hat{a}_{(\xi',\xi b)}, \qquad \hat{a}_{(\xi',\xi)}\alpha(b) = \hat{a}_{(\xi' b^*,\xi)}, \quad \hat{a}_{(\xi',\xi)}\beta(b) = \beta(b)\hat{a}_{(\xi',\xi)},$$

$$\hat{\beta}(b)a_{(\eta',\eta)} = a_{(\eta',\eta)}\hat{\beta}(b), \quad \alpha(b)a_{(\eta',\eta)} = a_{(\eta',\eta b)}, \quad \beta(b)a_{(\eta',\eta)} = a_{(\eta' b^*,\eta)}.$$

*Proof.* This follows from similar calculations as in the proof of Lemma 10.3.7, see [153, Lemma 4.3]. □

To shorten the notation, we denote the family $\mathcal{H}(B)$ by $\mathcal{B}$. Define $\hat{\beta}(\mathcal{B}) \subseteq \mathcal{L}_{\mathrm{id}}({}_{\hat{\beta}}E)$ and $\alpha(\mathcal{B}) \subseteq \mathcal{L}^{\mathrm{id}}({}_{\hat{\beta}}E)$ by

$$\hat{\beta}(\mathcal{B})^{\rho}_{\mathrm{id}} := \{\hat{\beta}(b) \mid b \in \mathcal{H}_{\rho}(B)\}, \quad \alpha(\mathcal{B})^{\mathrm{id}}_{\sigma} := \{\alpha(b) \mid b \in \mathcal{H}_{\sigma^*}(B)\},$$

and similarly define $\beta(\mathcal{B}) \subseteq \mathcal{L}_{\mathrm{id}}({}_{\beta}E)$, $\alpha(\mathcal{B}) \subseteq \mathcal{L}^{\mathrm{id}}({}_{\beta}E)$.

Given a right $C^*$-bimodule $F$ and a family $\mathcal{C} \subseteq \mathcal{L}(F)$, we denote by $\mathcal{C}' \subseteq \mathcal{L}(F)$ the family of all homogeneous operators that commute with all operators of $\mathcal{C}$.

**Proposition 11.4.3.** i) $[\widehat{\mathcal{A}}(V)\alpha(\mathcal{B})] = [\widehat{\mathcal{A}}(V)\hat{\beta}(\mathcal{B})] = \widehat{\mathcal{A}}(V) \subseteq \beta(\mathcal{B})'$. *If* $\widehat{\mathcal{A}}(V)$ *is a* $C^*$*-family, then it is a non-degenerate* $\mathcal{O}({}_{\hat{\beta}}E)$*-module.*

ii) $[\alpha(\mathcal{B})\mathcal{A}(V)] = [\beta(\mathcal{B})\mathcal{A}(V)] = \mathcal{A}(V) \subseteq \hat{\beta}(\mathcal{B})'$. *If* $\mathcal{A}(V)$ *is a* $C^*$*-family, then it is a non-degenerate* $\mathcal{O}({}_{\beta}E)$*-module.*

*Proof.* This follows easily from Lemma 11.4.2, see [153, Proposition 4.4]. □

The families $\widehat{\mathcal{A}}(V)$ and $\mathcal{A}(V)$ are non-degenerate in the following sense:

**Proposition 11.4.4.** i) $[\widehat{\mathcal{A}}(V)^* E] = E$ *if* ${}_{\beta}E$ *is decomposable.*

ii) $[\mathcal{A}(V)E] = E$ *if* ${}_{\hat{\beta}}E$ *is decomposable.*

iii) *If* ${}_{\beta}E$ *and* ${}_{\hat{\beta}}E$ *are decomposable, then* $[\widehat{\mathcal{A}}(V)^* \mathcal{H}({}_{\hat{\beta}}E)] = \mathcal{H}({}_{\hat{\beta}}E)$ *and* $[\mathcal{A}(V)\mathcal{H}({}_{\beta}E)] = \mathcal{H}({}_{\beta}E)$.

*Proof.* See [153, Proposition 4.5]. □

Next, we show that $\widehat{\mathcal{A}}(V)$ and $\mathcal{A}(V)$ are closed under multiplication. The proof involves the following observation. If ${}_{\beta}E$ is decomposable, then

$$\begin{aligned} [V[\mathcal{H}({}_{\beta}E) \otimes E]] &= [V\,\mathcal{H}({}_{\beta_{[1]}}(E_{\hat{\beta}} \otimes E))] && \text{(Proposition 11.2.20)} \\ &= \mathcal{H}({}_{\beta_{[1]}}(E \otimes {}_{\beta}E)) && \text{(Equation (10.6))} \\ &= [\mathcal{H}({}_{\beta}E) \otimes \mathcal{H}({}_{\beta}E)] && \text{(Proposition 11.2.19 ii)).} \end{aligned} \tag{11.6}$$

**Proposition 11.4.5.** i) $[\widehat{\mathcal{A}}(V)\widehat{\mathcal{A}}(V)] = \widehat{\mathcal{A}}(V)$ *if* ${}_{\beta}E$ *is decomposable.*

ii) $[\mathcal{A}(V)\mathcal{A}(V)] = \mathcal{A}(V)$ *if* ${}_{\hat{\beta}}E$ *is decomposable.*

*Proof.* We only prove assertion i). By definition, $[\widehat{\mathcal{A}}(V)\widehat{\mathcal{A}}(V)] \subseteq \mathcal{L}({}_{\hat{\beta}}E)$ is the family of closed subspaces spanned by all compositions of the form

$$\hat{a}_{(\xi',\xi)}\hat{a}_{(\zeta',\zeta)}\colon E \xrightarrow{|\zeta\rangle_{[2]}} E_{\hat{\beta}} \otimes E \xrightarrow{V} E \otimes_{\beta} E \xrightarrow{[\zeta'|_{[2]}} E \xrightarrow{|\xi\rangle_{[2]}} E_{\hat{\beta}} \otimes E \xrightarrow{V} E \otimes_{\beta} E \xrightarrow{[\xi'|_{[2]}} E,$$

where $\xi, \xi', \zeta, \zeta' \in {}_{\beta}E$ are homogeneous. Moving $[\zeta'|_{[2]}$ to the left and $|\xi\rangle_{[2]}$ to the right, we can write $\hat{a}_{(\xi',\xi)}\hat{a}_{(\zeta',\zeta)}$ in the form

$$E \xrightarrow{|\xi \otimes \zeta\rangle_{[2]}} E_{\hat{\beta}} \otimes (E \otimes_{\beta} E) \xrightarrow{V_{[13]}} (E_{\hat{\beta}} \otimes E) \otimes_{\beta} E \xrightarrow{V_{[12]}} E \otimes_{\beta} E \otimes_{\beta} E \xrightarrow{[\xi' \otimes \zeta'|_{[2]}} E.$$

Using the pentagon equation (10.7) and Proposition 11.2.19 ii), we find that the product $[\widehat{\mathcal{A}}(V)\widehat{\mathcal{A}}(V)]$ is equal to the family spanned by all compositions

$$E \xrightarrow{|\omega\rangle_{[2]}} E_{\hat{\beta}} \otimes (E \otimes_{\beta} E) \xrightarrow{V^*_{[23]}} E_{\hat{\beta}} \otimes E_{\hat{\beta}} \otimes E \xrightarrow{V_{[12]}} E \otimes_{\beta} E_{\hat{\beta}} \otimes E \xrightarrow{V_{[23]}} E \otimes_{\beta} E \otimes_{\beta} E \xrightarrow{[\omega'|_{[2]}} E,$$

where $\omega, \omega' \in {}_{\beta_{[1]}}(E \ominus {}_\beta E)$ are homogeneous. Now equation (11.6) implies that $[\widehat{\mathcal{A}}(V)\widehat{\mathcal{A}}(V)]$ is equal to the family spanned by all compositions

$$E \xrightarrow{|\vartheta\rangle_{[2]}} E_{\hat\beta}\ominus E \xrightarrow{\mathrm{id}_E \ominus |\eta\rangle_{[2]}} E_{\hat\beta}\ominus E_{\hat\beta}\ominus E \xrightarrow{V_{[12]}} E\ominus_\beta E_{\hat\beta}\ominus E \xrightarrow{\mathrm{id}_E \ominus \langle\eta'|_{[2]}} E\ominus_\beta E \xrightarrow{[\vartheta'|_{[2]}} E,$$

where $\vartheta, \vartheta' \in {}_\beta E$ are homogeneous and $\eta, \eta' \in E$ are arbitrary. Since

$$(\mathrm{id} \ominus \langle\eta'|_{[2]})V_{[12]} = V(\mathrm{id} \ominus \langle\eta'|_{[2]}), \quad (\mathrm{id} \ominus \langle\eta'|_{[2]})(\mathrm{id} \ominus |\eta\rangle_{[2]}) = \mathrm{id} \ominus \hat\beta(\langle\eta'|\eta\rangle),$$

the composition above is equal to

$$E \xrightarrow{|\vartheta\rangle_{[2]}} E_{\hat\beta}\ominus E \xrightarrow{\hat\beta_{[2]}(\langle\eta'|\eta\rangle)V} E\ominus_\beta E \xrightarrow{[\vartheta'|_{[2]}} E,$$

that is, to $\hat a_{(\vartheta',\vartheta'')}$, where $\vartheta'' = \hat\beta(\langle\eta'|\eta\rangle)\vartheta$. Note that $\vartheta'' \in {}_\beta E$ is homogeneous because $\hat\beta$ commutes with $\beta$. Using the fact that $E$ is full and that $\hat\beta$ is non-degenerate, we find that $[\widehat{\mathcal{A}}(V)\widehat{\mathcal{A}}(V)]$ is equal to the family spanned by all operators $\hat a_{(\vartheta',\vartheta'')}$, where $\vartheta', \vartheta'' \in {}_\beta E$ are homogeneous. This is $\widehat{\mathcal{A}}(V)$. □

**The comultiplications $\widehat{\boldsymbol\Delta}$ and $\boldsymbol\Delta$.** Denote by $\widehat{\mathcal{B}} \subseteq \mathcal{L}({}_{\hat\beta}E)$ and $\mathcal{B} \subseteq \mathcal{L}({}_\beta E)$ the $C^*$-families generated by $\widehat{\mathcal{A}}(V)$ and $\mathcal{A}(V)$, respectively. By Lemma 11.4.1, $\widehat{\mathcal{B}}$ and $\mathcal{B}$ commute with $\beta(B)$ and $\hat\beta(B)$, respectively, so that we can define morphisms

$$\widehat{\mathcal{B}} \to \mathcal{L}({}_{\hat\beta_{[2]}}(E \ominus {}_\beta E)), \ \hat a \mapsto 1 \ominus \hat a, \quad \mathcal{B} \to \mathcal{L}({}_{\beta_{[1]}}(E_{\hat\beta} \ominus E)), \ a \mapsto a \ominus 1,$$

(see Example 11.3.17 iii)). Composing with conjugation by $V^*$ or $V$, respectively, we obtain morphisms (see Example 11.3.17 ii) and equation (10.6))

$$\widehat\Delta = \widehat\Delta_V : \widehat{\mathcal{B}} \to \mathcal{L}({}_{\hat\beta_{[2]}}(E_{\hat\beta} \ominus E)), \quad \hat a \mapsto V^*(1 \ominus \hat a)V,$$
$$\Delta = \Delta_V : \mathcal{B} \to \mathcal{L}({}_{\beta_{[1]}}(E \ominus {}_\beta E)), \quad a \mapsto V(a \ominus 1)V^*.$$

**Proposition 11.4.6.** i) *If ${}_\beta E$ is decomposable and $[\widehat\Delta(\widehat{\mathcal{B}})(1 \ominus \widehat{\mathcal{B}})] = \widehat{\mathcal{B}} \ominus \widehat{\mathcal{B}} = [\widehat\Delta(\widehat{\mathcal{B}})(\widehat{\mathcal{B}} \ominus 1)]$, then $(\widehat{\mathcal{B}}, \widehat\Delta)$ is a flipped Hopf $C^*$-family.*

ii) *If ${}_{\hat\beta}E$ is decomposable and $[\Delta(\mathcal{B})(1 \ominus \mathcal{B})] = \mathcal{B} \ominus \mathcal{B} = [\Delta(\mathcal{B})(\mathcal{B} \ominus 1)]$, then $(\mathcal{B}, \Delta)$ is a Hopf $C^*$-family.*

*Proof.* We only prove assertion i); the proof of assertion ii) is similar. Let us make the assumptions stated in i). By Proposition 11.4.4, the $C^*$-family $\widehat{\mathcal{B}}$ is non-degenerate, and by the second assumption, $\widehat\Delta$ is a non-degenerate morphism $\widehat{\mathcal{B}} \to \mathcal{M}(\widehat{\mathcal{B}} \ominus \widehat{\mathcal{B}})$. It remains to show that $\widehat\Delta$ is coassociative. Let $\hat a \in \widehat{\mathcal{B}}^\rho_\sigma$, where $\rho, \sigma \in \mathrm{PAut}(B)$. By definition $\widehat\Delta(\hat a) = V^*(1 \ominus \hat a)V$, and hence

$$(\widehat\Delta \ominus \mathrm{id})(\widehat\Delta(\hat a)) = V^*_{[12]}V^*_{[23]}(1 \ominus 1 \ominus \hat a)V_{[23]}V_{[12]},$$

where $V_{[23]}V_{[12]}\colon E_{\hat{\beta}} \otimes E_{\hat{\beta}} \otimes E \to E \otimes_{\beta} E_{\hat{\beta}} \otimes E \to E \otimes {}_{\beta} E \otimes {}_{\beta} E$. We squeeze in conjugation $V_{[12]}$ and find

$$(\widehat{\Delta} \otimes \mathrm{id})(\widehat{\Delta}(\hat{a})) = V_{[12]}^{*}V_{[23]}^{*}V_{[12]}((1 \otimes 1) \otimes \hat{a})V_{[12]}^{*}V_{[23]}V_{[12]},$$

where $V_{[12]}^{*}(V_{[23]}V_{[12]})\colon E_{\hat{\beta}} \otimes E_{\hat{\beta}} \otimes E \to E \otimes {}_{\beta}E \otimes {}_{\beta}E \to (E_{\hat{\beta}} \otimes E) \otimes {}_{\beta}E$. By the pentagon equation (10.7), $V_{[12]}^{*}V_{[23]}V_{[12]}$ is equal to the composition $V_{[13]}V_{[23]}\colon E_{\hat{\beta}} \otimes E_{\hat{\beta}} \otimes E \to E_{\hat{\beta}} \otimes (E \otimes {}_{\beta}E) \to (E_{\hat{\beta}} \otimes E) \otimes {}_{\beta}E$. Therefore,

$$(\widehat{\Delta} \otimes \mathrm{id})(\widehat{\Delta}(\hat{a})) = V_{[23]}^{*}V_{[13]}^{*}((1 \otimes 1) \otimes \hat{a})V_{[13]}V_{[23]} = (\mathrm{id} \otimes \widehat{\Delta})(\widehat{\Delta}(\hat{a})). \qquad \square$$

When we pass from $V$ to $V^{\mathrm{op}}$, the legs get switched as follows:

**Proposition 11.4.7.** *We have*

$$\widehat{\mathcal{A}}(V^{\mathrm{op}}) = \mathcal{A}(V)^{*}, \quad \widehat{\Delta}_{V^{\mathrm{op}}} = \mathrm{Ad}_{\Sigma} \circ \Delta_{V},$$
$$\mathcal{A}(V^{\mathrm{op}}) = \widehat{\mathcal{A}}(V)^{*}, \quad \Delta_{V^{\mathrm{op}}} = \mathrm{Ad}_{\Sigma} \circ \widehat{\Delta}_{V}.$$

*Proof.* The proof is similar to the proof of Lemma 7.2.5. □

**Regularity and decomposability conditions on $V$.** Evidently, it is desirable to have simple criteria that tell whether the legs of a given pseudo-multiplicative unitary are Hopf $C^*$-families or not. We shall see that the regularity condition known from multiplicative unitaries (Section 7.3.1) can be adapted to that purpose, but has to be refined by additional assumptions.

First, let us explain how the regularity condition can be adapted to the present setting. As before, let $V\colon E_{\hat{\beta}} \otimes E \to E \otimes {}_{\beta}E$ be a pseudo-multiplicative unitary. For each $\xi, \eta' \in E$, we have adjointable operators

$$|\xi\rangle_{[2]}\colon E \to E_{\hat{\beta}} \otimes E,\ \zeta \mapsto \zeta \otimes \xi, \quad |\eta'\rangle_{[1]}\colon E \to E \otimes {}_{\beta}E,\ \zeta \mapsto \eta' \otimes \zeta.$$

Put $\langle\eta'|_{[1]} := |\eta'\rangle_{[1]}^{*}$. The regularity condition involves operators of the form

$$c_{(\eta',\xi)} := \langle\eta'|_{[1]}V|\xi\rangle_{[2]} \in \mathcal{L}_{B}(E)$$

and the space

$$\mathcal{C}(V) := \overline{\mathrm{span}}\{c_{(\eta',\xi)} \mid \xi, \eta' \in E\} = \big[\langle E|_{[1]}V|E\rangle_{[2]}\big].$$

**Definition 11.4.8.** We call a pseudo-multiplicative unitary $V\colon E_{\hat{\beta}} \otimes E \to E \otimes {}_{\beta}E$ *semi-regular* if $\mathcal{K}_{B}(E) \subseteq \mathcal{C}(V)$, and *regular* if $\mathcal{K}_{B}(E) = \mathcal{C}(V)$.

To ensure that the left leg $(\widehat{\mathcal{A}}(V), \widehat{\Delta})$ or the right leg $(\mathcal{A}(V), \Delta)$ of $V$ is a Hopf $C^*$-family, we need to impose stronger assumptions than regularity alone. The first step towards this refinement is the following observation:

**Lemma 11.4.9.** i) $c_{(\eta',\xi)} \in \mathcal{L}^{\theta}_{\mathrm{id}}({}_{\beta}E)$ *for all* $\xi \in E$, $\eta' \in \mathcal{H}_{\theta^*}({}_{\beta}E)$, $\theta \in \mathrm{PAut}(B)$.

ii) $c_{(\eta',\xi)} \in \mathcal{L}^{\theta}_{\mathrm{id}}({}_{\hat{\beta}}E)$ *for all* $\xi \in \mathcal{H}_{\theta}({}_{\hat{\beta}}E)$, $\eta' \in E$, $\theta \in \mathrm{PAut}(B)$.

*Proof.* i) If $\xi, \eta', \theta$ are as in i), then we have homogeneous operators

$$
{}_{\beta}E \xrightarrow[\text{(id,id)-hmg.}]{|\xi\rangle_{[2]}} {}_{\beta_{[1]}}(E_{\hat{\beta}} \otimes E) \xrightarrow[\text{(id,id)-hmg.}]{V} {}_{\beta_{[1]}}(E \otimes {}_{\beta}E) \xrightarrow[(\theta,\text{id})\text{-hmg.}]{\langle\eta'|_{[1]}} {}_{\beta}E
$$

by Proposition 11.2.16 and equation (11.1), and the composition $c_{(\eta',\xi)}$ is $(\theta, \mathrm{id})$-homogeneous by Proposition 11.2.8.

ii) If $\xi, \eta', \theta$ are as in ii), then we have homogeneous operators

$$
{}_{\hat{\beta}}E \xrightarrow[(\theta,\text{id})\text{-hmg.}]{|\xi\rangle_{[2]}} {}_{\hat{\beta}_{[2]}}(E_{\hat{\beta}} \otimes E) \xrightarrow[\text{(id,id)-hmg.}]{V} {}_{\hat{\beta}_{[2]}}(E \otimes {}_{\beta}E) \xrightarrow[\text{(id,id)-hmg.}]{\langle\eta'|_{[1]}} {}_{\hat{\beta}}E,
$$

and the claim follows as above. □

We define families $\widehat{\mathcal{C}}(V) \subseteq \mathcal{L}_{\mathrm{id}}({}_{\beta}E)$ and $\mathcal{C}(V) \subseteq \mathcal{L}_{\mathrm{id}}({}_{\hat{\beta}}E)$ by

$$
\widehat{\mathcal{C}}(V)^{\theta}_{\mathrm{id}} := \overline{\mathrm{span}}\{c_{(\eta',\xi)} \mid \xi \in E,\ \eta' \in \mathcal{H}_{\theta^*}({}_{\beta}E)\},
$$
$$
\mathcal{C}(V)^{\theta}_{\mathrm{id}} := \overline{\mathrm{span}}\{c_{(\eta',\xi)} \mid \xi \in \mathcal{H}_{\theta}({}_{\hat{\beta}}E),\ \eta' \in E\}
$$

for all $\theta \in \mathrm{PAut}(B)$. In the notation introduced before Lemma 11.4.2, we have $\widehat{\mathcal{C}}(V) = [\langle{}_{\beta}\mathcal{E}|_{[1]}V|E\rangle_{[2]}]$ and $\mathcal{C}(V) = [\langle E|_{[1]}V|_{\hat{\beta}}\mathcal{E}\rangle_{[2]}]$.

The following definition involves the $C^*$-families $\widehat{\mathcal{C}}(V)$ and $\mathcal{C}(V)$, the $C^*$-families $\mathcal{K}_{\mathrm{id}}({}_{\beta}E)$ and $\mathcal{K}_{\mathrm{id}}({}_{\hat{\beta}}E)$ that were defined before Proposition 11.2.19, the family $\mathcal{F}({}_{\beta}E) \subseteq \mathcal{H}({}_{\beta}E)$ defined by

$$
\mathcal{F}_{\theta}({}_{\beta}E) := \overline{\mathrm{span}}\{\xi_1\theta_3(\langle\xi_3|\xi_2\rangle) \mid \xi_i \in \mathcal{H}_{\theta_i}({}_{\beta}E), \theta_i \in \mathrm{PAut}(B),\ \theta_1\theta_2\theta_3^* \leq \theta\}
$$
$$
\subseteq \mathcal{H}_{\theta}({}_{\beta}E),
$$

and the family $\mathcal{F}({}_{\hat{\beta}}E) \subseteq \mathcal{H}({}_{\hat{\beta}}E)$ which is defined similarly.

**Definition 11.4.10.** We call a pseudo-multiplicative unitary $V\colon E_{\hat{\beta}} \otimes E \to E \otimes {}_{\beta}E$ *decomposably left regular* if the following conditions hold:

i) ${}_{\beta}E$ is decomposable, ii) $\widehat{\mathcal{C}}(V) = \mathcal{K}_{\mathrm{id}}({}_{\beta}E)$, iii) $\mathcal{F}({}_{\beta}E) = \mathcal{H}({}_{\beta}E)$.

We call $V$ *decomposably right regular* if the following conditions hold:

i) ${}_{\hat{\beta}}E$ is decomposable, ii) $\mathcal{C}(V) = \mathcal{K}_{\mathrm{id}}({}_{\hat{\beta}}E)$, iii) $\mathcal{F}({}_{\hat{\beta}}E) = \mathcal{H}({}_{\hat{\beta}}E)$.

**Theorem 11.4.11.** *Let $V\colon E_{\hat{\beta}} \otimes E \to E \otimes {}_{\beta}E$ be a pseudo-multiplicative unitary. If $V$ is decomposably left/right regular, then $(\widehat{\mathcal{A}}(V), \widehat{\Delta})/(\mathcal{A}(V), \Delta)$ is a flipped/ordinary Hopf $C^*$-family.*

*Proof.* See [152]. □

**The legs of the pseudo-multiplicative unitary of a groupoid.** Let us determine the legs of the pseudo-multiplicative unitary $W_G$ associated to a decomposable groupoid $G$ (see Example 11.1.5). We use the same notation as in Example 11.1.5 and in Proposition 11.2.32, and denote by $\pi\colon C_0(G) = B \to \mathcal{L}_B(L^2(G,\lambda))$ the representation given by multiplication operators.

**Proposition 11.4.12.** *Let $G$ be a locally compact, second countable, Hausdorff groupoid with left Haar system $\lambda$. If $G$ is decomposable, then the pseudo-multiplicative unitary $W_G$ is decomposably left regular and decomposably right regular.*

*Proof.* See [152]. □

**Proposition 11.4.13.** *Let $G$ be a locally compact, second countable, Hausdorff groupoid with left Haar measure $\lambda$.*

i) *The family $\widehat{\mathcal{A}}(W_G)$ is a $C^*$-family, and*

$$\widehat{\mathcal{A}}(W_G)^{\rho}_{\sigma} \subseteq \widehat{\mathcal{A}}(W_G)^{\mathrm{id}}_{\mathrm{id}} = \pi(C_0(G)) \quad \textit{for all } \rho,\sigma \in \mathrm{PAut}(B).$$

*For all $f \in C_0(G)$, $\zeta \in L^2(G,\lambda)_s \otimes L^2(G,\lambda)$ and $(x,y) \in G^2_{s,r}$,*

$$(\widehat{\Delta}(\pi(f))\zeta)(x,y) = f(xy)\zeta(x,y).$$

*In particular, $(\widehat{\mathcal{A}}(W_G),\widehat{\Delta})$ is a Hopf $C^*$-family.*

ii) *If $G$ is decomposable, then $\mathcal{A}(W_G)$ is a $C^*$-family, and*

$$\mathcal{A}(W_G)^{\rho}_{\sigma} = \overline{\mathrm{span}}\{L(f) \mid f \in C_c(U),\ U \subseteq G \textit{ open homogeneous}, \\ q_{U*} \leq \rho \textit{ and } q_{U*} \leq \sigma\}$$

*for all $\rho,\sigma \in \mathrm{PAut}(B)$. For every open homogeneous subset $U \subseteq G$ and all $f \in C_c(U)$, $\zeta \in L^2(G,\lambda) \otimes_r L^2(G,\lambda)$, $(x,y) \in G^2_{r,r}$,*

$$(\Delta(L(f))\zeta)(x,y) = \int_{G^{r_G(x)}} f(z)\zeta(z^{-1}x, z^{-1}y)\,d\lambda^{r_G(x)}(z).$$

*Proof.* This follows from Theorem 11.4.11, Proposition 11.4.12, and similar calculations as in the case where $G$ is a group (see Example 7.2.13). □

## 11.5 Coactions of Hopf $C^*$-families

This section gives a brief overview on coactions of Hopf $C^*$-families, which were introduced and studied in [152]. We state the pertaining definitions and explain how such coactions simultaneously generalize actions of groupoids on $C^*$-algebras and Fell bundles on decomposable groupoids. Let us mention that one can define reduced crossed products for coactions of Hopf $C^*$-families and prove an analogue of the Takesaki–Takai–Baaj–Skandalis duality theorem in a similar way as it was done for coactions of $C^*$-bialgebras in Chapter 9; see [152].

Throughout this section, let $A$ and $B$ be a $C^*$-algebra.

The concept of a coaction carries over to Hopf $C^*$-families as follows:

**Definition 11.5.1.** Let $(\mathcal{A}, \Delta)$ be a Hopf $C^*$-family on a right $C^*$-$B$-$B$-bimodule and $\mathcal{C}$ a non-degenerate $C^*$-family on a right $C^*$-$A$-$B$-bimodule.

A (*right*) *coaction* of $(\mathcal{A}, \Delta)$ on $\mathcal{C}$ is a non-degenerate morphism $\delta\colon \mathcal{C} \to \mathcal{M}(\mathcal{C} \ominus \mathcal{A})$ such that $[\delta(\mathcal{C})(1 \ominus \mathcal{A})] \subseteq \mathcal{C} \ominus \mathcal{A}$ and $(\delta \ominus \mathrm{id}) \circ \delta = (\mathrm{id} \ominus \Delta) \circ \delta$. We call $(\mathcal{C}, \delta)$ an $(\mathcal{A}, \Delta)$-*$C^*$-family* if $\delta$ is injective and $[\delta(\mathcal{C})(1 \ominus \mathcal{A})] = \mathcal{C} \ominus \mathcal{A}$.

Let $\mathcal{D}$ be a non-degenerate $C^*$-family on a right $C^*$-$A$-$B$-bimodule, and let $\delta_{\mathcal{C}}$ and $\delta_{\mathcal{D}}$ be coactions of $(\mathcal{A}, \Delta)$ on $\mathcal{C}$ and $\mathcal{D}$, respectively. A non-degenerate morphism $\phi\colon \mathcal{C} \to \mathcal{M}(\mathcal{D})$ is called *covariant* if $\delta_{\mathcal{D}} \circ \phi = (\phi \ominus \mathrm{id}) \circ \delta_{\mathcal{C}}$.

**Remark 11.5.2.** Similarly as above, one can define left coactions of Hopf $C^*$-families and right and left coactions of flipped Hopf $C^*$-families.

To consider coactions of Hopf $C^*$-families on $C^*$-algebras, we need to construct an internal tensor product of $C^*$-families with $C^*$-algebras. The necessary definitions are straightforward. Nevertheless, for Hopf $C^*$-families, it seems to be more natural to consider coactions on $C^*$-families than coactions on $C^*$-algebras.

**Definition 11.5.3.** A *$C^*$-$B$-algebra* is a $C^*$-algebra $C$ equipped with a non-degenerate $*$-homomorphism $B \to M(C)$. A *morphism* between $C^*$-$B$-algebras $C$ and $D$ is a non-degenerate $*$-homomorphism $\pi\colon C \to M(D)$ that satisfies $\pi(bc) = b\pi(c)$ for all $b \in B$ and $c \in C$. Given a $C^*$-$B$-algebra $C$, we define a family $\mathcal{H}(C) = (\mathcal{H}_\theta(C))_{\theta \in \mathrm{PAut}(B)}$ by

$$\mathcal{H}_\theta(C) := \{c \in [C\,\mathrm{Dom}(\theta)] \mid cb = \theta(b)c \text{ for all } b \in \mathrm{Dom}(\theta)\}, \quad \theta \in \mathrm{PAut}(B).$$

Let $C$ be a $C^*$-$B$-algebra. Then we can consider $C$ as a right $C^*$-$B$-$C$-bimodule, and each element $c \in \mathcal{H}_\theta(C)$, where $\theta \in \mathrm{PAut}(B)$, can be considered as a $(\theta, \mathrm{id})$-homogeneous operator via left multiplication. Thus we can identify $\mathcal{H}(C)$ with a subfamily of $\mathcal{L}_{\mathrm{id}}(C)$. Evidently $\mathcal{H}(C) \subseteq \mathcal{L}_{\mathrm{id}}(C)$ is a $C^*$-family.

**Definition 11.5.4.** Let $\mathcal{C}$ be a $C^*$-family on a $C^*$-$A$-$B$-bimodule $E$, and let $C$ be a $C^*$-$B$-algebra. The *internal tensor product* $\mathcal{C} \ominus C$ is the subspace of $\mathcal{L}_C(E \ominus C)$ given by

$$\mathcal{C} \ominus C := \overline{\mathrm{span}}\{c' \ominus c \mid c' \in \mathcal{C}^{\rho}_{\sigma},\ c \in \mathcal{H}_\theta(C), \text{ where } \rho \in \mathrm{PAut}(A),\ \sigma, \theta \in \mathrm{PAut}(B),\ \sigma \curlyvee \theta\}.$$

The internal tensor product constructed above has the following properties:

**Proposition 11.5.5.** *Let $\mathcal{C}$, $E$ and $C$ be as in the preceding definition.*

i) *The space $\mathcal{C} \ominus C$ is a $C^*$-algebra.*

ii) *If $B$ is decomposable and $\mathcal{C}$ is a non-degenerate $\mathcal{O}(E)$-module, then the natural representation $A \to \mathcal{L}_C(E \ominus C)$ turns $\mathcal{C} \ominus C$ into a $C^*$-$A$-algebra.*

iii) *For every morphism of $C^*$-$B$-algebras $\pi : C \to M(D)$, there exists a non-degenerate $*$-homomorphism*

$$\mathrm{id} \ominus \pi : \mathcal{C} \ominus C \to M(\mathcal{C} \ominus D)$$

*such that $(\mathrm{id} \ominus \pi)(c' \ominus c) = c' \ominus \pi(c)$ for all $c' \in \mathcal{C}^{\rho}_{\sigma}$, $c \in \mathcal{H}_{\theta}(C)$, where $\rho \in \mathrm{PAut}(A)$, $\sigma, \theta \in \mathrm{PAut}(B)$, $\sigma \curlyvee \theta$. If $B$ is decomposable and $\mathcal{C}$ is a non-degenerate $\mathcal{O}(E)$-module, then $\mathrm{id} \ominus \pi$ is a morphism of $C^*$-$A$-algebras.*

iv) *Let $\mathcal{D}$ be a $C^*$-family on some right $C^*$-$A$-$B$-bimodule $F$. Assume that $A$ and $B$ are decomposable and that $\mathcal{C}$ and $\mathcal{D}$ are non-degenerate $\mathcal{O}(E)$- or $\mathcal{O}(F)$-modules, respectively. Then every non-degenerate morphism of $C^*$-families $\psi : \mathcal{C} \to \mathcal{M}(\mathcal{D})$ induces a morphism of $C^*$-$A$-algebras*

$$\psi \ominus \mathrm{id} : \mathcal{C} \ominus C \to M(\mathcal{D} \ominus C)$$

*such that $(\psi \ominus \mathrm{id})(c' \ominus c) = \psi(c') \ominus c$ for all $c' \in \mathcal{C}^{\rho}_{\sigma}$, $c \in \mathcal{H}_{\theta}(C)$, where $\rho \in \mathrm{PAut}(A)$, $\sigma, \theta \in \mathrm{PAut}(B)$, $\sigma \curlyvee \theta$.*

v) *The internal tensor product is associative in the natural sense.*

*Proof.* See [152]. □

Similarly as above, one can define a flipped internal tensor product $C \otimes \mathcal{C}$ which has analogous properties. In the following definition, we use this flipped internal tensor product because we want to focus on right coactions:

**Definition 11.5.6.** Let $(\mathcal{A}, \Delta)$ be a Hopf $C^*$-family on a right $C^*$-$B$-$B$-bimodule $E$. Assume that $B$ is decomposable and that $\mathcal{A}$ is a non-degenerate $\mathcal{O}(E)$-module.

i) A (*right*) *coaction* of $(\mathcal{A}, \Delta)$ on a $C^*$-$B$-algebra $C$ is a morphism of $C^*$-$B$-algebras $\delta : C \to M(C \otimes \mathcal{A})$ such that $\delta(C)(1 \otimes \mathcal{A}) \subseteq C \otimes \mathcal{A}$ and $(\delta \otimes \mathrm{id}) \circ \delta = (\mathrm{id} \otimes \Delta) \circ \delta$.

ii) A morphism of $C^*$-$B$-algebras $\pi : C \to M(D)$ is called *covariant* with respect to coactions $\delta_C$ and $\delta_D$ of $(\mathcal{A}, \Delta)$ on $C$ and $D$, respectively, if $\delta_D \circ \pi = (\pi \otimes \mathrm{id}) \circ \delta_C$.

**Coactions, groupoid actions, and Fell bundles on groupoids.** Coactions of Hopf $C^*$-families generalize groupoid actions and Fell bundles on decomposable groupoids, very much like coactions of $C^*$-bialgebras generalize group actions and Fell bundles on groups (see Section 9.2). Let us briefly outline these relations.

Consider a locally compact, second countable, Hausdorff groupoid $G$ with left Haar system $\lambda$, and assume that $G$ is decomposable. The associated pseudo-multiplicative unitary $W_G$ (Example 11.1.5) gives rise to a flipped Hopf $C^*$-family $(\widehat{\mathcal{A}}(W_G), \widehat{\Delta})$ and a Hopf $C^*$-family $(\mathcal{A}(W_G), \Delta)$ (Proposition 11.4.13, 11.4.12). Note that the Hopf $C^*$-family $(\mathcal{A}(W_G), \Delta)$ can also be considered as a flipped Hopf $C^*$-family: We have a natural isomorphism

$$L^2(G,\lambda) \ominus {}_r L^2(G,\lambda) \cong L^2(G,\lambda)_r \oslash L^2(G,\lambda),$$

which induces an isomorphism $\mathcal{A}(W_G) \ominus \mathcal{A}(W_G) \cong \mathcal{A}(W_G) \oslash \mathcal{A}(W_G)$.

The concepts of a group action and of a Fell bundle generalize to groupoids easily, see, for example, [98] and [85].

**Theorem 11.5.7.** *Let G be a groupoid as above.*

i) *Injective coactions of the flipped Hopf $C^*$-family $(\widehat{\mathcal{A}}(W_G), \widehat{\Delta})$ on a $C^*$-$C_0(G^0)$-algebra $C$ correspond bijectively with continuous actions of the groupoid $G$ on $C$.*

ii) *Assume that $G$ is $r$-discrete. Then injective coactions of the flipped Hopf $C^*$-family $(\mathcal{A}(W_G), \Delta)$ on $C^*$-$C_0(G^0)$-algebras correspond bijectively (up to isomorphism) with upper semi-continuous Fell bundles on $G$.*

*Proof.* See [152]. □

# Chapter 12
# Appendix

## 12.1 $C^*$-algebras

Standard references on $C^*$-algebras are, for example, [34], [113], [121], [137], [149].

**Basic definitions and facts.** A $*$-*algebra* is a complex algebra $A$ with a conjugate-linear antiautomorphism $*\colon A \to A$ called the *involution* of $A$. A $*$-homomorphism between two $*$-algebras is a homomorphism of complex algebras that intertwines the respective involutions.

A *Banach algebra* is an algebra $A$ equipped with a norm $\|\cdot\|$ such that $A$ is complete and $\|ab\| \leq \|a\| \cdot \|b\|$ for all $a, b \in A$. A *Banach $*$-algebra* is a Banach algebra $A$ that is simultaneously a $*$-algebra such that $\|a\| = \|a^*\|$ for all $a \in A$. A Banach $*$-algebra $A$ is a $C^*$-*algebra* if $\|a^*a\| = \|a\|^2$ for all $a \in A$. A $C^*$-*(semi)norm* on a $*$-algebra $A$ is a (semi)norm that satisfies $\|a^*\| = \|a\|$ and $\|a^*a\| = \|a\|^2$ for all $a \in A$.

An *approximate unit* of a Banach algebra $A$ is a net $(u_\nu)_\nu$ of elements of $A$ that satisfies $\lim_\nu u_\nu a = a = \lim_\nu a u_\nu$ for all $a \in A$.

Let $A$ be a $C^*$-algebra. An element $a \in A$ is *central* if $ab = ba$ for all $b \in A$, a *partial isometry* if $aa^*a = a$, an *isometry* if $a^*a = 1$, a *unitary* if $a^*a = aa^* = 1$, *self-adjoint* if $a = a^*$, a *projection* if $a^2 = a = a^*$, *positive* if it can be written in the form $a = b^*b$ for some $b \in A$. The *natural order* on $A$ is given by $a \geq b :\Leftrightarrow a - b$ is positive.

Every $C^*$-algebra contains an approximate unit consisting of positive elements which have norm less than 1.

By an *ideal* of $A$ we mean a two-sided ideal in the algebraic sense that is closed with respect to the involution and with respect to the norm; thus, every ideal is a $C^*$-algebra again.

If $A$ is non-unital, then the *unitization* of $A$, denoted by $A^+$, is the $C^*$-algebra with underlying vector space $A \oplus \mathbb{C}$ and multiplication, involution, and norm given by

$$(a, \lambda) \cdot (b, \mu) := (ab + \lambda b + a\mu, \lambda\mu), \quad (a, \lambda)^* := (a^*, \bar{\lambda}),$$
$$\|(a, \lambda)\| := \sup\{\|ac + \lambda c\| \mid c \in A, \|c\| \leq 1\} \quad \text{for all } a, b \in A,\ \lambda, \mu \in \mathbb{C}.$$

The algebra $A$ is contained in $A^+$ as a two-sided ideal.

The *multiplier algebra* $M(A)$ of $A$ defined in Section 2.1.1 is a $C^*$-algebra with respect to the operator norm $\|T\| := \sup\{\|Ta\| \mid a \in A, \|a\| \leq 1\}$. The *strict topology* on $M(A)$ is the topology defined by the seminorms $\phi_a, \psi_a$ $(a \in A)$, where $\phi_a(T) = \|Ta\|$ and $\psi_a(T) = \|aT\|$ for all $a \in A$ and $T \in M(A)$. The $C^*$-algebra $A$ is dense in $M(A)$ with respect to the strict topology: if $(u_\nu)_\nu$ is a bounded approximate unit for $A$ and $T \in M(A)$, then $(Tu_\nu)_\nu$ is a net in $A$ that converges strictly to $T$.

Every $*$-homomorphism of $C^*$-algebras is automatically norm-decreasing. Let $A$ and $B$ be $C^*$-algebras. A $*$-homomorphism $\pi\colon A \to M(B)$ is called *non-degenerate* if $\pi(A)B$ is linearly dense in $B$. In that case, $\pi$ extends uniquely to a strictly continuous $*$-homomorphism $M(A) \to M(B)$.

For each unitary $U \in A$, we define $\mathrm{Ad}_U\colon A \to A$ by $a \mapsto UaU^*$; this is a $*$-automorphism.

**Commutative $C^*$-algebras and Gelfand duality.** The category of all commutative $C^*$-algebras with non-degenerate $*$-homomorphisms is equivalent to the category of all locally compact Hausdorff spaces with proper continuous maps. This correspondence is established as follows:

Let $X$ be a locally compact Hausdorff space. Then the algebra $C_0(X)$ of complex-valued functions on $X$ vanishing at infinity, equipped with the supremum norm, is a commutative $C^*$-algebra. The multiplier algebra $M(C_0(X))$ is canonically isomorphic to the $C^*$-algebra $C_b(X)$ of all bounded continuous functions on $X$. A map $\phi\colon X \to Y$ between locally compact Hausdorff spaces induces a non-degenerate $*$-homomorphism $\phi^*\colon C_0(Y) \to M(C_0(X))$ via $(\phi^* f)(x) = f(\phi(x))$. The image of $\phi^*$ is contained in $C_0(X)$ if and only if $\phi$ is proper.

Conversely, let $A$ be a commutative $C^*$-algebra. A *character* on $A$ is a non-zero $*$-homomorphism $A \to \mathbb{C}$. Denote by $\widehat{A}$ the set of all characters of $A$ endowed with the weakest topology that makes all functions of the form $\chi \mapsto \chi(a)$, where $a \in A$ and $\chi \in \widehat{A}$, continuous. Then $\widehat{A}$ is a locally compact Hausdorff space, and the Gelfand–Naimark theorem says that the natural map $A \to C_0(\widehat{A})$ is a natural isomorphism. Each $*$-homomorphism $\phi\colon A \to B$ of commutative $C^*$-algebras induces a proper continuous map $\phi^*\colon \widehat{B} \to \widehat{A}$ via $\chi \mapsto \chi \circ \phi$.

**Representations.** For every Hilbert space $H$, the space $\mathcal{L}(H)$ of all bounded linear operators on $H$ is a $C^*$-algebra, and the subspace $\mathcal{K}(H) \subseteq \mathcal{L}(H)$ of all compact linear operators is an ideal. Moreover, $\mathcal{L}(H) = M(\mathcal{K}(H))$.

A (*bounded*) *representation* of a $*$-algebra $A$ on a Hilbert space $H$ is a $*$-homomorphism $\pi\colon A \to \mathcal{L}(H)$. The representation $\pi$ is called *faithful* if $\pi$ is injective, and *non-degenerate* if $\pi(A)H$ is linearly dense in $H$. Note that if $A$ is a $C^*$-algebra, then every $*$-homomorphism $A \to \mathcal{L}(H)$ is automatically bounded.

**Positive functionals and states.** Let $A$ be a $C^*$-algebra. We denote by $A'$ the space of continuous linear functionals $A \to \mathbb{C}$. A linear map $\phi\colon A \to B$ of $C^*$-algebras is called *positive* if it preserves the order. A positive map is always continuous. More precisely, if $\phi \in A'$ is positive, then $\|\phi\| = \lim_\nu |\phi(u_\nu)| < \infty$ for every approximate unit $(u_\nu)_\nu$ of $A$. In particular, $\|\phi\| = \phi(1)$ if $A$ is unital. A *state* on a $C^*$-algebra $A$ is a positive linear functional of norm one.

By the Hahn–Banach theorem, there exists for each $a \in A$ a $\phi \in A'$ such that $\phi(a) = \|a\|$ and $\|\phi\| = 1$. For each $\omega \in A'$, there exist positive functionals $\omega_1, \dots, \omega_4 \in A'$ such that $\omega = \omega_1 - \omega_2 + i(\omega_3 - \omega_4)$.

**GNS-construction.** A *GNS-construction* for a state $\phi$ on a $C^*$-algebra $A$ consists of a Hilbert space $H$, a linear map $\Lambda\colon A \to H$ with dense image, and a representation $\pi\colon A \to \mathcal{L}(H)$, such that $\langle \Lambda(a)|\Lambda(b)\rangle = \phi(a^*b)$ and $\pi(a)\Lambda(b) = \Lambda(ab)$ for all $a, b \in A$. It is easy to see that the GNS-construction is unique up to a unitary transformation. For every GNS-construction $(H, \Lambda, \pi)$, there exists a unique vector $\zeta \in H$, called the *cyclic vector*, such that $\Lambda(a) = \pi(a)\zeta$ for all $a \in A$; moreover, $\|\zeta\| = 1$ and $\langle \zeta|\pi(a)\zeta\rangle = \phi(a)$ for all $a \in A$. If $(u_\nu)_\nu$ is a bounded approximate unit for $A$, then $(\pi(u_\nu)\xi)_\nu$ converges in norm to $\xi$ for each $\xi \in H$.

A particular GNS-construction for $\phi$ can be obtained as follows. By the Cauchy–Schwarz inequality, the set $N := \{a \in A \mid \phi(a^*a) = 0\} \subseteq A$ is a subspace (even a left ideal). Denote by $H$ the completion of the quotient $A/N$ with respect to the inner product $\langle a + N|b + N\rangle := \phi(a^*b)$. Then $H$ is a Hilbert space, the map $\Lambda\colon A \to H$, $a \mapsto a + N$, has dense image, and there exists a representation $\pi\colon A \to \mathcal{L}(H)$ such that $(H, \Lambda, \pi)$ is a GNS-construction.

The *universal GNS-representation* for a $C^*$-algebra $A$ is obtained as follows. For each state $\phi$ on $A$, choose a GNS-representation $(H_\phi, \Lambda_\phi, \pi_\phi)$. Put $H := \bigoplus_\phi H_\phi$, and define $\pi\colon A \to \mathcal{L}(H)$ by $(\pi(a)\xi)_\phi = \pi_\phi(a)\xi_\phi$ for every $a \in A$, $\xi = (\xi_\phi)_\phi \in H$, and every state $\phi$. Then $\pi$ is called the *universal GNS-representation* for $A$. This representation is always faithful; in particular, $A$ is isomorphic to $\pi(A) \subseteq \mathcal{L}(H)$.

**Enveloping $C^*$-algebra of a $*$-algebra.** Let $A$ be a $*$-algebra. If for every $a \in A$, the supremum

$$|a| := \sup\{\|\pi(a)\| \mid \pi\colon A \to B \text{ is a } *\text{-homomorphism, } B \text{ a } C^*\text{-algebra}\}$$

is finite, then $|\cdot|\colon a \to |a|$ is a $C^*$-seminorm on $A$. The completion of the quotient $A/\{a \in A \mid |a| = 0\}$ with respect to the norm induced by $|\cdot|$ is a $C^*$-algebra, called the *enveloping $C^*$-algebra* of $A$. Denote this $C^*$-algebra by $A_e$ and the natural map $A \to A_e$ by $q_e$. By definition, $A_e$ and $q_e$ have the following universal property: For every $*$-homomorphism $\pi\colon A \to B$, where $B$ is a $C^*$-algebra, there exists a unique $*$-homomorphism $\pi_e \colon A_e \to B$ such that $\pi = \pi_e \circ q_e$. Note that for each $a \in A$,

$$|a| = \sup\{\|\pi(a)\| \mid \pi \text{ is a bounded representation of } A \text{ on some Hilbert space } H\}.$$

**The minimal tensor product.** The *minimal tensor product* of two $C^*$-algebras $A$ and $B$ is defined as follows. Denote by $\pi_A \colon A \to \mathcal{L}(H_A)$ and $\pi_B \colon B \to \mathcal{L}(H_B)$ the universal GNS-representations of $A$ and $B$, respectively. Then the minimal tensor product $A \otimes B$ is the $C^*$-subalgebra of $\mathcal{L}(H_A \otimes H_B)$ generated by all operators of the form $\pi_A(a) \otimes \pi_B(b)$, where $a \in A$ and $b \in B$.

One has a canonical inclusion $M(A) \otimes M(B) \subseteq M(A \otimes B)$.

The minimal tensor product is functorial, that is, for each pair of $*$-homomorphisms $\phi_i \colon A_i \to B_i$ of $C^*$-algebras, where $i = 1, 2$, there exists a $*$-homomorphism $\phi_1 \otimes \phi_2 \colon A_1 \otimes A_2 \to B_1 \otimes B_2$ such that $(\phi_1 \otimes \phi_2)(a_1 \otimes a_2) = \phi_1(a_1) \otimes \phi_2(a_2)$ for all $a_i \in A_i$.

For each state $\phi$ on $A$ and each state $\psi$ on $B$, there exists a unique product state $\phi \otimes \psi$ on $A \otimes B$ such that $(\phi \otimes \psi)(a \otimes b) = \phi(a)\psi(b)$ for all $a \in A$ and $b \in B$. More generally, for each $\phi \in A'$ and $\psi \in B'$, there exists a unique functional $\phi \otimes \psi \in (A \otimes B)'$ such that $(\phi \otimes \psi)(a \otimes b) = \phi(a)\psi(b)$ for all $a \in A$, $b \in B$, and $\|\phi \otimes \psi\| \leq \|\phi\|\|\psi\|$ [149, Section IV.4, page 208].

**Complements on linear functionals.** The following results are well known, but it seems useful to include proofs.

**Proposition 12.1.1.** *Let $A$ be a $C^*$-algebra. Then each of the sets*

$$\{\phi(\cdot\, b) \mid \phi \in A', b \in A\}, \quad \{\phi(a\, \cdot) \mid \phi \in A', a \in A\},$$
$$\{\phi(a \cdot b) \mid \phi \in A', a, b \in A\}$$

*is equal to $A'$.*

*Proof.* We only prove the assertion concerning the first set; the remaining assertions follow similarly. The map $A \times A' \to A'$, $(b, \phi) \mapsto \phi(\cdot\, b)$, turns $A'$ into a Banach module over $A$. We shall show that $AA'$ is linearly dense in $A'$, and then the Cohen Factorization Theorem [62, Theorem 32.22], [36] implies $AA' = A'$.

Let $\phi \in A'$ be a state with GNS-construction $(H, \Lambda, \pi)$ and cyclic vector $\zeta$. Choose an approximate unit $(u_\nu)_\nu$ for $A$. Since $\lim_\nu \pi(u_\nu)\zeta = \zeta$, the functional $\phi(\cdot\, u_\nu) = \langle \zeta | \pi(\cdot)\pi(u_\nu)\zeta \rangle$ converges in norm to $\phi = \langle \zeta | \pi(\cdot)\zeta \rangle$ as $\nu$ tends to infinity. Using the fact that every functional can be written as a linear combination of states, we find that $AA'$ is dense in $A'$. $\square$

**Corollary 12.1.2.** *Let $A$ be a $C^*$-algebra and $\phi \in A'$. Then $\phi$ extends uniquely to a strictly continuous functional $\tilde{\phi} \in M(A)'$ of norm $\|\tilde{\phi}\| = \|\phi\|$. If $\phi$ is positive/a state, so is $\tilde{\phi}$.* □

**Proposition 12.1.3.** *Let $A$ be a $C^*$-algebra and $T \in M(A)$. Then $\|T\| = \sup\{\tilde{\omega}(T) \mid \omega \in A', \|\omega\| \leq 1\}$.*

*Proof.* Evidently, $\|T\| \geq \sup\{\tilde{\omega}(T) \mid \omega \in A', \|\omega\| \leq 1\}$. Let $(u_\nu)_\nu$ be an approximate unit for $A$ such that $\|u_\nu\| < 1$ for all $\nu$. Then $\|Tu_\nu\| \leq \|T\|\|u_\nu\| \leq \|T\|$ for all $\nu$ and $\lim_\nu Tu_\nu a = Ta$ for all $a \in A$, whence $\lim_\nu \|Tu_\nu\| = \|T\|$. Given $\epsilon > 0$, choose an index $\nu$ and an $\omega \in A'$ of norm 1 such that $|\|Tu_\nu\| - \|T\|| < \epsilon/2$ and $\|Tu_\nu\| = \omega(Tu_\nu)$. Then $\|T\| < \|Tu_\nu\| + \epsilon/2 = \omega(Tu_\nu) + \epsilon/2 \leq \tilde{\omega}(T) + \epsilon$. □

## 12.2 $C^*$-modules

An excellent reference on $C^*$-modules is [95]; other sources are, for example, [16, Section II.7], [57, Section 2.5], and [128, Chapter 2].

**Definition and basic facts.** Let $A$ be a $C^*$-algebra. A *(right) pre-$C^*$-module over $A$* or a *(right) pre-$C^*$-$A$-module* is a complex vector space $E$ equipped with a right module structure over $A$ and a sesquilinear map $\langle \cdot | \cdot \rangle \colon E \times E \to A$ such that

$$\langle \eta|\xi\rangle^* = \langle \xi|\eta\rangle, \quad \langle \eta|\xi a\rangle = \langle \eta|\xi\rangle a, \quad \langle \xi|\xi\rangle \geq 0 \quad \text{for all } \eta, \xi \in E,\ a \in A.$$

The assignment $\|\xi\| := \|\langle \xi|\xi\rangle\|^{1/2}$ defines a seminorm on $E$. Evidently, $\|\cdot\|$ is a norm if and only if $\langle \xi|\xi\rangle \neq 0$ for all $\xi \in E$, $\xi \neq 0$. If $\|\cdot\|$ is a norm and $E$ is complete with respect $\|\cdot\|$, then $E$ is called a *(right) $C^*$-module over $A$* or a *(right) $C^*$-$A$-module*.

To every pre-$C^*$-module $E$ over $A$, one can associate a $C^*$-module over $A$ as follows. The subset $N := \{\xi \in E \mid \|\xi\| = 0\}$ is easily seen to be a right submodule, and the quotient $E/N$ is a pre-$C^*$-module over $A$ with respect to the structure maps $\langle \eta + N|\xi + N\rangle := \langle \eta|\xi\rangle$ and $(\xi + N)a := \xi a + N$. The completion of $E/N$ with respect to the norm is a $C^*$-module over $A$.

If $E$ is a $C^*$-module over $A$, then $EA$ is linearly dense in $E$. Indeed, if $(u_\nu)_\nu$ is a bounded approximate unit for $A$, then for all $\xi \in E$,

$$\|\xi - \xi u_\nu\|^2 = \|\langle \xi - \xi u_\nu|\xi - \xi u_\nu\rangle\| = (1 - u_\nu^*)\langle \xi|\xi\rangle(1 - u_\nu) \xrightarrow{\nu\to\infty} 0.$$

The Cohen Factorization Theorem [62, Theorem 32.22], [36] implies that one even has $EA = E$.

A $C^*$-module $E$ over $A$ is called *full* if the set $\langle E|E\rangle$ is linearly dense in $A$.

A basic example of a $C^*$-module is the $C^*$-algebra $A$ itself, with inner product $\langle a|b\rangle := a^*b$ and the obvious right module structure. Another basic example: every Hilbert space can be considered as a $C^*$-module over $\mathbb{C}$.

**Operators on $C^*$-modules.** Let $E$ and $F$ be $C^*$-modules over a $C^*$-algebra $A$. An *adjoint* of a map $T: E \to F$ is a map $S: F \to E$ that satisfies $\langle \eta|T\xi\rangle = \langle S\eta|\xi\rangle$ for all $\eta \in F$ and $\xi \in E$. If it exists, the adjoint of $T$ is unique; it is denoted by $T^*$. In this case, $T$ and $T^*$ are bounded, $A$-linear in the sense that $T(\xi a) = (T\xi)a$, $T^*(\eta a) = (T^*\eta)a$ for all $\xi \in E$, $\eta \in F$, $a \in A$, and $T$ and $T^*$ satisfy $\|T\|^2 = \|T^*T\| = \|T^*\|^2$. The space of all adjointable operators from $E$ to $F$ is denoted by $\mathcal{L}_A(E,F)$. If $E = F$, this space is also denoted by $\mathcal{L}_A(E)$. Equipped with the natural operations and the operator norm, $\mathcal{L}_A(E)$ is a $C^*$-algebra. The space $\mathcal{L}_A(E,F)$, equipped with the inner product $\langle S|T\rangle := S^*T$ and the right module structure given by composition, is a $C^*$-module over $\mathcal{L}_A(E)$.

For each pair of elements $\xi \in E$ and $\eta \in F$, the map $|\eta\rangle\langle\xi|: E \to F$ given by $\zeta \mapsto \eta\langle\xi|\zeta\rangle$ defines an adjointable operator $E \to F$. Its adjoint is given by the map $|\xi\rangle\langle\eta|: F \to E$. An operator $T \in \mathcal{L}_A(E,F)$ is called *compact* if it can be approximated in norm by linear combinations of such elementary operators. The set of all compact operators is denoted by $\mathcal{K}_A(E,F)$; if $E = F$, this space is also denoted by $\mathcal{K}_A(E)$. The composition of a compact operator with an arbitrary operator is compact again, and $\mathcal{K}_A(E)$ is an ideal in $\mathcal{L}_A(E)$. Moreover, $\mathcal{L}_A(E) = M(\mathcal{K}_A(E))$. The space $\mathcal{K}_A(A,F)$ can be identified with $F$ via $|\eta\rangle\langle a| \equiv \eta a^*$ for all $\eta \in F, a \in A$.

The *strict topology* on $\mathcal{L}_A(E,F)$ is the topology generated by the family of seminorms $\phi_\xi$, $\xi \in E$, and $\psi_\eta$, $\eta \in F$, where $\phi_\xi(T) = \|T\xi\|$ and $\psi_\eta(T) = \|T^*\eta\|$ for all $\xi \in E$, $\eta \in F$, $T \in \mathcal{L}_A(E,F)$. The subspace $\mathcal{K}_A(E,F)$ is strictly dense in $\mathcal{L}_A(E,F)$.

A *representation* of a $C^*$-algebra $B$ on $E$ is a $*$-homomorphism $\pi: B \to \mathcal{L}_A(E)$. The representation is called *faithful* if it is injective, and *non-degenerate* if $\pi(B)E$ is linearly dense in $E$.

**Amplification.** Let $E$ be a $C^*$-module over a $C^*$-algebra $A$, and let $H$ be a Hilbert space. Then the algebraic tensor product $E \odot H$ is a pre-$C^*$-module over $A$ with respect to the structure maps

$$\langle \eta' \odot \xi'|\eta \odot \xi\rangle := \langle \eta'|\eta\rangle_E \cdot \langle \xi'|\xi\rangle_H \quad \text{and} \quad (\eta \odot \xi)a := \eta a \odot \xi$$

for all $\eta, \eta' \in E$, $\xi, \xi' \in H$, $a \in A$. The completion of this pre-$C^*$-module is a $C^*$-module over $A$, which we denote by $E \otimes H$. Similarly, we can define a $C^*$-$A$-module $H \otimes E$.

For each $S \in \mathcal{L}_A(E)$ and $T \in \mathcal{L}(H)$, there exists an operator $S \otimes T \in \mathcal{L}_A(E \otimes H)$ such that $(S \otimes T)(\eta \otimes \xi) = S\eta \otimes T\xi$ for all $\eta \in E$, $\xi \in H$. The map $\mathcal{L}_A(E) \odot \mathcal{L}(H) \to \mathcal{L}_A(E \otimes H)$ given by $S \odot T \mapsto S \otimes T$ extends to a $*$-homomorphism $\mathcal{L}_A(E) \otimes \mathcal{L}(H) \to \mathcal{L}_A(E \otimes H)$, which restricts to a $*$-isomorphism $\mathcal{K}_A(E) \otimes \mathcal{K}(H) \xrightarrow{\cong} \mathcal{K}_A(E \otimes H)$.

**Internal tensor product.** Let $E$ and $F$ be $C^*$-modules over $C^*$-algebras $A$ and $B$, respectively. Assume that we have fixed a representation $A \to \mathcal{L}_B(F)$. Then the algebraic tensor product $E \odot F$ is a pre-$C^*$-module over $B$ via

$$\langle \eta' \odot \xi' | \eta \odot \xi \rangle := \langle \xi' | \langle \eta' | \eta \rangle_E \xi \rangle_F \quad \text{and} \quad (\eta \odot \xi)b := \eta \odot \xi b$$

for all $\eta, \eta' \in E$, $\xi, \xi' \in F$, $b \in B$. The associated $C^*$-module over $B$ is denoted by $E \otimes_A F$ and called the *internal tensor product* of $E$ and $F$. For each $\eta \in E$ and $\xi \in F$, we denote by $\eta \otimes_A \xi$ the image of $\eta \odot \xi$ under the natural map $E \odot F \to E \otimes_A F$.

If $A = \mathbb{C}$ and the representation of $A = \mathbb{C}$ on $F$ is given by scalar multiplication, then $F$ is a Hilbert space and $F \otimes_{\mathbb{C}} E$ is equal to the $C^*$-module $F \otimes E$ introduced above.

If $E = A$ and the representation of $A$ on $F$ is non-degenerate, then there exists an isomorphism $A \otimes_A F \xrightarrow{\cong} F$, $a \otimes_A \xi \mapsto a\xi$.

We shall frequently use the following result [39, Proposition 1.34]:

**Proposition 12.2.1.** *For $i = 1, 2$, let $E_i$ be a $C^*$-module over a $C^*$-algebra $A$ and $F_i$ a $C^*$-module over a $C^*$-algebra $B$ with a fixed representation $A \to \mathcal{L}_B(F_i)$. Let $S \in \mathcal{L}_A(E_1, E_2)$ and $T \in \mathcal{L}_B(F_1, F_2)$. If $Ta\xi = aT\xi$ for all $a \in A$ and $\xi \in F_1$, then there exists a unique operator $S \otimes_A T \in \mathcal{L}_B(E_1 \otimes_A F_1, E_2 \otimes_A F_2)$ such that $(S \otimes_A T)(\eta \otimes_A \xi) = S\eta \otimes_A T\xi$ for all $\eta \in E_1$ and $\xi \in F_1$. Moreover, $(S \otimes_A T)^* = S^* \otimes_A T^*$.* □

The internal tensor product of $C^*$-modules is associative in an obvious sense.

## 12.3 Von Neumann algebras

Standard references on von Neumann algebras are, for example, [35], [75], [76], [137], [149], [150].

**Topologies on $\mathcal{L}(H)$.** Let $H$ be a Hilbert space. Then $\mathcal{L}(H)$ can be equipped with the several locally convex topologies:

- The family of seminorms $\phi_{\eta,\xi}$ given by $\phi_{\eta,\xi}(T) := |\langle \eta | T\xi \rangle|$, where $\eta, \xi \in H$, defines the *weak (operator) topology* on $\mathcal{L}(H)$.
- The family of seminorms $\psi_\eta$ given by $\psi_\eta(T) := \|T\eta\|$, where $\eta \in H$, defines the *strong (operator) topology* on $\mathcal{L}(H)$.

- The family of seminorms $\psi_\eta$, $\psi_\eta^*$, where $\psi_\eta^*$ is given by $\psi_\eta^*(T) := \|T^*\eta\|$ for each $\eta \in H$, defines the *strong-∗ (operator) topology* on $\mathcal{L}(H)$.
- The family of seminorms $\tilde{\phi}_{\eta,\xi}$ given by $\tilde{\phi}_{\eta,\xi}(T) := |\langle \eta|(T \otimes \mathrm{id})\xi\rangle|$, where $\eta, \xi \in H \otimes l^2(\mathbb{N})$, defines the *σ-weak (operator) topology* on $\mathcal{L}(H)$.
- The family of seminorms $\tilde{\psi}_\eta$ given by $\tilde{\psi}_\eta(T) := \|(T \otimes \mathrm{id})\eta\|$, where $\eta \in H \otimes l^2(\mathbb{N})$, defines the *σ-strong (operator) topology* on $\mathcal{L}(H)$.
- The family of seminorms $\tilde{\psi}_\eta$, $\tilde{\psi}_\eta^*$, where $\tilde{\psi}_\eta^*$ is given by $\tilde{\psi}_\eta^*(T) := \|(T^* \otimes \mathrm{id})\eta\|$ for each $\eta \in H \otimes l^2(\mathbb{N})$, defines the *σ-strong-∗ (operator) topology* on $\mathcal{L}(H)$.

These topologies are related as follows:

$$\begin{array}{ccccccc} \text{norm} & \supset & \sigma\text{-strong-}* & \supset & \sigma\text{-strong} & \supset & \sigma\text{-weak} \\ & & \cup & & \cup & & \cup \\ & & \text{strong-}* & \supset & \text{strong} & \supset & \text{weak,} \end{array}$$

where "⊃" means that the left-hand side is finer/stronger than the right-hand side.

**Definition and basic facts.** Let $H$ be a Hilbert space. The *commutant* of a subset $X \subseteq \mathcal{L}(H)$, usually denoted by $X'$, is the set of all $T \in \mathcal{L}(H)$ that commute with every element of $X$. Evidently, $X'$ is an algebra, and if $X^* = X$, then $X'$ is a $*$-algebra.

By von Neumann's double commutant theorem, the following conditions on a $C^*$-algebra $A \subseteq \mathcal{L}(H)$ containing $\mathrm{id}_H$ are equivalent: (i) $A$ is equal to the double commutant $A''$, (ii) $A$ is closed with respect to the weak topology. If these conditions hold, then $A$ is called a *von Neumann algebra*.

A linear functional on a von Neumann algebra is called *normal* if it is σ-weakly continuous. The space of all normal linear functionals on a von Neumann algebra $A$ is denoted by $A_*$. If $A$ is a von Neumann algebra on a Hilbert space $H$, then every normal linear functional $\omega \in A_*$ can be written in the form

$$\omega(x) = \sum_n \langle \eta_n | x\xi_n \rangle \quad \text{for all } x \in A,$$

where $(\eta_n)_n$ and $(\xi_n)_n$ are sequences of vectors in $H$ satisfying $\sum_n \|\eta_n\|^2 < \infty$ and $\sum_n \|\xi_n\|^2 < \infty$. Conversely, if $\omega$ has the form above, then it is normal.

For each element $x$ of a von Neumann algebra $A$, there exists a normal functional $\omega \in A_*$ such that $\|x\| = \omega(x)$ and $\|\omega\| = 1$.

More generally, a map $A \to B$ of von Neumann algebras is called *normal* if it is continuous with respect to the σ-weak topologies on $A$ and $B$, respectively.

**Tensor product.** Let $H$ and $K$ be Hilbert spaces. The *tensor product* of von Neumann algebras $N \subseteq \mathcal{L}(H)$ and $M \subseteq \mathcal{L}(K)$, which we denote by $N \mathbin{\bar{\otimes}} M$, is the von Neumann algebra on $H \otimes K$ generated by the operators $S \otimes T \in \mathcal{L}(H \otimes K)$, where $S \in N$ and $T \in M$. In particular, one easily finds $\mathcal{L}(H) \mathbin{\bar{\otimes}} \mathcal{L}(K) = \mathcal{L}(H \otimes K)$. More generally, $(N \mathbin{\bar{\otimes}} M)' = N' \mathbin{\bar{\otimes}} M'$ [149, IV, Theorem 5.9].

The tensor product of von Neumann algebras is functorial with respect to normal $*$-homomorphisms: For each pair of normal $*$-homomorphisms $\phi_i\colon M_i \to N_i$ of von Neumann algebras, where $i = 1, 2$, there exists a normal $*$-homomorphism $\phi_1 \mathbin{\bar{\otimes}} \phi_2\colon M_1 \mathbin{\bar{\otimes}} M_2 \to N_1 \mathbin{\bar{\otimes}} N_2$ such that $(\phi_1 \mathbin{\bar{\otimes}} \phi_2)(x_1 \mathbin{\bar{\otimes}} x_2) = \phi_1(x_1) \mathbin{\bar{\otimes}} \phi_2(x_2)$ for all $x_1 \in M_1$ and $x_2 \in M_2$.

For each $\phi \in N_*$ and $\psi \in M_*$, there exists a unique functional $\phi \mathbin{\bar{\otimes}} \psi \in (N \mathbin{\bar{\otimes}} M)_*$ such that $(\phi \mathbin{\bar{\otimes}} \psi)(a \mathbin{\bar{\otimes}} b) = \phi(a)\psi(b)$ for all $a \in N$, $b \in M$, and $\|\phi \mathbin{\bar{\otimes}} \psi\| \leq \|\phi\|\|\psi\|$ [149, Section IV.5]. If $\phi$ and $\psi$ are positive or states, then so is $\phi \mathbin{\bar{\otimes}} \psi$.

**Abstract von Neumann algebras.** A $C^*$-algebra $A$ is called a *$W^*$-algebra* if it is isomorphic as a Banach space to the dual of some other Banach space $A_*$. In this case, $A_*$ is uniquely determined (up to isomorphism) and called the *predual* of $A$. Every von Neumann algebra is a $W^*$-algebra, where the predual is the space of all normal linear functionals. Conversely, for every $W^*$-algebra, there exists a $*$-isomorphism onto a von Neumann algebra that is continuous with respect to the weak-$*$-topology.

In particular, for every von Neumann algebra $N$, the opposite algebra $N^{\mathrm{op}}$ is a $W^*$-algebra. Somewhat sloppily, we refer to $N^{\mathrm{op}}$ as a von Neumann algebra again.

## 12.4 Slice maps

Slice maps are an important tool for the theory of quantum groups in the setting of $C^*$-algebras and von Neumann algebras. We shall describe in detail how these slice maps are constructed and collect some frequently used formulas.

**Slice maps in the setting of $*$-algebras.** Let $A$ and $B$ be unital $*$-algebras with linear maps $\phi\colon A \to \mathbb{C}$, $\psi\colon B \to \mathbb{C}$. Then we can define linear *slice maps*

$$\begin{aligned} \phi \odot \mathrm{id}\colon A \odot B \to B, &\quad a \odot b \mapsto \phi(a)b, \\ \mathrm{id} \odot \psi\colon A \odot B \to B, &\quad a \odot b \mapsto a\psi(b). \end{aligned} \tag{12.1}$$

One easily verifies that for all $a, a' \in A$, $b, b' \in B$ and $x \in A \odot B$,

$$(\psi \circ (\phi \odot \mathrm{id}))(x) = (\phi \odot \psi)(x) = (\phi \circ (\mathrm{id} \odot \psi))(x) \tag{12.2}$$

and

$$\begin{aligned} b((\phi \odot \mathrm{id})(x))b' &= (\phi \odot \mathrm{id})((1 \odot b)x(1 \odot b')), \\ a((\mathrm{id} \odot \psi)(x))a' &= (\mathrm{id} \odot \psi)((a \odot 1)x(a' \odot 1)). \end{aligned} \tag{12.3}$$

If $\phi$ and $\psi$ are $*$-linear, so are $\phi \odot \mathrm{id}$ and $\mathrm{id} \odot \psi$. In general,

$$\begin{aligned} ((\phi \odot \mathrm{id})(x))^* &= (\bar{\phi} \odot \mathrm{id})(x^*), \quad \text{where } \bar{\phi}\colon A \to \mathbb{C},\ a \mapsto \overline{\phi(a^*)}, \\ ((\mathrm{id} \odot \psi)(x))^* &= (\mathrm{id} \odot \bar{\psi})(x^*), \quad \text{where } \bar{\psi}\colon B \to \mathbb{C},\ b \mapsto \overline{\psi(b^*)}. \end{aligned} \tag{12.4}$$

Given another $*$-algebra $C$, we can define a linear slice map

$$\phi \odot \mathrm{id} \odot \psi\colon A \odot C \odot B \to C, \quad a \odot c \odot b \mapsto \phi(a)c\psi(b),$$

and for all $x \in A \odot C$, $y \in C \odot B$,

$$\begin{aligned} (\phi \odot \mathrm{id})(x) \cdot (\mathrm{id} \odot \psi)(y) &= (\phi \odot \mathrm{id} \odot \psi)((x \odot 1)(1 \odot y)), \\ (\mathrm{id} \odot \psi)(y) \cdot (\phi \odot \mathrm{id})(x) &= (\phi \odot \mathrm{id} \odot \psi)((1 \odot y)(x \odot 1)). \end{aligned} \tag{12.5}$$

When we switch the order of the factors in the tensor product $A \odot C \odot B$, we obtain new slice maps that satisfy analogues of equation (12.5).

**Slice maps in the setting of $C^*$-algebras.** In the setting of $C^*$-algebras, one often uses slice maps that are defined on the minimal tensor product $A \otimes B$ of $C^*$-algebras $A$ and $B$ and on the multiplier algebra $M(A \otimes B)$.

**Proposition 12.4.1.** *Let $A$ and $B$ be $C^*$-algebras and $\phi \in A'$, $\psi \in B'$. Then $\phi \odot \mathrm{id}$ and $\mathrm{id} \odot \psi$ extend uniquely to linear maps*

$$\phi \otimes \mathrm{id}\colon M(A \otimes B) \to M(B), \quad \mathrm{id} \otimes \psi\colon M(A \otimes B) \to M(A)$$

*that are norm-continuous and strictly continuous on bounded subsets. We have $\|\phi \otimes \mathrm{id}\| = \|\phi\|$ and $\|\mathrm{id} \otimes \psi\| = \|\psi\|$. If $\phi$ or $\psi$ is positive, so is $\phi \otimes \mathrm{id}$ or $\mathrm{id} \otimes \psi$, respectively.*

**Remark 12.4.2.** The algebraic tensor product $A \odot B$ is strictly dense in $M(A \otimes B)$ because it is norm-dense in $A \otimes B$ and because $A \otimes B$ is strictly dense in $M(A \otimes B)$. Moreover, the unit ball of the algebraic tensor product $A \odot B$ is strictly dense in the unit ball of $M(A \otimes B)$. Consequently, $\phi \otimes \mathrm{id}$ and $\mathrm{id} \otimes \psi$ are uniquely determined by their restrictions to $A \odot B$. *Furthermore, it follows that the equations* (12.2)–(12.5) *and many more formulas satisfied by $\phi \odot \mathrm{id}$ and $\mathrm{id} \odot \psi$ extend to $\phi \otimes \mathrm{id}$ and $\mathrm{id} \otimes \psi$, respectively.* Finally, we find $(\phi \otimes \mathrm{id})(A \otimes B) \subseteq B$ and $(\mathrm{id} \otimes \psi)(A \otimes B) \subseteq A$.

*Proof of Proposition* 12.4.1. We only prove the assertions concerning $\phi \otimes \mathrm{id}$; for $\mathrm{id} \otimes \psi$, the arguments are similar. Uniqueness of $\phi \otimes \mathrm{id}$ was already observed above. To prove existence, we consider two cases.

*The case where $\phi$ is a state*: Let $(H_\phi, \Lambda_\phi, \pi_\phi)$ be a GNS-construction for $\phi$ with cyclic vector $\zeta_\phi$, and let $(H_B, \Lambda_B, \pi_B)$ be the universal GNS-construction for $B$. By definition of $A \otimes B$, there exists a non-degenerate $*$-homomorphism $\pi: A \otimes B \to \mathcal{L}(H_\phi \otimes H_B)$ such that $\pi(a \otimes b) = \pi_\phi(a) \otimes \pi_B(b)$ for all $a \in A$, $b \in B$. Extend $\pi$ to the multiplier algebra $M(A \otimes B)$ and consider the map

$$\Phi\colon M(A \otimes B) \to \mathcal{L}(H_B), \quad T \mapsto \langle \zeta_\phi|_{[1]} \circ \pi(T) \circ |\zeta_\phi\rangle_{[1]},$$

where $|\zeta_\phi\rangle_{[1]}\colon H_B \to H_\phi \otimes H_B$ is given by $\xi \mapsto \zeta_\phi \otimes \xi$, and $\langle \zeta_\phi|_{[1]} = |\zeta_\phi\rangle_{[1]}^*$. We have $\Phi(A \otimes B) \subseteq \pi_B(B)$ because for all $a \in A, b \in B$,

$$\Phi(a \otimes b) = \langle \zeta_\phi | \pi_\phi(a) \zeta_\phi \rangle \pi_B(b) = \phi(a) \pi_B(b).$$

Choose an approximate unit $(u_\nu)_\nu$ for $A$. Then $\lim_\nu \pi_\phi(u_\nu)\zeta_\phi = \zeta_\phi$. If $(T_\mu)_\mu$ is a norm-bounded net in $M(A \otimes B)$ that converges strictly to some $T \in M(A \otimes B)$, and if $b \in B$, then

$$\begin{aligned} \lim_\mu \Phi(T_\mu)\pi_B(b) &= \lim_{\mu,\nu} \langle \zeta_\phi|_{[1]} \circ \pi(T_\mu) \circ |\pi_\phi(u_\nu)\zeta_\phi\rangle_{[1]} \circ \pi_B(b) \\ &= \lim_{\mu,\nu} \langle \zeta_\phi|_{[1]} \circ \pi(T_\mu)(\pi_\phi(u_\nu) \otimes \pi_B(b)) \circ |\zeta_\phi\rangle_{[1]}. \end{aligned}$$

Using the relation $\pi(T_\mu)(\pi_\phi(u_\nu) \otimes \pi_B(b)) = \pi(T_\mu(u_\nu \otimes b))$ and the fact that $(T_\mu)_\mu$ converges strictly to $T$, we find

$$\lim_\mu \Phi(T_\mu)\pi_B(b) = \lim_\nu \langle \zeta_\phi|_{[1]} \circ \pi(T)(\pi_\phi(u_\nu) \otimes \pi_B(b)) \circ |\zeta_\phi\rangle_{[1]} = \Phi(T)\pi_B(b).$$

A similar argument shows that $\lim_\mu \pi_B(b)\Phi(T_\mu) = \pi_B(b)\Phi(T)$. Approximating any $T \in M(A \otimes B)$ strictly by a norm-bounded net in $A \otimes B$, we find that $\Phi(M(A \otimes B)) \subseteq \pi_B(M(B))$ and that $\Phi\colon M(A \otimes B) \to \pi_B(M(B))$ is strictly continuous on norm-bounded subsets.

Since $\pi_B\colon M(B) \to \pi_B(M(B))$ is an isomorphism, we can define $\phi \otimes \mathrm{id} := \pi_B^{-1} \circ \Phi$. By construction, $\phi \otimes \mathrm{id}$ is positive and has norm less than or equal to $\|\zeta_\phi\|^2 \|\pi\| = 1 = \|\phi\|$.

*General case:* Let $\phi \in A'$ be arbitrary. Then $\phi$ is a linear combinations of states, and all assertions except for the norm estimates follow easily from the special case considered above. Using Proposition 12.1.3, we find

$$\|\phi \otimes \mathrm{id}\| = \sup\{\|\omega \circ (\phi \otimes \mathrm{id})\| : \omega \in B' \subseteq M(B)',\ \|\omega\| \le 1\}.$$

But for $\omega$ as above, $\|\omega \circ (\phi \otimes \mathrm{id})\| = \|\phi \otimes \omega\| \le \|\phi\| \|\omega\| = \|\phi\|$, and hence $\|\phi \otimes \mathrm{id}\| \le \|\phi\|$. □

**Proposition 12.4.3.** *Let $A$, $B$ be $C^*$-algebras, $T \in M(A \otimes B)$, $\phi \in A'$, $\psi \in B'$.*

i) *If $(A \otimes 1)T(A \otimes 1) \subseteq A \otimes B$, then $(\phi \otimes \mathrm{id})(T) \in B$.*

ii) *If $(1 \otimes B)T(1 \otimes B) \subseteq A \otimes B$, then $(\mathrm{id} \otimes \psi)(T) \in A$.*

*Proof.* We only prove i), the proof of ii) is similar. By Proposition 12.1.1, we may assume that $\phi$ has the form $\omega(a_1 \cdot a_2)$, where $\omega \in A'$ and $a_1, a_2 \in A$. But then $(\phi \otimes \mathrm{id})(T) = (\omega \otimes \mathrm{id})((a_1 \otimes 1)T(a_2 \otimes 1)) \in (\omega \otimes \mathrm{id})(A \otimes B) \subseteq B$. □

**Slice maps in the setting of von Neumann algebras.** In the setting of von Neumann algebras, slice maps can be defined for normal functionals:

**Proposition 12.4.4.** *Let $A$ and $B$ be von Neumann algebras and $\phi \in A_*$, $\psi \in B_*$. Then the maps $\phi \odot \mathrm{id}$ and $\mathrm{id} \odot \psi$ extend uniquely to norm-continuous normal linear maps $\phi \mathbin{\bar\otimes} \mathrm{id}\colon A \mathbin{\bar\otimes} B \to B$ and $\mathrm{id} \mathbin{\bar\otimes} \psi\colon A \mathbin{\bar\otimes} B \to A$. We have $\|\phi \mathbin{\bar\otimes} \mathrm{id}\| = \|\phi\|$ and $\|\mathrm{id} \mathbin{\bar\otimes} \psi\| = \|\psi\|$. If $\phi$ or $\psi$ is positive, so is $\phi \mathbin{\bar\otimes} \mathrm{id}$ or $\mathrm{id} \mathbin{\bar\otimes} \psi$, respectively.*

**Remark 12.4.5.** Since $A \odot B$ is dense in $A \mathbin{\bar\otimes} B$ with respect to the $\sigma$-weak topology, the extensions $\phi \mathbin{\bar\otimes} \mathrm{id}$ and $\mathrm{id} \mathbin{\bar\otimes} \psi$ are uniquely determined by their restrictions to $A \odot B$. *Moreover, it follows that the equations* (12.2)–(12.5) *and many more formulas satisfied by $\phi \odot \mathrm{id}$ and $\mathrm{id} \odot \psi$ extend to $\phi \otimes \mathrm{id}$ and $\mathrm{id} \otimes \psi$, respectively.*

*Proof of Proposition* 12.4.4. We only prove the assertions concerning $\phi \mathbin{\bar\otimes} \mathrm{id}$. Uniqueness follows from the observation made above. We prove existence. Denote by $H$ and $K$ the underlying Hilbert spaces of $A$ and $B$, respectively. Since $\phi$ is normal, there exist sequences $(\eta_n)_n$ and $(\xi_n)_n$ of vectors in $H$ such that

$$C_\eta := \sum_n \|\eta_n\|^2 < \infty, \quad C_\xi := \sum_n \|\xi_n\|^2 < \infty,$$
$$\phi(a) = \sum_n \langle \eta_n | a \xi_n \rangle \quad \text{for all } a \in A.$$

For each $x \in A \mathbin{\bar\otimes} B$, the sesquilinear map $\sigma_x\colon K \times K \to \mathbb{C}$ given by

$$\sigma_x(\zeta, \zeta') := \sum_n \langle \eta_n \otimes \zeta | x(\xi_n \otimes \zeta') \rangle$$

is bounded: using the relation $|\langle \eta_n \otimes \zeta | x(\xi_n \otimes \zeta')\rangle| \le \|\eta_n\| \|\xi_n\| \|x\| \|\zeta\| \|\zeta'\|$ and the Cauchy–Schwarz inequality, we find

$$\|\sigma_x(\zeta, \zeta')\| \le C_\eta C_\xi \cdot \|x\| \cdot \|\zeta\| \cdot \|\zeta'\| \quad \text{for all } \zeta, \zeta' \in H.$$

Hence we can define $\phi \mathbin{\bar\otimes} \mathrm{id}\colon A \mathbin{\bar\otimes} B \to \mathcal{L}(K)$ by

$$\langle \zeta | (\phi \mathbin{\bar\otimes} \mathrm{id})(x) \zeta' \rangle := \sigma_x(\zeta, \zeta') \quad \text{for all } x \in A \mathbin{\bar\otimes} B,$$

and $\|\phi \mathbin{\bar{\otimes}} \mathrm{id}\| \leq C_\eta C_\xi$. Moreover,

$$\|\phi \mathbin{\bar{\otimes}} \mathrm{id}\| = \sup\{\|\omega \circ (\phi \mathbin{\bar{\otimes}} \mathrm{id})\| : \omega \in B_*,\ \|\omega\| \leq 1\},$$

and for $\omega$ as above, $\|\omega \circ (\phi \mathbin{\bar{\otimes}} \mathrm{id})\| = \|\phi \mathbin{\bar{\otimes}} \omega\| \leq \|\phi\|\|\omega\| = \|\phi\|$. Hence, $\|\phi \mathbin{\bar{\otimes}} \mathrm{id}\| \leq \|\phi\|$.

We claim that $\phi \mathbin{\bar{\otimes}} \mathrm{id}$ is normal. Let $\psi \in B_*$. Then there exist sequences $(\zeta_n)_n$ and $(\zeta'_n)_n$ of vectors in $K$ such that $C_\zeta := \sum_m \|\zeta_m\|^2 < \infty$, $C_{\zeta'} := \sum_m \|\zeta'_m\| < \infty$ and $\psi(b) = \sum_m \langle \zeta_m | b\zeta'_m \rangle$ for all $b \in B$. The relations

$$\sum_{n,m} \|\eta_n \otimes \zeta_m\|^2 \leq C_\eta C_\zeta < \infty, \quad \sum_{n,m} \|\xi_n \otimes \zeta'_m\|^2 \leq C_\xi C_{\zeta'} < \infty,$$

$$(\psi \circ (\phi \mathbin{\bar{\otimes}} \mathrm{id}))(x) = \sum_{n,m} \langle \eta_n \otimes \zeta_m | x(\xi_n \otimes \zeta'_m) \rangle \quad \text{for all } x \in A \mathbin{\bar{\otimes}} B$$

show that the composition $\psi \circ (\phi \mathbin{\bar{\otimes}} \mathrm{id})$ is normal. Since $\psi \in B_*$ was arbitrary, the claim follows.

Finally, a similar construction as in the proof of Proposition 12.4.1 shows that $\phi \mathbin{\bar{\otimes}} \mathrm{id}$ is positive if $\phi$ is a normal state, and consequently, $\phi \mathbin{\bar{\otimes}} \mathrm{id}$ is positive whenever $\phi$ is positive. □

## 12.5 Auxiliary results

**Proposition 12.5.1.** *Let $\pi : A \to B$ be a bounded homomorphism of Banach algebras. Assume that*

i) *$\pi$ is non-degenerate in the sense that $\overline{\mathrm{span}}\, \pi(A)B = B = \overline{\mathrm{span}}\, B\pi(A)$;*

ii) *$A$ has a bounded approximate unit, that is, a norm-bounded net $(u_\lambda)_\lambda$ of elements in $A$ such that $\lim_\lambda u_\lambda a = a = \lim_\lambda a u_\lambda$ for all $a \in A$.*

*Then $\pi$ extends uniquely to a homomorphism $\tilde{\pi} : M(A) \to M(B)$ such that*

$$\tilde{\pi}(T)\pi(a) = \pi(Ta) \quad \text{and} \quad \pi(a)\tilde{\pi}(T) = \pi(aT) \tag{12.6}$$

*for all $T \in M(A)$, $a \in A$.*

*Proof.* Let $T \in M(A)$ and put $v_\lambda := \pi(Tu_\lambda)$ for all $\lambda$. Then the net $(v_\lambda)_\lambda$ is norm-bounded because $\|v_\lambda\| \leq \|\pi\|\|T\|\|u_\lambda\|$ for all $\lambda$, and

$$\lim_\lambda v_\lambda \pi(a) = \lim_\lambda \pi(T(u_\lambda a)) = \pi(Ta),$$
$$\lim_\lambda \pi(a) v_\lambda = \lim_\lambda \pi((aT)u_\lambda) = \pi(aT)$$

for all $a \in A$. Since $(v_\lambda)_\lambda$ is norm-bounded and $\overline{\operatorname{span}}\, \pi(A)B = B = \overline{\operatorname{span}}\, B\pi(A)$, it follows that for every $b \in B$, the nets $(v_\lambda b)_\lambda$ and $(bv_\lambda)_\lambda$ converge in norm. Clearly, the assignments

$$\tilde{\pi}(T)b := \lim_\lambda v_\lambda b \quad \text{and} \quad b\tilde{\pi}(T) := \lim_\lambda bv_\lambda, \; b \in B,$$

define a multiplier $\tilde{\pi}(T) \in M(B)$ of norm

$$\|\tilde{\pi}(T)\| \leq \sup_\lambda \|v_\lambda\| \leq \|\pi\|\|T\| \sup_\lambda \|u_\lambda\|.$$

The map $T \mapsto \tilde{\pi}(T)$ is obviously linear and satisfies condition (12.6). Furthermore, by assumption i), the relation

$$\tilde{\pi}(S)\tilde{\pi}(T)\pi(a) = \tilde{\pi}(S)\pi(Ta) = \pi(STa) = \tilde{\pi}(ST)\pi(a), \quad a \in A,$$

implies that $\tilde{\pi}(S)\tilde{\pi}(T)b = \tilde{\pi}(ST)b$ for all $S, T \in M(A)$ and $b \in B$, and similarly $b\tilde{\pi}(S)\tilde{\pi}(T) = b\tilde{\pi}(ST)$. Therefore, $\tilde{\pi}$ is a homomorphism. □

**Corollary 12.5.2.** *Let $\pi\colon A \to B$ be a bounded $*$-homomorphism of Banach $*$-algebras that satisfies conditions* i) *and* ii) *of Proposition* 12.5.1. *Then $\pi$ extends to uniquely to a $*$-homomorphism $\tilde{\pi}\colon M(A) \to M(B)$ such that* (12.6) *holds.*

*Proof.* The fact that the extension $\tilde{\pi}$ provided by the previous proposition is a $*$-homomorphism follows easily from (12.6) and condition i). □

**Lemma 12.5.3.** *Let $V_1, \dots, V_n$ and $W$ be topological vector spaces, and let $F\colon V_1 \times \cdots \times V_n \to W$ be a multilinear map that is separately continuous in each component. Suppose that for each $i = 1, \dots, n$, we are given a closed subspace $Y_i \subseteq V_i$, and a subset $X_i \subseteq Y_i$ that is linearly dense. Then*

$$\begin{aligned}&\overline{\operatorname{span}}\{F(x_1, \dots, x_n) \mid x_i \in X_i \textit{ for all } i\}\\ &\quad = \overline{\operatorname{span}}\{F(y_1, \dots, y_n) \mid y_i \in Y_i \textit{ for all } i\}.\end{aligned}$$

*Proof.* Replacing $X_i$ by $\operatorname{span} X_i$, we may assume each $X_i$ is a subspace. Put

$$U := \overline{\operatorname{span}}\{F(x_1, \dots, x_n) \mid x_i \in X_i \text{ for all } i\}.$$

We prove by induction that for every $k = 0, \dots, n$,

$$F(y_1, \dots, y_k, x_{k+1}, \dots, x_n) \in U$$

for all $y_i \in Y_i$, $i = 1, \dots, k$, and $x_j \in X_j$, $j = k+1, \dots, n$. Note that for $k = n$, this statement implies the claim. For $k = 0$, the assertion is trivially satisfied. We assume that the statement holds for $k$ and prove it for $k+1$. Given

$y_i \in Y_i$, $i = 1, \dots, k+1$, and $x_j \in X_j$, $j = k+2, \dots, n$, we can approximate $y_{k+1}$ by some net $(x_{k+1,\nu})_\nu$ in $X_{k+1}$, and

$$F(y_1, \dots, y_k, x_{k+1,\nu}, x_{k+2}, \dots, x_n) \in U \quad \text{for every } \nu$$

by assumption. Since $U$ is closed and $F$ is continuous in the $(k+1)$st variable, we also have $F(y_1, \dots, y_k, y_{k+1}, x_{k+2}, \dots, x_n) \in U$. □

# Bibliography


[1] E. Abe, *Hopf algebras*. Cambridge Tracts in Math. 74, Cambridge University Press, Cambridge 1980.

[2] C. A. Akemann, G. K. Pedersen, and J. Tomiyama, Multipliers of $C^*$-algebras. *J. Funct. Anal.* **13** (1973), 277–301.

[3] S. Baaj, Multiplicateurs non bornés. Thèse 3ème cycle, Université Paris VI, Paris 1980.

[4] S. Baaj, Représentation régulière du groupe quantique $\mathrm{E}_\mu(2)$ de Woronowicz. *C. R. Acad. Sci. Paris Sér. I Math.* **314** (1992), 1021–1026.

[5] S. Baaj, Représentation régulière du groupe quantique des déplacements de Woronowicz. *Astérisque* **232** (1995), 11–48.

[6] S. Baaj and G. Skandalis, $C^*$-algèbres de Hopf et théorie de Kasparov équivariante. *$K$-Theory* **2** (1989), 683–721.

[7] S. Baaj and G. Skandalis, Unitaires multiplicatifs et dualité pour les produits croisés de $C^*$-algèbres. *Ann. Sci. École Norm. Sup.* (4) **26** (1993), 425–488.

[8] S. Baaj and G. Skandalis, Unitaires multiplicatifs commutatifs. *C. R. Acad. Sci. Paris Sér. I Math.* **336** (2003), 299–304.

[9] S. Baaj, G. Skandalis, and S. Vaes, Non-semi-regular quantum groups coming from number theory. *Comm. Math. Phys.* **235** (2003), 139–167.

[10] S. Baaj, G. Skandalis, and S. Vaes, Measurable Kac cohomology for bicrossed products. *Trans. Amer. Math. Soc.* **357** (2005), 1497–1524.

[11] S. Baaj and S. Vaes, Double crossed products of locally compact quantum groups. *J. Inst. Math. Jussieu* **4** (2005), 135–173.

[12] T. Banica, Théorie des représentations du groupe quantique compact libre $\mathrm{O}(n)$. *C. R. Acad. Sci. Paris Sér. I Math.* **322** (1996), 241–244.

[13] T. Banica, Le groupe quantique compact libre $\mathrm{U}(n)$. *Comm. Math. Phys.* **190** (1997), 143–172.

[14] T. Banica, Symmetries of a generic coaction. *Math. Ann.* **314** (1999), 763–780.

[15] T. Banica and B. Collins. Integration over quantum permutation groups. *J. Funct. Anal.* **242** (2007), 641–657.

[16] B. Blackadar, *Operator algebras*. Encyclopaedia Math. Sci. 122, Springer-Verlag, Berlin 2006.

[17] É. Blanchard, Déformations de $C^*$-algèbres de Hopf. *Bull. Soc. Math. France* **124** (1996), 141–215.

[18] R. J. Blattner, M. Cohen, and S. Montgomery, Crossed products and inner actions of Hopf algebras. *Trans. Amer. Math. Soc.* **298** (1986), 671–711.

[19] A. Borel, *Linear algebraic groups*. 2nd ed., Grad. Texts in Math. 126, Springer-Verlag, New York 1991.

[20] N. Bourbaki, *Éléments de mathématique. Fasc. XXXII. Théories spectrales*. Chap. 1 et 2: Algèbres normées. Groupes localement compacts commutatifs, Hermann, Paris 1967.

[21] O. Bratteli and D. W. Robinson, *Operator algebras and quantum statistical mechanics*. Vol. 1: $C^*$- and $W^*$-algebras, symmetry groups, decomposition of states, Texts Monogr. Phys., Springer-Verlag, New York 1979.

[22] T. Bröcker and T. tom Dieck, *Representations of compact Lie groups*. Grad. Texts in Math. 98, Springer-Verlag, New York 1995.

[23] K. A. Brown and K. R. Goodearl, *Lectures on algebraic quantum groups*. Adv. Courses Math. CRM Barcelona, Birkhäuser, Basel 2002.

[24] V. Chari and A. Pressley. *A guide to quantum groups*. Cambridge University Press, Cambridge 1994.

[25] C. Chevalley. *Theory of Lie groups I*. Princeton Math. Ser. 8, Princeton University Press, Princeton, N. J., 1946.

[26] A. Connes, On the spatial theory of von Neumann algebras. *J. Funct. Anal.* **35** (1980), 153–164.

[27] A. Connes, *Noncommutative geometry*. Academic Press Inc., San Diego 1994.

[28] J. B. Conway, *A course in functional analysis*. 2nd ed., Grad. Texts in Math. 96, Springer-Verlag, New York 1990.

[29] S. Dăscălescu, C. Năstăsescu, and Ş. Raianu, *Hopf algebras*, Monographs Textbooks Pure Appl. Math. 235, Marcel Dekker, New York 2001.

[30] M.-C. David, Paragroupe d'Adrian Ocneanu et algèbre de Kac. *Pacific J. Math.* **172** (1996), 331–363.

[31] J. De Cannière, Produit croisé d'une algèbre de Kac par un groupe localement compact. *Bull. Soc. Math. France* **107** (1979), 337–372.

[32] J. Dieudonné, *Treatise on analysis*. Vol. III, Pure Appl. Math. 10-III, Academic Press, New York 1972.

[33] J. Dieudonné, *Treatise on analysis*. Vol. VI, Pure Appl. Math. 10-VI, Academic Press, New York 1978.

[34] J. Dixmier, *Les $C^*$-algèbres et leurs représentations*. Cahiers Scientifiques, Fasc. XXIX, Gauthier-Villars & C$^{ie}$, Paris 1964.

[35] J. Dixmier, *Les algèbres d'opérateurs dans l'espace hilbertien* (*algèbres de von Neumann*). 2nd ed., Cahiers Scientifiques, Fasc. XXV, Gauthier-Villars, Paris 1969.

[36] R. S. Doran and J. Wichmann, *Approximate identities and factorization in Banach modules*. Lecture Notes in Math. 768, Springer-Verlag, Berlin 1979.

[37] V. G. Drinfeld, Quantum groups. In *Proc. Internat. Congr. Math.* (*Berkeley, Calif.*, 1986), Vol. 1, Amer. Math. Soc., Providence, RI, 1987, 798–820.

[38] M. J. Dupré and R. M. Gillette, *Banach bundles, Banach modules and automorphisms of* $C^*$*-algebras*. Pitman Res. Notes Math. Ser. 92, Pitman Advanced Publishing Program, Boston 1983.

[39] S. Echterhoff, S. Kaliszewski, J. Quigg, and I. Raeburn, A categorical approach to imprimitivity theorems for $C^*$-dynamical systems. *Mem. Amer. Math. Soc.* **180** (2006), no. 850.

[40] E. G. Effros and Z.-J. Ruan, Discrete quantum groups I: The Haar measure. *Internat. J. Math.* **5** (1994), 681–723.

[41] M. Enock, Produit croisé d'une algèbre de von Neumann par une algèbre de Kac. *J. Funct. Anal.* **26** (1977), 16–47.

[42] M. Enock, Quantum groupoids and pseudo-multiplicative unitaries. In *Locally compact quantum groups and groupoids* (*Strasbourg,* 2002), IRMA Lect. Math. Theor. Phys. 2, Walter de Gruyter, Berlin 2003, 17–47.

[43] M. Enock. Inclusions of von Neumann algebras and quantum groupoïds III. *J. Funct. Anal.* **223** (2005), 311–364.

[44] M. Enock, Quantum groupoids of compact type. *J. Inst. Math. Jussieu* **4** (2005), 29–133.

[45] M. Enock and R. Nest, Irreducible inclusions of factors, multiplicative unitaries, and Kac algebras. *J. Funct. Anal.* **137** (1996), 466–543.

[46] M. Enock and J.-M. Schwartz, Produit croisé d'une algèbre de von Neumann par une algèbre de Kac, II. *Publ. Res. Inst. Math. Sci.* **16** (1980), 189–232.

[47] M. Enock and J.-M. Schwartz, *Kac algebras and duality of locally compact groups*. Springer-Verlag, Berlin 1992.

[48] M. Enock and L. Vainerman, Deformation of a Kac algebra by an abelian subgroup. *Comm. Math. Phys.* **178** (1996), 571–596.

[49] M. Enock and J.-M. Vallin, $C^*$-algèbres de Kac et algèbres de Kac. *Proc. London Math. Soc.* (3) **66** (1993), 619–650.

[50] M. Enock and J.-M. Vallin, Inclusions of von Neumann algebras, and quantum groupoïds. *J. Funct. Anal.* **172** (2000), 249–300.

[51] M. Enock and J.-M. Vallin, Inclusions of von Neumann algebras and quantum groupoïds II. *J. Funct. Anal.* **178** (2000), 156–225.

[52] T. Falcone, $L^2$-von Neumann modules, their relative tensor products and the spatial derivative. *Illinois J. Math.* **44** (2000), 407–437.

[53] J. M. G. Fell and R. S. Doran, *Representations of* $^*$*-algebras, locally compact groups, and Banach* $^*$*-algebraic bundles*. Vol. 1: Basic representation theory of groups and algebras, Pure Appl. Math. 125, Academic Press, Boston, MA, 1988.

[54] J. M. G. Fell and R. S. Doran, *Representations of* $^*$*-algebras, locally compact groups, and Banach* $^*$*-algebraic bundles*. Vol. 2: Banach $^*$-algebraic bundles, induced representations, and the generalized Mackey analysis, Pure Appl. Math. 126, Academic Press, Boston 1988.

[55] W. Fulton and J. Harris, *Representation theory*. Grad. Texts in Math. 129, Springer-Verlag, New York 1991.

[56] V. V. Gorbatsevich, A. L. Onishchik, and E. B. Vinberg, *Foundations of Lie theory and Lie transformation groups*. Springer-Verlag, Berlin 1997.

[57] J. M. Gracia-Bondía, J. C. Várilly, and H. Figueroa, *Elements of noncommutative geometry*. Birkhäuser Advanced Texts, Birkhäuser, Boston 2001.

[58] U. Haagerup, The standard form of von Neumann algebras. *Math. Scand.* **37** (1975), 271–283.

[59] U. Haagerup, Operator valued weights in von Neumann algebras, I. *J. Funct. Anal.* **32** (1979), 175–206.

[60] U. Haagerup, Operator valued weights in von Neumann algebras, II. *J. Funct. Anal.* **33** (1979), 339–361.

[61] B. C. Hall, *Lie groups, Lie algebras, and representations*. Grad. Texts in Math. 222, Springer-Verlag, New York 2003.

[62] E. Hewitt and K. A. Ross, *Abstract harmonic analysis*. Vol. II: Structure and analysis for compact groups. Analysis on locally compact abelian groups, Grundlehren Math. Wiss. 152, Springer-Verlag, Berlin 1970.

[63] E. Hewitt and K. A. Ross, *Abstract harmonic analysis*. Vol. I: Structure of topological groups, integration theory, group representations, 2nd. ed., Grundlehren Math. Wiss. 115, Springer-Verlag, Berlin 1979.

[64] E. Hille, *Analytic function theory*. Vol. II, Ginn and Company, Boston 1962.

[65] G. P. Hochschild, *Basic theory of algebraic groups and Lie algebras*. Grad. Texts in Math. 75, Springer-Verlag, New York 1981.

[66] H. Hopf, Über die Topologie der Gruppen-Mannigfaltigkeiten und ihre Verallgemeinerungen. *Ann. of Math.* (2) **42** (1941), 22–52.

[67] A. Jacobs, The quantum $E(2)$ group. PhD thesis, Katholieke Universiteit Leuven, Leuven 2005. Available at http://hdl.handle.net/1979/154

[68] J. C. Jantzen, *Lectures on quantum groups*. Grad. Stud. Math. 6, Amer. Math. Soc., Providence, RI, 1996.

[69] K. K. Jensen and K. Thomsen, *Elements of KK-theory*. Birkhäuser, Boston 1991.

[70] M. Jimbo, A $q$-difference analogue of $U(\mathfrak{g})$ and the Yang-Baxter equation. *Lett. Math. Phys.* **10** (1985), 63–69.

[71] V. F. R. Jones, Index for subfactors. *Invent. Math.* **72** (1983), 1–25.

[72] V. Jones and V. S. Sunder, *Introduction to subfactors*, London Math. Soc. Lecture Note Ser. 234, Cambridge University Press, Cambridge 1997.

[73] A. Joyal and R. Street, An introduction to Tannaka duality and quantum groups. In *Category theory* (*Como*, 1990), Lecture Notes in Math. 1488, Springer-Verlag, Berlin 1991, 413–492.

[74] G. I. Kac, Extensions of groups to ring groups. *Mat. Sb. (N.S.)* **76** (1968), 473–496; English transl. *Math.-USSR Sb.* **5** (1968), 451–474.

[75] R. V. Kadison and J. R. Ringrose, *Fundamentals of the theory of operator algebras.* Vol. I: Elementary theory, Grad. Stud. Math. 15, Amer. Math. Soc., Providence, RI, 1997.

[76] R. V. Kadison and J. R. Ringrose, *Fundamentals of the theory of operator algebras.* Vol. II: Advanced theory, Grad. Stud. Math. 16, Amer. Math. Soc., Providence, RI, 1997.

[77] B.-J. Kahng, Non-compact quantum groups arising from Heisenberg type Lie bialgebras. *J. Operator Theory* **44** (2000), 303–334.

[78] B.-J. Kahng, Haar measure on a locally compact quantum group. *J. Ramanujan Math. Soc.* **18** (2003), 384–414.

[79] C. Kassel, *Quantum groups*, Grad. Texts in Math. 155, Springer-Verlag, New York 1995.

[80] A. Klimyk and K. Schmüdgen, *Quantum groups and their representations.* Texts Monogr. Phys., Springer-Verlag, Berlin 1997.

[81] E. Koelink and J. Kustermans, A locally compact quantum group analogue of the normalizer of $\mathrm{SU}(1,1)$ in $\mathrm{SL}(2,\mathbb{C})$. *Comm. Math. Phys.* **233** (2003), 231–296.

[82] E. Koelink and J. Kustermans, Quantum $\widetilde{\mathrm{SU}}(1,1)$ and its Pontryagin dual. In *Locally compact quantum groups and groupoids (Strasbourg,* 2002), IRMA Lect. Math. Theor. Phys. 2, Walter de Gruyter, Berlin 2003, 49–77.

[83] T. H. Koornwinder, Representations of the twisted $\mathrm{SU}(2)$ quantum group and some $q$-hypergeometric orthogonal polynomials. *Nederl. Akad. Wetensch. Indag. Math.* **51** (1989), 97–117.

[84] L. I. Korogodski and Y. S. Soibelman, *Algebras of functions on quantum groups: Part I.* Math. Surveys Monogr. 56, Amer. Math. Soc., Providence, RI, 1998.

[85] A. Kumjian, Fell bundles over groupoids. *Proc. Amer. Math. Soc.* **126** (1998), 1115–1125.

[86] J. Kustermans, One-parameter representations on C*-algebras. Technical report, Odense Universitet, 1997. Available at http://arxiv.org/abs/funct-an/9707009

[87] J. Kustermans, Locally compact quantum groups in the universal setting. *Internat. J. Math.* **12** (2001), 289–338.

[88] J. Kustermans, Locally compact quantum groups. In *Quantum independent increment processes I.* Lecture Notes in Math. 1865, Springer-Verlag, Berlin 2005, 99–180.

[89] J. Kustermans and L. Tuset, A survey of $C^*$-algebraic quantum groups, Part II. *Irish Math. Soc. Bull.* **44** (2000), 6–54.

[90] J. Kustermans and S. Vaes, Weight theory for C*-algebraic quantum groups. Technical report, Katholieke Universiteit Leuven and University College Cork, 1999. Available at http://arxiv.org/abs/math/9901063v1

[91] J. Kustermans and S. Vaes, Locally compact quantum groups. *Ann. Sci. École Norm. Sup.* (4) **33** (2000), 837–934.

[92] J. Kustermans and S. Vaes, The operator algebra approach to quantum groups. *Proc. Natl. Acad. Sci. USA* **97** (2000), 547–552.

[93] J. Kustermans and S. Vaes, Locally compact quantum groups in the von Neumann algebraic setting. *Math. Scand.* **92** (2003), 68–92.

[94] J. Kustermans and A. Van Daele, $C^*$-algebraic quantum groups arising from algebraic quantum groups. *Internat. J. Math.* **8** (1997), 1067–1139.

[95] E. C. Lance, *Hilbert $C^*$-modules. A toolkit for operator algebraists.* London Math. Soc. Lecture Note Ser. 210, Cambridge University Press, Cambridge 1995.

[96] M. B. Landstad, Quantizations arising from abelian subgroups. *Internat. J. Math.* **5** (1994), 897–936.

[97] M. B. Landstad, J. Phillips, I. Raeburn, and C. E. Sutherland, Representations of crossed products by coactions and principal bundles. *Trans. Amer. Math. Soc.* **299** (1987), 747–784.

[98] P.-Y. Le Gall, Théorie de Kasparov équivariante et groupoïdes I. *K-Theory* **16** (1999), 361–390.

[99] F. Lesieur, Groupoïdes quantiques mesurés: axiomatique, étude, dualité, exemples. PhD thesis, Universite d'Orléans, Orléans 2003. Available at http://www.univ-orleans.fr/mapmo/publications/lesieur/these.php

[100] S. Levendorski and Y. Soibelman, Algebras of functions on compact quantum groups, Schubert cells and quantum tori. *Comm. Math. Phys.* **139** (1991), 141–170.

[101] R. Longo, A duality for Hopf algebras and for subfactors. I. *Comm. Math. Phys.* **159** (1994), 133–150.

[102] R. Longo and J. E. Roberts, A theory of dimension. *K-Theory*, **11** (1997), 103–159.

[103] G. Lusztig, *Introduction to quantum groups*. Progr. Math. 110, Birkhäuser, Boston 1993.

[104] S. Mac Lane, *Categories for the working mathematician*. 2nd ed., Grad. Texts in Math. 5, Springer-Verlag, New York 1998.

[105] A. Maes and A. Van Daele, Notes on compact quantum groups. *Nieuw Arch. Wisk.* (4) **16** (1998), 73–112.

[106] S. Majid, Hopf–von Neumann algebra bicrossproducts, Kac algebra bicrossproducts, and the classical Yang–Baxter equations. *J. Funct. Anal.* **95** (1991), 291–319.

[107] S. Majid, *Foundations of quantum group theory*. Cambridge University Press, Cambridge 1995.

[108] T. Masuda, K. Mimachi, Y. Nakagami, M. Noumi, and K. Ueno, Representation of quantum groups. In *Mappings of operator algebras* (*Philadelphia, PA,* 1988), Progr. Math. 84, Birkhäuser, Boston 1990, 119–128.

[109] T. Masuda, K. Mimachi, Y. Nakagami, M. Noumi, and K. Ueno, Representations of the quantum group $SU_q(2)$ and the little $q$-Jacobi polynomials. *J. Funct. Anal.* **99** (1991), 357–386.

[110] T. Masuda, Y. Nakagami, and S. L. Woronowicz, A $C^*$-algebraic framework for quantum groups. *Internat. J. Math.* **14** (2003), 903–1001.

[111] S. Montgomery, *Hopf algebras and their actions on rings*. CBMS Regional Conf. Ser. in Math. 82, Amer. Math. Soc., Providence, RI, 1993.

[112] S. A. Morris, *Pontryagin duality and the structure of locally compact abelian groups*. 29 London Math. Soc. Lecture Note Ser. 29, Cambridge University Press, Cambridge 1977.

[113] G. J. Murphy, *$C^*$-algebras and operator theory*. Academic Press, Boston 1990.

[114] A. Ocneanu, Quantized groups, string algebras and Galois theory for algebras. In *Operator algebras and applications*, Vol. 2: Mathematical physics and subfactors, London Math. Soc. Lecture Note Ser. 136, Cambridge University Press, Cambridge 1988, 119–172.

[115] A. Ocneanu, Operator algebras, topology and subgroups of quantum symmetry – construction of subgroups of quantum groups. In *Taniguchi Conference on Mathematics Nara '98*, Adv. Stud. Pure Math. 31, Math. Soc. Japan, Tokyo 2001, 235–263.

[116] M. O'uchi, Pseudo-multiplicative unitaries on Hilbert $C^*$-modules. *Far East J. Math. Sci. (FJMS)* **2001**, Special Volume, Part II, 229–249.

[117] A. Pal, On some quantum groups and their representations. PhD thesis, Indian Statistical Institute, New Delhi 1995. Available at http://www.isid.ac.in/~arup/papers/thesis.pdf

[118] A. Pal, Haar measure on $E_q(2)$. *Pacific J. Math.* **176** (1996), 217–233.

[119] A. L. T. Paterson, *Amenability*. Math. Surveys Monogr. 29, Amer. Math. Soc., Providence, RI, 1988.

[120] A. L. T. Paterson, *Groupoids, inverse semigroups, and their operator algebras*. Progr. Math. 170, Birkhäuser, Boston 1999.

[121] G. K. Pedersen, *$C^*$-algebras and their automorphism groups*. London Math. Soc. Monogr. 14, Academic Press, London 1979.

[122] L. Pontrjagin, *Topological groups*. Princeton Math. Ser. 2, Princeton University Press, Princeton 1939.

[123] W. Pusz, Quantum $GL(2,\mathbb{C})$ group as double group over "$az+b$" quantum group. *Rep. Math. Phys.* **49** (2002), 113–122.

[124] W. Pusz and P. M. Sołtan, Functional form of unitary representations of the quantum '$az+b$' group. *Rep. Math. Phys.* **52** (2003), 309–319.

[125] W. Pusz and S. L. Woronowicz, A quantum $GL(2,\mathbb{C})$ group at roots of unity. *Rep. Math. Phys.* **47** (2001), 431–462.

[126] W. Pusz and S. L. Woronowicz, A new quantum deformation of '$ax+b$' group. *Comm. Math. Phys.* **259** (2005), 325–362.

[127] W. Pusz and S. L. Woronowicz, On a quantum group of unitary operators. The quantum '$az + b$' group. In *Twenty years of Bialowieza: a mathematical anthology*, World Sci. Monogr. Ser. Math. 8, World Scientific Publishing, Singapore 2005, 229–261.

[128] I. Raeburn and D. P. Williams, *Morita equivalence and continuous-trace $C^*$-algebras*. Math. Surveys Monogr. 60, Amer. Math. Soc., Providence, RI, 1998.

[129] J. Renault, *A groupoid approach to* C*-*algebras*. Lecture Notes in Math. 793, Springer-Verlag, Berlin 1980.

[130] M. A. Rieffel, Some solvable quantum groups. In *Operator algebras and topology* (*Craiova,* 1989), Pitman Res. Notes Math. Ser. 270, Longman Sci. Tech., Harlow 1992, 146–159.

[131] M. A. Rieffel, Compact quantum groups associated with toral subgroups. In *Representation theory of groups and algebras*, Contemp. Math. 145, Amer. Math. Soc., Providence, RI, 1993, 465–491.

[132] M. A. Rieffel, Non-compact quantum groups associated with abelian subgroups. *Comm. Math. Phys.* **171** (1995), 181–201.

[133] M. Rosso, Comparaison des groupes $SU(2)$ quantiques de Drinfel'd et de Woronowicz. *C. R. Acad. Sci. Paris Sér. I Math.* **304** (1987), 323–326.

[134] M. Rosso, Algèbres enveloppantes quantifiées, groupes quantiques compacts de matrices et calcul différentiel non commutatif. *Duke Math. J.* **61** (1990), 11–40.

[135] M. Rowicka, Exponential equations related to the quantum '$ax + b$' group. *Comm. Math. Phys.* **244** (2004), 419–453.

[136] W. Rudin, *Functional analysis*. 2nd ed., McGraw-Hill, New York 1991.

[137] S. Sakai, *$C^*$-algebras and $W^*$-algebras*. Ergeb. Math. Grenzgeb. 60, Springer-Verlag, Berlin 1971.

[138] J.-L. Sauvageot, Sur le produit tensoriel relatif d'espaces de Hilbert. *J. Operator Theory* **9** (1983), 237–252.

[139] J.-L. Sauvageot, Produits tensoriels de $\mathcal{Z}$-modules et applications. In *Operator algebras and their connections with topology and ergodic theory* (*Buşteni,* 1983), Lecture Notes in Math. 1132, Springer-Verlag, Berlin 1985, 468–485.

[140] S. Shnider and S. Sternberg, *Quantum groups: From coalgebras to Drinfeld algebras*. International Press, Cambridge, MA, 1993.

[141] P. M. Sołtan, New quantum "$az + b$" groups. *Rev. Math. Phys.* **17** (2005), 313–364.

[142] P. M. Sołtan and S. L. Woronowicz, A remark on manageable multiplicative unitaries. *Lett. Math. Phys.* **57** (2001), 239–252.

[143] T. A. Springer, *Linear algebraic groups*. 2nd ed., Progr. Math. 9, Birkhäuser, Boston 1998.

[144] Ş. Strătilă, *Modular theory in operator algebras*. Editura Academiei, Bucharest; Abacus Press, Tunbridge Wells, Kent, 1981.

[145] M. E. Sweedler, *Hopf algebras*. W. A. Benjamin, New York 1969.

[146] W. Szymański, Finite index subfactors and Hopf algebra crossed products. *Proc. Amer. Math. Soc.* **120** (1994), 519–528.

[147] M. Takesaki, *Tomita's theory of modular Hilbert algebras and its applications*. Lecture Notes in Math. 128, Springer-Verlag, Berlin 1970.

[148] M. Takesaki, Duality and von Neumann algebras. In *Lectures on operator algebras*, Tulane Univ. Ring and Operator Theory Year, 1970–1971, Vol. II, Lecture Notes in Math. 247, Springer, Berlin 1972, 665–786.

[149] M. Takesaki, *Theory of operator algebras I*. Encyclopaedia Math. Sci. 124, Springer-Verlag, Berlin 2002.

[150] M. Takesaki, *Theory of operator algebras II*. Encyclopaedia Math. Sci. 125, Springer-Verlag, Berlin 2003.

[151] T. Tannaka, *Sôtai genri* (Japanese) [The principle of duality]. Iwanami Shoten, Tokyo 1951.

[152] T. Timmermann, Pseudo-multiplicative unitaries and pseudo-Kac systems on $C^*$-modules. PhD thesis, Universität Münster, Münster 2005. Available at http://www.math.uni-muenster.de/sfb/about/publ/heft394.ps

[153] T. Timmermann, Pseudo-multiplicative unitaries on C*-modules and Hopf C*-families I. *J. Noncommut. Geom.* **1** (2007), 497–542.

[154] T. Timmermann, $C^*$-pseudo-multiplicative unitaries. SFB 478 - Geometrische Strukturen in der Mathematik, Münster, Technical Report No. 481, September 2007. Available at www.arxiv.org/abs/0709.2995.

[155] T. Timmermann, $C^*$-pseudo-Kac systems and duality for coactions of concrete Hopf $C^*$-bimodules. SFB 478 - Geometrische Strukturen in der Mathematik, Münster, Technical Report No. 484, October 2007. Available at www.arxiv.org/abs/0709.4617.

[156] T. Timmermann, Finite-dimensional Hopf-$C^*$-bimodules and $C^*$-pseudo-multiplicative unitaries. SFB 478 - Geometrische Strukturen in der Mathematik, Münster, November 2007. Available at www.arxiv.org/abs/0711.1420.

[157] S. Vaes, Examples of locally compact quantum groups through the bicrossed product construction. In *XIIIth International congress on mathematical physics* (*London*, 2000), International Press, Boston 2001, 341–348.

[158] S. Vaes, Locally compact quantum groups. PhD thesis, Katholieke Universiteit Leuven, Leuven 2001. Available at http://wis.kuleuven.be/analyse/stefaan/PhD.html

[159] S. Vaes, The unitary implementation of a locally compact quantum group action. *J. Funct. Anal.* **180** (2001), 426–480.

[160] S. Vaes, A new approach to induction and imprimitivity results. *J. Funct. Anal.* **229** (2005), 317–374.

[161] S. Vaes, Strictly outer actions of groups and quantum groups. *J. Reine Angew. Math.* **578** (2005), 147–184.

[162] S. Vaes and L. Vainerman, Extensions of locally compact quantum groups and the bicrossed product construction. *Adv. Math.* **175** (2003), 1–101.

[163] S. Vaes and L. Vainerman, On low-dimensional locally compact quantum groups. In *Locally compact quantum groups and groupoids* (*Strasbourg,* 2002), IRMA Lect. Math. Theor. Phys. 2, Walter de Gruyter, Berlin 2003, 127–187.

[164] S. Vaes and A. Van Daele, Hopf C*-algebras. *Proc. London Math. Soc.* (3) **82** (2001), 337–384.

[165] L. Vainerman, 2-cocycles and twisting of Kac algebras. *Comm. Math. Phys.* **191** (1998), 697–721.

[166] L. Vainerman, The bicrossed product construction for locally compact quantum groups. *Bull. Kerala Math. Assoc.* **2005**, Special issue, 99–136.

[167] L. Vainerman and G. I. Kac, Nonunimodular ring groups and Hopf-von Neumann algebras. *Dokl. Akad. Nauk SSSR* **211** (1973), 1031–1034; English transl. *Soviet. Math. Dokl.* **14** (1973), 1144–1148.

[168] L. Vainerman and G. I. Kac, Nonunimodular ring groups and Hopf-von Neumann algebras. *Mat. Sb.* (*N.S.*) **94** (1974), 194–225; English transl. *Math. USSR Sb.* **23** (1974), 185–214.

[169] L. L. Vaksman and Y. S. Soibelman, Algebra of functions on the quantum group SU(2). *Funktsional. Anal. i Prilozhen.* **22** (1988), 1–14; English transl. *Funct. Anal. Appl.* **22** (1988), 170–181.

[170] J.-M. Vallin, C*-algèbres de Hopf et C*-algèbres de Kac. *Proc. London Math. Soc.* (3) **50** (1985), 131–174.

[171] J.-M. Vallin, Bimodules de Hopf et poids opératoriels de Haar. *J. Operator Theory* **35** (1996), 39–65.

[172] J.-M. Vallin, Unitaire pseudo-multiplicatif associé à un groupoïde. Applications à la moyennabilité. *J. Operator Theory* **44** (2000), 347–368.

[173] A. Van Daele, Quantum deformation of the Heisenberg group. In *Current topics in operator algebras* (*Nara,* 1990), World Scientific Publishing, Singapore 1991, 314–325.

[174] A. Van Daele, Multiplier Hopf algebras. *Trans. Amer. Math. Soc.* **342** (1994), 917–932.

[175] A. Van Daele, The Haar measure on a compact quantum group. *Proc. Amer. Math. Soc.* **123** (1995), 3125–3128.

[176] A. Van Daele, Discrete quantum groups, *J. Algebra* **180** (1996), 431–444.

[177] A. Van Daele, An algebraic framework for group duality. *Adv. Math.* **140** (1998), 323–366.

[178] A. Van Daele, The Haar measure on some locally compact quantum groups. Technical report, Katholieke Universiteit Leuven, Leuven 2001. Available at http://arxiv.org/abs/math/0109004

[179] A. Van Daele and S. Wang, Universal quantum groups. *Internat. J. Math.* **7** (1996), 255–263.

[180] A. Van Daele and S. L. Woronowicz, Duality for the quantum $E(2)$ group. *Pacific J. Math.* **173** (1996), 375–385.

[181] V. S. Varadarajan, *Lie groups, Lie algebras, and their representations*. Grad. Texts in Math. 102, Springer-Verlag, New York 1984.

[182] R. Vergnioux, KK-théorie équivariante et opérateur de Julg-Valette pour les groupes quantiques. PhD thesis, Universite Paris VII, Paris 2002. Available at http://www.institut.math.jussieu.fr/theses/2002/vergnioux/

[183] N. Ja. Vilenkin and A. U. Klimyk, *Representation of Lie groups and special functions*. Volume 3: Classical and quantum groups and special functions, Math. Appl. 75 (Soviet Series), Kluwer Academic Publishers, Dordrecht 1992.

[184] D. V. Voiculescu, K. J. Dykema, and A. Nica, *Free random variables*. CRM Monogr. Ser. 1, Amer. Math. Soc., Providence, RI, 1992.

[185] S. Wang, General constructions of compact quantum groups. PhD thesis, University of California, Berkeley 1993.

[186] S. Wang, Free products of compact quantum groups. *Comm. Math. Phys.* **167** (1995), 671–692.

[187] S. Wang, Tensor products and crossed products of compact quantum groups. *Proc. London Math. Soc.* (3) **71** (1995), 695–720.

[188] S. Wang, Deformations of compact quantum groups via Rieffel's quantization. *Comm. Math. Phys.* **178** (1996), 747–764.

[189] S. Wang, Quantum symmetry groups of finite spaces. *Comm. Math. Phys.* **195** (1998), 195–211.

[190] A. Weil, *L'intégration dans les groupes topologiques et ses applications*. Hermann, Paris 1940.

[191] D. P. Williams, *Crossed products of $C^*$-algebras*. Math. Surveys Monogr. 134, Amer. Math. Soc., Providence, RI, 2007.

[192] S. L. Woronowicz. On the purification of factor states. *Comm. Math. Phys.* **28** (1972), 221–235.

[193] S. L. Woronowicz, Compact matrix pseudogroups. *Comm. Math. Phys.* **111** (1987), 613–665.

[194] S. L. Woronowicz, Twisted $SU(2)$ group. An example of a noncommutative differential calculus. *Publ. Res. Inst. Math. Sci.* **23** (1987), 117–181.

[195] S. L. Woronowicz, Tannaka-Krein duality for compact matrix pseudogroups. Twisted $SU(N)$ groups. *Invent. Math.* **93** (1988), 35–76.

[196] S. L. Woronowicz, Quantum $E(2)$ group and its Pontryagin dual. *Lett. Math. Phys.* **23** (1991), 251–263.

[197] S. L. Woronowicz, Unbounded elements affiliated with $C^*$-algebras and noncompact quantum groups. *Comm. Math. Phys.* **136** (1991), 399–432.

[198] S. L. Woronowicz, Operator equalities related to the quantum $E(2)$ group. *Comm. Math. Phys.* **144** (1992), 417–428.

[199] S. L. Woronowicz. Quantum $SU(2)$ and $E(2)$ groups. Contraction procedure. *Comm. Math. Phys.* **149** (1992), 637–652.

[200] S. L. Woronowicz, $C^*$-algebras generated by unbounded elements. *Rev. Math. Phys.* **7** (1995), 481–521.

[201] S. L. Woronowicz, From multiplicative unitaries to quantum groups. *Internat. J. Math.* **7** (1996), 127–149.

[202] S. L. Woronowicz, Compact quantum groups. In *Symétries quantiques* (*Les Houches*, 1995), North-Holland, Amsterdam 1998, 845–884.

[203] S. L. Woronowicz, Quantum exponential function. *Rev. Math. Phys.* **12** (2000), 873–920.

[204] S. L. Woronowicz, Quantum "$az + b$" group on complex plane. *Internat. J. Math.* **12** (2001), 461–503.

[205] S. L. Woronowicz, Haar weights on some quantum groups. In *Group 24: Physical and mathematical aspects of symmetries*, Proceedings of the 24th international colloquium on group theoretical methods in physics, Paris, 15–20 July 2002, Conference Series 173, Institute of Physics Publishing, Bristol 2002, 763–772.

[206] S. L. Woronowicz and K. Napiórkowski, Operator theory in the $C^*$-algebra framework. *Rep. Math. Phys.* **31** (1992), 353–371.

[207] S. L. Woronowicz and S. Zakrzewski, Quantum '$ax + b$' group. *Rev. Math. Phys.* **14** (2002), 797–828.

[208] T. Yamanouchi, Bicrossproduct Kac algebras, bicrossproduct groups and von Neumann algebras of Takesaki's type. *Math. Scand.* **71** (1992), 252–260.

# Symbol Index

# Index